AF389144

# Volatile Compounds and Smell Chemicals (Odor and Aroma) of Food

# Volatile Compounds and Smell Chemicals (Odor and Aroma) of Food

Editor

**Eugenio Aprea**

MDPI • Basel • Beijing • Wuhan • Barcelona • Belgrade • Manchester • Tokyo • Cluj • Tianjin

*Editor*
Eugenio Aprea
University of Trento/Fondazione Edmund Mach
Italy

*Editorial Office*
MDPI
St. Alban-Anlage 66
4052 Basel, Switzerland

This is a reprint of articles from the Special Issue published online in the open access journal *Molecules* (ISSN 1420-3049) (available at: https://www.mdpi.com/journal/molecules/special_issues/Molecules_Compounds_Food).

For citation purposes, cite each article independently as indicated on the article page online and as indicated below:

LastName, A.A.; LastName, B.B.; LastName, C.C. Article Title. *Journal Name* **Year**, *Article Number*, Page Range.

ISBN 978-3-03943-412-1 (Hbk)
ISBN 978-3-03943-413-8 (PDF)

# Contents

# About the Editor

**Eugenio Aprea** has been an Associate Professor of Food Chemistry at University of Trento since 2019 (https://webapps.unitn.it/du/it/Persona/PER0000552/Pubblicazioni) and is also affiliated with the Research and Innovation Center of the Edmund Mach Foundation, where he has been a researcher for 19 years. His main scientific interests are the characterization of food product properties related to sensory perceptions and the relationship linking these two aspects: volatile compounds, color, texture, food composition, and sensory perception. His research interest includes data analysis and multivariate data analysis. He has published, until September 2020, 92 papers in Web of Science Core Collection Journals, mainly in the fields of food science and analytical chemistry, which have been cited more than 2100 times with an h-index of 29 (https://orcid.org/0000-0003-3648-0459).

# Preface to "Volatile Compounds and Smell Chemicals (Odor and Aroma) of Food"

Volatile organic compounds (VOCs) are molecules characterized by a low boiling point responsible for their volatility at room temperature. It means that at ordinary room temperature, they are diffused into the surrounding environment. VOCs are from different origins and are generated from different chemical and biochemical pathways. Among the constituents of food, VOCs are important, interesting and informative molecules. Indeed, several of them give rise to odor and aroma but can also be used by analysts to monitor ripening, senescence, and decay in fruit and vegetables, as well as to monitor and control the changes during food processing and storage (i.e., preservation, fermentation, cooking, and packaging). This Special Issue aims to attract up-to-date original contributions covering all aspects of volatile compounds research in the food sector.

**Eugenio Aprea**
*Editor*

*Editorial*

# Special Issue "Volatile Compounds and Smell Chemicals (Odor and Aroma) of Food"

**Eugenio Aprea** [1,2]

1   Center Agriculture Food Environment, University of Trento/Fondazione Edmund Mach, via E. Mach 1, 38010 San Michele all'Adige (TN), Italy; eugenio.aprea@unitn.it

2   Department of Food Quality and Nutrition, Research and Innovation Centre, Fondazione Edmund Mach (FEM), via E. Mach 1, 38010 San Michele all'Adige (TN), Italy

Academic Editor: Derek J. McPhee
Received: 11 August 2020; Accepted: 20 August 2020; Published: 21 August 2020

**Keywords:** odour; aroma profiling; flavour; food processing; Maillard reaction; oxidation; fermentation; food decay; solid-phase micro-extraction; gas chromatography; spectrometry; gas sensors; fingerprinting; volatilome

Among the constituents of food, volatile compounds are a particularly intriguing group of molecules, because they give rise to odour and aroma. Indeed, olfaction is one of the main aspects influencing the appreciation or dislike of particular food items. Volatile compounds are perceived through the smelling sensory organs of the nasal cavity, and evoke numerous associations and emotions, even before the food is tasted. Such a reaction occurs because the information from these receptors is directed to the hippocampus and amygdala, the key regions of the brain involved in learning and memory.

In addition to identifying the odour-active compounds, the analysis of the volatile compounds in food is also applied for detecting the ripening, senescence, and decay in fruit and vegetables, as well as monitoring and controlling the changes during food processing and storage (i.e., preservation, fermentation, cooking, and packaging). Nineteen research papers and two review papers cover the topics of this Special Issue. The majority of the papers investigated volatile compounds from vegetable origin products (16) and four papers reported volatile compounds from animal origin products (dairy products).

Six of the research papers were mainly focused on the sensory aspects related to volatile compounds [1–6], and another six papers investigated aspects related to processing [7–9], five papers were more oriented to the quality of the products [10–14], cultivation practice [15] and storage [16,17], two further papers were more focused on measurement techniques and procedures [18,19]. Lastly, the two review papers reviewed the sensory characteristics (flavour and texture) linked to a physiological disorder of apple, namely watercore [20] and the influence of ruminant diet on the volatile flavour compounds in cheese [21].

Muñoz-González et al. [1] studied the effects of ethanol concentration on the dynamics of oral aroma release after wine consumption. They found that a subject's ethanol concentration and nature of the aroma compound immediately or after 4 min from wine intake influenced oral aroma release. Just after wine intake, a higher release was observed for the more polar compounds and a lower release for more apolar compounds, while after 4 min, all the six esters investigated increased in the oral cavity, increasing ethanol content in wine. The latter effect could be related to an increase in the fruity aroma persistence in the wines.

Endrizzi et al. [2] studied the effect of artificial flavourings on consumer liking of apples. By means of a conjoint experiment, external information (use of claims) was investigated as well. Depending on their personal liking, an apple consumers' preference is affected by different flavouring treatments.

The external information ("traditional" or "selected flavour") influenced apple acceptability for some groups of consumers depending on their food approach (attitude toward natural food interest and food neophobia).

Deuscher et al. [3] used gas chromatography–olfactometry to classify dark chocolates differing in organoleptic properties by the key aroma compounds. Thirty-eight discriminant key odorants were identified, 13 of which were described for the first time in a cocoa product.

Liberto et al. [4] explored the possibility to use HS-SPME-MS-Enose coupled with chemometrics in order to predict in-cup coffee sensory quality in routine controls. This preliminary study has demonstrated that the proposed approach requires a number of compromises in terms of model robustness and acceptance of the errors in prediction.

In the work of Genovese et al., [5] it was shown that high levels of extravirgin olive oil (EVOO) phenolic compounds influence the in-mouth perceived intensity of both flavours and off-flavours. While the high levels of EVOO phenolic compounds did not produce any effect on the headspace concentration of volatile compounds, they reduced the perceptions of the positive fruity score and of the score of the "fusty–muddy" defect while increased the score of "winey–vinegary" defect.

Yan et al. [6] used fast volatile fingerprinting to differentiate the grades of olive oils (extra virgin, refined and pomace) and indicating the mass peaks more intense for the different olive oils grades.

Fatty acids and their derived volatile compounds were studied in "Thompson Seedless" Raisins during air-drying and sun-drying processes by Wang et al. [7]. The study shows that air-drying is more favourable for the production of fatty acids, whereas sun-drying is more advantageous in terms of fatty acid-derived volatiles.

The work of Ma et al. [8] evaluated the influence of clarification treatments on the volatile composition and aromatic attributes of "Italian Riesling" Icewine. The study shows that while all treatments improved the limpidity of wines, a decreased concentration of aroma compounds was observed as well as a decrease of colour intensity. The aroma and taste properties of the wine samples were influenced more by bentonite fining, while membrane filtration mainly influenced colour and aroma. On the other hand, soybean protein and centrifugation treatments achieved better sensory quality.

Vidal et al. [9] in their study, applied the response surface methodology approach in order to determine the optimum extraction conditions for maximizing the content of volatile compounds and pigments in extra virgin olive oil obtained from Arbequina, Koroneiki, and Arbosana Cultivars. The authors showed that extra-virgin olive oil from irrigated crops and from the Arbequina cultivar had the highest content of volatile compounds.

Chitarrini et al. [10] combined gas chromatography–mass spectrometry and sensory analysis to investigate the aroma of new and standard apple varieties grown at two altitudes. The study indicates twelve volatile organic compounds changing in relationship with pedoclimatic locations, independent of the variety, and significant interactions between variety and altitude on sensory parameters were reported as well.

In a further work, Chitarrini and co-workers [11] used the same approach (gas chromatography–mass spectrometry and sensory analysis) to describe four different honey wines: multi-floral blossom honey and a forest honeydew honey with and without the addition of black currant during fermentation.

Smith and Peterson [12] combined gas chromatography/mass spectrometry/olfactometry with sensory recombination and descriptive analysis to identify aroma differences in refined and whole-grain extruded maize puffs. Maillard reaction products were reported as the main responsible for perceived differences.

Ianni et al. [13] studied the proteolytic process in Caciocavallo cheese obtained from Friesian cows fed zinc, selenium, and iodine supplementation and the possible effect on volatile compounds formation. In particular, it was observed that selenium negatively influenced the biochemical processes leading to the formation of 3-methyl butanol.

A further study on the effect of animal feeding on volatile compounds in dairy products was undertaken by Bennato and co-workers [14]. These authors showed that milk from goats fed a dietary supplementation with olive leaves produced a yogurt, with characteristics potentially indicative of an improvement in nutritional properties and flavour.

Slaghenaufi et al. [15] studied how different modalities of grape withering may influence volatile compounds in young and aged Corvina wines. The fermentative volatile metabolites responsible for wine aroma modulation were affected by the withering process, and in particular, the on-vine withering with blocked xylem is an interesting alternative to conventional fruit withering for the production of wines where a mild withering is requested.

Yang et al. [16] studied the effect of cold storage on the volatile component in jujube fruits by using headspace-gas chromatography–ion mobility spectrometry showing the possibility to determine the different storage periods.

Lobo-Prieto et al. [17] tracked the sensory characteristics of virgin olive oils during storage, showing some disagreements between the sensory assessment and the oxidative stability index analysed by Rancimat.

In the paper of Coelho and co-workers [18], a liquid–liquid microextraction method combined with GC-MS was proposed for the analysis of compounds in fermented beverages and spirits. The method was validated for a set of compounds typically found in fermented beverages.

In the work of Capozzi et al. [19] two complementary analytical techniques, Headspace-Solid Phase Microextraction-Gas Chromatography-Mass Spectrometry and Proton-Transfer Reaction–Mass Spectrometry coupled to a Time of Flight mass analyser were coupled to study the volatile organic profile of Mascarpone cheese. Mascarpone is a base ingredient in industrial, culinary, and homemade preparations (e.g., it is a key constituent of a widely appreciated Italian dessert "Tiramisù") which has been scarcely investigated for its aroma profile.

Tanaka and co-workers [20], in their review paper, summarized the recent studies related to the physiology of watercore, a physiological disorder of the apple, with special focus on "Fuji" and related cultivars.

Ianni et al., [21] in their review paper, summarized the major families of volatile compounds most commonly found in cheese obtained from lactating dairy ruminants fed experimental diets at various ripening stages, describing, in greater detail, the role of the animal diet in influencing the cheese flavour.

In conclusion, volatile compounds are responsible for the aroma of food and may be influenced by processing, storage, harvesting or animal feeding, and they require a dedicated analytical approach for the identification and the monitoring All these aspects were covered by the manuscripts submitted and published in the Special Issue "Volatile Compounds and Smell Chemicals (Odor and Aroma) of Food". These manuscripts contributed—with their topics and their quality—to the success of this Special Issue.

**Funding:** This research received no external funding.

**Acknowledgments:** The Guest Editor wishes to thank all the authors for their contributions to this Special Issue, all the reviewers for their work in evaluating the submitted articles and the editorial staff of Molecules, especially Larry Li, for their kind help in making this Special Issue.

## References

1. Muñoz-González, C.; Pérez-Jiménez, M.; Criado, C.; Pozo-Bayón, M.Á. Effects of Ethanol Concentration on Oral Aroma Release After Wine Consumption. *Molecules* **2019**, *24*, 3253. [CrossRef] [PubMed]
2. Endrizzi, I.; Aprea, E.; Betta, E.; Charles, M.; Zambanini, J.; Gasperi, F. Investigating the Effect of Artificial Flavours and External Information on Consumer Liking of Apples. *Molecules* **2019**, *24*, 4306. [CrossRef]
3. Deuscher, Z.; Gourrat, K.; Repoux, M.; Boulanger, R.; Labouré, H.; Le Quéré, J.-L. Key Aroma Compounds of Dark Chocolates Differing in Organoleptic Properties: A GC-O Comparative Study. *Molecules* **2020**, *25*, 1809. [CrossRef]

*Molecules* **2020**, *25*, 3811

4.    Liberto, E.; Bressanello, D.; Strocchi, G.; Cordero, C.; Ruosi, M.R.; Pellegrino, G.; Bicchi, C.; Sgorbini, B. HS-SPME-MS-Enose Coupled with Chemometrics as an Analytical Decision Maker to Predict In-Cup Coffee Sensory Quality in Routine Controls: Possibilities and Limits. *Molecules* **2019**, *24*, 4515. [CrossRef] [PubMed]

5.    Genovese, A.; Mondola, F.; Paduano, A.; Sacchi, R. Biophenolic Compounds Influence the In-Mouth Perceived Intensity of Virgin Olive Oil Flavours and Off-Flavours. *Molecules* **2020**, *25*, 1969. [CrossRef]

6.    Yan, J.; Alewijn, M.; van Ruth, S.M. From Extra Virgin Olive Oil to Refined Products: Intensity and Balance Shifts of the Volatile Compounds versus Odor. *Molecules* **2020**, *25*, 2469. [CrossRef] [PubMed]

7.    Wang, D.; Javed, H.U.; Shi, Y.; Naz, S.; Ali, S.; Duan, C.-Q. Impact of Drying Method on the Evaluation of Fatty Acids and Their Derived Volatile Compounds in 'Thompson Seedless' Raisins. *Molecules* **2020**, *25*, 608. [CrossRef] [PubMed]

8.    Ma, T.-Z.; Gong, P.-F.; Lu, R.-R.; Zhang, B.; Morata, A.; Han, S.-Y. Effect of Different Clarification Treatments on the Volatile Composition and Aromatic Attributes of 'Italian Riesling' Icewine. *Molecules* **2020**, *25*, 2657. [CrossRef]

9.    Vidal, A.M.; Alcalá, S.; De Torres, A.; Moya, M.; Espínola, J.M.; Espínola, F. Fresh and Aromatic Virgin Olive Oil Obtained from Arbequina, Koroneiki, and Arbosana Cultivars. *Molecules* **2019**, *24*, 3587. [CrossRef]

10.    Chitarrini, G.; Dordevic, N.; Guerra, W.; Robatscher, P.; Lozano, L. Aroma Investigation of New and Standard Apple Varieties Grown at Two Altitudes Using Gas Chromatography-Mass Spectrometry Combined with Sensory Analysis. *Molecules* **2020**, *25*, 3007. [CrossRef]

11.    Chitarrini, G.; Debiasi, L.; Stuffer, M.; Ueberegger, E.; Zehetner, E.; Jaeger, H.; Robatscher, P.; Conterno, L. Volatile Profile of Mead Fermenting Blossom Honey and Honeydew Honey with or without *Ribes nigrum*. *Molecules* **2020**, *25*, 1818. [CrossRef] [PubMed]

12.    Smith, K.; Peterson, D.G. Identification of Aroma Differences in Refined and Whole Grain Extruded Maize Puffs. *Molecules* **2020**, *25*, 2261. [CrossRef] [PubMed]

13.    Ianni, A.; Bennato, F.; Martino, C.; Grotta, L.; Franceschini, N.; Martino, G. Proteolytic Volatile Profile and Electrophoretic Analysis of Casein Composition in Milk and Cheese Derived from Mironutrient-Fed Cows. *Molecules* **2020**, *25*, 2249. [CrossRef] [PubMed]

14.    Bennato, F.; Innosa, D.; Ianni, A.; Martino, C.; Grotta, L.; Martino, G. Volatile Profile in Yogurt Obtained from Saanen Goats Fed with Olive Leaves. *Molecules* **2020**, *25*, 2311. [CrossRef]

15.    Slaghenaufi, D.; Boscaini, A.; Prandi, A.; Dal Cin, A.; Zandonà, V.; Luzzini, G.; Ugliano, M. Influence of Different Modalities of Grape Withering on Volatile Compounds of Young and Aged Corvina Wines. *Molecules* **2020**, *25*, 2141. [CrossRef]

16.    Yang, L.; Liu, J.; Wang, X.; Wang, R.; Ren, F.; Zhang, Q.; Shan, Y.; Ding, S. Characterization of Volatile Component Changes in Jujube Fruits during Cold Storage by Using Headspace-Gas Chromatography-Ion Mobility Spectrometry. *Molecules* **2019**, *24*, 3904. [CrossRef]

17.    Lobo-Prieto, A.; Tena, N.; Aparicio-Ruiz, R.; Morales, M.T.; García-González, D.L. Tracking Sensory Characteristics of Virgin Olive Oils During Storage: Interpretation of Their Changes from a Multiparametric Perspective. *Molecules* **2020**, *25*, 1686. [CrossRef]

18.    Coelho, E.; Lemos, M.; Genisheva, Z.; Domingues, L.; Vilanova, M.; Oliveira, J.M. Validation of a LLME/GC-MS Methodology for Quantification of Volatile Compounds in Fermented Beverages. *Molecules* **2020**, *25*, 621. [CrossRef]

19.    Capozzi, V.; Lonzarich, V.; Khomenko, I.; Cappellin, L.; Navarini, L.; Biasioli, F. Unveiling the Molecular Basis of Mascarpone Cheese Aroma: VOCs analysis by SPME-GC/MS and PTR-ToF-MS. *Molecules* **2020**, *25*, 1242. [CrossRef]

20.    Tanaka, F.; Hayakawa, F.; Tatsuki, M. Flavor and Texture Characteristics of 'Fuji' and Related Apple (*Malus domestica* L.) Cultivars, Focusing on the Rich Watercore. *Molecules* **2020**, *25*, 1114. [CrossRef]

21.    Ianni, A.; Bennato, F.; Martino, C.; Grotta, L. Giuseppe Martino Volatile Flavor Compounds in Cheese as Affected by Ruminant Diet. *Molecules* **2020**, *25*, 461. [CrossRef] [PubMed]

 *molecules*

*Article*

# Effects of Ethanol Concentration on Oral Aroma Release After Wine Consumption

Carolina Muñoz-González, María Pérez-Jiménez, Celia Criado and María Ángeles Pozo-Bayón *

Instituto de Investigación en Ciencias de la Alimentación (CIAL), Consejo Superior de Investigaciones Científicas (CSIC)-Universidad Autónoma de Madrid (UAM), Campus de Excelencia Científica, 28049 Madrid, Spain
* Correspondence: m.delpozo@csic.es; Tel.: +34-91-0017-961; Fax: +34-91-0017-905

Received: 25 July 2019; Accepted: 3 September 2019; Published: 6 September 2019

**Abstract:** This paper evaluates, for the first time, the effects of ethanol concentration on the dynamics of oral (immediate and prolonged) aroma release after wine consumption. To do this, the intraoral aroma release of 10 panelists was monitored at two sampling points (0 and 4 min) after they rinsed their mouths with three rosé wines with different ethanol content (0.5% $v/v$, 5% $v/v$ and 10% $v/v$) that were aromatized with six fruity esters (ethyl butanoate, isoamyl acetate, ethyl pentanoate, ethyl hexanoate, ethyl octanoate and ethyl decanoate). Overall, the results indicated that the extent of the effects of ethanol content on the oral aroma release were influenced by the subject, the ethanolconcentration and the type of aroma compound. This effect was also different in the immediate than in the prolonged aroma release. In the first in-mouth aroma monitoring, an increase in the ethanol content provoked a higher release of the more polar and volatile esters (ethyl butanoate, ethyl pentanoate), but a lower release for the more apolar and less volatile esters (ethyl octanoate, ethyl decanoate). Regarding the prolonged oral aroma release, an increase of ethanol content in wine increased the oral aroma release of the six esters, which might also increase the fruity aroma persistence in the wines. Future works with a higher number of individuals will be needed to understand the mechanisms behind this phenomenon.

**Keywords:** wine; ethanol; intra-oral SPME; oral aroma release; aroma persistence

---

## 1. Introduction

Ethanol, the most abundant volatile compound in wines and the second most abundant after water, is a determinant of its sensory quality [1]. The presence of ethanol has been related to different sensations, such as burning feeling or palate warming [2], physical and perceived viscosity [2] or sourness and sweetness balance [3], among others. Apart from its contribution per se, it has been suggested that ethanol concentration plays a significant role in the overall aroma perception in wines [4–7], although the mechanisms behind these effects are not completely known. However, the increasing interest of consumers for light, fruity, and low alcohol beverages has increased the necessity to deeper understand this phenomenon in order to promote new types of low alcoholic wines.

The studies about the effects of ethanol concentration on aroma compounds have been addressed through sensory and/or chemical approaches. From a sensory point of view, studies performed orthonasally (when a wine is smelled) have found that an increase in ethanol content provokes a decrease of the fruity and floral notes [4–7], while an increase in wood, pepper, or chemical notes [7], which would produce an odor imbalance. In this regard, in vitro static studies have shown that the presence of ethanol might change wine polarity which would modify the distribution of volatile compounds between the gas and liquid phases according to their physico-chemical properties (for a review on this topic see [1]). Although very relevant, these studies may not give a clear understanding of how the wine is actually perceived when it is consumed.

When a wine is consumed, a fraction of aroma compounds travel from the mouth to the olfactory receptors via retronasal after having undergone oral processing (swallowing, breathing, interaction with saliva, adsorption with mucus, etc.) [8–11]. This generates a pulse of aroma known as immediate aroma perception. Moreover, another aroma fraction remains in the oral cavity when the wine is no longer in the mouth, and being released over time, which is responsible for the prolonged aroma perception or aroma persistence. Therefore, the mechanisms involved in retronasal release and perception are dynamic and more complex than those behind orthonasal perception. In this regard, some retronasal sensory dynamic studies have found that ethanol increased the duration and intensity of floral notes [12]. In another study, the effect of ethanol on the olfactory intensity of specific wine volatile phenols has also been observed [13]. However, to our knowledge, there are no in vivo studies that have measured the effects of ethanol content on the partitioning and release of aroma compounds in the mouth considering the dynamics of wine consumption. This could be due, among others, to analytical problems related to the high abundance of ethanol compared to other volatiles of interest. To the authors' knowledge, only one in vivo study has measured the effects of ethanol concentration on the retronasal aroma release in model wines [14] showing that the presence of ethanol increased the overall in vivo aroma release. This might be explained by changes in surface tension increases the surface area of wine coating the mouth during consumption, as well as by the Marangoni effect and the Rayleigh-Bénard convection described by Tsachaki and collaborators [15,16] using in vitro approaches. However, all these studies have been performed in model wines and have not taken into account the complexity of the wine matrix composition, which could have affected the observed effects. In this regard, Tsachaki and collaborators [16] showed that the in vitro aroma release in real wines was closer to that of an aqueous solution than to model wines. Therefore, to understand the effects of ethanol content on aroma release and perception, it is important to consider the real wine matrix composition and to use closer approaches to that of wine consumption.

With these antecedents in mind, the objective of this work was to study how ethanol concentration affects the dynamic (immediate and prolonged) oral aroma release in real wines. To this end, a dealcoholized rosé wine (0.5% *v/v* ethanol content) dosed with increasing amounts of ethanol (5% and 10%) was used. The three wines, aromatized with six fruity esters (4 mg/L), were evaluated by 10 panelists at two time points (30 sec "immediate aroma release" and 4 min after wine expectoration "prolonged oral release") following the intra-oral SPME procedure for oral monitoring. These results will provide more information about how aroma release changes during consumption, and will be important to understand the role of ethanol on aroma persistence.

## 2. Results and Discussion

To evaluate the effect of ethanol content on the in vivo aroma release during wine consumption, the oral aroma release of 10 panelists was monitored at two sampling points (Figure 1) after they rinsed their mouths with three rosé wines aromatized with six fruity esters (Table 1) presenting different ethanol content (0.5% *v/v*, 5% *v/v* and 10% *v/v*).

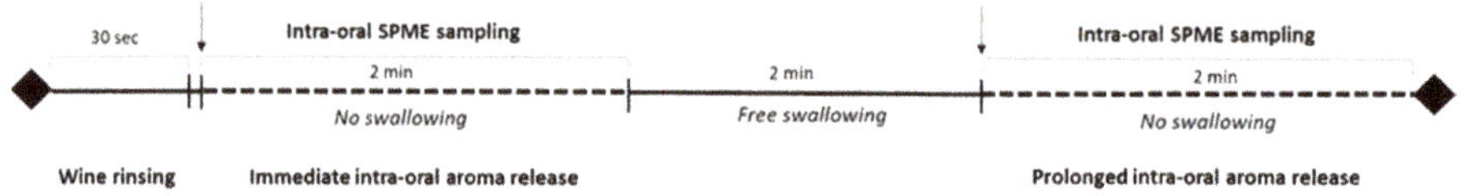


**Figure 1.** Sampling procedure followed for oral aroma monitoring.

**Table 1.** Physicochemical properties of the aroma compounds employed in this study.

| Aroma Compounds | CAS Number | MW [a] (g/mol) | log P [b] | BP [c] (°C) | Solubility [d] (mg/L) | Aroma Descriptor [e] |
| --- | --- | --- | --- | --- | --- | --- |
| Ethyl butyrate | 105-54-4 | 116.16 | 1.85 | 125.79 | 4900 | Pineapple Strawberry |
| Isoamyl acetate | 123-92-2 | 130.19 | 2.26 | 134.87 | 2000 | Banana |
| Ethyl pentanoate | 539-82-2 | 130.19 | 2.34 | 148.37 | 2210 | Fruity |
| Ethyl hexanoate | 123-66-0 | 144.22 | 2.83 | 170.05 | 309 | Apple |
| Ethyl octanoate | 106-32-1 | 172.27 | 3.81 | 210.70 | 70 | Fruity apricot |
| Ethyl decanoate | 110-38-3 | 200.32 | 4.79 | 247.43 | 16 | Grape |

[a] Molecular weight. [b] Hydrophobic constant estimated using molecular modeling software EPI Suite (U.S. EPA 2000–2007). [c] Boiling point estimated using molecular modeling software EPI Suite (U.S. EPA 2000–2007). [d] Solubility in water estimated using molecular software EPI Suite (U.S. EPA 2000–2007). [e] From Flavornet database (http://www.flavornet.org; accessed October 2009) or PubChem (https://pubchem.ncbi.nlm.nih.gov/).

## 2.1. Effects of Ethanol Concentration and Individual Differences on the Immediate Oral Aroma Release

Immediate aroma release refers to the aroma compounds trapped in a SPME fiber immediately after the panelists rinsed their mouths with the wine samples and expectorated (Figure 1). Data obtained for the 10 volunteers in each of the wines were submitted to two-way ANOVA analyses considering individuals and ethanol content as factors and their interactions. Results (Table 2) showed that the effect of subjects on aroma release was highly significant ($p < 0.0001$) for all the assayed compounds (Table 2), which indicated a high interindividual variability on the immediate oral release of esters. The variability on the aroma released observed among individuals equally trained in the same consumption procedure and using the optimized SPME conditions is not surprising and it could be related to differences in oral physiological factors, such as, salivary flow rate and composition or oral cavity volume, among others, as it has been previously described in different works [8–11].

**Table 2.** Results of the two-way ANOVA performed to compare the statistical significance of the ethanol content on the immediate intra-oral aroma release (n = 10). Each sample was analyzed in triplicate.

| | | Ethyl Butyrate | Isoamyl Acetate | Ethyl Pentanoate | Ethyl Hexanoate | Ethyl Octanoate | Ethyl Decanoate |
| --- | --- | --- | --- | --- | --- | --- | --- |
| Mean | | 4,596,854.41 | 7,107,442.85 | 6,254,852.57 | 8,892,546.68 | 26,049,220.47 | 12,679,314.42 |
| SEM [a] | | 355,864.23 | 509,112.21 | 453,805.55 | 744,257.30 | 1,704,957.45 | 850,164.95 |
| $R^2$ [b] | | 0.94 | 0.94 | 0.91 | 0.87 | 0.95 | 0.93 |
| F [c] | | 25.27 | 24.43 | 17.228 | 11.47 | 30.10 | 21.29 |
| Pr > F [d] | | <0.0001 | <0.0001 | <0.0001 | <0.0001 | <0.0001 | <0.0001 |
| Subject | F | 56.14 | 59.49 | 40.55 | 31.37 | 78.78 | 58.32 |
| | Pr > F | <0.0001 | <0.0001 | <0.0001 | <0.0001 | <0.0001 | <0.0001 |
| Ethanol (%) | F | 4.04 | 1.06 | 3.73 | 0.57 | 6.75 | 4.29 |
| | Pr > F | 0.024 | 0.356 | 0.031 | 0.567 | 0.003 | 0.019 |
| Subject*Ethanol (%) | F | 12.30 | 9.57 | 7.17 | 3.31 | 8.54 | 4.65 |
| | Pr > F | <0.0001 | <0.0001 | <0.0001 | 0.000 | <0.0001 | <0.0001 |

[a] Standard error of mean. [b] Coefficient of determination. [c] F statistic. [d] Probability values.

Regarding the effect of ethanol content on the immediate oral aroma release, it seemed that this effect was less important than the subject effect and aroma compound and ethanol concentration-dependent (Table 2). To better visualize these results, data were plotted, and shown in Figure 2. To do so, data were normalized considering the values obtained for the 0.5% *v/v* as 0% and calculating the ratio for the other two wines. As can be seen, the oral release of two compounds (isoamyl acetate andethyl hexanoate) was not significantly affected by variations in the ethanol content. Moreover, certain compounds (ethyl butyrate and ethyl pentanoate) showed a significantly higher immediate oral release after the consumption of the wines with high ethanol content (5% *v/v* and 10% *v/v*) compared to 0.5% *v/v*, while others (ethyl octanoate and ethyl decanoate) showed the opposite behavior. The differences observed among compounds are not unexpected, and they could be related to the physicochemical properties of the aroma compounds, such as hydrophobicity (log P) or boiling point (BP), as it has been previously suggested by different authors [5,15,17–19]. Interestingly, the less polar and volatile compounds of this study (ethyl octanoate and ethyl decanoate) were less released in the wines with high ethanol content. This could be due to the fact that these compounds could be more soluble in the wine matrix as ethanol concentration increases. Therefore, the higher the ethanol content, the higher the retention of these compounds in the wine matrix and the lower the immediate oral release. On the contrary, the more polar and volatile compounds (ethyl butyrate and ethyl pentanoate) were more released in the wines with high ethanol content. These compounds would have had less affinity for the wine matrix as ethanol concentration increased, and therefore, they could have been more intra-orally released. However, the immediate release of the compound isoamyl acetate (showing a similar log P and BP values than those of ethyl pentanoate) was not significantly affected by the ethanol content. Interestingly, this is the only compound with a non-linear structure of the compounds assayed, which means that not only the phsysicochemical characteristics, but the structure of the compound could be important to explain oral aroma release.

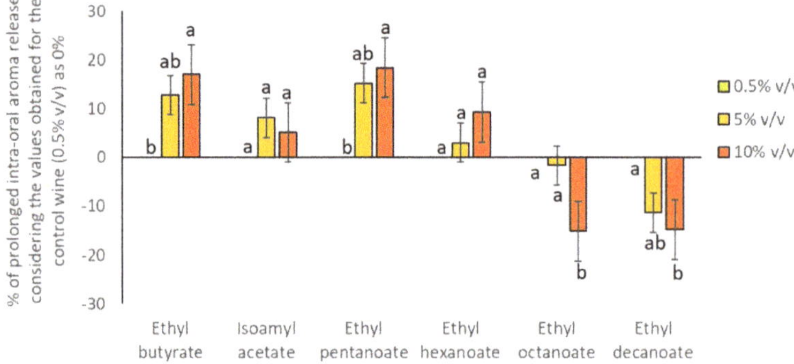


**Figure 2.** Percentage of immediate oral aroma release obtained for the volunteers (n = 10) after rinsing their mouths with the wines (considering the values obtained for the W 0.5% as 0% and comparing this value with the amount determined for the rest of wines). Different letters across the different compounds denote statistical differences after the application of the Tukey test for means comparison ($p < 0.05$).

Finally, the interaction between subject and ethanol content was significant for all the aroma compounds assayed (Table 2). This means that the effect of ethanol on the immediate oral aroma release is individual dependent, and therefore, not all the panelists would be affected in the same way by changes in ethanol concentration. Figure 3 shows an example of the release values obtained for the compound ethyl decanoate by the 10 panelists. As can be seen, the variation in the ethanol content only significantly affected the oral aroma release of ethyl decanoate for three of the ten subjects (S2, S3 and S4). Moreover, the effects of ethanol content on oral aroma release showed opposite behaviors depending on the subject. While the presence of ethanol significantly reduced the release of ethyl decanoate for S2 and S4, it increased the release for S3. The different effects of ethanol concentration-depending on the subject are interesting, and it should be studied in future works considering a high number of individuals.

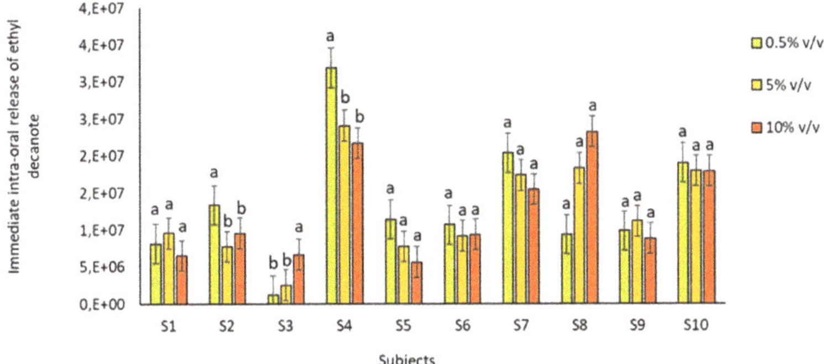


**Figure 3.** Immediate oral ethyl decanoate release (absolute peak areas) obtained for the volunteers (n = 10) after rinsing their mouths with the wines. Different letters across the different compounds denote statistical differences (*p* value < 0.05) after the application of the Tukey test for means comparison.

## 2.2. Effects of Ethanol and Individual Differences on the Prolonged Oral Aroma Release

To globally understand the effects of ethanol during wine consumption it is important to consider not only the aroma released immediately after wine intake, but also the oral aroma release that lingers in the mouth once the wine is no longer in the oral cavity. Therefore, after monitoring the immediate aroma release, the panelists were asked to keep their mouths closed and 2 min later, a second oral monitoring was carried out in the same conditions as described above (Figure 1). With these data, a two-way ANOVA analysis was performed in order to understand the effect of individual differences and ethanol content on the prolonged oral aroma release. Results are shown in Table 3. Similar to what happened in the immediate oral aroma release, a high variability among individuals and in the interactions between individuals and ethanol content was observed for all the aroma compounds assayed (Table 3).

**Table 3.** Results of the two-way ANOVA performed to compare the statistical significance of the ethanol content on the prolonged intra-oral aroma release (n = 10). Each sample was analyzed in triplicate.

| | | Ethyl Butyrate | Isoamyl Acetate | Ethyl Pentanoate | Ethyl Hexanoate | Ethyl Octanoate | Ethyl Decanoate |
|---|---|---|---|---|---|---|---|
| Mean | | 133,838.96 | 321,509.22 | 199,024.90 | 633,104.58 | 5,761,616.25 | 6,940,717.29 |
| SEM [a] | | 31,033.08 | 43,941.31 | 36,020.31 | 105,826.43 | 519,018.38 | 515,250.95 |
| $R^2$ [b] | | 0.83 | 0.96 | 0.86 | 0.90 | 0.91 | 0.91 |
| F [c] | | 9.02 | 45.07 | 11.32 | 21.13 | 19.41 | 18.74 |
| Pr > F [d] | | <0.0001 | <0.0001 | <0.0001 | <0.0001 | <0.0001 | <0.0001 |
| Subject | F | 19.80 | 121.52 | 31.67 | 63.78 | 55.75 | 48.91 |
| | Pr > F | <0.0001 | <0.0001 | <0.0001 | <0.0001 | <0.0001 | <0.0001 |
| Ethanol | F | 8.91 | 12.64 | 7.44 | 2.56 | 3.76 | 7.55 |
| (%) | Pr > F | 0.000 | <0.0001 | 0.001 | 0.087 | 0.030 | 0.001 |
| Subject*Ethanol | F | 3.62 | 10.38 | 1.66 | 1.98 | 2.82 | 4.54 |
| (%) | Pr > F | 0.000 | <0.0001 | 0.079 | 0.028 | 0.002 | <0.0001 |

[a] Standard error of mean. [b] Coefficient of determination. [c] F statistic. [d] Probability values.

Regarding the effects of ethanol content on the prolonged oral aroma release, it can be observed that the presence of ethanol significantly increased the oral release for five of the six esters assayed. Moreover, the release of ethyl hexanoate was also increased by the ethanol content, but the effect did not reach the significance level (*p*-value = 0.087) (Table 3). Once again, to better visualize these results, data were plotted, and shown in Figure 4. Here, data were normalized considering the values obtained for the 0.5% *v/v* as 0% and calculating the ratio for the other two wines. As can be observed, all the esters were more released at higher ethanol content. Moreover, the more apolar compounds (ethyl octanoate and ethyl decanoate) were also significantly more released in the 5% *v/v* than in 0.5% *v/v*. Interestingly, these results showed the same trend as the previous studies and dynamic sensory studies performed on this topic which showed an enhancement of aroma release in the presence of ethanol [14,20], and an increase in the duration of fruity/floral notes as the ethanol concentration increases [12].

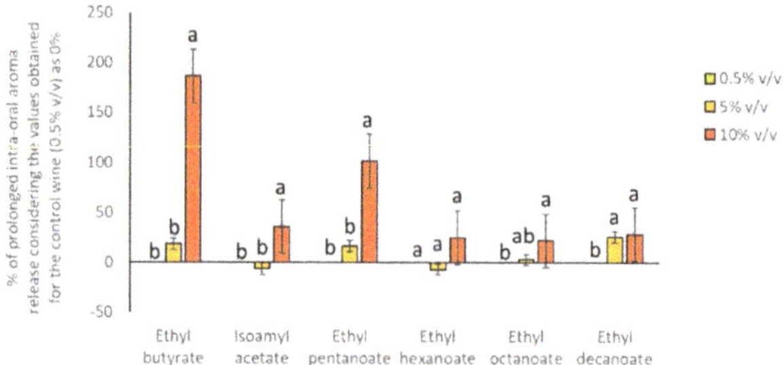


**Figure 4.** Percentage of prolonged oral aroma release obtained for the volunteers (n = 10) after rinsing their mouths with the wines (considering the values obtained for the W0.5% as 0% and comparing this value with the amount determined for the rest of wines). Different letters across the different compounds denote statistical differences after the application of the Tukey test for means comparison ($p < 0.05$).

Although additional studies are required in order to elucidate the mechanisms of action of ethanol content on the persistence of aroma compounds after wine intake, the higher prolonged oral aroma release observed in the presence of ethanol (from 5% *v/v* onwards) could be due to different facts. On the one hand, ethanol might induce changes in surface tension affecting the distribution of the wine in the mouth and pharynx during consumption, allowing the sample to better spread out and favoring the formation of a larger surface in the oral cavity for volatile release [20]. On the other hand, in the presence of ethanol, esters could be more solubilized in the layer of wine that remains in the mouth after wine rinsing and, therefore, these compounds could be released more slowly over time. Finally, the mixture of wine lingering in the oral cavity could be submitted to the Marangoni and Rayleigh convection phenomenon described in vitro by Tsachaki and co-authors [15]. In this case, the presence of ethanol would help aroma compounds be released more easily from the layer of wine to the oral headspace. Additionally, other mechanisms related to the interactions of ethanol with salivary proteins or enzymes could not be discarded. In this regard, it could be possible that the presence of ethanol could have inhibited certain salivary enzymes (e.g., esterases), more prone to metabolize long chain esters [21], like ethyl octanoate and ethyl decanoate. Additionally, the presence of ethanol (in higher concentrations than those of the aroma compounds) could have occupied the binding sites of the salivary proteins modifying the interactions between aroma compounds and oral proteins as suggested previously in the presence of sugar [22]. Therefore, additional studies are required in order to corroborate these hypotheses.

## 3. Material and Methods

### 3.1. Wine Samples

A dealcoholized rosé wine (Matarromera, Pesquera de Duero, Spain) from the Tempranillo grape variety with an ethanol concentration of 0.5% *v/v* (control wine) was selected for this study. To this wine, different amounts of ethanol content (Panreac Quimica S.A., Barcelona, Spain) were added to obtain two additional wines with 5% *v/v* and 10% *v/v* of ethanol content.

To reinforce the aroma profile of the wines, six food-grade esters (Sigma-Aldrich, Steinheim, Germany) were employed in this study. They present different physicochemical properties and are associated with fruity notes in wines (Table 1). Independent stock solutions of the compounds were prepared in food-grade ethanol (Panreac Quimica S.A., Barcelona, Spain). From here, each aroma compound was added to the wines immediately before each session to obtain a final concentration of 4 mg/L. It was previously determined by HS-SPME-GC/MS [23] that the amount of these compounds present in the original wine was negligible compared to the amount of aroma added (Table S1).

### 3.2. Intra-Oral SPME Sampling

#### 3.2.1. Subjects

Ten young, healthy volunteers (four male and six female) between 18–36 years old participated in this study. The sampling procedures were explained in detail to the subjects who also provided written consent to participate. The Bioethical Committee of the Spanish National Council of Research (CSIC) approved this study.

#### 3.2.2. Intra-Oral SPME Procedure for Aroma Monitoring

To study how ethanol affects the oral processing of typical wine esters, subjects rinsed their mouths with three aromatized wines containing different ethanol content (0.5% *v/v*, 5% *v/v* and 10% *v/v*), and then, the aroma released into the oral cavity after wine expectoration was monitored following the intra-oral SPME procedure previously developed and validated by Esteban-Fernandez and collaborators [24]. The representation of the analytical procedure is schematized in Figure 1. Briefly, subjects were instructed to introduce the aromatized wine (15 mL) into their oral cavity, kept it for 30 s and then spit it off. During rinsing, special care was taken to keep the lips closed, not to swallow and not to open the velum-tongue border prior to expectoration. At 30 s after expectoration, a DVB/CAR/PDMS (Divinylbenzene/Carboxen/Polydimethylsiloxane 50/30 μm thickness, 2 cm length) coated SPME fiber (Supelco, Bellefonte, PA, USA) with a home-made adaptor, was introduced into the oral cavity of the volunteer. After 2 min of oral aroma extraction, in which swallowing was not allowed, the fiber was removed from the oral cavity and immediately placed into the split/splitless injector (splitless mode) of the Gas Chromatograph (GC) Agilent 6890N coupled to a quadrupole Mass Detector (MS) Agilent 5973 (Santa Clara, CA, USA). This measurement is referred to as an immediate oral release. Four minutes after wine expectoration, a second fiber was introduced in the oral cavity following the exact protocol described previously (Figure 1). This second measurement is referred to as a prolonged oral aroma release.

After each injection, the fiber was cleaned for 10 min to avoid any memory effect. Analyses were performed three times with each wine type (0.5% *v/v*, 5% *v/v* and 10% *v/v*) by the ten volunteers (3 wine × 10 volunteers × 3 repetitions = 90 injections). An inter-fiber repeatability study was previously carried out. This allowed the selection of the most similar fibers (RSD values <10% for the extraction of the same aroma compound) to complete the study. Moreover, the use of a short sampling time also known as "true headspace" [25], avoids any problems derived from the use of SPME, such as the changes in the distribution of the molecules between the liquid and gas phases at equilibrium, due to adsorption phenomena onto the fiber itself, or to problems of saturation, due to an excess of ethanol absorbed onto the fiber compared to the compounds of interest [25]. In our case, the equilibrium is not reached, the fiber is not saturated, and therefore, the detected composition can be considered as representative of the real intraoral-headspace composition.

#### 3.2.3. GC/MS Analyses

Volatile compounds were separated on a DB-Wax polar capillary column (60 m × 0.25 mm i.d. × 0.50 um film thickness) from Agilent (J&W Scientific, Folsom, USA). Helium was the carrier gas at a flow rate of 1 mL min$^{-1}$. The oven temperature was initially held at 40 °C for 2 min, then increased at 8 °C each minute to 240 °C and held for 15 min.

For the MS system (Agilent 5973N), the temperature of the transfer line, quadrupole and ion source were 270, 150 and 230 °C respectively. Electron impact mass spectra were recorded at 70 eV ionization voltages, and the ionization current was 10 μA. The acquisitions were performed in a scan (from 35 to 350 amu) and SIM modes. The identification of the six target compounds was based on the comparison of retentions times and mass spectra. The mass spectra were compared with those from NIST 2.0 database and with those from reference compounds analysed in the same conditions. Since no internal standard was used during the oral SPME extractions, absolute peak areas (APAs) were obtained. The use of APAs to express aroma release was sufficient for this type of analyses as the aim of the work was to compare the extent of aroma release between wine samples.

### 3.2.4. Statistical Analyses

Two-way ANOVA and the Tukey test were used to determine significant differences in oral release parameters of the six aroma compounds considering subjects and wine type (ethanol concentration) as factors. The significance level was $p < 0.05$ throughout the study. The XL-Stat 2017.01 program was used for data processing (Addinsoft, Paris, France).

### 4. Conclusions

The present results showed that the effects of ethanol content on oral aroma release after wine consumption are dependent on the subjects, the ethanol concentration and their interactions and the nature of the aroma compound assayed.

The effects of ethanol content on oral aroma release were different immediately after wine consumption than four minutes later. In the first in-mouth aroma monitoring, an increase in the ethanol content provoked a higher release of the more polar and volatile esters (ethyl butanoate, ethyl pentanoate), but a lower release for the more apolar and less volatile esters (ethyl octanoate, ethyl decanoate). Regarding the prolonged oral aroma release, an increase of ethanol content in wine, increased the oral aroma release of the six esters. This effect was greater when the concentration of alcohol rose from 0.5% to 10% *v/v*, and it could be related to an increase of the fruity aroma persistence in the wines.

Moreover, important interindividual differences have been noted in this work, which could be related to variation in oral physiology parameters among subjects. Interestingly, high interaction between subject and ethanol was also found, meaning that in each subject, retronasal aroma release is differently affected by ethanol content. Future studies with a high number of individuals are needed in order to validate these results and to investigate the mechanisms behind the observed effects. Nonetheless, it is important to state that this research has shown the effect of ethanol on aroma release within the oral cavity but no information on the travel of aroma compounds to the receptors in the nose has been provided, which could be also of interest for new studies. Overall, these findings could help winemakers produce new low alcohol wines considering the effect of ethanol on aroma release.

**Supplementary Materials:** The following are available online, Table S1: Endogenous concentration of the target aroma compounds in the control wine (0.5% *v/v*).

**Author Contributions:** M.Á.P.-B. conceived the experiment. M.P.-J. and C.C. conducted the analyses. C.M.-G. performed the data analysis. C.M.-G. and M.Á.P.-B. wrote the manuscript. All authors reviewed the manuscript.

**Funding:** This research was funded by the Spanish MINECO through the AGL201678936-R (AEI/FEDER, UE) Project.

**Acknowledgments:** C.M.-G. and M.P.-J. thank the JdC and FPI programs (MINECO) for her respective postdoctoral and PhD research contracts (MINECO). The authors thank the volunteers who participated in this study and Francisco Javier Barroso for his technical assistance.

## References


1. Ickes, C.M.; Cadwallader, K.R. Effects of Ethanol on Flavor Perception in Alcoholic Beverages. *Chemosens. Percept.* **2017**, *10*, 119–134. [CrossRef]
2. Gawel, R.; Van Sluyter, S.; Waters, E.J. The effects of ethanol and glycerol on the body and other sensory characteristics of Riesling wines. *Aust. J. Grape Wine Res.* **2007**, *13*, 38–45. [CrossRef]
3. Zamora, M.C.; Goldner, M.C.; Galmarini, M.V. Sourness? Sweetness Interactions in Different Media: White Wine, Ethanol and Water. *J. Sens. Stud.* **2006**, *21*, 601–611. [CrossRef]
4. Escudero, A.; Campo, E.; Fariña, L.; Cacho, J.; Ferreira, V. Analytical Characterization of the Aroma of Five Premium Red Wines. Insights into the Role of Odor Families and the Concept of Fruitiness of Wines. *J. Agric. Food Chem.* **2007**, *55*, 4501–4510. [CrossRef]
5. Goldner, M.C.; Zamora, M.C.; Lira, P.D.L.; Gianninoto, H.; Bandoni, A. Effect of ethanol level in the perception of aroma attributes and the detection of volatile compounds in red wine. *J. Sens. Stud.* **2009**, *24*, 243–257. [CrossRef]
6. Guth, H. Quantitation and Sensory Studies of Character Impact Odorants of Different White Wine Varieties. *J. Agric. Food Chem.* **1997**, *45*, 3027–3032. [CrossRef]
7. King, E.S.; Dunn, R.L.; Heymann, H. The influence of alcohol on the sensory perception of red wines. *Food Qual. Prefer.* **2013**, *28*, 235–243. [CrossRef]
8. Doyennette, M.; Déléris, I.; Feron, G.; Guichard, E.; Souchon, I.; Trelea, I. Main individual and product characteristics influencing in-mouth flavour release during eating masticated food products with different textures: Mechanistic modelling and experimental validation. *J. Theor. Biol.* **2014**, *340*, 209–221. [CrossRef]
9. Gierczynski, I.; Guichard, E.; Laboure, H. Aroma perception in dairy products: the roles of texture, aroma release and consumer physiology. A review. *Flavour Fragr. J.* **2011**, *26*, 141–152. [CrossRef]
10. Muñoz-González, C.; Martín-Álvarez, P.J.; Moreno-Arribas, M.V.; Pozo-Bayón, M. Ángeles Impact of the Nonvolatile Wine Matrix Composition on the In Vivo Aroma Release from Wines. *J. Agric. Food Chem.* **2013**, *62*, 66–73. [CrossRef]
11. Salles, C.; Chagnon, M.-C.; Feron, G.; Guichard, E.; Labouré, H.; Morzel, M.; Semon, E.; Tarrega, A.; Yven, C. In-Mouth Mechanisms Leading to Flavor Release and Perception. *Crit. Rev. Food Sci. Nutr.* **2010**, *51*, 67–90. [CrossRef]
12. Baker, A.K.; Ross, C.F. Wine finish in red wine: The effect of ethanol and tannin concentration. *Food Qual. Prefer.* **2014**, *38*, 65–74. [CrossRef]
13. Petrozziello, M.; Asproudi, A.; Guaita, M.; Borsa, D.; Motta, S.; Panero, L.; Bosso, A. Influence of the matrix composition on the volatility and sensory perception of 4-ethylphenol and 4-ethylguaiacol in model wine solutions. *Food Chem.* **2014**, *149*, 197–202. [CrossRef]
14. Muñoz-González, C.; Rodríguez-Bencomo, J.J.; Moreno-Arribas, M.V.; Pozo-Bayón, M.Á. Feasibility and application of a retronasal aroma-trapping device to study in vivo aroma release during the consumption of model wine-derived beverages. *Food Sci. Nutr.* **2014**, *2*, 361–370. [CrossRef]
15. Tsachaki, M.; Linforth, R.S.T.; Taylor, A.J. Dynamic Headspace Analysis of the Release of Volatile Organic Compounds from Ethanolic Systems by Direct APCI-MS. *J. Agric. Food Chem.* **2005**, *53*, 8328–8333. [CrossRef]
16. Tsachaki, M.; Linforth, R.S.T.; Taylor, A.J. Aroma Release from Wines under Dynamic Conditions. *J. Agric. Food Chem.* **2009**, *57*, 6976–6981. [CrossRef]
17. Aznar, M.; Tsachaki, M.; Linforth, R.S.; Ferreira, V.; Taylor, A.J. Headspace analysis of volatile organic compounds from ethanolic systems by direct APCI-MS. *Int. J. Mass Spectrom.* **2004**, *239*, 17–25. [CrossRef]
18. Boelrijk, A.; Basten, W.; Burgering, M.; Gruppen, H.; Voragen, A.; Smit, G. The effect of co-solvent on the release of key flavours in alcoholic beverages; comparing in vivo with artificial mouth-MS nose measurements, Flavour research at the dawn of the twenty-first century. In Proceedings of the 10th Weurman Flavour Research Symposium, Beaune, France, 25–28 June 2002; pp. 204–207.
19. Tsachaki, M.; Gady, A.-L.; Kalopesas, M.; Linforth, R.S.T.; Atheès, V.; Marin, M.; Taylor, A.J. Effect of Ethanol, Temperature, and Gas Flow Rate on Volatile Release from Aqueous Solutions under Dynamic Headspace Dilution Conditions. *J. Agric. Food Chem.* **2008**, *56*, 5308–5315. [CrossRef]
20. Clark, R.; Linforth, R.; Bealin-Kelly, F.; Hort, J. Effects of ethanol, carbonation and hop acids on volatile delivery in a model beer system. *J. Inst. Brew.* **2011**, *117*, 74–81. [CrossRef]

21. Buettner, A. Influence of Human Salivary Enzymes on Odorant Concentration Changes Occurring in Vivo. 1. Esters and Thiols. *J. Agric. Food Chem.* **2002**, *50*, 3283–3289. [CrossRef]
22. Friel, E.N.; Taylor, A.J. Effect of Salivary Components on Volatile Partitioning from Solutions. *J. Agric. Food Chem.* **2001**, *49*, 3898–3905. [CrossRef]
23. Rodríguez-Bencomo, J.J.; Muñoz-González, C.; Andújar-Ortiz, I.; Martín-Álvarez, P.J.; Moreno-Arribas, M.V.; Pozo-Bayón, M.Á. Assessment of the effect of the non-volatile wine matrix on the volatility of typical wine aroma compounds by headspace solid phase microextraction/gas chromatography analysis. *J. Sci. Food Agric.* **2011**, *91*, 2484–2494. [CrossRef]
24. Esteban-Fernández, A.; Rocha-Alcubilla, N.; Muñoz-González, C.; Moreno-Arribas, M.V.; Pozo-Bayon, M.A. Intra-oral adsorption and release of aroma compounds following in-mouth wine exposure. *Food Chem.* **2016**, *205*, 280–288. [CrossRef]
25. Robinson, A.L.; Ebeler, S.E.; Heymann, H.; Boss, P.K.; Solomon, P.S.; Trengove, R.D. Interactions between Wine Volatile Compounds and Grape and Wine Matrix Components Influence Aroma Compound Headspace Partitioning. *J. Agric. Food Chem.* **2009**, *57*, 10313–10322. [CrossRef]

**Sample Availability:** Samples of the compounds are not available from the authors.

Article

# Investigating the Effect of Artificial Flavours and External Information on Consumer Liking of Apples

Isabella Endrizzi [1,*], Eugenio Aprea [1,2], Emanuela Betta [1], Mathilde Charles [3],
Jessica Zambanini [1] and Flavia Gasperi [1,2]

[1] Department of Food Quality and Nutrition, Research and Innovation Centre, Fondazione Edmund
Mach (FEM), Via E. Mach 1, 38010 San Michele all'Adige, Italy; eugenio.aprea@fmach.it (E.A.);
emanuela.betta@fmach.it (E.B.); jessica.zambanini@fmach.it (J.Z.); flavia.gasperi@fmach.it (F.G.)

[2] Center Agriculture Food Environment University of Trento/Fondazione Edmund Mach, via E. Mach 1,
38010 San Michele all'Adige, Italy

[3] Sensory and Behaviour Sciences research group, Sportslab, Decathlon SA, 59665 Villeneuve d'Ascq, France;
mathildeccharles@gmail.com

[*] Correspondence: isabella.endrizzi@fmach.it; Tel.: +39-0461-615388

Academic Editor: Henryk H. Jeleń
Received: 31 October 2019; Accepted: 24 November 2019; Published: 26 November 2019

**Abstract:** In this paper, the influence of flavour modification, artificially induced, on consumer acceptability of apple fruit is studied. The method consists of modifying the flavour of a real food matrix dipping apples into flavour solutions. Two flavouring compounds (linalool and anethole) that were responsible of "floral" and "anise" aroma descriptors, respectively, were considered here. The effectiveness of flavouring treatments was confirmed by instrumental analysis of volatile compounds profile using solid-phase microextraction gas chromatography/mass spectrometry (SPME/GC-MS) and by discriminative and descriptive sensory analyses. The effect of flavour-impact was evaluated in an informed test on the two flavoured 'Fuji' apples: the consumers were asked to evaluate the global liking of the treated and non-treated apples with information regarding the aromatic features. Participants' additional data on the characteristics on their "ideal apple", attitudes toward natural food, food neophobia, and demographic data were also recorded by specific questionnaires. A statistically significant effect on liking was found for the flavour factor, whereas external information only affected apple acceptance for subgroups of consumers, depending on their attitude towards food.

**Keywords:** flavouring treatment; SPME/GC-MS; triangle test; sensory profile; apple acceptability; external information

---

## 1. Introduction

Several authors investigated the influence of sensory attributes, both being related to texture and flavour, on consumer liking of apples. For example, the positive correlation between liking and texture parameters like crispness, hardness or juiciness has been demonstrated [1–3]. Symoneaux and colleagues [4] reported that crunchiness and sweetness are the main sensory preference key drivers in apple. Endrizzi and colleagues also found that an increase in some texture parameters has less influence on the sweetest apples and a greater influence on the least sweet [5]. The papers reported in the literature suggest that consumers classify apples according to a texture dimension (soft to firm) and a taste dimension (sweet to acid), which suggests that there are two main apple consumer categories: those who prefer firm, juicy, and quite acid apples, and those who like sweeter, but less firm, apples [6–10].

### *1.1. Studying Flavour in Apples*

It seems that apple flavour, as the retro-nasal perception of volatile compounds, is a factor of secondary importance for consumers or important just for a limited number of consumers [11,12]. This could be because crunchiness, juiciness, and the ratio between acid and sweet are effectively dominant characters in apples; in other words, these characteristics stimulate our senses first when we eat an apple [13]. Consumer experience when tasting an apple is the result of the multisensory integration of olfactory, taste, and tactile sensations, being modulated by the dynamic evolution of the tasting event and the location of sensory stimuli in the mouth [14]. For these reasons, consumers are commonly not educated to distinguish between taste and flavour and, therefore, their definition of sweet or acid actually includes a set of gustatory and olfactory characteristics. However, the acceptability test results also depend strongly on the variety of apples included in comparison during the study, and thus the selection of samples plays a fundamental role [5,12]. Sometimes, fruits that are developed during breeding programs, with positive flavour attribute, have to be excluded from a study, because available fruits are not enough to conducts the tests (generally one to three trees are available) or because they are considered to be unpleasant for other sensory aspects and cannot be proposed to the consumer [15]. Nevertheless, flavour continues to remain a primary focus in apple breeding [16,17], and it is known that it is an important factor because it also affects other dominant characteristics, such as sweetness [18,19]. In the past, the role of fruit volatiles and tastants on sensory perception has been studied while using model solutions [20], juices [21,22], fruit pulps [23,24], and the injection of flavours into fruit pieces [14,25]. More recently [13], the influence of aroma perception on taste and texture in apple was studied in our laboratory by dipping fruit pieces (cylinders) in a flavour solution.

### *1.2. Studying Intrinsic and Extrinsic Factors*

In recent years, the number of publications that investigates consumer choice by simultaneously evaluating intrinsic and extrinsic factors by means of rating-based conjoint experiments has increased (see among others [26,27]). Nevertheless, far fewer studies investigated taste as a factor in a conjoint framework, mainly measuring the effect of different levels of sweetness or sourness, sometimes combined by texture attributes [5,28–31]. As far as the authors' know, there are no conjoint studies that examine the effect of different flavours/aromas. It is also known that, external information, as claimed, has a role in consumers' perception influencing food choices. Several studies have been performed investigating the effect of nutrition and health claims (see among others [5,32,33], information regarding origin [34], production method [35], quality labels [36], or sustainability labels [37]).

### *1.3. Objective of the Study*

Here, for the first time as far as the author knows, a conjoint study including tasting was carried out on apple, measuring the effect on liking of different flavours and of external information regarding the flavour of fruit under evaluation. To do that, the effect of flavour-impact was artificially induced in the real food matrix to compare consumer acceptability of unmodified versus flavoured Fuji apples. The effectiveness of the flavouring treatment was evaluated by instrumental and sensory methods. After that, a consumer test was performed to investigate the overall liking of the treated apples. Therefore, the objective of this work is two-fold: to check the applicability and the effectiveness of the flavouring method proposed and evaluate the effect on acceptability of the chosen flavours in apple, in informed conditions.

## 2. Results

The evaluation of the effectiveness of flavouring treatment was achieved by instrumental (Section 2.1.1) and sensory perception verification (Sections 2.1.2 and 2.1.3). Afterwards, consumers assessed the overall liking of apples (Section 2.2).

### 2.1. Verification of Flavouring Treatment Effectiveness

#### 2.1.1. Quantification of Volatile Organic Compounds

Calibration curves, based on solid-phase microextraction gas chromatography/mass spectrometry (SPME/GC-MS) analysis, were built for the quantification of linalool and anethole present in the apple pulp. The coefficient of variation of the quantification method, which was calculated over 10 consecutive injections, was lower than 14% for linalool and lower than 11% for anethole. Table 1 reports an average amount of flavouring and ethanol penetrating in the apples, as calculated over 10 different fruits per each treatment. Linalool and anethole isomer (estragole) are common constituents that are found in apple headspace [38], generally at low concentration, and they may increase depending on several factors, including ripening and storage. The average amount of linalool and anethole naturally present in the pealed apples used in this study was 0.030 (min 0.026, max 0.031) and 0.019 (min 0.008, max 0.038) mg/kg, respectively. After flavouring treatment, the amount of linalool that was found in the apple pulp was on average 1.05 ± 0.24 mg/kg. Nevertheless, the amount of anethole was 1.02 ± 0.14 mg/kg. The ethanol naturally occurring in the used apples was, on average, 67.1 ± 16.0 mg/kg and only marginally influenced by the soaking treatment (from 52.7 to 81.2 mg/kg). In the Supplementary Materials, we also report the spatial (radial) adsorption trend of linalool and anethole in the apple flesh (Figures S1 and S2). In conclusion, the volatile organic compounds analysis by SPME/GC-MS confirmed that the treatment allowed for the penetration of the flavouring agent into the apple flesh. Furthermore, the two concentrations of flavouring agent employed allowed for obtaining apples with two levels (statistically different) of linalool and anethole.

**Table 1.** Average amount (mg/kg) and standard deviation (in parenthesis) of ethanol, linalool and anethole in non-flavoured (R) and flavoured (L, A) apples that were treated with two different concentrations (1 and 2.5 g/L), calculated over 10 different fruits per each treatment by solid-phase microextraction gas chromatography/mass spectrometry (SPME/GC-MS).

|          | R             | L_1           | L_2.5         | A_1           | A_2.5         |
|----------|---------------|---------------|---------------|---------------|---------------|
| ethanol  | 67.11 (15.99) | 77.16 (25.57) | 73.96 (23.91) | 81.23 (31.88) | 52.68 (17.35) |
| linalool | 0             | 0.35 (0.07)   | 1.05 (0.24)   | 0             | 0             |
| anethole | 0             | 0             | 0             | 0.49 (0.16)   | 1.02 (0.14)   |

#### 2.1.2. Triangle Test

Table 2 reports the results of the triangle tests. Apples that were flavoured with anethole, in both concentrations, were perceived to be statistically different from those not treated, while those that were flavoured with linalool were perceived as being statistically different from those that were not treated just in the preparation with the highest concentration. Both compounds were targeted to be presented at the same concentration in the apples after the treatment; nevertheless, the compounds may not elicit the same intensity of aroma at equal concentrations, as confirmed by the data reported in Table 2. We expected a higher rate of correct responses for linalool at low concentration, but the detection thresholds in apple could be different from those found in water, according to the anethole (0.073 mg/L) and linalool (0.00017 mg/L) thresholds in water [39]. At the tested higher concentrations, linalool is more clearly perceived than anethole, perhaps because the floral scent that is given by linalool in apples is recognised as an incongruous sensation, while anethole is not perceived as such: indeed, some varieties are characterised by this attribute in aged apples [40].

**Table 2.** Total number of responses, number and percentage of correct responses and the related p-value of each triangle comparison. Statistically significant comparisons (*p*-value < 0.05) are reported in bold.

| Comparison | Description | Total Responses | Correct Responses | % of Correct Responses | *p*-Value |
| --- | --- | --- | --- | --- | --- |
| 1 | R vs. L_1 | 38 | 18 | 47 | 0.051 |
| 2 | R vs. L_2.5 | 38 | 27 | 71 | **<0.001** |
| 3 | R vs. A_1 | 38 | 22 | 58 | **0.002** |
| 4 | R vs. A_2.5 | 38 | 19 | 50 | **0.025** |

### 2.1.3. Descriptive Sensory Analysis

Descriptive profiling data of the six apples (R, A_1, A_2.5, L_1, and L_2.5) were submitted to PCA to explore similarities and differences among the treated and untreated apples. Figure 1 reports the model on all of the sensory attributes. The analysis revealed that the first three PCs accounted for 93% of the variability. PC1 accounted for 44% of the variability and clearly apple A_2.5 is separated from others along this dimension, whereas sample L_2.5 differs from others along PC2 (25%). Thus, anise flavoured apples are characterised by anise odour and flavour, while apples that are treated with linalool are described with "floral honey" odour, "floral", and "pineapple" odours and flavours. This is consistent with [25], where fruit pieces that were injected with linalool were mainly described as floral/citrus. In Figure 1b, where PC2 and PC3 (24%) are plotted together, apples that are treated with the solution at the lower concentration (L_1 and A_1) are represented closer and seem to be characterised by higher levels of sweetness. Overall, the results from descriptive sensory analysis revealed that the flavouring method was successful in modifying the samples for their flavour component, but not the other sensory properties.

(**a**)

**Figure 1.** *Cont.*

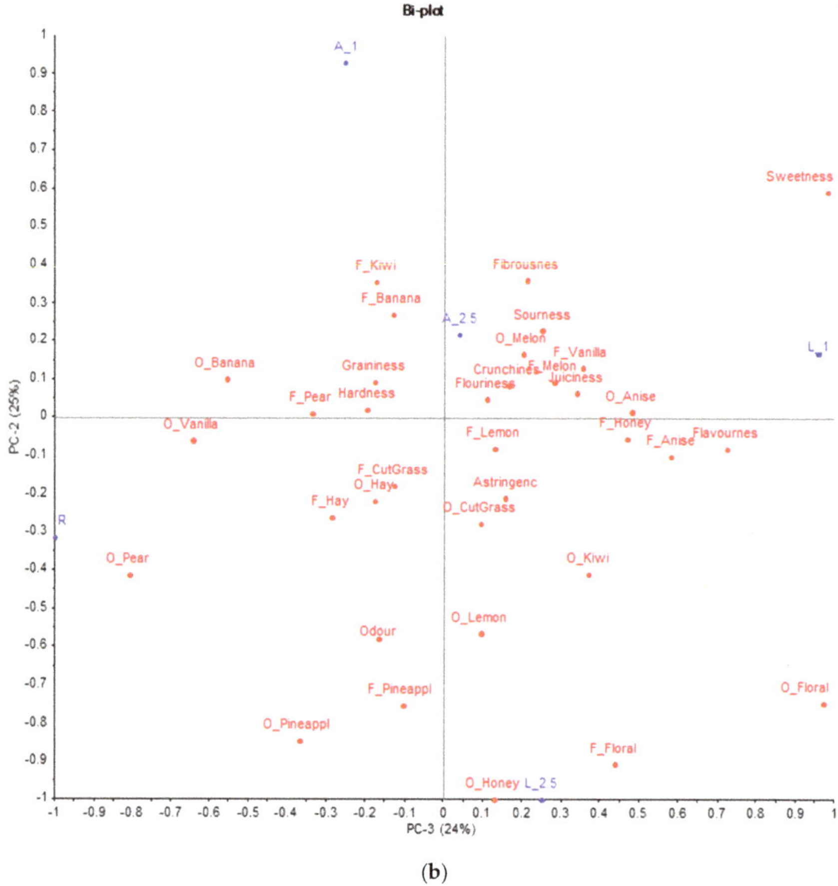



**(b)**

**Figure 1.** Bi-plot of PCA model on all sensory variables: PC1 vs. PC2 (**a**) and PC3 vs. PC2 (**b**). The six apples, two varieties (Golden Delicious and Fuji) in the three treatments (R, A, and L), are written in bolt.

## 2.2. Consumer Study

### 2.2.1. Description of Consumer Panels

Fifty-one percent of the panel ($N$ = 207) was composed of male, 20% being childless, and 43% living in a family. Seven percent of the panel has a postgraduate degree, 65% was non-smoker, and 64% did sports more than twice a week. For the majority of the 207 consumers an ideal apple should be very crunchy (61%), very juicy (67%), very aromatic (63%), but just fairly acid (92%) and fairly sweet (76%). The panel was segmented in three groups according to its attitude towards natural food interest (NFI): low (31%), medium (38%), and high (31%). A further segmentation on the basis of the food neophobia scale (FNS) was also performed: low (34%), medium (32%), and high food neophobia (34%).

### 2.2.2. Informed Testing

The results of the ANOVA mixed model showed significant main effects on liking for consumer and flavour ($p$ < 0.0001 and $p$ = 0.007, respectively) and a non-significant main effect for external information ($p$ = 0.108). All of the interactions are highly significant ($p$ < 0.0001). The interaction between flavour and information (F × I) had the strongest effect (MS = 13.00), followed by the flavour level (MS = 12.77), with anise flavoured apple (A; M = 6.31, SD = 1.64) being the least preferred sample statistically different from floral flavoured apple, which is the most preferred (L; M = 6.65, SD = 1.73). The sample with no added flavour had a different evaluation, depending on submitted information (Figure 2): awarded under the claim 'traditional' (R; M = 6.83, SD = 1.58), but penalised if it was labelled as 'chosen for its intense flavour' (R; M = 6.29, SD = 1.69). Table 3 reports the results of the ANOVA model (1), recalculated for specific subgroups identified by FNS (FNS_3 = 70 consumers with high food neophobia), NFI (NFI_2 = 77 consumers with moderate interest towards natural food), age (on Age_1 = 67 consumers under 33 years of age), and gender (105 males and 102 females). The significant main effect of flavour is reconfirmed in people under 33 years of age (Age_1) and in males (Figure 3). Additionally, there are significant effects on the liking of external information regarding flavour: females, subjects with a moderate interest towards natural food and subjects with high level of food neophobia evaluated apples that were labelled as 'traditional' higher than those claimed as 'chosen for its intense flavour' (Figure 4). The finding concerning the F × I interaction was also confirmed in several sub-groups of consumers: males ($p = 0.011$) and females ($p = 0.016$), for those who have a low interest in natural foods ($p = 0.008$) and low neophobia ($p = 0.015$).

**Table 3.** *p*-values of ANOVA mixed model on the effects of consumer and conjoint factors on liking (All = 207 consumers), on FNS_3 = 70 consumers with high food neophobia, on NFI_2 = 77 consumers with moderate interest towards natural food, on Age_1 = 67 consumers under 33 years of age, on males (105) and females (102). Statistically significant effects (*p*-value < 0.05) are reported in bold.

| Source of Variation | Effect | All | FNS_3 | NFI_2 | Age_1 | Males | Females |
|---|---|---|---|---|---|---|---|
| Consumer (C) | Random | **0.000** | **0.001** | **0.013** | **0.014** | **0.000** | **0.000** |
| Flavour (F) | Fixed | **0.007** | 0.175 | 0.147 | **0.003** | **0.004** | 0.098 |
| Info (I) | Fixed | 0.108 | **0.007** | **0.042** | 0.885 | 0.953 | **0.041** |
| C × F | Random | **0.000** | **0.000** | **0.003** | **0.000** | **0.000** | **0.000** |
| C × I | Random | **0.000** | **0.006** | 0.065 | 0.210 | **0.015** | **0.008** |
| F × I | Fixed | **0.000** | 0.091 | 0.168 | 0.063 | **0.011** | **0.016** |

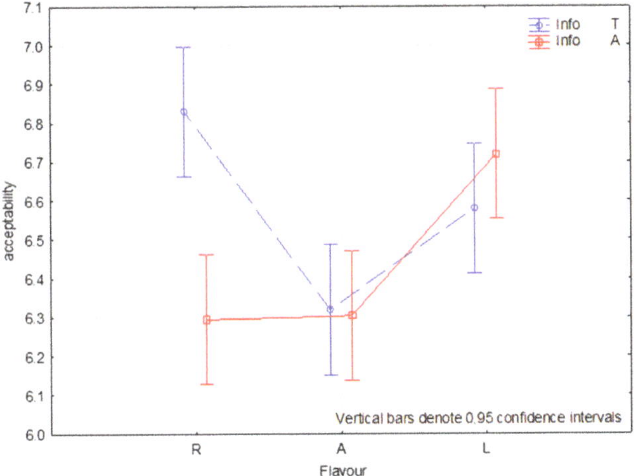


**Figure 2.** Representation of interaction effect between Flavour (R = reference, A = anethole, L = linalool) and Information (T = traditional, A = chosen for its intense aroma) on liking in the two-way ANOVA mixed effect model.

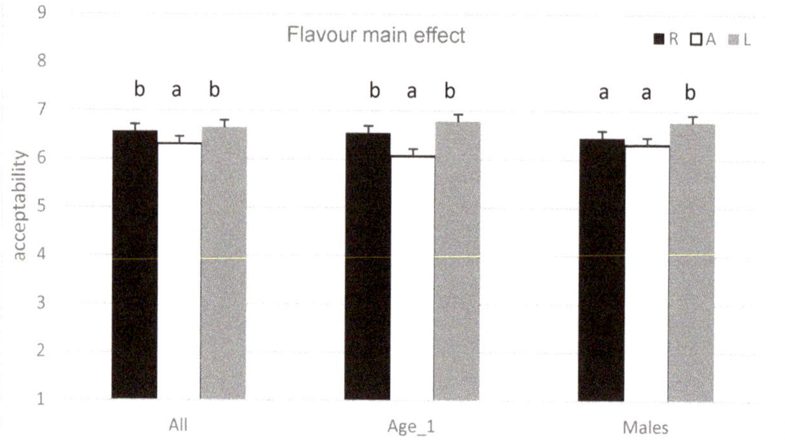

**Figure 3.** Main effect of flavour (R = reference, A = anethole, L = linalool) on liking of the whole consumer sample (All) and of specific subgroups are reported (Age_1 = 67 consumers under 33 years of age, 105 males). Within the same group, mean values with different letters are significantly different ($p < 0.05$).

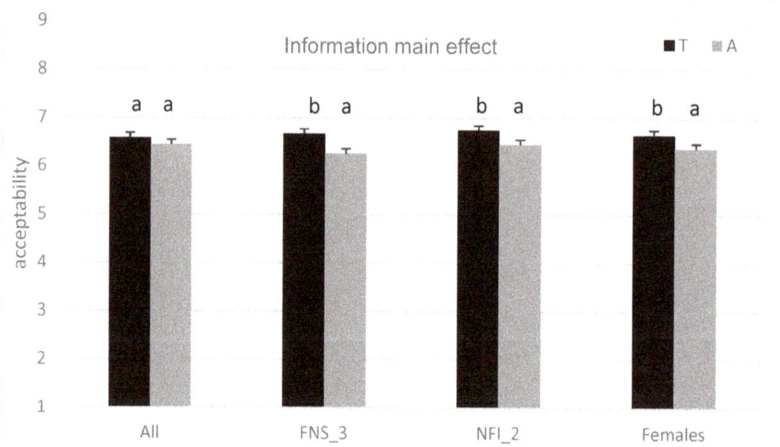

**Figure 4.** Main effect of information (T = traditional, A = chosen for its intense aroma) on liking of the whole consumer sample (All) and on specific subgroups of it: FNS_3 = 70 consumers with high food neophobia, NFI_2 = 77 consumers with moderate interest towards natural food, 102 consumer females. Within the same group, mean values with different letters are significantly different ($p < 0.05$).

## 3. Discussion

### 3.1. Flavouring Method

In the present paper, a new way to investigate flavour impact in apples was proposed. The method used has the advantage of not modifying the apple in its shape, which allows for it to be consumed as a whole fruit if necessary, or in apple slices, as presented here. In previous studies, participants tasted fruit cylinders, cubes, or quartered pieces [13,15,25]. This approach was applied on the Fuji apple, one of the most popular apple varieties in Italy and Europe. It is characterised by appreciated taste and texture attributes with a mild flavour profile [10,41], and this is the reason why we have chosen it. Anethole and linalool have been chosen between the shortlist of those volatile compounds that are already naturally present in apples in low amount [38]. Even if anethole has been shown being noticeable in specific varieties as Ambrosia and in increasing concentration by age [39]. The research was successful in achieving the aim of developing a method that was able to alter the flavour profile of fruit while using apple fruit as an example. The results demonstrate that the compounds reproducibly penetrate into the fruit tissue. This finding is also supported by the perception of the panellists who were able first to discriminate the samples and then to describe them as different in terms of the aromatic profile. The flavouring method should only change the ortho- and retro-olfactory perception and not the other characteristics in order to be effective. This makes sense only from the instrumental point of view, because any different perception that we record could be the result of both a real modification and a multisensory interaction (among the others [42,43]). In our case, the perception of texture attributes remains unchanged between the flavoured samples and the reference, whereas flavoured samples that were prepared with the lowest concentration solution showed a higher level of sweetness (Figure 1). This could be the result of a sensory interaction. It is known that the perception of sweetness in apples is not due only to the sugar content, but also to a strong aromatic component [19]. Here, it seems that when the aroma is perceived clearly and is therefore above the recognition threshold, it is correctly associated and measured; otherwise, it seems to contribute to the evaluation of sweetness. However, further investigations are needed to validate this thesis.

### 3.2. Effect of the Flavour

The goal of this work is to study the impact of flavour on consumer when information is present. Here, the evidence that consumers' hedonic scores changed in response to the different flavour treatment was demonstrated: the anise-flavoured apple is always the least appreciated, while the floral flavoured is the most appreciated one. The smell of anise is undoubtedly a smell/flavour that divides people according to their liking, even if there is no information regarding that in the literature. Jaeger et al. [25], who injected essences (linalool among the others) into pieces of kiwi, found that their samples that were treated with high linalool were very high in the intensity of floral/citrus attribute and this contributed to diminished acceptability. Indeed, in their study, the sample that was injected with high linalool was the least appreciated. The significant effect of the flavour found for the whole panel (All in Figure 3) is only reconfirmed for male and younger consumers. This is a result that is contrary to the expectations for the authors, who expected mature people and women to be more discriminative because they are generally more open to novelties. Previous works in literature, in fact, described that younger participants showed a higher degree of food neophobia [44] and that women are generally more open than men [45], even if, for gender, some authors described the contrary [46].

### 3.3. Effect of Information

Giving information about "traditional" or "selected flavour" profile while also considering the total mean data did not influence apple acceptance. This result is in accordance with Endrizzi et al. [5], who found that external information are not significant for the whole panel, but rather for specific consumer groups, depending on their sensitivity to the information given. If we associate 'traditional' information with the concept of domestic, locally grown, this result is in contrast with those found in the literature where consumers are demonstrated to prefer locally grown apples [47–50]. This is might be because if, in the cited papers, the concept 'local' is always contraposed to 'non-local' and thus related to 'freshness'. Here, we are measuring a different aspect that is more related with flavour profile. However, information regarding 'traditional apple' increased the acceptance of apples for specific subgroups of people, depending on consumer characteristics (Figure 4): for females, for highly food neophobic consumers, and for consumers moderately interested towards natural food. These results are not surprising for neophobic consumers who are more comfortable with familiar products and, for females, being generally less neophobic in accordance with the findings that were found by Demattè et al. [44]. With regard to the group of people moderately interested in natural food, as compared to highly NFI people, they evaluate products with reduced or non-fat content as less healthy and those with added sweeteners more healthy [51]. We would have expected the positive effect of traditional information on liking, probably linked to the concept of unmodified, in highly NFI people instead, who consider the importance of eating organic, not processed without additives foods. A further result concerns the interaction between Flavour and Information (F × I), which was found to be significant for the whole panel (Figure 2), but also in the subgroup of consumers with a high interest towards natural food and for the group of neophobics. From this, it emerges that the effect of external information only affects the sample without adding flavour, for which there is, on average, a higher rating if it is presented as 'traditional'. It seems that the external information only has effect on a product when it is appreciated, penalizing it if presented with the claim 'chosen for its intense flavour' since the sample R, the one without added flavour, is undoubtedly the preferred sample. The information does not affect the rating when the product has an unpleasant (unexpected or incongruent) flavour.

## 4. Design and Methods

### 4.1. Flavouring Treatment

#### 4.1.1. Preliminary Trials

The method for modifying the flavour, as already presented by Charles et al. [13] for fruit pieces, consisted here of dipping the whole (unpeeled) fruit in a flavour solution with the objective of modifying the apple flavour with negligible alteration of other sensory attributes. In the preliminary phase of this study, several flavour compounds that were compatible with the apple matrix at different concentration levels were screened and tested (data not presented here). A focus group of researchers, as combined to sensory and instrumental analyses, evaluated the effectiveness of the "dipping" method in relation to the apple variety, the compound used, the concentration to be reached, the impact of other ingredients in the flavouring solution, and to evaluate the congruence between the selected flavour and the apple matrix global perception in order to obtain a realistic product. The focus group started evaluating seven compounds (limonene, cinnamaldehyde linalool, eugenol, anethole, butyl acetate, and benzaldehyde). Just for few of them, the "dipping" method produced a perceptible change in the flavour profile of the apples. Among these, those that induced a realistic flavour in the apples were chosen.

#### 4.1.2. The Flavouring Solutions

As a result of the preliminary trials, two odorants at two levels of concentration were chosen to flavour a batch of 'Fuji' apple: linalool (L) at 1 and 2.5 g/L responsible for floral flavour and anethole (A) at 1 and 2.5 g/L responsible for anise flavour. The flavouring solutions were prepared dissolving pure food grade linalool (95%, Sigma-Aldrich, St. Louis, MO, USA) and anethole (99%, Sigma-Aldrich) in a 90% ethanol-water solution, respectively. The 90% ethanol-water solution was chosen to favour the complete dissolution of the flavouring agent, since anethole has weak solubility in water.

No flavouring treatment was considered and the same base solution (90% ethanol-water) without any flavouring agent was used for the preparation of the reference sample (R).

#### 4.1.3. Sample Preparation

The Fuji fruits were harvested in 2013 at a commercial maturity from orchards that were located in Trentino region (North of Italy). A batch of 60 fruits that was homogeneous as possible in ripening, size, and without any visible external damage was bought from local retailers.

As the penetration of the flavouring compound could depend on the size of the whole fruit, the fruits were sorted for weight and divided into three groups on the basis of these measures, and then stored at room temperature (18 ± 1 °C) for 24 h prior to the flavouring treatment. The flavouring treatment was then carried out without distinction on the three groups. The fruits were immersed in the flavouring solution for 1 min. and then removed and placed on a drip grid inserted in a 20 L plastic container. The containers were sealed with parafilm M$^{®}$ and stored for 72 h at 20 °C in dark room prior the tests.

According to this protocol, the following five apple samples were obtained: two "floral" samples flavoured by linalool at 1 and 2.5 g/L (L_1 and L_2,5); two "anise" samples flavoured by anethole at 1 and 2.5 g/L (A_1 and A_2,5), and one "reference" sample (R).

The penetration of the flavour compounds in apple flesh was verified and quantified by SPME/GC-MS analysis, being briefly described in the following Section (Section 4.2.1). Two sensory panels performed discrimination analysis and descriptive sensory analysis to verify the differences between reference and treated apples in terms of human perception (Section 4.2.2).

Only the concentration of 2.5 g/L for both flavourings was chosen for the following consumer test based on the results of these analyses.

### 4.2. Verification of Flavouring Treatment Effectiveness

#### 4.2.1. Quantification of Volatile Organic Compounds

Ten whole fruits for each flavouring treatment (L and A) at each concentration (1 and 2.5 g/L) and further ten whole fruits as reference (R) were submitted to quantitative analysis by SPME/GC-MS in order to investigate the effectiveness and the variability of the flavouring method. Seventy grams from the peeled fruits were diced while using a commercial cutter and immediately inserted in a glass vessel where they were mixed with 75 mL of deionised water, 30 g of sodium chloride, 250 mg of ascorbic acid, and 250 mg of citric acid [38]. From the homogenised samples, a 5 g aliquot was inserted into a 20 mL screw cap vial, suitable for volatile analysis, and it was spiked with 50 μL of 2-octanol (2.5 mg/L) used as internal standards. The vials were placed in the thermostated autosampler tray at 4 °C before the HS-SPME/GC-MS analysis. Calibration curves considering seven calibration points (including blank) were built for the quantification of ethanol (from 0 to 120 mg/kg), linalool (from 0 to 1.00 mg/kg), and anethole (from 0 to 1.10 mg/kg), dissolving known concentrations of the chemicals in apple puree. Volatile compounds from the headspace were extracted and then concentrated on a 2 cm Solid Phase Microextration fibre coated with divinylbenzene/carboxen/polydimethylsiloxane 50/30 μm (DBV/CAR/PDMS, Supelco, Bellefonte, PA, USA). The choice of fibre, as well as analysis parameters, was based on previous work [37]. The fibre was exposed to the apple headspace for 30 min. with the samples being equilibrated at 40 °C. Volatile compounds that adsorbed on the SPME fibre were desorbed at 250 °C in the injector port of a GC that was interfaced with a mass detector, which operated in electron ionization mode (EI, internal ionization source; 70 eV) with a scan range from *m/z* 35–350 (GC Clarus 500, PerkinElmer, Norwalk CT, USA). Separation was achieved on a HP-Innowax fused-silica capillary column (30 m, 0.32 mm ID, 0.5 μm film thickness; Agilent Technologies, Palo Alto, CA, USA). The GC oven temperature program consisted of 40 °C for 3 min., and then 40–220 °C at 4 °C/min., stable at 220 °C for 1 min., and then 220–250 at 10 °C/min., and finally 250 °C for 1 min. Helium was used as the carrier gas with a constant column flow rate of 1.5 mL/min.

#### 4.2.2. Sensory Analysis

All of the sensory tests were conducted in the FEM Sensory Laboratory compliant to the EN ISO standards 8589 [52], which was equipped with 22 individual booths while using FIZZ 2.46A software (Biosystemes, Couternon, France) to collect the responses.

#### Triangle Test

The possible unspecific sensory differences in treated apples were investigated while using the triangle method [53] to determine whether flavouring treatment induced perceptible sensory differences in apples. Four consecutive triangle tests were organized to compare the floral and anise flavoured apples at two different concentration levels (1 and 2.5 g/L) with the reference apple (R vs. L_1; R vs. L_2,4; R vs. A_1; R vs. A_2,5). Thirty-eight judges, who were employed at FEM (Fondazione Edmund Mach), with previous experience in discriminative or descriptive sensory analysis, were invited to perform the tests. The apple samples were first peeled, cut using an apple-slicer-corer (12 slices), and then dipped in an antioxidant solution (0.2% citric acid, 0.2% ascorbic acid, 0.5% calcium chloride) [41] preliminarily validated by Corollaro et al. [54]. One slice per sample was served in plastic cups that were labelled with three-digit random codes. Sample presentation order was randomised and counterbalanced over the panel. The judges were asked to evaluate by both smelling and tasting the samples under red light and identify the different sample in each of the four triad randomly presented.

Descriptive Analysis

A panel of 10 experienced judges performed the sensory profile of apples according to the descriptive analysis method, while using a consensus lexicon that was developed by Corollaro et al. [41]. For the description of specific odours and flavours, the panel used a list of 11 sensory attributes that were classified in four different categories: fruity (O/F Pear, O/F Banana, O/F Lemon, O/F Kiwi, O/F Pineapple, and O/F Melon), vegetal (O/F Cut grass and O/F Hay), spicy (O/F Vanilla and O/F Anise), and floral (O/F Floral Honey) [13]. In this experiment, in which linalool-flavoured samples are evaluated, the judges were allowed to describe sample odours or flavours while using individual extra attributes not contained in the list because just one floral attribute was included in sensory list. These free evaluations were collected in O/F_Floral, as all extra attributes reported were referable to this category. For each sample (R, A_1, A_2.5, L_1, and L_2.5), randomly presented in the test session, eight apple cylinders (1.8 cm diameter, 1.2 cm high each) were cut, dipped in an antioxidant solution, and then served in a plastic cup that was labelled with three-digit numbers and presented in a balanced order over the panel. The samples were evaluated under red light. Refer to Corollaro et al. for further details regarding the selection of the panel, its performance, general lexicon development, and sensory test procedures [41]. For specific odours and flavours attribute sensory definition, evaluation procedure, and references refer to Charles et al. [13].

*4.3. Consumer Study*

The consumer study was performed in the FEM sensory laboratory by 207 consumers who attended the "open door" event for the FEM's 140th anniversary (51% males; age: M = 41, SD = 14, Min. = 16, Max = 74). All subjects, who were not paid, declared to like apples and voluntarily joined the sensory evaluations. In addition to socio-demographic data, the participants also provided information regarding the characteristics of their ideal apple, choosing one of the three options (very, fairly, or little) for each ideal feature investigated (crunchiness, juiciness, sweetness, acidity, and aroma). They also provide their attitude toward natural food interest [51] and food neophobia scale [55]: the participants rated their degree of agreement with a series of positive and negative statements conveniently translated in Italian while using a nine-point scale rather than the original seven-point scale [5,24].

4.3.1. Conjoint Test

The test consisted of an experiment that was evaluated in 'informed' conditions combining conjoint analysis with the tasting of the two treated apples (floral and anise flavoured) and the untreated one. Each consumer received six apples in total according to a complete factorial design with three treatments and two information levels: three apples (R, A_2.5 and L_2.5) presented twice every time together with a different claim: 'traditional' or 'chosen for its intense flavour. The claims regarding apple flavour were submitted to consumers on the computer screen (Figure 5) just before tasting the samples. Consumers rated the overall liking of the six apples on a nine-point scale from 1 = "Dislike extremely" to 9 = "Like extremely". No verbal instructions were given to the consumers prior to testing: the consumers were told to pay attention and carefully read all of the instructions provided during the test.

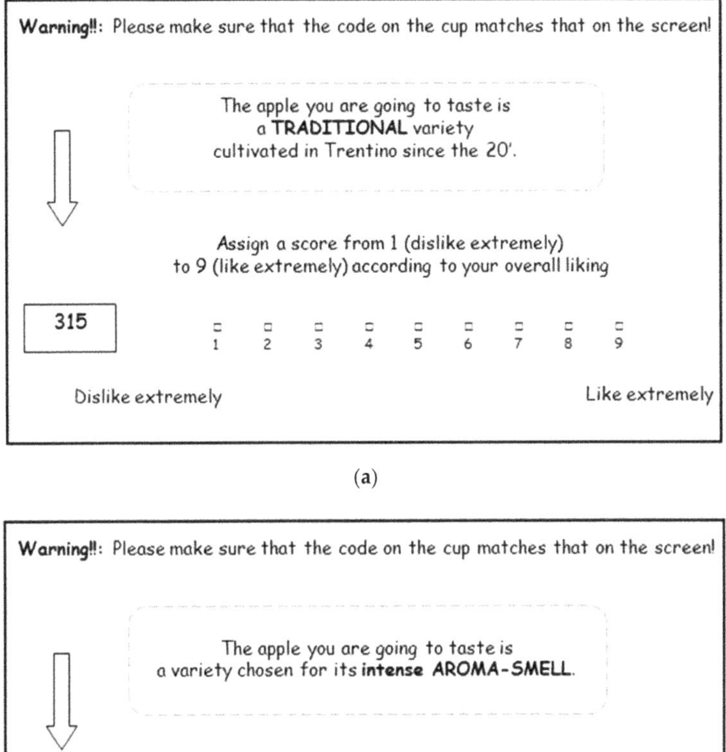



(a)



(b)

**Figure 5.** Examples of screen used in the conjoint study: (**a**) the information about traditional apple; and, (**b**) the information about aromatic apple.

### 4.3.2. Sample Preparation

All of the samples were peeled, cut using an apple-slicer-corer (12 slices), dipped in an antioxidant solution, and one slice per sample was served in plastic cups labelled with three-digit random codes and then evaluated under white light. The heaviest apples from G3 ($M = 213 \pm 6$ g, $N = 84$) were submitted to the test first, then those from G2 ($M = 197 \pm 4$ g, $N = 73$), and finally those from G1 ($M = 183 \pm 5$ g, $N = 50$): in this way, each consumer tasted apple slices that were obtained from apples of the same weight range.

### 4.4. Statistical Analyses

For triangle tests, the difference between the two products was considered to be statistically significant when the error was less than or equal to 5% (alpha = 0.05), which corresponded to a level of confidence that was greater than or equal to 95%. Standardised data of descriptive profiling were submitted to PCA (without any weighting option) to explore the similarities and differences among the treated and untreated apples. The liking data that were obtained in the consumer testing were first verified to measure the effect of apple weight (G1, G2, and G3, as mentioned in Section 4.3.2) with a one-way anova on acceptability. The liking data were further analysed using the following model (Equation (1)) with main effects and two-factor interactions for the design variables plus random effect of consumer because no effect was present:

$$y_{ijk} = \mu + \alpha_i + \beta_j + C_k + \alpha\beta_{ij} + \alpha C_{ik} + \beta C_{jk} + \varepsilon_{ijk}, \; I = 1, I, j = 1, J, k = 1, K, \quad (1)$$

Here, $y_{ijk}$ is the (ijk)th observation, $\mu$ is the general mean, $\alpha_i$, $\beta_j$ are the main effects of the two conjoint factors flavour and information, respectively. $\alpha\beta_{ij}$. is the fixed interaction effect. $C_k$ represents the main effects of consumers, whereas $\alpha C_{ik}$ and $\beta C_{jk}$ are its interaction effects with the design variables, and $\varepsilon_{ijk}$ is the independent random noise. All of these random effects are assumed to be independent and homoscedastic. The same ANOVA model (Equation (1)) was recalculated in specific demographic subgroups of consumers identified by gender (105 males and 102 females), age, natural food interest (NFI), and food neophobia scale (FNS) in order to identify which groups of people were more influenced by the information. According to the 33rd and 66th percentile points, consumers were classified in three groups: for age (Age_1 = 67 consumers under 33 years of age, Age_2 = 70 consumers between 33 and 48 years, and Age_3 = 70 consumers over 48 years), NFI score (NFI_1 = 65 consumers with low interest towards natural food, NFI_2 = 77 consumers with moderate interest, NFI_3 = 65 consumers with high interest), and FNS score (FNS_1 = 71 consumers with low level of food neophobia, FNS_2 = 66 consumers with medium level, and FNS_3 = 70 consumers with high level).

Summary statistics, analyses of variance (ANOVA), post-hoc Tukey's test, and correlation analysis for the sensory parameters were performed while using Statistica 13.1 (StatSoft, Inc., Tulsa, OK, USA). The Unscrambler X 10.4.1 (CAMO Software AS., Oslo Science Park, Gaustadalléen 21, 0349 Oslo, Norway) was used for the principal component analysis (PCA).

## 5. Conclusions

This study showed that the model system for modifying the flavour of a real structure that is achieved by dipping the whole fruit in flavour solutions is suitable for flavouring the fruits without altering their nature. Overall, tasted apples were acceptable showing scores between 6 and 7. It seems the claim, especially the indication of 'traditional', has a positive effect on liking. Different flavouring treatments significantly affect consumers' preference, depending on their personal liking, while external information only affected apple acceptability for some groups of consumers depending on their food approach. This confirms that flavour is an important factor in apple consumer preference and it could negatively influence the preference as it happened here. External information seems to be less relevant in general, but it can be important for some people to be more receptive to the information provided. The work presented here is an example of how consumer science can contribute to the effectiveness of fruit breeding programs by providing clear consumer advice and a way to early test novel flavour in product development.

**Supplementary Materials:** The following are available online. Figure S1: Linalool gradient from the peel to the core of the apple after flavouring treatment, Figure S2: Anethole gradient from the peel to the core of the apple after flavouring treatment.

**Author Contributions:** Conceptualization, I.E., E.A. and F.G.; methodology, I.E., E.A. and F.G.; formal analysis, I.E.; investigation, I.E., E.A., E.B., M.C., J.Z. and F.G.; writing—original draft preparation, I.E.; data curation, I.E. and E.A.; writing—review and editing, I.E., E.A., M.C. and F.G.; visualization, I.E.; supervision, I.E., E.A. and F.G.; funding acquisition, F.G.

**Funding:** This work was supported by the Autonomous Province of Trento, Italy.

**Acknowledgments:** The authors are grateful for the cooperation of FEM panellists.

**Conflicts of Interest:** The authors declare no conflict of interest. The funders had no role in the design of the study; in the collection, analyses, or interpretation of data; in the writing of the manuscript, or in the decision to publish the results.

## References


1. Péneau, S.; Brockhoff, P.B.; Hoehn, E.; Escher, F.; Nuessli, J. Relating consumer evaluation of apple freshness to sensory and physico-chemical measurements. *J. Sens. Stud.* **2006**, *22*, 313–335. [CrossRef]
2. Hampson, C.R.; Quamme, H.A.; Hall, J.W.; Mac Donald, R.A.; King, M.C.; Cliff, M.A. Sensory evaluation as selection tool in apple breeding. *Euphytica* **2000**, *111*, 79–90. [CrossRef]
3. Harker, R.F.; Kupfermanb, E.M.; Marinc, A.B.; Gunsona, F.A.; Triggs, C.M. Eating quality standards for apples based on consumer preferences. *Postharvest Biol. Technol.* **2008**, *50*, 70–78. [CrossRef]
4. Symoneaux, R.; Galmarini, M.V.; Mehinagic, E. Comment analysis of consumer's likes and dislikes as an alternative tool to preference mapping. A case study on apples. *Food Qual. Preference* **2012**, *24*, 59–66. [CrossRef]
5. Endrizzi, I.; Torri, L.; Corollaro, M.L.; Demattè, M.L.; Aprea, E.; Charles, M. A conjoint study on apple acceptability: Sensory characteristics and nutritional information. *Food Qual. Preference* **2015**, *40*, 39–48. [CrossRef]
6. Daillant-Spinnler, B.; MacFie, H.; Beyts, P.; Hedderley, D. Relationships between perceived sensory properties and major preference directions of 12 varieties of apples from the southern hemisphere. *Food Qual. Preference* **1996**, *7*, 113–126. [CrossRef]
7. Tomala, K.; Baryłko-Pikielna, N.; Jankowski, P.; Jezioreka, K.; Wasiak-Zysb, G. Acceptability of scab-resistant versus conventional apple cultivars by Polish adult and young consumers. *J. Sci. Food Agric.* **2009**, *89*, 1035–1045. [CrossRef]
8. Nestrud, M.A.; Lewless, H.T. Perceptual mapping of apples and cheeses using projective mapping and sorting. *J. Sens. Stud.* **2010**, *25*, 390–405. [CrossRef]
9. Seppä, L.; Railio, J.; Vehkalahti, K.; Tahvonen, R.; Tuorila, H. Hedonic responses and individual definitions of an ideal apple as predictors of choice. *J. Sens. Stud.* **2013**, *28*, 346–357. [CrossRef]
10. Bonany, J.; Brugger, C.; Buehler, A.; Carbó, J.; Codarin, S.; Donati, F.; Echeverria, G.; Egger, S.; Guerra, W.; Hilaire, C.; et al. Preference mapping of apple varieties in Europe. *Food Qual. Preference* **2014**, *32*, 317–329. [CrossRef]
11. Jaeger, S.R.; Andani, Z.; Wakeling, I.N.; Macfie, H.J.H. Consumer preferences for fresh and aged apples: A cross-cultural comparison. *Food Qual. Preference* **1998**, *9*, 355–366. [CrossRef]
12. Charles, M. Contribution to Methodological Reflections Related to Consumer Preference Study and Sensory Interactions—An Application on Apples. Unpublished Ph.D. Thesis in Food Science, LUNAM University, Angers, France, 2013. Available online: https://tel.archives-ouvertes.fr/tel-00931819 (accessed on 31 October 2019).
13. Charles, M.; Endrizzi, I.; Aprea, E.; Zambanini, J.; Betta, E.; Gasperi, F. Dynamic and static sensory methods to study the role of aroma on taste and texture: A multisensory approach to apple perception. *Food Qual. Preference* **2017**, *62*, 17–30. [CrossRef]
14. Spence, C. The multisensory perception of flavour. *Psychologist* **2010**, *23*, 720–723.
15. Altisent, R.; Jaeger, S.; Johnston, J.W.; Harker, R. Injection of flavor essences into fruit pieces. A new approach for exploring consumer preferences for novel flavors of apple fruit. *J. Sens. Stud.* **2013**, *28*, 405–413. [CrossRef]
16. Harker, F.R.; Gunson, F.A.; Jaeger, S.R. The case for fruit quality: An interpretive review of consumer attitudes, and preferences for apples. *Postharvest Biol. Technol.* **2003**, *28*, 333–347. [CrossRef]
17. Wismer, W.V.; Harker, F.R.; Gunson, F.A.; Rossiter, K.L.; Lau, K.; Seal, A.G.; Lowe, R.G.; Beatson, R. Identifying flavour targets for fruit breeding: A kiwifruit example. *Euphytica* **2005**, *141*, 93–104. [CrossRef]
18. Echeverría, G.; Graell, J.; Lara, I.; Lòpez, M.L.; Puy, J. Panel consonance in the sensory evaluation of apple attributes: Influence of mealiness on sweetness perception. *J. Sens. Stud.* **2008**, *23*, 656–670. [CrossRef]
19. Aprea, E.; Charles, M.C.; Endrizzi, I.; Corollaro, M.L.; Betta, E.; Biasioli, F.; Gasperi, F. Sweet taste in apple: The role of sorbitol, individual sugars, organic acids and volatile compounds. *Sci. Rep.* **2017**, *7*, 44950. [CrossRef]

20. Young, H.; Gilbert, J.M.; Murray, S.N.; Ball, R.D. Causal effects of aroma compounds on Royal Gala apple flavour. *J. Sci. Food Agric.* **1996**, *22*, 49–64. [CrossRef]

21. Ball, R.D.; Murray, S.H.; Young, H.; Gilbert, J.M. Statistical analysis relating analytical and consumer panel assessments of kiwifruit flavor compounds in a model juice base. *Food Qual. Preference* **1998**, *9*, 255–266. [CrossRef]

22. Gasperi, F.; Aprea, E.; Biasioli, F.; Carlin, S.; Endrizzi, I.; Pirretti, G.; Spilimbergo, S. Effects of supercritical $CO_2$ and $N_2O$ pasteurisation on the quality of fresh apple juice. *Food Chem.* **2009**, *115*, 129–136. [CrossRef]

23. Rossiter, K.L.; Young, H.; Walker, S.B.; Miller, M.; Dawson, D.M. Effect of sugars and acids on consumer acceptability of kiwifruit. *J. Sens. Stud.* **2000**, *15*, 241–250. [CrossRef]

24. Marsh, K.B.; Friel, E.N.; Gunson, A.; Lund, C.; MacRae, E. Perception of flavour in standardised fruit pulps with additions of acids or sugars. *Food Qual. Preference* **2006**, *17*, 376–386. [CrossRef]

25. Jaeger, S.R.; Harker, R.; Trigg, C.M.; Gunson, A.; Campbell, R.L.; Jackmam, R.; Requejo-Jackman, C. Determining consumer purchase intensions: The importance of dry matter, size, and price of kiwifruit. *J. Food Sci.* **2011**, *76*, 177–184. [CrossRef] [PubMed]

26. De Pelsmaeker, S.; Dewettinck, K.; Gellynck, X. The possibility of using tasting as a presentation method for sensory stimuli in conjoint analysis. *Trends Food Sci. Technol.* **2013**, *29*, 108–115. [CrossRef]

27. Asioli, D.; Varela, O.; Herseleth, M.; Almli, V.L.; Olsen, N.V.; Næs, T. A discussion of recent methodologies for combining sensory and estrinsic product properties in consumer studies. *Food Qual. Preference* **2017**, *56*, 266–273. [CrossRef]

28. Cerda, A.A.; Garcia, L.Y.; Ortega-Farias, S.; Ubilla, A.M. Consumer preferences and willingness topay for organic apples. *Cienc. Investig. Agrar.* **2012**, *39*, 47–59. [CrossRef]

29. Del Carmen, D.R.; Esguerra, E.B.; Absulio, W.L.; Maunahan, M.V.; Masilungan, G.D.; Collins, R.; Sun, T. Understanding consumers preferences for fresh table-ripe papaya using survey and conjoint methods of analyses. *Acta Hortic.* **2013**, *1012*, 1379–1386. [CrossRef]

30. Rødbotten, M.; Martinsen, B.K.; Borge, G.I.; Mortvedt, H.S.; Knutsen, S.H.; Lea, P.; Næs, T. A cross-cultural study of preference for apple juice with different sugar and acid contents. *Food Qual. Preference* **2009**, *20*, 277–284. [CrossRef]

31. Johansen, S.B.; Næs, T.; Øyaas, J.; Hersleth, M. Acceptance of calorie-reduced yoghurt: Effects of sensory characteristics and product information. *Food Qual. Preference* **2010**, *21*, 13–21. [CrossRef]

32. Ares, G.; Gimènez, A.; Gàmbaro, A. Influence of nutritional knowledge on perceived healthiness and willingness to try functional foods. *Appetite* **2008**, *51*, 663–668. [CrossRef] [PubMed]

33. Arruda Pinto, V.R.; de Oliveira Freitas, T.B.; de Souza Dantas, M.I.; Della Luicia, S.M.; Fernandes Melo, L.; Rodriges Minim, V.P.; Bressan, J. Influence of package and health-related claims on perception and sensory acceptability of snack bars. *Food Res. Int.* **2017**, *101*, 103–113. [CrossRef] [PubMed]

34. Ha, C.-H.; Lee, S.M.; Lee, E.-K.; Kim, K.-O. Effect of flour information (origin and organic) and consumer attitude to health and natural product on bread acceptability of Korean consumers. *J. Sens. Stud.* **2017**, *32*, e12281. [CrossRef]

35. Samant, S.S.; Seo, H.-S. Effects of label understanding level on consumers' visual attention toward sustainability and process-related label claims found on chicken meat products. *Food Qual. Preference* **2016**, *50*, 48–56. [CrossRef]

36. Bernebéu, R.; Rabadan, A.; El Orche, N.E.; Diaz, M. Influence of quality labels on the formation of preferences of lamb meat consumers. A Spanish case study. *Meat Sci* **2018**, *135*, 129–133. [CrossRef] [PubMed]

37. Van Loo, E.J.; Caputo, V.; Nayga, R.M., Jr.; Verbeke, W. Consumers' valuation of sustainability labels on meat. *Food Policy* **2014**, *49*, 137–150. [CrossRef]

38. Aprea, E.; Corollaro, M.L.; Betta, E.; Endrizzi, I.; Dematte, M.L.; Biasioli, F.; Gasperi, F. Sensory and instrumental profiling of 18 apple cultivars to investigate the relation between perceived quality and odour and flavour. *Food Res. Int.* **2012**, *49*, 677–686. [CrossRef]

39. Czerny, M.; Christlbauer, M.; Christlbauer, M.; Fischer, A.; Granvogl, M.; Hammer, M.; Schieberle, P. Re-investigation on odour thresholds of key food aroma compounds and development of an aroma language based on odour qualities of defined aqueous odorant solutions. *Eur. Food Res. Technol.* **2008**, *228*, 265–273. [CrossRef]

40. Khomenko, J.; Cappellin, L.; Charles, M.; Marini, F.; Gasperi, F.; Biasioli, F. Prediction of anise flavour in apple from GC-MS and PTR-MS volatile profiling. In Proceedings of the 4th MS Food Day, Foggia, Italy, 7–9 October 2015; Società Chimica Italiana: Roma, Italy, 2015.

41. Corollaro, M.L.; Endrizzi, I.; Bertolini, A.; Aprea, E.; Demattè, M.L.; Costa, F.; Biasioli, F.; Gasperi, F. Sensory profiling of apple: Methodological aspects, cultivar characterisation and postharvest changes. *Postharvest Biol. Technol.* **2013**, *77*, 111–120. [CrossRef]

42. Poinot, P.; Arvisenet, G.; Ledauphin, J.; Gaillard, J.-L.; Prost, C. How can aroma–related cross–modal interactions be analysed? A review of current methodologies. *Food Qual. Preference* **2013**, *28*, 304–316. [CrossRef]

43. Thomas-Danguin, T.; Sinding, C.; Tournier, C.; Saint-Eve, A. Multimodal interactions. In *Flavor, from Food to Behaviors, Wellbeing and Health*; Étiévant, P., Guichard, E., Salles, C., Voilley, A., Eds.; Woodhead Publishing: Cambridge, UK, 2016; pp. 121–141.

44. Demattè, M.L.; Endrizzi, I.; Biasioli, F.; Corollaro, M.L.; Pojer, N.; Zampini, M.; Aprea, E.; Gasperi, F. Food neophobia and its relation with olfactory ability in common odour identification. *Appetite* **2013**, *68*, 112–117. [CrossRef] [PubMed]

45. Tuorila, H.M.; Lähteenmäki, L.; Pohjalainen, L.; Lotti, L. Food neophobia among the Finns and related responses to familiar and unfamiliar foods. *Food Qual. Preference* **2001**, *12*, 29–37. [CrossRef]

46. Demattè, M.L.; Endrizzi, I.; Gasperi, F. Food neophobia and its relation with olfaction. *Front Psychol.* **2013**, *5*, 127. [CrossRef] [PubMed]

47. Novotorova, N.K.; Mazzocco, M.A. Consumer preferences and trade-offs for locally grown and genetically modified apples: A conjoint analysis approach. *Int. Food Agribus. Man.* **2008**, *11*, 31–53.

48. Dentoni, D.; Tonsor, G.; Calantone, R.; Peterson, C. The direct and indirect effect of locally-grown attributes on consumers' attitudes towards a product. *Agric. Resour. Econ. Rev.* **2009**, *38*, 384–396. [CrossRef]

49. Wang, Q.; Sun, J.; Parsons, R. Consumer preferences and willingness to pay for locally grown organic apples: Evidence from a conjoint study. *Hort. Sci.* **2010**, *45*, 376–381. [CrossRef]

50. Skreli, E.; Imami, D. Analyzing consumers' preferences for apple attributes in Tirana, Albania. *Int. Food Agribus. Man.* **2012**, *15*, 137–156.

51. Roininen, K.; Lahteenmaki, L.; Tuorila, H. Quantification of consumer attitudes to health and hedonic characteristics of foods. *Appetite* **1999**, *33*, 71–88. [CrossRef]

52. EN ISO. *International Standard 8589. Sensory Analysis—General Guidance for the Design of Test Rooms*; International Organisation of Standardization: Genève, Switzerland, 2014; Available online: https://www.iso.org/standard/36385.html (accessed on 31 October 2019).

53. ISO. *International Standard 4120. Sensory Analysis-Methodology-Triangle Test*; International Organisation of Standardization: Genève, Switzerland, 2004; Available online: https://www.iso.org/standard/33495.html (accessed on 31 October 2019).

54. Corollaro, M.L.; Aprea, E.; Endrizzi, I.; Betta, E.; Demattè, M.L.; Charles, M.; Bergamaschi, M.; Costa, F.; Biasioli, F.; Corelli Grappatelli, L.; et al. A combined sensory-instrumental tool for apple quality evaluation. *Postharvest Biol. Technol.* **2014**, *96*, 135–144. [CrossRef]

55. Pliner, P.; Hobden, K. Development of a Scale to Measure Neophobia in Humans the Trait of Food. *Appetite* **1992**, *19*, 105–120. [CrossRef]

**Sample Availability:** Samples of the compounds are not available from the authors.

*Article*

# Key Aroma Compounds of Dark Chocolates Differing in Organoleptic Properties: A GC-O Comparative Study

Zoé Deuscher [1,2], Karine Gourrat [1,3], Marie Repoux [4], Renaud Boulanger [2,5], Hélène Labouré [1] and Jean-Luc Le Quéré [1,*]

1   Centre des Sciences du Goût et de l'Alimentation (CSGA), AgroSup Dijon, CNRS, INRAE, Université de Bourgogne Franche-Comté, F-21000 Dijon, France
2   CIRAD, UMR Qualisud, F-34398 Montpellier, France
3   ChemoSens Platform, CSGA, F-21000 Dijon, France
4   Valrhona, 14 av. du Président Roosevelet, F-26602 Tain l'Hermitage, France
5   Qualisud, Univ Montpellier, CIRAD, Montpellier SupAgro, Univ d'Avignon, Univ de La Réunion, F-34000 Montpellier, France
*   Correspondence: jean-luc.le-quere@inrae.fr

Academic Editor: Eugenio Aprea
Received: 31 March 2020; Accepted: 12 April 2020; Published: 15 April 2020

**Abstract:** Dark chocolate samples were previously classified into four sensory categories. The classification was modelled based on volatile compounds analyzed by direct introduction mass spectrometry of the chocolates' headspace. The purpose of the study was to identify the most discriminant odor-active compounds that should characterize the four sensory categories. To address the problem, a gas chromatography-olfactometry (GC-O) study was conducted by 12 assessors using a comparative detection frequency analysis (cDFA) approach on 12 exemplary samples. A nasal impact frequency (NIF) difference threshold combined with a statistical approach (Khi$^2$ test on k proportions) revealed 38 discriminative key odorants able to differentiate the samples and to characterize the sensory categories. A heatmap emphasized the 19 most discriminant key odorants, among which heterocyclic molecules (furanones, pyranones, lactones, one pyrrole, and one pyrazine) played a prominent role with secondary alcohols, acids, and esters. The initial sensory classes were retrieved using the discriminant key volatiles in a correspondence analysis (CA) and a hierarchical cluster analysis (HCA). Among the 38 discriminant key odorants, although previously identified in cocoa products, 21 were formally described for the first time as key aroma compounds of dark chocolate. Moreover, 13 key odorants were described for the first time in a cocoa product.

**Keywords:** dark chocolate; key odorant; key aroma; gas chromatography-olfactometry (GC-O); comparative detection frequency analysis (cDFA); nasal impact frequency (NIF); correspondence analysis (CA); hierarchical cluster analysis (HCA); heatmap

## 1. Introduction

Dark chocolate may contain 35%, and up to 85%–99% for high cocoa content samples, of the ingredients originating from cocoa (cocoa solids and cocoa butter). The appreciation of dark chocolate is mainly related to its sensory properties, which are greatly influenced by the cocoa beans' aroma and by the complex manufacturing process [1] that gives rise to the final chocolate product. The volatile composition of cocoa beans and of the resulting dark chocolate has been the subject of many gas chromatography-mass spectrometry (GC-MS) studies, with the aim of characterizing the (i) chocolate quality attributes, (ii) variety and origin of cocoa beans, and (iii) the process, including the fermentation and drying of cocoa beans, roasting, and conching.

Thus, quality attributes of dark chocolate produced from Vietnamese cocoa were recently investigated [2], as well as the influence of the cocoa variety on the fermentation step studied [3–5], influence of the cocoa origin on the cocoa flavor examined [6–9], and the link between the origin and process searched for [10]. The process, starting with fermentation [11–16] and followed by drying [17] and roasting [18–23] steps, has been the subject of many studies. Specialized reviews and treatises gathered such knowledge [1,24–26].

Among these investigations, studies aiming to identify the only aroma-active compounds in dark chocolate (the dark chocolate key aroma compounds) are scarce. Nevertheless, some important gas chromatography-olfactometry (GC-O) studies, completed by GC-MS, allowed identification of the major dark chocolate key aroma compounds. Thus, the influence of conching was examined, with the aim of identifying the key odorant compounds in dark chocolate [27]. Aroma-active compounds in dark and milk chocolate in relation to sensory perception were investigated [28], and recently, key aroma compounds in two commercial dark chocolates with high cocoa contents were characterized [29]. As already cited, the major results are reported in reviews on cocoa and cocoa products [24–26] as well as in a specialized treatise dedicated to chocolate [1].

Direct injection mass spectrometry (DIMS) methods, such as proton transfer reaction mass spectrometry (PTR-MS), have been used in some studies conducted on dark chocolates. Their objective was generally to classify the chocolates according to the variety and/or the origin of the transformed cocoa beans by measuring and comparing their relative volatile organic compounds (VOCs) patterns [30,31]. Investigations into the relationships between VOC patterns obtained using a DIMS method and the organoleptic properties of foodstuffs obtained as sensory profiles measured by a panel are rather scarce [32]. Dark chocolates of diverse cocoa origins and cultivars, but manufactured using the same standard process at the pilot level at an industrial plant, could be classified into four sensory categories [32]. This classification was based on a quantitative descriptive analysis (QDA) protocol conducted by an internal expert panel: The panelists rated 36 flavor attributes, among which 33 were aroma descriptors. The classification into the four sensory categories is, in this case, the basis of a quality control procedure to define the ultimate use of merchantable cocoa lots as they are received at the factory [32]. In a recent study, we were able to model this sensorial classification by deciphering the volatilome of 206 dark chocolate samples using the DIMS method PTR-MS [32]. This approach, combined with chemometrics and variable selection procedures, allowed us to propose a highly significant prediction model of the sensory categories (poles) based on a limited number of ions (10 to 22 depending on the selection method used) [32]. Some of them were tentatively identified as volatile compounds on the basis of their mass formulae determined thanks to the time-of-flight (TOF) mass analyzer used, and literature data [32]. However, some of these ions represented ion fragments and most of the supposedly molecular ions represented many possible isobaric compounds or isomers. None of them were securely identified and their aroma activities in the chocolates were not determined.

Comparative GC-O studies conducted so far have been aimed at emphasizing key odorants that could contrast samples differing in terms of cultivars, origins, processes, or brands. Distinguishing the same types of samples, only categorized in different classes based on sensory criteria, using a comparative GC-O methodology appears challenging. Therefore, the aim of the present study was to determine the discriminant key aroma compounds that should allow the four previously characterized and modelled sensory poles to be distinguished. To achieve this, three exemplary samples of each sensory pole were chosen on the basis of the availability and respective position in the sensory space of the 206 samples described above. A combined GC-MS and GC-O approach was conducted on each of the selected 12 samples. The key aroma compounds were determined by the detection frequency analysis (DFA) method [33,34], using a panel of 12 assessors. The results obtained for each sample were compared in order to emphasize the discriminant key aroma molecules in the chocolates. Finally, a correspondence analysis (CA) and a hierarchical cluster analysis (HCA) were conducted in order to find potential relationships between the discriminant features and the 12 samples.

## 2. Results

In a previous study [32], four sensory categories (named sensory poles) of dark chocolates, essentially based on aroma evaluation, were characterized and modelled using the headspace fingerprints of 206 samples obtained with a direct introduction mass spectrometry technique (PTR-MS). Key aroma compounds that should discriminate these categories were searched for using 12 samples representative of the sensory poles. The equilibrated positions of these 12 samples (three in each pole) in the sensory space of the 206 chocolates defined in a QDA are illustrated on the principal component analysis (PCA) planes of the sensory data displayed in Figure A1 (Appendix A).

To identify the impact aroma compounds, a GC-O approach, completed by GC-MS, was conducted on the 12 samples by a panel of 12 assessors. The discriminant odorants were determined using a comparative detection frequency analysis (cDFA). The chocolate aroma extracts under study were obtained in triplicate by hydro-distillation under vacuum in a solvent-assisted flavor evaporation (SAFE) device [35], completed by headspace (HS) extracts obtained using solid-phase microextraction (SPME) to account for the most volatile odorants. The reliability and repeatability of the hydro-distillation step were checked by the use of an internal standard. SPME conditions were optimized in order to get the same GC response and odor intensity as in the SAFE method for a volatile "reference" peak (butane-2,3-dione), which was detected by the entire panel in both methods.

### 2.1. Determination of Impact Compounds by GC-O: Comparative Detection Frequency Analysis (cDFA)

The aim of any GC-O experiment is to screen the volatiles isolated from a particular food in order to determine their relative odor potency and to prioritize the most potent odorants, the key odorants, for subsequent identification. Different GC-O methods have been developed within three main paradigms. Dilution techniques (aroma extract dilution analysis, AEDA; and combined hedonic aroma measurement, CHARM) analyze aroma extracts through several successive dilutions as far as no odorant is perceived [36,37]. Both are valuable screening methods that use a large number of dilutions, but generally, a very small number of assessors. Therefore, both methods are not amenable to statistics, and, being based only on individual detection thresholds, they are associated with inherent drawbacks, such as the sensitivity of the small panel of sniffers, with potential inattention and/or specific anosmia, and time-consuming successive sessions [38–41]. The method named Osme (odor-specific magnitude estimation) is supposed to measure the odor intensity of the eluting species [42]. The third method, detection frequency analysis (DFA), only determines when an odorant is detected [33,34]. Osme requires trained panelists familiar with the odor intensity notion whereas DFA results in an easier task. Both Osme and DFA employ a single optimized extract dilution but a larger number of sniffers; they use the sensitivity of several assessors to average and mitigate inter-individual variation and allow statistical evaluation of the data. Different studies showed that the three methods give similar and correlated results in terms of screening impact odorants [41,43–45]. In the present study, DFA was used with a panel of 12 assessors sniffing the 12 samples SAFE extracts (10 assessors for SPME extracts). The triplicate SAFE extracts of each sample were pooled for the GC-O analysis and HS-SPME extracts optimized as described above (see also the experimental section for details).

The GC-O experiments allowed the detection of 8480 odor events all together, which were grouped for each sample in olfactive areas (OAs) on the basis of their linear retention indices' (LRIs) closeness. Therefore, each OA was characterized by the number of individual odor events detected by the panel members; this number defined its detection frequency, expressed as a percentage as the nasal impact frequency (NIF %) [33]. A detection threshold filter was applied to remove noise and retain only the most intense and significant OAs. No definite rule exists to determine such a threshold. A minimal 12.5% NIF value was set as being necessary to build an OA [46]. However, in the current literature, values varied from this lower level to a 50% NIF threshold [34,47,48]. Owing to the number of odor events detected by the panelists in the 12 samples, illustrating the odorous richness and sensory diversity of the dark chocolates under study, and the number of replicates per sample (12 or 10), a 50% threshold was applied, as also chosen by others [47–49]. This meant an OA was finally retained as significant only if its detection frequency was ≥ 50% in at least one sample. By applying this 50% NIF threshold, 96 OAs were finally considered (Table 1), i.e., a rather great number of OAs despite the high threshold level used.

Some specificities were evidenced as some of these OAs were found at high levels in all the samples, whereas some were found more specifically in some samples. However, the total number of significant OAs detected per sample was quite similar, with an average of 43 (min 35–max 53). Thus, among the OAs commonly detected at high levels in all samples and identified as butane-2,3-dione (OA n°7), ethyl 2-methylbutanoate (10), oct-1-en-3-one (19), dimethyltrisulfide (24), trimethylpyrazine (26), 5-ethyl-2,3-dimethylpyrazine and 3-methylthiopropanal (34, coeluted but distinguishable), phenylacetaldehyde (45), 2- and 3-methylbutanoic acid (47), phenylmethanol (59), and 4-vinylguaiacol (80), most of them were previously identified in cocoa mass or dark chocolate or in other cocoa categories (Table 1). Vanillin (OA n°93) was also of this type but was not further considered as vanillin was added in the recipe as a flavoring ingredient. Only a few of these OAs were detected by almost all the assessors (NIF ≈ 100%) in all the samples (Table 1). They are limited to ubiquitous molecules, such as butane-2,3-dione (OA n°7), dimethyltrisulfide (24), 3-methylthiopropanal (34), phenylacetaldehyde (45), 2- and 3-methylbutanoic acid (47), and 4-vinylguaiacol (80). The number of such OAs, not discriminative because they were detected by all the panelists and delivered the highest NIF values in all the samples, was inferior to 10% of the detected OAs, as recommended by Etievant and Chaintreau [41] to allow olfactive discrimination between samples in GC-O, thus validating the extract concentrations chosen for our study [40]. Other OAs were also found in all the samples at common lower levels, for instance, 2-methylpropanal (OA n°2), ethyl propanoate (6), 3-hydroxybutan-2-one (18), 2-ethenyl-6-methylpyrazine (35), acetylpyrazine (43), 1-phenylethyl acetate (48, coeluted with another ester: methyl 2-methylpentanoate), 2,3,5-trimethyl-6-(3-methylbutyl)pyrazine (49), 1*H*-pyrrole-2-carbaldehyde (69), and nonanoic acid (78). Most of them were also previously identified in cocoa or chocolate categories (Table 1). Together with the previous ones, these compounds were found to be important for the overall aroma of the samples and may represent a common aromatic background of dark chocolate, which is not able to discriminate the samples according to their differing sensory properties.

Numerous OAs were found with very different NIF values between samples (Table 1). To consider a significant difference between samples, it was assumed that an NIF value difference of 30% between the lowest and the highest values is at least necessary when working with a panel of eight assessors [33]. By applying this difference threshold for our panel of 12 (or 10 for HS) assessors, an NIF difference > 30% between at least two samples, i.e., a difference of at least 4 assessors (for the SAFE extracts), 73 OAs were considered discriminant among the 96 initially retained (Table 1). Amongst these discriminant odorants, most of them were previously identified as volatile compounds of dark chocolate or cocoa mass (Table 1) and many of them determined as key odorants of dark chocolate. Thus, the Strecker aldehydes 2- and 3-methylbutanal (OAs n°3 and 4) with their characteristic cocoa and chocolate olfactive notes were found together with other key chocolate aldehydes: 2-methylbut-2-enal (OA n°12), heptanal (15), oct-2-enal (30), non-2-enal (39), and 2-phenylbut-2-enal (61) [1,27,29,50,51]. The key alcohols heptan-2-ol (OA n°20), 2-phenylethanol (60), phenol (67), 4-methyphenol (72) [1], butane-2,3-diol (40) [50,52], and guaiacol (58) [29,50] were found to be discriminant together with other alcohols, generally present in cocoa products (Table 1), but for some of them, these have never been described as dark chocolate key aroma compounds: Ethanol (OA n°5), butan-2-ol (8), 2-methylbut-3-en-2-ol (9), 3-methylbutan-1-ol (16), 3-ethoxypropan-1-ol (23), octan-2-ol (28), 3-ethyl-4-methylpentan-1-ol (37), 1-phenylethanol (54), farnesol (88), and octadecan-1-ol (95). Dark chocolate key esters were also found as discriminative features: Ethyl 2- and 3-methylbutanoate (OA n°11 and 12), isoamyl acetate (13), pentyl acetate (14), ethyl phenylacetate (52), 2-phenylethyl acetate (55), and ethyl cinnamate (74). Other esters, for some of them this is the first time being described as key aroma compounds of dark chocolate (Table 1), were also considered as discriminant: Hept-2-yl acetate (OA n°17), ethyl nonanoate (39), 3-hydroxypropyl acetate (or propane-1,3-diol monoacetate, 50), pentan-2-yl benzoate and ethyl dodecanoate (56), methyl tetradecanoate (65), isopropyl palmitate (82), and 2-phenylethyl lactate (86). Some key aroma ketones were also found to be discriminant: Nonan-2-one (OA n°25) already characterized in dark chocolate (Table 1), and others, while cited in cocoa products (Table 1), were described for the first time as key aromas in dark chocolate: Acetophenone (OA n°46) and heptadecan-2-one (81). The hydroxyketone 3-hydroxy-4-phenylbutan-2-one (OA n°84) has never been described in cocoa products but was produced from phenylalanine in the Maillard reaction in a study on roast aroma formation [53]. Discriminant carboxylic acids were also found among dark chocolate key aromas: Acetic acid (OA n°33), butanoic acid (44), and phenylacetic acid (94). Octanoic acid (OA n°71) and dodecanoic acid (92), while found in cocoa products (Table 1), are formally cited for the first time as key odorants of dark chocolate.

Numerous lactones were found to be discriminant, some of them as part of dark chocolate's key aroma compounds: δ-octenolactone (OA n°66), γ-nonalactone (68), γ-decalactone (77), and δ-decalactone (79); δ-octalactone (63), δ-decenolactone (83), and γ-dodecalactone (89) are new dark chocolate key odorants whereas δ-pentalactone (OA n°53) is newly described in cocoa products. Some important heterocycles were also found amongst the 73 discriminant odorants: The furanones furaneol (4-hydroxy-2,5-dimethylfuran-3(2*H*)-one, OA n°70), a key aroma of dark chocolate; dihydroactinidiolide (4,4,7a-trimethyl-5,6,7,7a-tetrahydro-1-benzofuran-2(4*H*)-one, OA n°87), found as a new dark chocolate key odorant; and 5-[(2*Z*)oct-2-en-1-yl]dihydrofuran-2(3*H*)-one (90), newly described as a key aroma of a cocoa product. Pyrans and pyranones were also part of the discriminant odorants, together with some pyrroles: The pyranol *trans*-linalool-3,7-oxide (6-ethenyl-2,2,6-trimethyltetrahydro-2*H*-pyran-3-ol, OA n°51); the pyranones maltol (OA n°63, 3-hydroxy-2-methyl-4*H*-pyran-4-one), 3-hydroxy-2,3-dihydromaltol (85), newly cited in dark chocolate's key compounds; dihydromaltol (57), newly determined in a cocoa product; and the pyrroles 2-acetylpyrrole (OA n°64), 5-methyl-1*H*-pyrrole-2-carbaldehyde (73), and 1*H*-indole (91), all three already described as key odorants of dark chocolate (Table 1).

Among the heterocycles, pyrazines were found in numerous key compound examples, and many of them were found to be discriminant. Thus, being already established as key odorants of dark chocolate (Table 1), 2,5-dimethylpyrazine (OA n°21), ethylpyrazine (22), 3-ethyl-2,5-dimethylpyrazine, 2-ethyl-3,5-dimethylpyrazine (32), and 3-isobutyl-2,5-dimethylpyrazine (38) were discriminative features, together with the pyrazines newly identified as key odorants of dark chocolate (Table 1): 2,6-diethylpyrazine (OA n°31), 2-ethenyl-5-methylpyrazine (36), 2-isobutyl-3,5-dimethylpyrazine (38), and 2-isobutyl-3,5,6-trimethylpyrazine (41).

Finally, among the key sulfur aromas, only methanethiol (OA n°1), for the first time being described as a key odorant of dark chocolate, was found to be discriminant to distinguish the 12 samples (Table 1).

**Table 1.** Characterization of olfactive areas (OAs) of the 12 chocolate samples (1A to 4C) for which the nasal impact frequency (NIF) ≥ 50% in at least one sample. [a] OA number, OA 1 to 10 obtained in HS-SPME-GC-MS, OA from 11 obtained with SAFE extracts; [b] Linear retention indices (LRIs) calculated in GC-MS on DB-FFAP and DB-5 columns, "n.d." = not detected, "-" = irrelevant; [c] NIF values in each sample (%); [d] Odor attributes given by the panel; [e] Discriminant OA: x = NIF difference > 30% between at least 2 samples, xx = NIF difference > 50% between at least 2 samples (see text); [f] Reliability of identification: *1* = MS, LRIs and odor identical to published data and/or data found in databases, *2* = 1 with additional MW confirmed by CI, *3* = 2 with additional injection of standards on equivalent DB-5 and/or DB-FFAP columns, *4* = after MDGC-MS/O; [g] Compounds identified in references related to cocoa and/or dark chocolate (not exhaustive list) and according to the database Volatile Compounds in Food (VCF): (http://www.vcf-online.nl) [54].

| OA [a] | LRI [b] (GC-MS) | | NIF [c] (%) | | | | | | | | | | | | Odor Attributes [d] | Disc. [e] | Compound Identification | Reliability of Identification [f] | References [g] |
|---|---|---|---|---|---|---|---|---|---|---|---|---|---|---|---|---|---|---|---|
| | DB-FFAP | DB-5 | 1A | 1B | 1C | 2A | 2B | 2C | 3A | 3B | 3C | 4A | 4B | 4C | | | | | |
| 1 | 701 | n.d. | 100 | 100 | 90 | 80 | 90 | 80 | 50 | 70 | 20 | 90 | 60 | 70 | cheese, cabbage, sulfur | xx | methanethiol | 1 | [32,55] |
| 2 | 817 | n.d. | 60 | 60 | 30 | 60 | 60 | 70 | 60 | 60 | 50 | 60 | 50 | 60 | chocolate, cocoa, roasted | | 2-methylpropanal | 1 | [1,27,28] |
| 3 | 920 | n.d. | 60 | 40 | 50 | 60 | 90 | 70 | 50 | 60 | 80 | 80 | 100 | 40 | cocoa, chocolate | xx | 2-methylbutanal | 3 | [1,27,29,50,52] |
| 4 | 923 | n.d. | 80 | 70 | 60 | 40 | 30 | 50 | 80 | 60 | 40 | 30 | 30 | 60 | cocoa | x | 3-methylbutanal | 3 | [1,27–29,50–52] |
| 5 | 942 | n.d. | 30 | 40 | 40 | 40 | 50 | 30 | 40 | 50 | 30 | 30 | 80 | 40 | fruity, solvent | x | ethanol | 3 | [5,11,12,16,56–58] |
| 6 | 961 | n.d. | 50 | 50 | 70 | 70 | 50 | 30 | 60 | 40 | 40 | 50 | 60 | 60 | fruity, floral | | ethyl propanoate | 1 | [7,54] |
| 7 | 991 * | n.d. | 100 | 100 | 100 | 100 | 100 | 100 | 100 | 100 | 90 | 100 | 100 | 100 | butter | | butane-2,3-dione | 3 | [1,27,29,50–52] |
| 8 | 1025 | n.d. | 10 | 10 | 80 | 10 | 10 | 0 | 30 | 20 | 10 | 20 | 20 | 0 | rubber | xx | butan-2-ol | 1 | [7,26,54,59] |
| 9 | 1040 | n.d. | 50 | 20 | 10 | 20 | 10 | 20 | 20 | 0 | 30 | 40 | 20 | 10 | fruity, floral | x | 2-methylbut-3-en-2-ol | 2 | [7,56,60] |
| 10 | 1054 ** | n.d. | 90 | 70 | 60 | 70 | 70 | 60 | 90 | 80 | 80 | 60 | 90 | 60 | fruity | | ethyl 2-methylbutanoate | 1 | [17,26,29,61] |
| 11 | 1072 | 849 | 33 | 75 | 42 | 50 | 67 | 42 | 67 | 67 | 83 | 42 | 67 | 33 | fruity, floral | x | ethyl 3-methylbutanoate | 2 | [17,26,29,50,52,62] |
| 12 | 1108 | n.d. | 25 | 25 | 8 | 17 | 42 | 75 | 0 | 8 | 8 | 8 | 8 | 0 | hot plastic | xx | (E)-2-methylbut-2-enal | 2 | [1,7,24,27,63] |
| 13 | 1127 | 874 | 8 | 33 | 0 | 17 | 42 | 58 | 8 | 8 | 0 | 17 | 8 | 0 | fruity, candy | xx | isoamyl acetate | 3 | [29,50,57,60,62,64,65] |
| 14 | 1183 | n.d. | 0 | 33 | 17 | 8 | 50 | 58 | 8 | 8 | 8 | 8 | 8 | 0 | fruity | xx | pentyl acetate | 2 | [12,29,50,52,60,65] |
| 15 | 1196 | 903 | 42 | 42 | 17 | 8 | 75 | 50 | 42 | 42 | 42 | 67 | 42 | 17 | fruity, floral | xx | heptanal | 3 | [1,27,60,63,65] |
| 16 | 1211 | 750 | 17 | 33 | 0 | 42 | 50 | 33 | 17 | 25 | 0 | 33 | 50 | 8 | cheesy | x | 3-methylbutan-1-ol | 3 | [20,56,57,60] |
| 17 $ | 1267 | 891 | 17 | 42 | 0 | 8 | 92 | 75 | 58 | 67 | 75 | 83 | 67 | 42 | fruity, flowery | xx | styrene | 2 | [3,12,65] |
| | | 1038 | | | | | | | | | | | | | | | hept-2-yl acetate | 2 | [66] |
| 18 | 1296 | 742 | 58 | 75 | 83 | 50 | 75 | 50 | 58 | 58 | 75 | 58 | 75 | 50 | butter | | 3-hydroxybutan-2-one | 3 | [1,26,63,67] |
| 19 | 1309 | 975 | 92 | 83 | 100 | 100 | 100 | 100 | 92 | 75 | 83 | 92 | 100 | 75 | mushroom | | oct-1-en-3-one | 2 | [20,22,28,29,51,54,61] |
| 20 | 1323 | 902 | 8 | 42 | 17 | 33 | 58 | 50 | 42 | 33 | 25 | 58 | 58 | 42 | fruity, mushroom, vegetal | x | heptan-2-ol | 3 | [1,3,26,57,59,60,65,68] |
| 21 | 1330 | 914 | 25 | 25 | 17 | 0 | 50 | 33 | 25 | 25 | 25 | 50 | 25 | 8 | roasted, chocolate | x | 2,5-dimethylpyrazine | 2 | [1,27,28,50,52] |
| 22 | 1346 | 916 | 67 | 67 | 50 | 75 | 92 | 75 | 42 | 75 | 67 | 42 | 58 | 42 | roasted cereals, peanut | x | ethylpyrazine | 2 | [1,7,24,27,59,67] |
| 23 $ | 1383 | 1128 | 67 | 0 | 42 | 58 | 17 | 0 | 0 | 0 | 0 | 0 | 0 | 0 | metallic, musty | xx | *allo*-ocimene | 2 | [19,65,68] |
| | | 798 | | | | | | | | | | | | | | | 3-ethoxypropan-1-ol | 2 | [65] |
| 24 | 1387 | 969 | 83 | 100 | 92 | 92 | 100 | 83 | 83 | 100 | 100 | 100 | 100 | 83 | sulfur, cabbage | | dimethyltrisulfide | 3 | [27–29,50–52] |
| 25 | 1397 | 1091 | 0 | 50 | 0 | 17 | 33 | 42 | 33 | 33 | 67 | 33 | 42 | 25 | fruity, floral, vegetal | xx | nonane-2-one | 3 | [6,28,60,64,65] |
| 26 | 1410 | 1001 | 83 | 92 | 92 | 100 | 92 | 92 | 83 | 100 | 75 | 92 | 100 | 100 | roasted, vegetal, earthy | | trimethylpyrazine | 3 | [27–29,50–52] |
| 27 | 1419 | - | 25 | 58 | 25 | 17 | 58 | 50 | 8 | 8 | 8 | 17 | 8 | 8 | fruity | x | unknown | | - |
| 28 | 1427 | n.d. | 0 | 33 | 0 | 0 | 0 | 25 | 8 | 50 | 42 | 92 | 75 | 67 | fruity, floral, candy | xx | octan-2-ol | 1 | [7,11,56,57,65,68] |
| 29 | 1431 | - | 67 | 75 | 42 | 92 | 50 | 67 | 58 | 58 | 75 | 75 | 67 | 83 | roasted, nutty | x | unknown | | - |
| 30 | 1438 | 1058 | 67 | 92 | 92 | 92 | 83 | 75 | 92 | 42 | 92 | 100 | 83 | 67 | vegetal, earthy | xx | (E)-oct-2-enal | 1 | [3,16,26,29,68] |
| 31 | 1440 | 1075 | 8 | 25 | 17 | 0 | 33 | 33 | 17 | 75 | 17 | 33 | 42 | 33 | vegetal | xx | 2,6-diethylpyrazine | 2 | [62] |
| 32 $ | 1450 | 1076 | 25 | 0 | 25 | 0 | 33 | 50 | 8 | 25 | 8 | 8 | 17 | 17 | vegetal, roasted | x | 3-ethyl-2,5-dimethylpyrazine and/or | 2 | [27,28] |
| | | | | | | | | | | | | | | | | | 2-ethyl-3,5-dimethylpyrazine | | [1,27,29,50,51] |
| 33 | 1462 | n.d. | 100 | 58 | 92 | 58 | 83 | 67 | 25 | 42 | 67 | 58 | 25 | 0 | vinegar | xx | acetic acid | 3 | [1,28,29,50,52] |
| 34 $ | 1466 | 1078 | 100 | 100 | 100 | 100 | 92 | 92 | 100 | 100 | 100 | 100 | 100 | 100 | roasted, then potato | | 5-ethyl-2,3-dimethylpyrazine | 2 | [1,50,63,69,70] |
| | | 907 | | | | | | | | | | | | | | | 3-methylthiopropanal | 3 | [1,27,63,71,72] |

**Table 1.** *Cont.*

| OA [a] | LRI [b] (GC-MS) | | NIF [c] (%) | | | | | | | | | | | | Odor Attributes [d] | Disc. [e] | Compound Identification | Reliability of Identification [f] | References [g] |
| --- | --- | --- | --- | --- | --- | --- | --- | --- | --- | --- | --- | --- | --- | --- | --- | --- | --- | --- |
| | DB-FFAP | DB-5 | 1A | 1B | 1C | 2A | 2B | 2C | 3A | 3B | 3C | 4A | 4B | 4C | | | | | |
| 35 | 1497 | 1016 | 67 | 42 | 42 | 58 | 67 | 58 | 42 | 58 | 67 | 92 | 92 | 50 | vegetal, earthy, roasted | xx | 2-ethenyl-6-methylpyrazine | 2 | [1,27,62,63,73] |
| 36 | 1508 | 1021 | 0 | 33 | 0 | 0 | 33 | 42 | 33 | 58 | 58 | 67 | 67 | 25 | vegetal, earthy, roasted | xx | 2-ethenyl-5-methylpyrazine | 2 | [54] |
| 37 | 1514 | 1004 | 58 | 17 | 58 | 25 | 42 | 17 | 17 | 33 | 58 | 42 | 42 | 17 | flowery, vegetal | x | 3-ethyl-4-methylpentan-1-ol | 1 | - |
| 38 $ | 1530 | 997<br>1011 | 50 | 83 | 50 | 75 | 83 | 83 | 75 | 92 | 92 | 83 | 83 | 83 | vegetal, pepper | x | 3-isobutyl-2,5-dimethylpyrazine<br>and/or 2-isobutyl-3,5-dimethylpyrazine | 2<br>2 | [1,24,27,70]<br>[27,70] |
| 39 $ | 1542 | 1293<br>1159 | 92 | 92 | 75 | 67 | 75 | 67 | 67 | 75 | 75 | 75 | 58 | 42 | vegetal, cardboard, flowery | x | ethyl nonanoate<br>(E)-non-2-enal | 3<br>3 | [5,54]<br>[16,29,51,61,65] |
| 40 | 1552 | 792 | 0 | 67 | 8 | 17 | 50 | 50 | 50 | 75 | 67 | 58 | 67 | 50 | flowery | xx | butane-2,3-diol | 3, 4 | [5,26,50,52,56,60,65,67] |
| 41 | 1594 | 1273 | 83 | 33 | 58 | 58 | 50 | 50 | 25 | 25 | 33 | 50 | 25 | 33 | vegetal, cucumber | xx | 2-isobutyl-3,5,6-trimethylpyrazine | 2 | [54,70] |
| 42 | 1597 | 988 | 17 | 25 | 25 | 17 | 25 | 25 | 33 | 50 | 42 | 33 | 50 | 0 | vegetal, earthy | | 3-hydroxybutanoic acid | 1 | - |
| 43 | 1635 | 1063 | 50 | 58 | 67 | 83 | 67 | 67 | 58 | 92 | 58 | 50 | 75 | 58 | roasted | | acetylpyrazine | 2 | [29,54,65,71,74] |
| 44 | 1638 | 788 | 67 | 50 | 50 | 58 | 58 | 92 | 50 | 67 | 58 | 50 | 50 | 33 | cheese | xx | butanoic acid | 3 | [16,20,29,51,59,61,68] |
| 45 | 1653 | 1046 | 100 | 100 | 100 | 92 | 92 | 83 | 100 | 100 | 100 | 100 | 100 | 92 | flowery | | phenylacetaldehyde | 3 | [1,27,28,50,51] |
| 46 | 1660 | 1066 | 0 | 17 | 8 | 33 | 17 | 17 | 8 | 50 | 33 | 25 | 17 | 33 | floral, fruity | x | acetophenone | 3 | [2,60,62,64,65,68,74] |
| 47 | 1676 | 886<br>876 | 100 | 100 | 92 | 100 | 100 | 100 | 92 | 100 | 100 | 92 | 100 | 92 | melted cheese | | 2-methylbutanoic acid<br>3-methylbutanoic acid | 3<br>3 | [12,29,51,61,67,75]<br>[28,29,50–52] |
| 48 $ | 1710 | 1188<br>842 | 58 | 50 | 58 | 58 | 58 | 67 | 58 | 67 | 75 | 67 | 67 | 50 | vegetal, roasted, fruity | | 1-phenylethyl acetate<br>methyl 2-methylpentanoate | 2<br>2 | [2]<br>- |
| 49 | 1726 | 1386 | 75 | 67 | 67 | 92 | 67 | 67 | 58 | 58 | 83 | 67 | 75 | 83 | floral, anise, minty | | 2,3,5-trimethyl-6-(3-methylbutyl)pyrazine | 2 | [54] |
| 50 | 1748 | 955 | 17 | 25 | 8 | 42 | 33 | 33 | 17 | 33 | 33 | 33 | 58 | 50 | unpleasant | x | 3-hydroxypropyl acetate | 2 | - |
| 51 | 1766 | 1175 | 67 | 33 | 0 | 17 | 33 | 50 | 17 | 50 | 50 | 33 | 33 | 33 | fruity, roasted, vegetal | xx | trans-linalool-3,7-oxide | 2 | [1,12,16,27,50,52,60,63,68] |
| 52 | 1795 | 1242 | 0 | 92 | 0 | 8 | 75 | 67 | 58 | 50 | 50 | 50 | 42 | 25 | floral | xx | ethyl phenylacetate | 3 | [29,50,51,57,60,62,65,68] |
| 53 | 1818 | 958 | 75 | 75 | 58 | 75 | 83 | 83 | 83 | 92 | 67 | 83 | 100 | 75 | roasted, vegetal | x | δ-pentalactone | 2 | - |
| 54 | 1825 | 1061 | 67 | 92 | 75 | 50 | 100 | 92 | 92 | 100 | 100 | 92 | 100 | 67 | floral, rose, fruity | x | 1-phenylethanol | 3 | [17,26,56,59,60,65,68] |
| 55 | 1828 | 1255 | 8 | 17 | 0 | 42 | 25 | 33 | 33 | 33 | 42 | 33 | 42 | 50 | earthy, moldy | x | 2-phenylethyl acetate | 3 | [1,28,29,50,51] |
| 56 $ | 1848 | 1391<br>1590 | 0 | 33 | 17 | 0 | 17 | 50 | 8 | 33 | 42 | 42 | 17 | 17 | roasted, nut, spicy | x | pentan-2-yl benzoate<br>ethyl dodecanoate | 2<br>3 | [12]<br>[3,56,57,59,76] |
| 57 | 1869 | 1064 | 17 | 58 | 75 | 50 | 75 | 83 | 33 | 67 | 83 | 42 | 50 | 58 | roasted, caramel, fruity | xx | dihydromaltol | 2 | - |
| 58 | 1872 | 1087 | 58 | 67 | 92 | 58 | 58 | 50 | 83 | 42 | 50 | 67 | 75 | 50 | roasted, smoked, sweet | x | guaiacol | 3, 4 | [16,20,24,29,50,59] |
| 59 | 1892 | 1036 | 75 | 67 | 75 | 58 | 83 | 67 | 67 | 67 | 67 | 75 | 75 | 75 | sweet, fruity, floral | | phenylmethanol | 2 | [17,20,65,77] |
| 60 | 1921 | 1116 | 67 | 75 | 83 | 100 | 92 | 58 | 83 | 67 | 75 | 67 | 92 | 83 | floral, rose | x | 2-phenylethanol | 3 | [1,27–29,50,51] |
| 61 | 1944 | 1269 | 8 | 25 | 17 | 17 | 58 | 42 | 33 | 42 | 50 | 17 | 42 | 25 | floral | x | 2-phenylbut-2-enal | 2 | [1,17,24,26,27,50,57,60,63,65,68] |
| 62 | 1976 | - | 50 | 50 | 50 | 58 | 75 | 33 | 67 | 67 | 75 | 58 | 58 | 33 | roasted, fruity, spicy | x | unknown | | - |
| 63 $ | 1980 | 1279<br>1081 | 67 | 33 | 17 | 0 | 17 | 17 | 0 | 0 | 17 | 0 | 8 | 42 | fruity, sweet | x | δ-octalactone<br>maltol | 2, 4 | [65,72,77]<br>[1,24,26,28,63,65] |
| 64 | 1985 | 1069 | 17 | 33 | 17 | 17 | 25 | 42 | 33 | 42 | 50 | 42 | 17 | 8 | hot plastic | x | 2-acetylpyrrole | 2 | [24,27,28,57,60,65,67,68] |
| 65 | 2011 | 1721 | 8 | 50 | 25 | 50 | 25 | 42 | 0 | 25 | 67 | 42 | 25 | 17 | vegetal, metallic | xx | methyl tetradecanoate | 2, 4 | [54,59] |
| 66 | 2015 | 1261 | 25 | 50 | 50 | 42 | 75 | 58 | 50 | 83 | 58 | 42 | 58 | 50 | sweet, vegetal | xx | δ-octenolactone | 2, 4 | [20,26,29,51,61] |
| 67 | 2018 | 982 | 50 | 25 | 0 | 0 | 17 | 25 | 33 | 17 | 50 | 33 | 17 | 33 | floral, fruity | x | phenol | 3, 4 | [1,17,24,26,63,68] |
| 68 | 2039 | 1358 | 25 | 50 | 33 | 25 | 50 | 25 | 67 | 67 | 58 | 67 | 42 | 33 | fruity, sweet | x | γ-nonalactone | 3 | [20,29,54,61,77] |
| 69 | 2043 | 1012 | 67 | 58 | 75 | 58 | 83 | 50 | 67 | 50 | 67 | 50 | 83 | 75 | sweet, fruity | | 1H-pyrrole-2-carbaldehyde | 2 | [1,7,27,50,59,63,67] |

**Table 1.** *Cont.*

| OA [a] | LRI [b] (GC-MS) | | NIF [c] (%) | | | | | | | | | | | | Odor Attributes [d] | Disc. [e] | Compound Identification | Reliability of Identification [f] | References [g] |
|---|---|---|---|---|---|---|---|---|---|---|---|---|---|---|---|---|---|---|---|
| | DB-FFAP | DB-5 | 1A | 1B | 1C | 2A | 2B | 2C | 3A | 3B | 3C | 4A | 4B | 4C | | | | | |
| 70 [$] | 2046 | 1063 | 58 | 50 | 17 | 42 | 25 | 67 | 0 | 50 | 0 | 0 | 0 | 58 | caramel, strawberry | xx | furaneol | 3, 4 | [1,6,20,24,28,29,51,63] |
| 71 | 2075 | 1170 | 42 | 50 | 33 | 42 | 42 | 25 | 0 | 42 | 50 | 25 | 33 | 42 | unpleasant | x | octanoic acid | 3 | [5,7,57,59,70,78] |
| 72 | 2096 | 1064 | 42 | 75 | 58 | 83 | 67 | 83 | 83 | 92 | 83 | 83 | 75 | 50 | animal, unpleasant, urine | x | 4-methylphenol | 3 | [1,24,27,29,63,71,77] |
| 73 | 2122 | 1124 | 33 | 17 | 25 | 0 | 17 | 25 | 0 | 58 | 42 | 33 | 67 | 42 | floral, spicy, fruity | xx | 5-methyl-1*H*-pyrrole-2-carbaldehyde | 2 | [28] |
| 74 [$] | 2139 | 1466 | 50 | 17 | 25 | 17 | 33 | 25 | 17 | 17 | 17 | 8 | 8 | 33 | fruity, vegetal | x | (*E*)-ethyl cinnamate | 2 | [1,17,26,29,59,60,63] |
| 75 [$] | 2142 | - | 25 | 33 | 50 | 33 | 25 | 33 | 17 | 17 | 17 | 8 | 8 | 33 | floral, sweet, fruity | | unknown | | - |
| 76 [$] | 2147 | - | 33 | 25 | 8 | 8 | 42 | 8 | 8 | 42 | 58 | 0 | 0 | 0 | roasted, spicy | xx | unknown | | - |
| 77 | 2156 | 1465 | 33 | 17 | 25 | 17 | 8 | 25 | 58 | 33 | 33 | 17 | 42 | 42 | sweet, fruity, peach | x | γ-decalactone | 1 | [22,28,51,61,72,79] |
| 78 | 2197 | 1265 | 50 | 58 | 33 | 33 | 33 | 42 | 33 | 58 | 50 | 42 | 60 | 58 | animal, unpleasant | | nonanoic acid | 1 | [5,7,80] |
| 79 | 2205 | 1490 | 17 | 42 | 8 | 0 | 50 | 25 | 50 | 42 | 42 | 17 | 25 | 25 | fruity, floral, woody | x | δ-decalactone | 2 | [20,22,28,72,77] |
| 80 | 2212 | 1308 | 100 | 83 | 100 | 92 | 83 | 100 | 100 | 83 | 100 | 100 | 92 | 92 | curry, licorice, clove, spicy | | 4-vinylguaiacol | 3 | [66] |
| 81 | 2234 | 1900 | 58 | 58 | 50 | 33 | 83 | 50 | 67 | 92 | 67 | 58 | 92 | 75 | floral, fruity, vegetal | xx | heptadecan-2-one | 2 | [56] |
| 82 | 2240 | 2022 | 0 | 17 | 42 | 25 | 17 | 17 | 33 | 8 | 8 | 8 | 58 | 50 | floral, fruity | xx | isopropyl palmitate | 1 | - |
| 83 | 2246 | 1471 | 17 | 0 | 17 | 25 | 33 | 25 | 67 | 42 | 67 | 25 | 0 | 8 | fruity, sweet, coconut | xx | δ-decenolactone | 2 | [20,22,54,77] |
| 84 | 2272 | 1343 | 25 | 33 | 17 | 50 | 25 | 33 | 25 | 33 | 50 | 17 | 58 | 58 | unpleasant, dust | xx | 3-hydroxy-4-phenylbutan-2-one | 2 | - |
| 85 | 2278 | 1143 | 17 | 25 | 17 | 17 | 17 | 17 | 17 | 50 | 8 | 0 | 33 | 8 | roasted, chicory coffee | x | 3-hydroxy-2,3-dihydromaltol | 1 | [54] |
| 86 | 2338 | 1501 | 42 | 50 | 25 | 0 | 50 | 50 | 0 | 17 | 8 | 0 | 0 | 0 | woody, vegetal | x | 2-phenylethyl lactate | 1 | - |
| 87 | 2353 | 1527 | 50 | 33 | 17 | 33 | 25 | 25 | 25 | 25 | 8 | 25 | 8 | 8 | dust | x | dihydroactinidiolide | 1 | - |
| 88 | 2365 | n.d. | 8 | 8 | 42 | 25 | 33 | 50 | 33 | 17 | 42 | 50 | 42 | 42 | floral | x | farnesol | 1 | - |
| 89 | 2387 | 1675 | 0 | 8 | 8 | 17 | 0 | 25 | 42 | 25 | 33 | 8 | 58 | 75 | fruity, peach | xx | γ-dodecalactone | 3 | [29,72] |
| 90 [$] | 2412 | 1652 | 8 | 58 | 17 | 25 | 42 | 58 | 67 | 75 | 50 | 50 | 83 | 75 | fruity, floral | xx | 5-[(2Z)oct-2-en-1-yl]dihydrofuran-2(3H)-one | 2 | - |
| | | 1214 | | | | | | | | | | | | | rubber, medicinal | | 4-vinylphenol | 3 | [27,29] |
| 91 | 2464 | 1290 | 33 | 67 | 25 | 33 | 58 | 67 | 42 | 50 | 42 | 42 | 42 | 58 | unpleasant, floral | x | 1*H*-indole | 3 | [1,27,29,63] |
| 92 | 2511 | 1565 | 33 | 42 | 50 | 17 | 42 | 50 | 42 | 50 | 33 | 42 | 25 | 0 | unpleasant, animal, leather | x | dodecanoic acid | 3 | [5,17,26,54] |
| 93 [***] | 2587 | 1392 | 83 | 92 | 92 | 92 | 92 | 83 | 92 | 83 | 50 | 83 | 83 | 92 | vanilla, sweet, cocoa | | vanillin | 3 | [1,27–29,51,54] |
| 94 | 2591 | 1245 | 67 | 42 | 33 | 25 | 42 | 50 | 33 | 50 | 58 | 0 | 0 | 0 | floral, unpleasant | xx | phenylacetic acid | 2 | [26,29,51,54,59,70,77] |
| 95 | 2602 | 2086 | 17 | 25 | 58 | 8 | 50 | 17 | 0 | 33 | 42 | 8 | 8 | 17 | floral | xx | octadecan-1-ol | 1 | - |
| 96 | 2628 | - | 8 | 8 | 33 | 17 | 0 | 8 | 0 | 8 | 0 | 50 | 25 | 17 | floral | | unknown | | - |

* also identified in the SAFE extract at LRI 995; ** also identified in the SAFE extract at the same LRI; *** vanillin, not further considered (see text); [$] coelution not resolved on DB-FFAP column, either deconvoluted or resolved in MDGC-MS-O.

Some of these OAs found to be discriminant were characteristic of only one or a few samples. Thus, OAs n°9 (2-methylbut-3-en-2-ol), 63 (δ-octalactone and maltol), 74 (ethyl cinnamate), and 87 (dihydroactinidiolide) reached an NIF value ≥ 50% in sample 1A only. OA n°8 (butan-2-ol) reached this NIF value only in sample 1C, while OAs n°12 (2-methylbut-2-enal), 13 (isoamyl acetate), 32 (3- and/or 2-ethyldimethylpyrazines), and 56 (pentan-2-yl benzoate and ethyl dodecanoate) attained this level only in sample 2C. OAs n°64 (2-acetylpyrrole) and 76 (unknown) seemed characteristic of sample 3C, while OA n°55 (2-phenylethyl acetate) reached this NIF value only in sample 4C. The NIF values of these OAs stayed below 50% in the other samples in rather equilibrated proportions (Table 1). They could be more specific of the respective corresponding samples, where their NIFs reached a value ≥ 50%. Moreover, some OAs could reach an NIF value ≥ 50% in a particular sample, while attaining values just below 50% in other samples. Thus, OA n°77 (γ-decalactone) reached 58% in sample 3A and 42% (i.e., a difference of only two assessors detecting the component) in samples 4B and 4C. When using this type of OA, differentiating samples from sensory poles 3 and 4 will be difficult. On the contrary, some OAs may reflect a strong specificity. Thus, OA n°28 (octan-2-ol) reached a 92% NIF value in sample 4A while being almost not detected in poles 1 and 2 samples. The same applied for OA n°33 (acetic acid), with a value of 100% in sample 1A, while it was not detected in sample 4C. Besides, OA n°52 (ethyl phenylacetate) appeared particular, with a 92% NIF value in sample 1B while it was not detected in the other samples of pole 1 (Table 1).

In order to rationalize the data, correspondence analysis (CA) was used to study the potential relationships between the 73 discriminant OAs and 12 samples through the NIF values gathered in Table 1. This multivariate exploratory analysis appeared suitable for the nature of the data that exhibited frequencies of detection. While highly significant (Khi$^2$ independence test: $p$-value < 0.0001) and allowing a rather clear separation between groups of samples (Figure S1, Supplementary data), the analysis revealed many variables (OAs) that were poorly represented (i.e., localized in the center of the CA plot), exhibiting no real change in the detection frequencies between samples. They represented common key impact compounds but were not able to participate in the differentiation of the different samples (Figure S1). Moreover, a parametric analysis (comparison of k proportions) conducted on the OAs (Khi$^2$ test) delivered insignificant $p$-values ($\alpha = 0.05$) for most of them (Table A1, Appendix B). To remove this noise in the CA, the NIF difference threshold between at least two samples was increased from > 30% to > 50%, meaning that a difference of at least six assessors (for the SAFE extracts) was judged necessary to define a discriminant OA. This more drastic threshold retained 34 discriminant OAs (Table 1) for which most of the $p$-values in the Khi$^2$ test of the k proportions comparison were also highly significant (Table 2). Therefore, the selection of the significant variables on the detection frequency basis of the GC-O analyses (difference threshold > 50%) revealed a good accord with the parametric comparison of k proportions. A CA was realized with these 34 significant OAs (Figures 1 and A2, Appendix B), resulting in a highly significant analysis (Khi$^2$ independence test: $p$-value < 0.0001), meaning that some relationships between the 34 OAs and the 12 samples should exist. As expected, the center of the CA plots was clarified with fewer ill-represented variables.

**Table 2.** Key discriminant aroma compounds that characterize the four sensory poles as determined in a cDFA GC-O analysis using a 50% discriminative threshold (see text for a complete explanation).[a] OA number, as in Table 1; [b] LRI on DB-FFAP, as in Table 1; [c] Odor attributes given by the panel; [d] Identification (refer to Table 1); [e] Chemical Abstracts Service registry number; [f] Mass formula; [g] Molecular Weight; [h] Pertinent odor attributes found in the databases VCF and The Good Scents Company (http://www.thegoodscentscompany.com/); [i] $p$-value of the Khi$^2$ test ($\alpha = 0.05$) obtained for the parametric test comparison of k proportions using the data of Table 1.

| OA [a] | LRI [b] | Odor [c] | Identification [d] | CAS [e] | Formula [f] | MW [g] | Lit. Odor [h] | $p$-value [i] |
|---|---|---|---|---|---|---|---|---|
| 1 | 701 | cabbage, sulfur | **methanethiol** * | 74-93-1 | $CH_4S$ | 48.1 | cabbage, sulfur | <0.001 |
| 3 | 920 | cocoa, chocolate | 2-methylbutanal | 96-17-3 | $C_5H_{10}O$ | *86.1* ** | cocoa, nutty | 0.081 |
| 8 | 1025 | rubber | **butan-2-ol** * | 78-92-2 | $C_4H_{10}O$ | 74.1 | medicine, solvent | <0.001 |
| 12 | 1108 | hot plastic | (*E*)-2-methylbut-2-enal | 1115-11-3 | $C_5H_8O$ | *84.1* | solvent, ethereal | <0.0001 |
| 13 | 1127 | fruity, candy | isoamyl acetate | 123-92-2 | $C_7H_{14}O_2$ | *130.2* | fruity, banana | <0.001 |
| 14 | 1183 | fruity | pentyl acetate | 628-63-7 | $C_7H_{14}O_2$ | *130.2* | fruity, banana | <0.001 |
| 15 | 1196 | fruity, floral | heptanal | 111-71-7 | $C_7H_{14}O$ | *114.2* | fresh, green | 0.037 |
| 17 | 1267 | fruity, flowery | **hept-2-yl acetate** * | 5921-82-4 | $C_9H_{18}O_2$ | *158.2* | fruity | <0.0001 |
| 23 | 1383 | metallic, musty | ***allo*-ocimene** * | 673-84-7 | $C_{10}H_{16}$ | *136.2* | herbal, peppery | <0.0001 |
| 25 | 1397 | fruity, floral, vegetal | nonan-2-one | 821-55-6 | $C_9H_{18}O$ | *142.2* | sweet, herbal, fruity | 0.020 |
| 28 | 1427 | fruity, floral, candy | **octan-2-ol** * | 123-96-6 | $C_8H_{18}O$ | *130.2* | fruit, fresh, green | <0.0001 |
| 30 | 1438 | vegetal, earthy | (*E*)-oct-2-enal | 2548-87-0 | $C_8H_{14}O$ | 126.2 | green, herbal, leaf | 0.017 |
| 31 | 1440 | vegetal | **2,6-diethylpyrazine** * | 13067-27-1 | $C_8H_{12}N_2$ | *136.2* | green | 0.011 |
| 33 | 1462 | vinegar | acetic acid | 64-19-7 | $C_2H_4O_2$ | 60.1 | vinegar, pungent | <0.0001 |
| 36 | 1508 | vegetal, earthy, roasted | **2-ethenyl-5-methylpyrazine** * | 13925-08-1 | $C_7H_8N_2$ | *120.1* | coffee | <0.001 |
| 40 | 1552 | flowery | butane-2,3-diol | 513-85-9 | $C_4H_{10}O_2$ | *90.1* | floral | <0.001 |
| 41 | 1594 | vegetal, cucumber | 2-isobutyl-3,5,6-trimethylpyrazine * | 46187-37-5 | $C_{11}H_{18}N_2$ | *178.3* | - | 0.107 |
| 44 | 1638 | cheese | butanoic acid | 107-92-6 | $C_4H_8O_2$ | 88.1 | cheese | 0.463 |
| 51 | 1766 | fruity, vegetal, roasted | *trans*-linalool-3,7-oxide | 39028-58-5 | $C_{10}H_{18}O_2$ | *170.2* | floral, woody, wintergreen | 0.061 |
| 52 | 1795 | floral | ethyl phenylacetate | 101-97-3 | $C_{10}H_{12}O_2$ | *164.2* | floral | <0.0001 |
| 57 | 1869 | caramel, fruity, roasted | **dihydromaltol** * | 38877-21-3 | $C_6H_8O_3$ | *128.1* | - | 0.019 |
| 65 | 2011 | vegetal, metallic | **methyl tetradecanoate** * | 124-10-7 | $C_{15}H_{30}O_2$ | *242.4* | orris, petal, waxy | 0.020 |
| 66 | 2015 | sweet, vegetal | δ-octenolactone | 16400-69-4 | $C_8H_{12}O_2$ | *140.2* | coconut $ | 0.337 |
| 70 | 2046 | caramel, strawberry | furaneol | 3658-77-3 | $C_6H_8O_3$ | *128.1* | caramel, strawberry | <0.0001 |
| 73 | 2122 | floral, spicy, fruity | 5-methyl-1*H*-pyrrole-2-carbaldehyde | 1192-79-6 | $C_6H_7NO$ | *109.1* | - | 0.005 |
| 76 | 2147 | roasted, spicy | unknown | - | - | - | - | <0.001 |
| 81 | 2234 | floral, fruity, vegetal | **heptadecan-2-one** * | 2922-51-2 | $C_{17}H_{34}O$ | 254.5 | - | 0.076 |
| 82 | 2240 | floral, fruity | **isopropyl palmitate** * | 142-91-6 | $C_{19}H_{38}O_2$ | 298.5 | fatty, oily | 0.010 |
| 83 | 2246 | fruity, sweet, coconut | **δ-decenolactone** * | 54814-64-1 | $C_{10}H_{16}O_2$ | *168.2* | coconut, fruity | <0.001 |
| 84 | 2272 | unpleasant, dust | **3-hydroxy-4-phenylbutan-2-one** * | 5355-63-5 | $C_{10}H_{12}O_2$ | *164.2* | burnt plastic | 0.280 |
| 89 | 2387 | fruity, peach | **γ-dodecalactone** * | 2305-05-7 | $C_{12}H_{22}O_2$ | *198.3* | peach, fruit | <0.0001 |
| 90 | 2412 | fruity, floral | 5-[(2*Z*)oct-2-en-1-yl]dihydrofuran-2(3*H*)-one * | 156318-46-6 | $C_{12}H_{20}O_2$ | *196.3* | - | 0.001 |
| 94 | 2591 | floral, unpleasant | phenylacetic acid | 103-82-2 | $C_8H_8O_2$ | 136.1 | floral, urine | <0.001 |
| 95 | 2602 | floral | **octadecan-1-ol** * | 112-92-5 | $C_{18}H_{38}O$ | 270.5 | oily | 0.010 |

* although most of them previously identified in cocoa products (see Table 1), to the authors' knowledge, these compounds (in bold character) are formally described for the first time as key aroma compounds of dark chocolate; ** MW in bold italic means MW confirmed by CI; $ odor description in [22]; -: not described or not relevant.

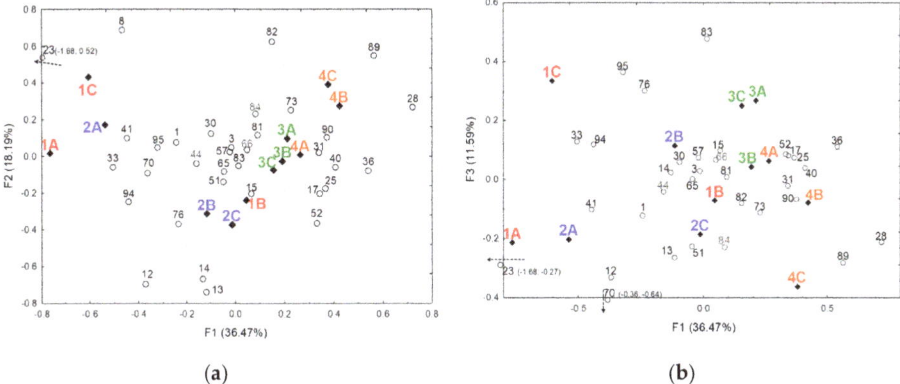

**Figure 1.** Correspondence analysis (CA) between the 12 samples and the 34 discriminant OAs defined by their NIF values. OAs (light circles) are plotted according to their NIF in samples (plain diamonds) in the dimensions 1 and 2 that gathered 54.66% of cumulative inertia (**a**), and 1 and 3 (**b**), respectively. The CA plots are zoomed in for clarity, and the coordinates of extra variables (23 and 70) indicated in brackets with their direction. The OA numbers are those found in Table 1. The sample names are colored for illustrative purpose, with pole 1 samples appearing in red, pole 2 samples in blue, pole 3 samples in green, and pole 4 ones in orange. A 3D plot (dimensions 1, 2, and 3) of the CA may be found in Appendix B (Figure A2). CA independence test: Khi$^2$ = 5444 (critical value 408, $\alpha$ = 0.05, degrees of freedom = 363), $p < 0.0001$.

CA plots (Figures 1 and A2) were used to study potential proximities between samples on the one hand, and between samples and OAs on the other hand. Factor 1 (36.47% of inertia) clearly separates poles 1 and 2 samples from poles 3 and 4 ones, positioned on the negative and the positive sides of the factor, respectively (Figure 1a). Factor 3 (11.59% of inertia) allows a better separation between poles 3 and 4 samples (Figure 1b). Sample 2A is found in proximity with samples 1A and 1C (on the negative side of F1) while sample 1B is close to samples 2B and 2C, near the center of F1 and on the negative side of F2. These findings were already pinpointed in the related previous experiment, where samples belonging to poles 1 and 2 presented large intra-class distances in a PCA conducted on the samples' volatilome data [32], a phenomenon also apparent in Figure A1. Meanwhile, samples belonging to poles 3 and 4 were found close together, being obviously very similar in terms of the volatiles composition, as previously noticed [32]. However, as in the PLS-DA previously conducted on the volatilome data [32], samples belonging to poles 3 and 4 were better distinguished on the third factor F3 (Figure 1b). The CA finally clearly distinguished four groups of three chocolates through their proximities on the plots: Sample groups {1A, 1C, 2A}, {2B, 2C, 1B}, {3A, 3B, 3C}, and {4A, 4B, 4C}, respectively. The fact that sample 2A was classified with samples 1A and 1C, and sample 1B was classified with samples 2B and 2C, respectively, illustrated, as already outlined, the large intra-class variability of the corresponding sensory poles 1 and 2, which partially overlapped (Figure A1, Appendix A). The proximity of the sensory poles 3 and 4 with partial overlap was also apparent (Figure A1).

OAs more associated with particular samples are clearly visible on the CA plots (Figures 1 and A2). Thus, OAs n°23 (*allo*-ocimene), 8 (butan-2-ol), 70 (furaneol), 33 (acetic acid), and 41 (2-isobutyl-3,5,6-trimethylpyrazine), found on the negative side of factor F1, distinguish the sample group {1A, 1C, 2A}. Opposed on the positive side of F1, OAs n°28 (octan-2-ol), 36 (2-ethenyl-5-methylpyrazine), 89 (γ-dodecalactone) distinguish group {4A, 4B, 4C}, and to a lesser extent group {3A, 3B, 3C}, together with OAs n°40 (butane-2,3-diol), 90 (5-[oct-2-en-1-yl]dihydrofuran-2(*3H*)-one), 31 (2,6-diethylpyrazine), and to a lesser extent OAs n°17 (hept-2-yl acetate), 25 (nonane-2-one), and 52 (ethyl phenylacetate), seem more specific of group {3}.

On the negative side of factor F2, OAs n°12 (2-methylbut-2-enal), 13 (isoamyl acetate), 14 (pentyl acetate), 76 (unknown), and 15 (heptanal) are associated with the sample group {2B, 2C, 1B}, in which OA n°82 (isopropyl palmitate), found on the positive side of F2, is less present. It is noteworthy that, except for OA n°41 related to the sample group {1A, 1C, 2A}, all the OAs with a non-significant *p*-value in the Khi$^2$ test of the comparison of k proportions (Table 2) are displayed in the center of the CA plots (Figure 1). They were poorly represented in the correspondence analysis and did not participate in the differentiation of the samples. This was particularly true for the OAs n°44 (butanoic acid, *p*-value 0.463), 66 (δ-octenolactone, *p*-value 0.337), and 84 (3-hydroxy-4-phenylbutan-2-one, *p*-value 0.280). This again revealed a good agreement between both variable selection methods, one based on sensory results inferred from the GC-O difference threshold in detection frequencies, and the other one based on statistics that are more conventional.

In order to go deeper into the data presented in the CA plots and objectively define the relationships that exist between the 34 discriminant OAs and the 12 samples, a heatmap was constructed using the NIF data found in Table 1. This heatmap (Figure 2) independently classified variables (OAs) and individuals (samples) thanks to a hierarchical cluster analysis (HCA) centered on Euclidian distances. The resulting samples' clustering largely confirmed the correspondence analysis and the evidenced relationships. Thus, four clusters were clearly defined (see also Figure A3, Appendix B): The sample groups were {1A, 1C, 2A} and {1B, 2B, 2C}, reflecting the intra-class variability of sensory poles 1 and 2, as already outlined; and {4B, 4C, 3A, 4A} showing the proximity of sample 3A with pole 4 samples, and particularly with sample 4A, and {3B, 3C}. Four to six clusters of variables could also be clearly seen (Figure 2). The first sample cluster {1A, 1C, 2A} was particularly defined by very low NIF values for a series of compounds grouped together in the HCA. Thus, the low NIFs of OAs n°40 (butan-2,3-diol), 36 (2-ethenyl-5-methylpyrazine), 25 (nonan-2-one), 90 (5-[oct-2-en-1-yl]dihydrofuran-2(3*H*)-one), 28 (octan-2-ol), 17 (hept-2-yl acetate), 15 (heptanal), and 52 (ethyl phenylacetate) characterized this cluster, together with, to a lesser extent, the low NIFs of OAs n°81 (heptadecan-2-one), 73 (5-methyl-1*H*-pyrrole-2-carbaldehyde), 66 (δ-octenolactone), and 31 (2,6-diethylpyrazine). It was also defined by high NIFs of three OAs clustered in an HCA branch: 23 (*allo*-ocimene), 33 (acetic acid), and 41 (2-isobutyl-3,5,6-trimethylpyrazine). The sample group {3B, 3C}, opposed to the first one, was defined by high NIFs of OAs n°94 (phenylacetic acid), 76 (unknown), 51 (linalool-3,7-oxide), and 83 (δ-decenolactone) clustered in an HCA branch, and 95 (octadecan-1-ol) and 57 (dihydromaltol), grouped in another branch. It was also defined by low NIFs of OAs n°1 (methanethiol), 41, 23, and 82 (isopropyl palmitate). The third sample cluster {1B, 2B, 2C} was characterized by medium to high NIFs for the branch grouping OAs n°12 (2-methylbut-2-enal), 13 (isoamyl acetate), and 14 (pentyl acetate); the group 15, 17, and 52; the cluster 1, 70 (furaneol), and 33 (acetic acid); and medium to low NIF values for the cluster 82, 84 (3-hydroxy-4-phenylbutan-2-one), and 89 (γ-dodecalactone). Finally, the last sample cluster {4B, 4C, 3A, 4A} was the less homogeneous, and could be better interpreted by considering the two sub-groups {4B, 4C} and {3A, 4A} defined in the HCA. The first sub-group displayed high NIFs for the OA clusters 73-81, 82-84-89, and for OAs 28 and 90, and medium to high values for OAs 31 and 40. Both sub-clusters shared medium to low values for OA groups 12-13-14, 51-76-94, and for acetic acid (OA 33) and *allo*-ocimene (23). The proximity of samples 3A and 4A in the second sub-group was characterized by medium to low NIFs for cluster dihydromaltol (OA 57)-octadecan-1-ol (OA 95), for furaneol (OA 70) and nonan-2-one (OA 25), and medium to high values for the OA cluster 15-17-52 (Figure 2).

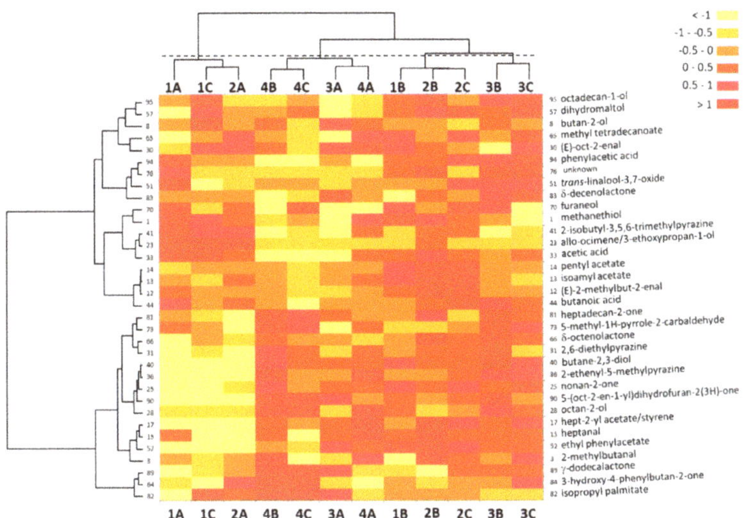


**Figure 2.** Heatmap displaying the results of a hierarchical cluster analysis (HCA) conducted independently on both samples' and variables' (OAs) dimensions, for the 34 discriminant OAs. NIF values' importance varies from >1 (highest value, in red) to < −1 (lowest values, in yellow). OA numbers are those found in Table 1. An HCA conducted on only samples showing the 4 distinctive clusters displayed here may be found in Appendix B (Figure A3) for clarity purposes. The data were centered and scaled; dissimilarity Euclidian distances were used with the Ward amalgamation method.

## 2.2. Identification of Impact Compounds

Ninety-six OAs reached the 50% NIF threshold used in the DFA and were considered as significant impact odorants of the chocolate samples under study. Among them, only 28 were defined by a single well-resolved GC-MS peak, and they were easily and unambiguously identified by their EI and CI mass spectra and their LRIs on DB-FFAP by comparison with authentic standard data (Table 1). Some other compounds, although present in co-eluted peaks, displayed clear EI mass spectra, sometimes after deconvolution using the AMDIS or PARADISE [81] software packages. Thus, seven more compounds (3-methylthiopropanal in OA n°34, ethyl nonanoate and non-2-enal in OA 39, 2- and 3-methylbutanoic acid in OA 47, ethyl dodecanoate in OA 56, and 4-vinylphenol in OA 90) could be unambiguously identified (Table 1). For 3-methylthiopropanal and 4-vinylphenol, their respective characteristic odor notes detected by the assessors in the descending part of the GC peaks (potato and medicinal, respectively) also aided their identification. Using the same procedure, 58 compounds were tentatively identified by comparison of their MS, LRI on DB-FFAP, and odor to data found in published literature and/or found in libraries. Injections of the sample extracts on a DB-5 column allowed confirmation of most of the identified peaks after determining their LRIs, which were compared to published data using the column and/or to LRI data found in databases. Among the 93 aroma compounds identified so far (35 unambiguously and 58 tentatively) in 83 OAs, only 17 molecular weights were not confirmed by chemical ionization (CI) using methane and ammonia as reagent gases. CI was a successful method to confirm identification when limited information was present in MS databases and/or when EI mass spectra were ambiguous. For example, MW of OA n°69, tentatively identified by its impure mass spectrum to 1*H*-pyrrole-2-carbaldehyde (MW = 95) on the basis of the similarity index using the Wiley 11th Editition/NIST 2017 database (Figure 3a), was confirmed by methane- and ammonia-CI (Figure 3b).

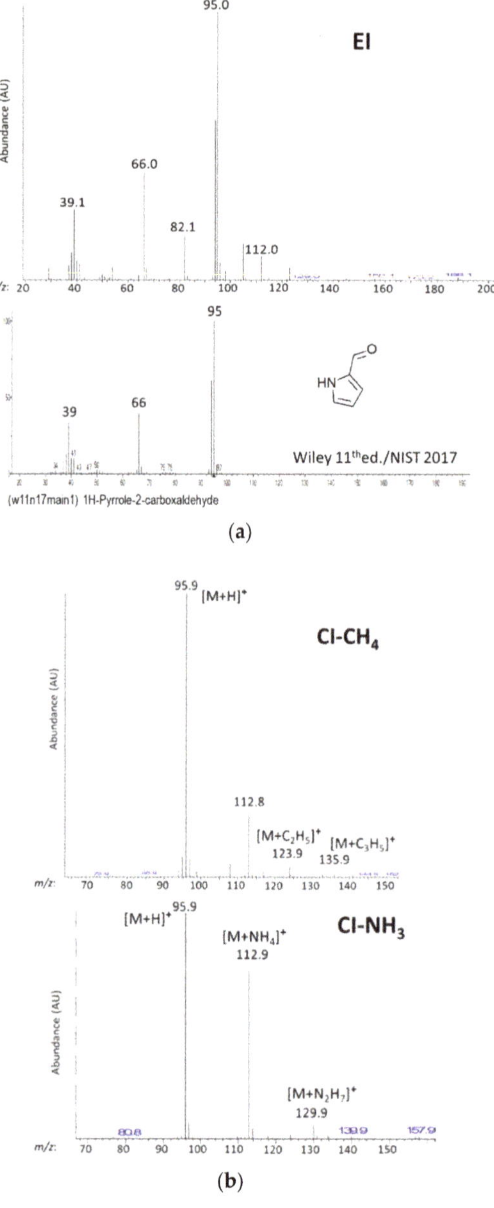


**Figure 3.** Mass spectra of 1H-pyrrole-2-carbaldehyde (OA n°69, MW 95) obtained in EI compared to the Wiley 11th Edition/NIST 2017 database reference spectrum (**a**) and in CI with methane and ammonia as reagent gases (**b**). Diagnostic ions on both CI spectra are indicated (**b**).

Thus, the methane-CI spectrum displayed the diagnostic ions $[M + H]^+$ at *m/z* 96, $[M + 29]^+ = [M + C_2H_5]^+$ at *m/z* 124, and $[M + 41]^+ = [M + C_3H_5]^+$ at *m/z* 136. This was confirmed in ammonia-CI by the diagnostic ions $[M + H]^+$ at *m/z* 96, $[M + 18]^+ = [M + NH_4]^+$ at *m/z* 113, probably enhanced by an impurity found at *m/z* 112 in the EI mass spectrum, and $[M + 35]^+ = [M + N_2H_7]^+$ at *m/z* 130 (Figure 3).

For EI and CI mass spectra data acquisitions, basic/neutral and acidic fractions obtained after chemical fractionation of the chocolate extracts were checked when needed, in order to clarify some co-elutions. For instance, γ-nonalactone (OA n°68), just preceding OA 69 by four LRI units, was more clearly identified in the basic/neutral fraction. Odor descriptions given by the 12 assessors in the DFA experiment were also compared to odor attributes found in databases to aid the identification task. Most of the time, this comparison confirmed the identifications inferred from the MS and LRI data (Table A1, Appendix B). Seven OAs remained problematic in terms of the odor description and/or identification because they exhibited co-eluting species that were clearly visible in EI and CI mass spectra obtained using the DB-FFAP column. Therefore, heart-cutting MDGC-MS/O was used to resolve these problems with the DB-FFAP column in the first dimension and a DB-5 one in the second dimension. Three OAs were thus clearly identified and unambiguously confirmed by MS and LRI data of standards obtained on both column types: Butane-2,3-diol (OA n°40), guaiacol (OA 58), and furaneol (OA 70), with the odor attributes also comparable to published data (Table A1). OA n°63 was tentatively determined as a mixture of δ-octalactone and maltol. As their respective odors, in agreement with the published data, are similar (Table A1), the fruity-sweet note of OA 63 could be due to one of them or to the mixture. Finally, a heart-cut of the OAs n°65, 66, and 67 grouped in a single MDGC run allowed the identification of phenol (67) and a tentative identification of methyl tetradecanoate (65) and δ-octenolactone (66). For OA n°32, it was not possible to differentiate 3-ethyl-2,5-dimethylpyrazine from 2-ethyl-3,5-dimethylpyrazine as these molecules shared the same mass spectra, the same LRIs on both DB-FFAP and DB-5 columns, and the same odor descriptions (Table A1). Moreover, both have been described in dark chocolate (Table 1). Therefore, OA 32 was due to either one of these pyrazines or to a mixture of both volatiles. Finally, within the 96 OAs retained as significant impact components of the dark chocolates under investigation, by applying a 50% SNIF threshold in DFA, 101 odorous compounds were identified (39) or tentatively identified (62) with rather good confidence, and 6 remained as unknown.

## 3. Discussion

The main objective of the study was to identify the most potent odorants in chocolates, and particularly the key odorants that could discriminate the samples, and potentially the predefined sensory poles. Clearly, as usual in GC-O studies, the potent odorants were not the most abundant volatiles in the extracts. Thus, the most abundant compounds found in common in all the samples were acetoin (3-hydroxybutan-2-one, LRI$_{DB-FFAP}$ 1296), trimethylpyrazine (LRI 1410), tetramethylpyrazine (LRI 1480), 3-methylbutanoic acid (LRI 1676), the two diastereoisomers of butane-2,3-diol monoacetate LRIs 1575 and 1587), phenylacetaldehyde (LRI 1653), phenylethanol (LRI 1921), and 2-acetylpyrrole (LRI 1985). In pole 1 samples, acetic acid (LRI 1462) was also found to be particularly abundant. As expected by the powerful aromatic nature of dark chocolate, a large number of odorous compounds were detected by the GC-O panel. Applying a high 50% NIF threshold to the data, 96 olfactive areas were finally retained that revealed 107 active odorants, among which six remained unidentified (Table 1). This rather important retained number, despite the application of a demanding threshold, equals or even surpasses the number of OAs found in highly odorous products, like alcoholic beverages, such as Cognac, for example [40], or even chocolate [29]. Identification of most of the impact compounds were based on classical extract handling and instrumental means, GC-MS in electron and chemical ionization, with the help of chemical fractionation of the extracts and MDGC-MS/O. However, some of them appearing in the co-eluted peaks were tentatively identified by complementary comparison of the odor attributes used by the panel to published odor descriptors. Thus, the odor attributes given by the panel to AO n°17 (fruity, flowery) suggested hept-2-yl acetate, known as fruity, rather than styrene, which imparts a plastic note. The same applied for OA n°23 (metallic, musty) attributed to *allo*-ocimene rather than 3-ethoxypropan-1-ol reported as fruity (Table 2, and Table A1). OA n°90, most often described as fruity and floral, was tentatively attributed to the lactone 5-(oct-2-en-1-yl)dihydrofuran-2(3*H*)-one rather than to 4-vinylphenol only detected by fewer panelists in the descending GC peak as rubber and medicinal

(Table 2, and Table A1). OA n°32 was not fully resolved as both candidates 3-ethyl-2,5-dimethylpyrazine and 2-ethyl-3,5-dimethylpyrazine were not separated on DB-FFAP nor on DB5 columns (Table 1) and have both been described with the same vegetal, roasted notes (Table A1). The same applied for OA n°38, attributed to the positional isomer candidates 3-isobutyl-2,5-dimethylpyrazine and 2-isobutyl-3,5-dimethylpyrazine, only separated on the DB5 column but imparting the same vegetal, pepper olfactive note that was not described in consulted databases (Tables 1 and A1).

Most of the identified key odorants have been found previously in cocoa products, including cocoa mass or liquor, and/or dark chocolate (Table 1). However, to the authors' knowledge, some of them were described here formally for the first time as key odorants of dark chocolate: Methanethiol (OA n°1), ethanol (5), ethyl propanoate (6), butan-2-ol (8), 2-methylbut-3-en-2-ol (9), 3-methylbutan-1-ol (16), hept-2-yl acetate (17), *allo*-ocimene (23), octan-2-ol (28), 2,6-diethylpyrazine (31), 2-ethenyl-5-methylpyrazine (36), 2-isobutyl-3,5-dimethylpyrazine (38), 2-isobutyl-3,5,6-trimethylpyrazine (41), acetophenone (46), 1-phenylethyl acetate (48), 2,3,5-trimethyl-6-(3-methylbutyl)pyrazine (49), 1-phenylethanol (54), ethyl dodecanoate (56), phenylmethanol (59), δ-octalactone (63), methyl tetradecanoate (65), octanoic acid (71), nonanoic acid (78), heptadecan-2-one (81), δ-decenolactone (83), 3-hydroxy-2,3-dihydromaltol (85), γ-dodecalactone (89), and OA n°92, dodecanoic acid (Table 1, and Table A1). Moreover, to the best of the authors' knowledge, 13 key odorants are described for the first time in the composition of a cocoa product (Table 1, and Table A1). However, all of them have been previously described in foodstuffs or beverages. Thus, 3-ethyl-4-methylpentan-1-ol (OA n°37) was previously described in brandies [54]; 3-hydroxybutanoic acid (42) was described in various fruits, wine and honey [54]; methyl 2-methylpentanoate (48) in potato and tea [54]; 3-hydroxypropyl acetate (50) in bread and wines [54]; δ-pentalactone (53) in various foods and beverages [54]; dihydromaltol (57) in milk products and wine [54]; isopropyl palmitate (82) in various food products [54]; 3-hydroxy-4-phenylbutan-2-one (84) in honey and wines [54]; 2-phenylethyl lactate (86) in cheddar cheese [54]; dihydroactinidiolide (87) in a lot of foodstuffs, beverages, and seeds [54]; farnesol (88) in a lot of foods and beverages [54]; 5-(oct-2-en-1-yl)dihydrofuran-2(3*H*)-one (90) in chicken [54]; and octadecan-1-ol (95) in a lot of products, including milk products, fruits, and tea [54].

In order to determine the discriminative features that should allow samples to be distinguished, based on the work of Pollien et al. [33], firstly a GC-O comparative approach where a 30% difference threshold was considered in the DFA data, i.e., an NIF difference > 30% between at least two samples, was attempted. Among the initial 96 potent OAs, this procedure revealed 73 OAs in which an NIF difference > 30% between at least two samples exists (Table 1). To understand the discriminative variables better, a correspondence analysis was conducted to visualize the proximities between OAs and samples. Despite its statistical significance, this CA displayed rather noisy plots, where many variables (OAs) poorly represented in the center of the CA plots bore little correspondence information (Figure S1). To look more objectively at the data, a statistical comparison of k proportions (Khi$^2$ test) was used on the whole NIF dataset of Table 1. The results clearly confirmed the non-discriminant OAs (*p*-values highly non-significant, $\alpha$ = 0.05) and revealed non-significant *p*-values for most of the OAs ill-defined on the CA plots (Table A1). Therefore, a more demanding difference threshold (50%), i.e., an NIF difference > 50% between at least two samples, was applied to the NIF data. This more drastic difference threshold selected 34 OAs (Table 1) for which most of the *p*-values obtained in the Khi$^2$ test were also highly significant (Table 2). These 34 OAs defined by 34 odorants, among which only one remained unknown (Table 2), were considered the discriminative features that allowed the samples to be distinguished. Noteworthy, most of their main odor qualities cited by the panelists corresponded generally to odor attributes that were found in the literature and databases (Table 2). The CA conducted using these 34 key odorants revealed significant proximities between particular odorants and the samples (Figure 1). Finally, the CA distinguished four groups of three samples: {1A, 1C, 2A}, {2B, 2C, 1B}, {3A, 3B, 3C}, and {4A, 4B, 4C}. These groups represented a clear image of the four sensory poles, with each sensory pole being characterized by particular key odorants (see results). These groups and their respective proximities also reflected the intra-class variability of sensory poles 1 and 2 [32] (sample 2A grouped with 1A and 1C, and sample 1B grouped with 2B and 2C, respectively), with the concomitant difficulties encountered in sampling pertinent exemplary chocolates considering their partial overlapping evidenced in Figure A1 and in [32]. They also reflected the similarities of poles 3 and 4 [32], albeit distinguishable (Figure 1). The heatmap produced with the NIF data of the 34 discriminant odorants (Figure 2) largely confirmed the CA. The sample clusters defined by HCA showed the same tendencies: Variability of sensory poles 1 and 2, proximity of poles 3 and 4 with sample 3A grouped with pole 4 samples, and particularly with sample 4A. One advantage of such a heatmap based on HCA is the clustering of explanatory variables, thus evidenced in a better manner. For instance, a cluster of OAs with very low NIF values characterized the samples more related to pole 1 (with also sample 2A), which included high NIF values for a cluster composed of acetic acid (OA n°33), *allo*-ocimene (23), 2-isobutyl-3,5,6-trimethylpyrazine (41), methanethiol (1), and furaneol (70) to a lesser extent. The heatmap appeared complementary to the correspondence analysis for the treatment of GC-O data, with the aim of discriminating chocolate samples differentiated on sensory criteria, with the association of discriminant key odorants. Within these 34 discriminant key odorants, 17 are described formally for the first time as key flavor compounds of dark chocolate (Table 2). The criterion based on the NIF difference threshold introduced by Pollien et al. [33] for discriminating samples in GC-O using a comparative analysis based on the detection frequency (named here cDFA) appeared to be in very good accordance with the statistical approach, which used the Khi$^2$ test calculated in the comparison of k proportions (Table 2). However, a few discrepancies were noticed that merit discussion.

Three key odorants out of the 34 retained discriminant ones were not at all significant with $p$-values > 0.15 (Table 2). These compounds, butanoic acid (OA n°44, $p$-value = 0.463), δ-octenolactone (66, $p$-value = 0.337), and 3-hydroxy-4-phenylbutan-2-one (84, $p$-value = 0.280), as already outlined, were situated near the origin in the CA plots, and therefore, were not well represented in the correspondence analysis (Figure 1). They did not belong to the same cluster on the heatmap (Figure 2). However, butanoic acid was found with similar medium NIF values in all the samples except a high value in sample 2C and a low value in chocolate 4C (Table 1); this behavior explained both the retained 50% NIF difference threshold and the non-significant Khi$^2$ test. The same applied for δ-octenolactone (high NIF value in the only 3B sample vs. low NIF value in the single 1A one), and to a lesser extent for 3-hydroxy-4-phenylbutan-2-one. Therefore, the three compounds can hardly be considered as discriminant features, as clearly indicated by the Khi$^2$ test ($p$ > 0.15). A heatmap conducted with the remaining 31 discriminant features using a classical non-specific filtering of 50% on the standard deviation (std) criterion (i.e., eliminating 50% of the variables with the lowest std for clarity purpose) revealed interesting features (Figure S2, Supplementary data). Particularly, by removing the non-significant variables (based on the Khi$^2$ test) and the variables with the lowest std (both types contributing to background noise), the samples' clustering appeared in good conformity with the initial sensory classification, with the clusters {3B, 3A, 3C} and {4C, 4A, 4B}, corresponding to sensory poles 3 and 4, well defined on discriminant key odorants (Figure S2).

Besides, seven of the OAs not retained as significant based on the 50% NIF difference threshold had significant $p$-values in the Khi$^2$ test (Table A1). Thus, 3-methylbutan-1-ol (OA n°16, $p$-value = 0.030), OA n°27 (unknown, $p$-value = 0.003), 1-phenylethanol (54, $p$-value = 0.002), OA n°56 (pentan-2-yl benzoate/ethyl dodecanoate, $p$-value = 0.035), OA n°63 (δ-octalactone/maltol, $p$-value < 0.001), 2-phenylethyl lactate (86, $p$-value < 0.0001), and OA n°96 (unknown, $p$-value = 0.012) should be considered. It is noteworthy that all these but one (OA n°96) satisfied the 30% NIF difference threshold criterion and were retained in the initial 73 discriminant OAs (Table 1). Their NIF values in the samples were of two types (Table 1): Most of them (six out of seven) had generally low NIF values, with no detection (NIF = 0) in some samples, and were very often characteristic of a particular sensory pole. Thus, OA n°27 was more clearly detected in poles 1 and 2, OA n°63 seemed to characterize pole 1, and 2-phenylethyl lactate (86) was not detected at all in pole 4 and characterized poles 1 and

2, which was contrary to OA n°96 that seemed significantly detected only in pole 4. The remaining 1-phenylethanol (OA n°54) had generally very high NIF values except in one sample (2A). All these behaviors explained both the retained 30% NIF difference threshold and the significant *p*-values in the Khi$^2$ test. Therefore, it sounded reasonable to include them as significant variables in the differentiation of the chocolates. A heatmap was calculated using the 38 'discriminant' variables (31 + 7) based on both the NIF difference threshold and Khi$^2$ test. For clarity purposes and to highlight the most significant variables that could discriminate the samples, a 50% non-specific filtering on the std criterion was again applied, therefore resulting in only 19 variables being displayed (Figure 4). However, an HCA conducted with the complete set of 38 variables was also performed and resulted in the same sample clustering (Appendix B, Figure A4).

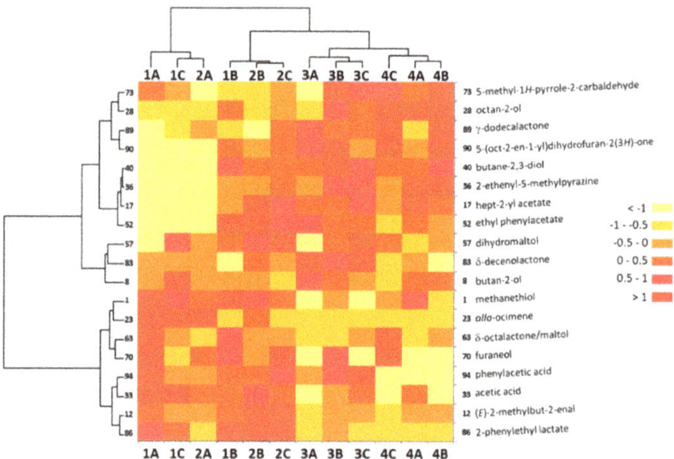



**Figure 4.** Heatmap displaying the results of a hierarchical cluster analysis (HCA) conducted independently on both samples and variables (OAs) dimensions, for the 38 "discriminant" OAs determined by both NIF difference threshold and khi$^2$ test (see text). NIF values importance varies from >1 (highest value, in red) to < −1 (lowest values, in yellow). OA numbers are those found in Table 1. The data were centered and scaled; dissimilarity Euclidian distances were used with the Ward amalgamation method; %(std) non-specific filtering was used with a 50% threshold, resulting in the display of the only 19 variables with the highest std. An HCA conducted on the samples with the 38 variables, showing the same four distinctive clusters displayed here, may be found in Appendix B (Figure A4) for clarity purposes.

Noteworthy, the grouped samples were respectively gathered in four well-separated clusters {1A, 1C, 2A}, {1B, 2B, 2C}, {3A, 3B, 3C}, and {4C, 4A, 4B} corresponding to the initially defined sensory poles, with the limit of the misclassification of samples 1B and 2A already outlined. Clustered variables allowed qualification of the sample groups (Figure 4). Thus, 5-methyl-1*H*-pyrrole-2-carbaldehyde (OA 73) and octan-2-ol (28) were more perceived in poles 3 and 4 samples. A group of key odorants was clearly less perceived in pole 1 represented by the cluster {1A, 1C, 2A}: γ-dodecalactone (OA 89), OA n°90 5-(oct-2-en-1-yl)dihydrofuran-2(3*H*)-one, butane-2,3-diol (40), 2-ethenyl-5-methylpyrazine (36), hept-2-yl acetate (17), and ethyl phenylacetate (52). Butan-2-ol (OA 8) and δ-decenolactone (83) were more perceived in pole 3. Methanethiol (1) and *allo*-ocimene (23) characterized poles 1 and 2 together with OA n°63 (δ-octalactone/maltol) and furaneol (70) while the last compounds were less perceived in poles 3 and 4. Acetic acid (OA 33) and phenylacetic acid (94) had higher NIF values in poles 1 and 2, phenylacetic acid being particularly absent from pole 4 samples (Table 1). Finally, 2-methylbut-2-enal (OA 12) and 2-phenylethyl lactate (86) were less perceived in poles 3 and 4, the latter characterizing particularly pole 2 chocolates (Figure 4). These 19 particularly significant key odorants that allowed discrimination of the 12 chocolates in agreement with the sensory poles could not be related to the cocoa variety and/or origin as the initial classification was only based on sensory properties [32]. Moreover, the dark chocolate key odorant compounds constitute a flavor balance that is the result of many factors, including the cocoa variety, post-harvest treatments linked to origin, and a complex processing that includes roasting. For instance, acetic acid and phenylacetic acid are final degradation products of the amino acids alanine and phenylalanine, respectively, which accumulate from the fermentation of cocoa beans to the final product [1], but acetic acid is also a marker of the cocoa variety Criollo [31,82]. Heterocyclic compounds like lactones, pyrazines, pyrroles, pyranones, furanones, and the Strecker aldehydes, formed in abundance in the Maillard reaction during the roasting step, are already present in fermented cocoa beans [1,11,17,77,83]. Thus, it was recently reported that interactions between cocoa botanical and geographical origin, formulation, and process showed difficulties in identifying individual markers linked to the different steps along the supply chain [31].

Most of the key odorants identified in the present study were potential candidates for the molecular ions identified in the PTR-ToF-MS analyses of the samples' headspace volatiles [32]. However, the volatiles with higher molecular weights were only identified in the present study, illustrating the fact that headspace analyses are less sensitive than vacuum extraction procedures. Among the 38 'discriminant' key odorants identified here, only 6 were found in the discriminant ions that allowed classification of the initial 206 chocolate samples [32]: Butan-2-ol (OA n°8), 2-methylbut-2-enal (12), 3-methylbutan-1-ol (16), 2,6-diethylpyrazine (31), butane-2,3-diol (40), and 1-phenylethanol (54). This result reflects the different criteria retained to classify the samples. In the PTR-MS study, the relative abundance of the volatile components (represented by 143 ions) were used after headspace extraction; furthermore, the variables (ions) were highly correlated [32]. In the present study, the discriminative features were determined for their odor quality as key odorants in a comparative GC-O experiment, i.e., in a sequential discrete detection mode. While being impact odorants, they were sometimes found in very low abundance, and a lot of them (with the higher molecular weights) were simply not detected in the samples' headspace.

The comparative GC-O conducted here used the detection frequency analysis method with the data expressed in nasal impact frequencies. Although this are not a direct measurement of the perceived odor intensities, it can be demonstrated that NIF values increase as a function of the concentration [33], and consequently with odor intensities. It was assumed that working with a panel of 8–10 assessors, an NIF difference of 30% would generally indicate a significant concentration difference for individual perceived odorants [33]. In the present study, we worked with a panel of 10–12 assessors and finally assessed an NIF difference of 50% between at least two samples as being necessary for an odorant to differentiate them with high significance on this component. This assessment was largely confirmed by the Khi$^2$ test of comparing k proportions, which was made usable with the type of data obtained using DFA. Therefore, if the NIF values we obtained did not exactly measure the concentrations, at least they gave a good idea of the relative abundances of the key odorants in the samples, which were finally retained as discriminative features.

## 4. Materials and Methods

### 4.1. Chocolate Samples

All the dark chocolates under study were produced in an industrial pilot plant using the same 'standard' transformation process, with the same mass of cocoa (65%) from different origins and varieties, of sugar, of soy lecithin, and of vanillin. They were supplied by the Valrhona Company, chocolate producer in Tain L'Hermitage (France). Twelve chocolates, three in each of the four sensory poles previously defined at the industrial level, were chosen among 206 chocolate samples that were used to build a predictive model [32]. Being representative of the sensory categories, they were chosen according to four decision criteria: Availability (sufficient quantity available to conduct all the experiments of the project), uniform distribution in the four sensory poles, coverage of the maximum acceptable variability within each sensory pole, and distinct origins. Their positions in the sensory space are highlighted in the PCA planes of the sensory data of the 206 samples in Figure A1 (Appendix A) for illustrative purposes. In the following, they will be noted xA, xB, and xC, x being the sensory pole (x = 1 to 4). The samples were stored under vacuum at −20 °C before their analysis.

### 4.2. Extraction of the Volatiles

After being thawed at room temperature, each sample of dark chocolate was cut into small cubes (ca. 1 cm$^3$). Suspended in 100 mL of ultra-pure water (MilliQ system, Millipore, Bedford, MA, USA), the sample was placed in the sample flask of a solvent-assisted flavor evaporation (SAFE) glassware [35], where, together with a magnetic stirrer, 300 µL of an aqueous solution at 93 mg/L of 2-methylheptan-3-one (CAS 13019-20-0; 99% pure; Sigma-Aldrich, St Louis, MO, USA) used as internal standard were added. The resulting standard concentration was 0.28 mg/L. The round-bottom flask was placed in a water bath at 37 °C (just above the chocolate melting point) and the stirred slurry with melted chocolate was distilled under vacuum in the SAFE apparatus at 1 Pa. After a distillation time of 1h30min, the frozen hydro-distillate (ca. 100 mL) was thawed at room temperature, and then a liquid-liquid extraction was conducted with methylene chloride ($CH_2Cl_2$) as solvent (Carlo Erba, Val de Reuil, France; purity > 99.9%, distilled just before use). Three successive 15-min extraction steps were realized under agitation using 3 × 15 mL $CH_2Cl_2$ in a water-ice bath (ca. 4 °C) and the recovered organic extracts were pooled and dried over anhydrous $Na_2SO_4$ (5 g). The extract was then filtered through glass wool before being concentrated to 400 µL (adjusted volume with $CH_2Cl_2$) using two successive Kuderna-Danish apparatuses (Merck, Darmstadt, Germany) of decreasing size equipped with a Snyder column. The extracts obtained in triplicate for each chocolate sample were stored at −20 °C before use.

### 4.3. Determination of Impact Compounds by GC-O Comparative Detection Frequency Analysis (cDFA)

The 12 chocolate extracts (pooled triplicates of each extraction) were submitted to GC-O using the detection frequency analysis (DFA) method [33,34]. The extracts' concentration was optimized in dummy assays conducted with three assessors to follow the recommendation of Etiévant and Chaintreau [41] to avoid overexpressing the number of odorants that could be detected by all the panelists (thereafter not discriminant).

Analyses were performed on a 6890A GC (Agilent Technologies, Santa Clara, CA, USA) equipped with an FID and an in-house sniffing port using a DB-FFAP column (30 m × 0.32 mm i.d., 0.5 mm film thickness; Agilent Technologies). He4e, 1 µL of extract was injected using a splitless/split injector in splitless mode for 0.5 min, then switched to split mode (25 mL/min) at a temperature of 240 °C. The initial oven temperature was set to 40 °C and then increased at 4 °C/min to a final temperature of 240 °C held for 10 min. Analyses were performed in constant flow mode at a carrier gas (He) velocity of 44 cm/s. At its end, the column was connected to a Y-type seal glass, and the effluent was split into two equal parts (50% to FID, 50% to sniffing port) by two deactivated capillaries (both 1.1 m, 0.32 mm i.d.). The FID and transfer line to the sniffing port were heated at 240 °C. Humidified air (25 mL/min) was added to the transfer line to prevent nasal mucosa dehydration. Linear retention indices (LRIs) were calculated by a weekly injection of a reference solution of n-alkanes ($C_7$ to $C_{30}$; Sigma-Aldrich) according to van den Dool and Kratz [84].

Twelve assessors belonging to the CSGA staff (8 women and 3 men, 21 to 61 years old, nine of them with previous experience in GC-O) participated after having been informed and having signed a consent form. Each of them sniffed the 12 extracts once, in a randomized order using a Williams Latin square design, for a period of ca. 40 min starting 3.8 min after the injection (solvent delay). Data were acquired by the OpenLab software (6850/6890 GC System, V2.3, Agilent Technologies) for the chromatographic part and by the ODP recorder (Gerstel, Mülheim an der Ruhr, Germany) for the descriptors citing part (a button and a microphone allowed the recording of odor events and their vocal description). DFA is based on the determination of olfactive areas (OAs) in a sample by gathering all the odor events detected by the panel on the basis of their LRI closeness, grouped if the difference is inferior to a few LRI values. A threshold of 50% for the detection frequency, also known as the nasal impact frequency (NIF) [33], was set as necessary to retain an OA [34,48]. This threshold equals a minimum of six odor events detected by the panel in a sample to retain an OA, i.e., an OA was retained only when six assessors detected it in at least one sample [34,48].

cDFA was performed to obtain a first impression of the odorants, which may contribute to the overall aroma of the dark chocolates and to highlight differences between them. Although they are not a direct measurement of the perceived odor intensities, NIFs increase with concentration [33]. Therefore, they can be used to compare peak intensities between aromagrams. According to Pollien et al. [33], a difference in NIF values of at least 30% (between the lowest and the highest values of one OA) is assumed to be a significant concentration difference. Therefore, to be considered as a discriminant OA, a 30% difference (corresponding to a difference of four odor events) between at least two samples for the given OA was applied. To highlight very discriminant OAs, a 50% difference (six differing odor events between at least two samples) was also applied in a second time.

To determine the very volatile impact compounds whose retention times do not allow separation of them from the solvent peak on the DB-FFAP column (generally for LRIs $\leq$ 1000), a headspace (HS) technique was used. A solid-phase microextraction (SPME) method was optimized (chocolate sample size, addition of water or not, equilibration time and temperature, extraction time, desorption time, type of SPME fiber) in order to get the same GC response and odor intensity as in the SAFE method for a volatile reference peak, clearly identified as butane-2,3-dione (LRI: 995 on DB-FFAP, odor descriptor: butter) and detected by the entire panel (NIF: 100%). Thus, 2 g of chocolate cut in small cubes (ca. 0.5 cm$^3$) were suspended in 1 mL of purified water within a 20-mL sample vial containing a magnetic stirrer. The vial, closed by a PTFE-lined screw cap, was equilibrated under agitation (250 rpm) at 60 °C for 15 min in a water bath. Then, the extraction was realized with a triple-phase divinylbenzene/carboxen/polydimethylsiloxane (DVB/CAR/PDMS) SPME 2-cm fiber (Supelco Sigma-Aldrich) for 30 min at the same temperature. Then, the SPME fiber was desorbed for 5 min in the GC injector maintained at 240 °C (splitless mode). As only the most volatile compounds were sniffed in that case, the GC oven set at an initial temperature of 40 °C was programmed at 4 °C/min to 80 °C and then to 240 °C (maintained for 10 min) at 20 °C/min. Other GC and signal acquisition parameters were the same as the ones mentioned earlier, except the sniffing period that lasted 5 min only, and 10 assessors from the initial 12 ones participated (two were not available). As previously stated, a weekly injection of n-alkanes was used for LRI calculation, this time after adsorption on the same SPME fiber. The GC-O data were processed the same way as previously stated.

All through the GC-O procedure, the quality of the GC column was checked for repeatability (retention times, peak heights and peak areas) weekly by injecting a reference solution (Grob Test Mix, Sigma-Aldrich).

### 4.4. Identification of the Impact Compounds

The compounds responsible for OAs were identified by GC-MS.

The triplicate SAFE extracts of the 12 chocolate samples were analyzed on a 7890A GC coupled to a 5975C mass selective detector (MSD, Agilent Technologies) using the same column as in the GC-O study. GC-MS data of SPME extracts were also obtained in duplicate on the DB-FFAP column, using the same conditions as those used for the GC-O experiments. A complementary study was performed on a DB-5MS column (30 m × 0.32 mm i.d., 0.5 μm film thickness, Agilent Technologies) to confirm the identifications by obtaining MS and LRIs on a second column with a different polarity, and thus avoiding overlooking possibly coeluting compounds. The used GC conditions were the same. The data were obtained on the DB-5 column on a pooled solution of the triplicate SAFE extracts of the 12 chocolates. Analyses were conducted using the same chromatographic parameters with a solvent delay of 3.5 min, except for SPME analyses, and LRIs were calculated as previously described. Electron ionization (EI) spectra were obtained with electron energy of 70 eV at a rate of 4 scans/s, covering the *m/z* range 29–350 with a source temperature of 230 °C. Data were acquired using the ChemStation software (ver. A.03.00, Agilent Technologies). The reliability of the compound identification was first assured by comparison of the experimental mass spectra to mass spectral data contained in various databases: NIST 2017/Wiley 11th Edition, MassBank (https://masbank.eu/MassBank), Pherobase (https://www.pherobase.com), and our in-house database INRAMass containing more than 10,000 mass spectra of volatiles. The software packages AMDIS (ver. 2.73, NIST) and PARADISE [81] (ver. 2.92, http://www.models.life.ku.dk/paradise) were used for mass spectra deconvolution of coeluted peaks. Besides spectral information, compound identification was confirmed by comparison of the experimental LRIs to published data and to data found in the following online databases: NIST Chemistry WebBook (http://webbook.nist.gov/chemistry), Volatile Compounds in Food (http://www.vcf-online.nl) [54], the Pherobase, and the LRI & Odour database (http://www.odour.org.uk). When standards were available in our collection of aroma compounds, identifications were confirmed by comparing their MS and LRI obtained on equivalent DB-FFAP and/or DB-5 columns.

Chemical ionization (CI) was also carried out with methane and ammonia as reagent gases on the pooled triplicates of each sample. CI analyses were conducted with a source pressure of 0.1 kPa for both gases at a source temperature of 150 °C and with an electron energy of 240 eV. Molecular weights (MWs) were determined by observing diagnostic ions depending on chemical classes [85,86].

To aid compound identification, a chemical fractionation of the pooled triplicate extracts of each chocolate sample was also performed to separate the basic/neutral fraction from the acidic one. An aliquot (200 µL) of each $CH_2Cl_2$ extract was diluted in 100 mL of purified water. The aqueous solution was adjusted to pH 9 with NaOH (0.045 M) and agitated for one hour. The basic/neutral fraction was recovered by extraction with $CH_2Cl_2$ (3 × 10 mL). The remaining aqueous solution was adjusted to pH 2 with aqueous HCl (18%), stirred for one hour, and the acidic fraction was recovered by extraction with $CH_2Cl_2$ (3 × 10 mL). Both organic fractions were dried, filtered, and concentrated as previously described, and analyzed by GC-MS on both DB-FFAP and DB-5 columns.

Finally, to separate some co-eluting species not clearly resolved by the use of the columns of different polarities, a two-dimensional GC-MS/O system (MDGC-MS/O) was used. The first gas chromatograph (GC1) was a 7890A GC (Agilent Technologies) equipped with FID as a monitoring detector and a DB-FFAP column (30 m × 0.25 mm i.d., 0.5 µm film thickness, Agilent Technologies). The second GC (GC2) was also a 7890A GC equipped with a DB-5MS column (30 m × 0.25 mm i.d., 0.5 µm film thickness, Agilent Technologies) and coupled to a 5975C MSD (Agilent Technologies) and to a sniffing port (ODP 3, Gerstel). The connection between GC1 and GC2 was provided by a Deans switch (Agilent Technologies) followed by a cryotrap system (CTS, Gerstel) cooled down to −100 °C by liquid nitrogen. Fractions transported by the Deans switch (heart-cuts) from GC1 to GC2 were released to GC2 by a rapid heating (ca. 20 °C/s) of the CTS trap to 240 °C. GC ovens were successively temperature programmed from 40 to 240 °C at a rate of 4 °C/min. All other parameters were fixed as previously described. After the second column separation in GC2, 2/3 of the flow was diverted to the ODP and 1/3 to the MSD via two deactivated capillaries of adequate dimensions (0.83 m × 0.18 mm i.d. and 0.50 m × 0.10 mm i.d., respectively) via a capillary flow purged splitter (Agilent Technologies).

*4.5. Statistical Data Analysis*

All the statistical data treatments were performed using the software packages XLSTAT (Addinsoft, Paris, France) and/or Statistica (ver. 13.3, TIBCO Software Inc., Tulsa, OK, USA).

## 5. Conclusions

The aim of the present study was to identify the discriminant key odorants that should allow four previously characterized sensory categories of dark chocolates to be distinguished, which were modelled using the volatilome of 206 samples [32]. To address the question, a GC-O study was conducted by 12 assessors using a comparative detection frequency analysis (cDFA) on 12 samples chosen on availability and exemplariness criteria. A nasal impact frequency (NIF) difference of 50% for a key odorant between at least two samples was retained to differentiate the samples. A correspondence analysis (CA) revealed a classification that could be related to the sensory categories initially defined, through the proximities found between the most discriminant key odorants and the chocolate samples. The approach was confirmed and completed by a statistical analysis (Khi$^2$ test on proportions) made feasible with the DFA data. Finally, 38 key odorants discriminated the samples and allowed retrieval of the sensory categories thanks to a hierarchical cluster analysis (HCA). The discriminative relationships were illustrated in a heatmap, where the 19 most significant key odorants were identified.

**Supplementary Materials:** The following are available online, Figure S1: Correspondence analysis (CA) between the 12 samples and the 73 discriminant OAs defined by their NIF values using a NIF difference threshold >30%; Figure S2: Heatmap displaying the 15 most "discriminant" OAs of 31 significant OAs initially retained on NIF difference threshold >50%.

**Author Contributions:** Conceptualization, methodology, and validation, J.-L.L.Q.; investigation and analyses, Z.D.; data curation, Z.D. and J.-L.L.Q.; formal analysis, Z.D., H.L., and J.-L.L.Q.; resources, K.G.; writing—original draft preparation, Z.D. and J.-L.L.Q.; writing—review and editing, Z.D., J.-L.L.Q., H.L., K.G., M.R, and R.B.; supervision, J.-L.L.Q.; project administration, J.-L.L.Q., H.L., M.R., and R.B. All authors have read and agreed to the published version of the manuscript.

**Funding:** This research was funded by Valrhona and Agropolis Foundation (Grant N°ID 1505-003, Investissement d'Avenir programme, Labex Agro ANR-10LABX-0001-01)

**Acknowledgments:** The authors are indebted to Pierre Costet and Florent Coste (Valrhona, Tain l'Hermitage, France) for giving them the opportunity of this study (CHAMAN project), and for the chocolate samples. The authors thank Sébastien Preys (Ondalys, Clapiers, France) for analyzing preliminary data in order to choose the samples under study and for interesting discussions, and Géraldine Lucchi (CSGA, ChemoSens Platform) for her help with PARADISE and MDGC-MS/O. Rolande Koumbangoye and Shamsia Pithon are thanked for their technical help (chemical fractionation and MDGC-MS/O respectively). INRAE, Regional Council of Bourgogne Franche-Comté and the European Regional Development Fund (ERDF) are thanked for equipment funding.

**Conflicts of Interest:** The authors declare no conflict of interest. The funders had no role in the design of the study; in the collection, analyses, or interpretation of data; in the writing of the manuscript, or in the decision to publish the results.

## Appendix A

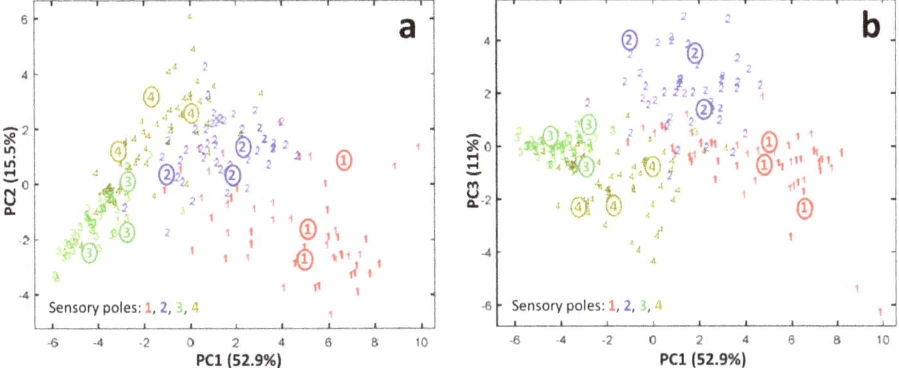


**Figure A1.** PCA score plots (**a**: PC1 vs. PC2 and **b**: PC1 vs. PC3, respectively) of the 206 chocolates calculated from the average intensities of 36 flavor attributes determined by quantitative descriptive analysis. The data were normalized by the sum of the intensities. For illustrative purposes, the colored numbers correspond to the four different sensory poles. The twelve samples under investigation (three in each pole) are circled.

## Appendix B

**Table A1.** Key aroma compounds in dark chocolate (12 samples) determined in a cDFA GC-O analysis using a 50% NIF threshold (12 assessors, see text for a complete explanation).[a] OA number, as in Table 1; [b] LRI on DB-FFAP, as in Table 1; [c] Odor attributes given by the panel; [d] Identification (refer to Table 1); [e] Chemical Abstracts Service registry number; [f] Mass formula; [g] Molecular Weight; [h] Pertinent odor attributes found in the databases VCF [54] and The Good Scents Company (http://www.thegoodscentscompany.com/); [i] $p$-value of the Khi$^2$ test ($\alpha = 0.05$) obtained for the parametric test comparison of k proportions using the data of Table 1.

| OA [a] | LRI [b] | Odor [c] | Identification [d] | CAS [e] | Formula [f] | MW [g] | Lit. Odor [h] | $p$-value [i] |
|---|---|---|---|---|---|---|---|---|
| 1 | 701 | cheese, cabbage, sulfur | methanethiol * | 74-93-1 | $CH_4S$ | 48.1 | cabbage, sulfur | <0.001 |
| 2 | 817 | chocolate, cocoa, roasted | 2-methylpropanal | 78-84-2 | $C_4H_8O$ | 72.1 | cocoa, malt, nut, caramel | 0.959 |
| 3 | 920 | cocoa, chocolate | 2-methylbutanal | 96-17-3 | $C_5H_{10}O$ | 86.1 | cocoa, nutty | 0.081 |
| 4 | 923 | cocoa | 3-methylbutanal | 590-86-3 | $C_5H_{10}O$ | 86.1 | cocoa, malt | 0.168 |
| 5 | 942 | fruity, solvent | ethanol * | 64-17-5 | $C_2H_6O$ | 46.1 | sweet, ripe apple, ethereal | 0.630 |
| 6 | 961 | fruity, floral | ethyl propanoate * | 105-37-3 | $C_5H_{10}O_2$ | 102.1 | apple, grape, sweet | 0.837 |
| 7 | 991 | butter | butane-2,3-dione | 431-03-8 | $C_4H_6O_2$ | 86.1 | butter | 0.436 |
| 8 | 1025 | rubber | butan-2-ol * | 78-92-2 | $C_4H_{10}O$ | 74.1 | medicine, solvent | <0.001 |
| 9 | 1040 | fruity, floral | 2-methylbut-3-en-2-ol * | 115-18-4 | $C_5H_{10}O$ | 86.1 | herb | 0.315 |
| 10 | 1054 | fruity | ethyl 2-methylbutanoate | 7452-79-1 | $C_7H_{14}O_2$ | 130.2 | fruit, apple, kiwi | 0.666 |
| 11 | 1072 | fruity, floral | ethyl 3-methylbutanoate | 108-64-5 | $C_7H_{14}O_2$ | 130.2 | fruity, sweet, apple | 0.151 |
| 12 | 1108 | hot plastic | (E)-2-methylbut-2-enal | 1115-11-3 | $C_5H_8O$ | 84.1 | solvent, ethereal | <0.0001 |
| 13 | 1127 | fruity, candy | isoamyl acetate | 123-92-2 | $C_7H_{14}O_2$ | 130.2 | fruity, banana | <0.001 |
| 14 | 1183 | fruity | pentyl acetate | 628-63-7 | $C_7H_{14}O_2$ | 130.2 | fruity, banana | <0.001 |
| 15 | 1196 | fruity, floral | heptanal | 111-71-7 | $C_7H_{14}O$ | 114.2 | fresh, green | 0.037 |
| 16 | 1211 | cheesy | 3-methylbutan-1-ol * | 123-51-3 | $C_5H_{12}O$ | 88.1 | fermented, fusel | 0.030 |
| 17 | 1267 | fruity, flowery | hept-2-yl acetate * | 5921-82-4 | $C_9H_{18}O_2$ | 158.2 | fruity | <0.0001 |
| 18 | 1296 | butter | 3-hydroxybutan-2-one | 513-86-0 | $C_4H_8O_2$ | 88.1 | butter, cream | 0.696 |
| 19 | 1309 | mushroom | oct-1-en-3-one | 4312-99-6 | $C_8H_{14}O$ | 126.2 | mushroom, earth | 0.176 |
| 20 | 1323 | fruity, mushroom, vegetal | heptan-2-ol | 543-49-7 | $C_7H_{16}O$ | 116.2 | mushroom, coconut, green | 0.184 |
| 21 | 1330 | roasted, chocolate | 2,5-dimethylpyrazine | 123-32-0 | $C_6H_8N_2$ | 108.1 | cocoa, roasted nuts | 0.213 |
| 22 | 1346 | roasted cereals, peanut | ethylpyrazine | 13925-00-3 | $C_6H_8N_2$ | 108.1 | roasted, peanut butter | 0.203 |
| 23 | 1383 | metallic, musty | allo-ocimene * | 673-84-7 | $C_{10}H_{16}$ | 136.2 | herbal, peppery | <0.0001 |
| 24 | 1387 | sulfur, cabbage | dimethyltrisulfide | 3658-80-8 | $C_2H_6S_3$ | 126.3 | cabbage, sulfur | 0.330 |
| 25 | 1397 | fruity, floral, vegetal | nonane-2-one | 821-55-6 | $C_9H_{18}O$ | 142.2 | sweet, herbal, fruity | 0.020 |
| 26 | 1410 | roasted, vegetal, earthy | trimethylpyrazine | 14667-55-1 | $C_7H_{10}N_2$ | 122.2 | cocoa, roast, earth | 0.451 |
| 27 | 1419 | fruity | unknown | - | - | - | - | 0.003 |
| 28 | 1427 | fruity, floral, candy | octan-2-ol * | 123-96-6 | $C_8H_{18}O$ | 130.2 | fruit, fresh, green | <0.0001 |
| 29 | 1431 | roasted, nutty | unknown | - | - | - | - | 0.385 |
| 30 | 1438 | vegetal, earthy | (E)-oct-2-enal | 2548-87-0 | $C_8H_{14}O$ | 126.2 | green, herbal, leaf | 0.017 |
| 31 | 1440 | vegetal | 2,6-diethylpyrazine * | 13067-27-1 | $C_8H_{12}N_2$ | 136.2 | green | 0.011 |
| 32 | 1450 | vegetal, roasted | 3-ethyl-2,5-dimethylpyrazine | 13360-65-1 | $C_8H_{12}N_2$ | 136.2 | roast, potato | 0.062 |
| | | | and/or 2-ethyl-3,5-dimethylpyrazine | 13925-07-0 | $C_8H_{12}N_2$ | 136.2 | roast, potato, burnt | |
| 33 | 1462 | vinegar | acetic acid | 64-19-7 | $C_2H_4O_2$ | 60.1 | vinegar, pungent | <0.0001 |
| 34 | 1466 | roasted, | 5-ethyl-2,3-dimethylpyrazine | 15707-34-3 | $C_8H_{12}N_2$ | 136.2 | burnt, popcorn, roast | 0.518 |
| | | then, potato | 3-methylthiopropanal | 3268-49-3 | $C_4H_8OS$ | 104.2 | cooked potato | |
| 35 | 1497 | vegetal, earthy, roasted | 2-ethenyl-6-methylpyrazine | 13925-09-2 | $C_7H_8N_2$ | 120.1 | roasted, hazelnut | 0.128 |
| 36 | 1508 | vegetal, earthy, roasted | 2-ethenyl-5-methylpyrazine * | 13925-08-1 | $C_7H_8N_2$ | 120.1 | coffee | <0.001 |
| 37 | 1514 | flowery, vegetal | 3-ethyl-4-methylpentan-1-ol * | 38514-13-5 | $C_8H_{18}O$ | 130.2 | - | 0.112 |

**Table A1.** *Cont.*

| OA [a] | LRI [b] | Odor [c] | Identification [d] | CAS [e] | Formula [f] | MW [g] | Lit. Odor [h] | p-value [i] |
|---|---|---|---|---|---|---|---|---|
| 38 | 1530 | vegetal, pepper | 3-isobutyl-2,5-dimethylpyrazine | 32736-94-0 | $C_{10}H_{16}N_2$ | 164.2 | - | 0.193 |
| | | | and/or **2-isobutyl-3,5-dimethylpyrazine** * | 70303-42-3 | $C_{10}H_{16}N_2$ | 164.2 | - | |
| 39 | 1542 | vegetal, cardboard, flowery | **ethyl nonanoate** * | 123-29-5 | $C_{11}H_{22}O_2$ | 186.3 | soapy, waxy | 0.376 |
| | | | (E)-non-2-enal | 18829-56-6 | $C_9H_{16}O$ | 140.2 | paper, cut grass, cucumber | |
| 40 | 1552 | flowery | butane-2,3-diol | 513-85-9 | $C_4H_{10}O_2$ | 90.1 | floral | <0.001 |
| 41 | 1594 | vegetal, cucumber | **2-isobutyl-3,5,6-trimethylpyrazine** * | 46187-37-5 | $C_{11}H_{18}N_2$ | 178.3 | - | 0.107 |
| 42 | 1597 | vegetal, earthy | **3-hydroxybutanoic acid** * | 300-85-6 | $C_4H_8O_3$ | 104.1 | butter | 0.264 |
| 43 | 1635 | roasted | acetylpyrazine | 22047-25-2 | $C_6H_6N_2O$ | 122.1 | roasted, toasted | 0.582 |
| 44 | 1638 | cheese | butanoic acid | 107-92-6 | $C_4H_8O_2$ | 88.1 | cheese | 0.463 |
| 45 | 1653 | flowery | phenylacetaldehyde | 122-78-1 | $C_8H_8O$ | 120.1 | geranium, hyacinth | 0.347 |
| 46 | 1660 | floral, fruity | **acetophenone** * | 98-86-2 | $C_8H_8O$ | 120.1 | mimosa, acacia, sweet | 0.171 |
| 47 | 1676 | melted cheese | 2-methylbutanoic acid | 116-53-0 | $C_5H_{10}O_2$ | 102.1 | cheese, fermented | 0.693 |
| | | | 3-methylbutanoic acid | 503-74-2 | $C_5H_{10}O_2$ | 102.1 | cheese, sweat | |
| 48 | 1710 | vegetal, roasted, fruity | **1-phenylethyl acetate** * | 93-92-5 | $C_{10}H_{12}O_2$ | 164.2 | green, leafy, rose, fruit | 0.990 |
| | | | **methyl 2-methylpentanoate** * | 2177-77-7 | $C_7H_{14}O_2$ | 130.2 | fruity, apple | |
| 49 | 1726 | floral, anise, minty | **2,3,5-trimethyl-6-(3-methylbutyl)pyrazine** * | 10132-43-1 | $C_{12}H_{20}N_2$ | 192.3 | anise-like, floral | 0.805 |
| 50 | 1748 | unpleasant | **3-hydroxypropyl acetate** * | 36678-05-4 | $C_5H_{10}O_3$ | 118.1 | - | 0.353 |
| 51 | 1766 | fruity, roasted, vegetal | *trans*-linalool-3,7-oxide | 39028-58-5 | $C_{10}H_{18}O_2$ | 170.2 | floral, woody, wintergreen | 0.061 |
| 52 | 1795 | floral | ethyl phenylacetate | 101-97-3 | $C_{10}H_{12}O_2$ | 164.2 | floral | <0.0001 |
| 53 | 1818 | roasted, vegetal | **δ-pentalactone** * | 542-28-9 | $C_5H_8O_2$ | 100.1 | sweet | 0.567 |
| 54 | 1825 | floral, rose, fruity | **1-phenylethanol** * | 98-85-1 | $C_8H_{10}O$ | 122.2 | floral, rose | 0.002 |
| 55 | 1828 | earthy, moldy | 2-phenylethyl acetate | 103-45-7 | $C_{10}H_{12}O_2$ | 164.2 | tobacco, honey | 0.239 |
| 56 | 1848 | roasted, nut, spicy | **pentan-2-yl benzoate** * | 39180-02-4 | $C_{12}H_{16}O_2$ | 192.3 | - | 0.035 |
| | | | **ethyl dodecanoate** * | 106-33-2 | $C_{14}H_{28}O_2$ | 228.4 | nut, leaf | |
| 57 | 1869 | roasted, caramel, fruity | **dihydromaltol** * | 38877-21-3 | $C_6H_8O_3$ | 128.1 | - | 0.019 |
| 58 | 1872 | roasted, smoked, sweet | guaiacol | 90-05-1 | $C_7H_8O_2$ | 124.1 | smoke, bacon, wood, vanilla | 0.327 |
| 59 | 1892 | sweet, fruity, floral | **phenylmethanol** * | 100-51-6 | $C_7H_8O$ | | floral, fruity, balsam | 0.993 |
| 60 | 1921 | floral, rose | 2-phenylethanol | 60-12-8 | $C_8H_{10}O$ | 122.2 | rose, rose water | 0.342 |
| 61 | 1944 | floral | 2-phenylbut-2-enal | 4411-89-6 | $C_{10}H_{10}O$ | 146.2 | narcissus | 0.190 |
| 62 | 1976 | roasted, fruity, spicy | unknown | - | - | - | - | 0.509 |
| 63 | 1980 | fruity, sweet | **δ-octalactone** * | 698-76-0 | $C_8H_{14}O_2$ | 142.2 | sweet, coconut, tropical | <0.001 |
| | | | maltol | 118-71-8 | $C_6H_6O_3$ | 126.1 | sweet, caramel, fruity | |
| 64 | 1985 | hot plastic | 2-acetylpyrrole | 1072-83-9 | $C_6H_7NO$ | 109.1 | coumarinic, licorice | 0.376 |
| 65 | 2011 | vegetal, metallic | **methyl tetradecanoate** * | 124-10-7 | $C_{15}H_{30}O_2$ | 242.4 | orris, petal, waxy | 0.020 |
| 66 | 2015 | sweet, vegetal | δ-octenolactone | 16400-69-4 | $C_8H_{12}O_2$ | 140.2 | coconut-like | 0.337 |
| 67 | 2018 | floral, fruity | phenol | 108-95-2 | $C_6H_6O$ | 94.1 | medicine, phenolic | 0.067 |
| 68 | 2039 | fruity, sweet | γ-nonalactone | 104-61-0 | $C_9H_{16}O_2$ | 156.2 | peach, coconut, sweet | 0.177 |
| 69 | 2043 | sweet, fruity | 1*H*-pyrrole-2-carbaldehyde | 1003-29-8 | $C_5H_5NO$ | 95.1 | musty | 0.649 |
| 70 | 2046 | caramel, strawberry | furaneol | 3658-77-3 | $C_6H_8O_3$ | 128.1 | caramel, strawberry | <0.0001 |
| 71 | 2075 | unpleasant | **octanoic acid** * | 124-07-2 | $C_8H_{16}O_2$ | 144.2 | rancid, waxy, sweat | 0.442 |
| 72 | 2096 | animal, unpleasant, urine | 4-methylphenol | 106-44-5 | $C_7H_8O$ | 108.1 | horse, smoke, stable | 0.136 |
| 73 | 2122 | floral, spicy, fruity | 5-methyl-1*H*-pyrrole-2-carbaldehyde | 1192-79-6 | $C_6H_7NO$ | 109.1 | - | 0.005 |
| 74 | 2139 | fruity, vegetal | (E)-ethyl cinnamate | 103-36-6 | $C_{11}H_{12}O_2$ | 176.2 | floral, fruit, sweet | 0.449 |
| 75 | 2142 | floral, sweet, fruity | unknown | - | - | - | - | 0.472 |
| 76 | 2147 | roasted, spicy | unknown | - | - | - | - | <0.001 |
| 77 | 2156 | sweet, fruity, peach | γ-decalactone | 706-14-9 | $C_{10}H_{18}O_2$ | 170.2 | peach, sweet, apricot | 0.327 |
| 78 | 2197 | animal, unpleasant | **nonanoic acid** * | 112-05-0 | $C_9H_{18}O_2$ | 158.2 | dirty, cheese, waxy | 0.847 |

**Table A1.** *Cont.*

| OA [a] | LRI [b] | Odor [c] | Identification [d] | CAS [e] | Formula [f] | MW [g] | Lit. Odor [h] | *p*-value [i] |
|---|---|---|---|---|---|---|---|---|
| 79 | 2205 | fruity, floral, woody | δ-decalactone | 705-86-2 | $C_{10}H_{18}O_2$ | 170.2 | fruity, sweet | 0.092 |
| 80 | 2212 | curry, licorice, clove, spicy | **4-vinylguaiacol *** | 7786-61-0 | $C_9H_{10}O_2$ | 150.2 | clove, curry, smoke | 0.384 |
| 81 | 2234 | floral, fruity, vegetal | **heptadecan-2-one *** | 2922-51-2 | $C_{17}H_{34}O$ | 254.5 | - | 0.076 |
| 82 | 2240 | floral, fruity | **isopropyl palmitate *** | 142-91-6 | $C_{19}H_{38}O_2$ | 298.5 | fatty, oily | 0.010 |
| 83 | 2246 | fruity, sweet, coconut | **δ-decenolactone *** | 54814-64-1 | $C_{10}H_{16}O_2$ | 168.2 | coconut, fruity | <0.001 |
| 84 | 2272 | unpleasant, dust | 3-hydroxy-4-phenylbutan-2-one * | 5355-63-5 | $C_{10}H_{12}O_2$ | 164.2 | burnt plastic | 0.280 |
| 85 | 2278 | roasted, chicory coffee | **3-hydroxy-2,3-dihydromaltol *** | 28564-83-2 | $C_6H_8O_4$ | 144.1 | roast, earth | 0.214 |
| 86 | 2338 | woody, vegetal | **2-phenylethyl lactate *** | 10138-63-3 | $C_{11}H_{14}O_3$ | 194.2 | rose | <0.0001 |
| 87 | 2353 | dust | **dihydroactinidiolide *** | 15356-74-8 | $C_{11}H_{16}O_2$ | 180.2 | ripe, woody | 0.449 |
| 88 | 2365 | floral | **farnesol *** | 4602-84-0 | $C_{15}H_{26}O$ | 222.4 | floral | 0.278 |
| 89 | 2387 | fruity, peach | **γ-dodecalactone *** | 2305-05-7 | $C_{12}H_{22}O_2$ | 198.3 | peach, fruit | <0.0001 |
| 90 | 2412 | fruity, floral | **5-[(2Z)oct-2-en-1-yl]dihydrofuran-2(3H)-one *** | 156318-46-6 | $C_{12}H_{20}O_2$ | 196.3 | - | 0.001 |
| 91 | 2464 | unpleasant, floral | 1H-indole | 120-72-9 | $C_8H_7N$ | 117.1 | animal, fecal, floral | 0.559 |
| 92 | 2511 | unpleasant, animal, leather | **dodecanoic acid *** | 143-07-7 | $C_{12}H_{24}O_2$ | 200.3 | fat, wax, oil | 0.280 |
| 93 | 2587 | vanilla, sweet, cocoa | vanillin | 121-33-5 | $C_8H_8O_3$ | 152.1 | vanilla, chocolate | 0.236 |
| 94 | 2591 | floral, unpleasant | phenylacetic acid | 103-82-2 | $C_8H_8O_2$ | 136.1 | floral, urine | <0.001 |
| 95 | 2602 | floral | **octadecan-1-ol *** | 112-92-5 | $C_{18}H_{38}O$ | 270.5 | oily | 0.010 |
| 96 | 2628 | floral | unknown | - | - | - | - | 0.012 |

* although most of them previously identified in cocoa products (see Table 1), to the authors' knowledge, these compounds (in bold character) are formally described for the first time as key aroma compounds of dark chocolate; -: not described or not relevant.

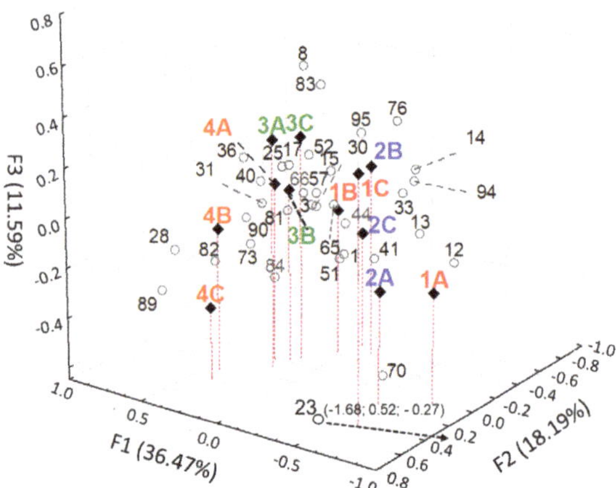

**Figure A2.** 3-D plot (dimensions 1, 2, and 3) of a correspondence analysis (CA) between the 12 samples and the 34 discriminant OAs defined by their NIF values. OAs (light circles) are plotted according to their NIF in samples (plain diamonds), respectively. The CA plot is zoomed in for clarity, and the coordinates of the extra variable (23) indicated in brackets with its direction. The OA numbers are those found in Table 1. The sample names are colored for illustrative purpose, with pole 1 samples appearing in red, pole 2 samples in blue, pole 3 samples in green, and pole 4 ones in orange. CA independence test: Khi$^2$ = 5444 (critical value 408, α = 0.05), degrees of freedom = 363, *p* < 0.0001.

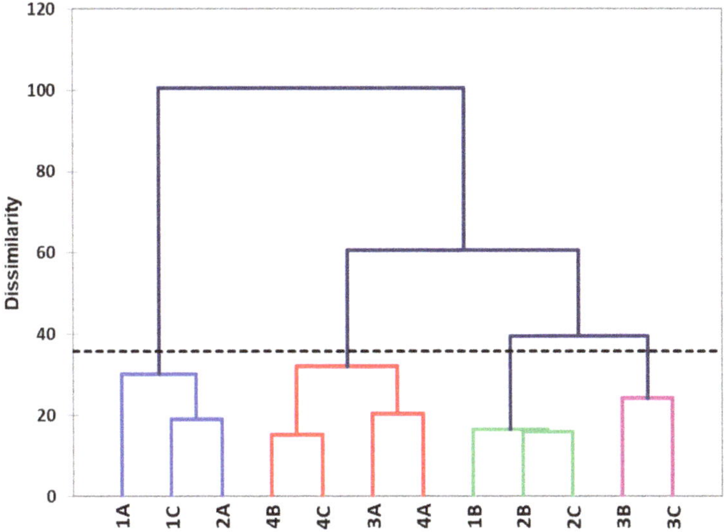

**Figure A3.** Hierarchical cluster analysis (HCA) conducted on the samples with the NIF values of the 34 OAs determined as discriminant based on 50% NIF difference threshold, displaying four clusters. The data were centered and scaled; dissimilarity Euclidian distances were used with the Ward amalgamation method.

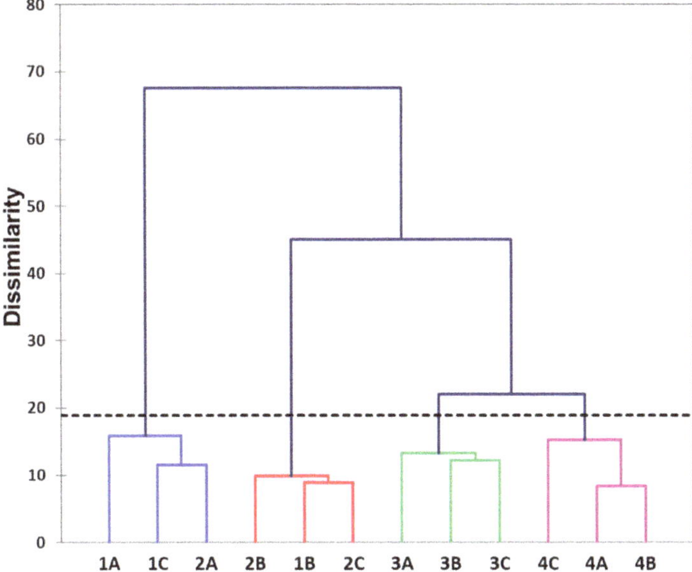

**Figure A4.** Hierarchical cluster analysis (HCA) conducted on the samples with the NIF values of the 38 'discriminant' OAs, with discrimination based on both NIF difference threshold and $Khi^2$ test (see discussion), displaying four clusters. The data were centered and scaled; dissimilarity Euclidian distances were used with the Ward amalgamation method.

## References


1. Afoakwa, E.O. *Chocolate Science and Technology*; Wiley Blackwell: Chichester, UK, 2016.
2. Tran, P.D.; Van Durme, J.; Van de Walle, D.; de Winne, A.; Delbaere, C.; de Clercq, N.; Phan, T.T.Q.; Phuc Nguyen, C.-H.; Tran, D.N.; Dewettinck, K. Quality Attributes of Dark Chocolate Produced from Vietnamese Cocoa Liquors. *J. Food Qual.* **2016**, *39*, 311–322. [CrossRef]
3. Bastos, V.S.; Uekane, T.M.; Bello, N.A.; de Rezende, C.M.; Flosi Paschoalin, V.M.; Del Aguila, E.M. Dynamics of volatile compounds in TSH 565 cocoa clone fermentation and their role on chocolate flavor in Southeast Brazil. *J. Food Sci. Technol.* **2019**, *56*, 2874–2887. [CrossRef]
4. Magalhães da Veiga Moreira, I.; Gabriela da Cruz Pedrozo Miguel, M.; Lacerda Ramos, C.; Ferreira Duarte, W.; Efraim, P.; Freitas Schwan, R. Influence of Cocoa Hybrids on Volatile Compounds of Fermented Beans, Microbial Diversity during Fermentation and Sensory Characteristics and Acceptance of Chocolates. *J. Food Qual.* **2016**, *39*, 839–849. [CrossRef]
5. Menezes, A.G.T.; Batista, N.N.; Ramos, C.L.; de Andrade e Silva, A.R.; Efraim, P.; Pinheiro, A.C.M.; Schwan, R.F. Investigation of chocolate produced from four different Brazilian varieties of cocoa (*Theobroma cacao* L.) inoculated with Saccharomyces cerevisiae. *Food Res. Int.* **2016**, *81*, 83–90. [CrossRef]
6. Liu, M.; Liu, J.; He, C.; Song, H.; Liu, Y.; Zhang, Y.; Wang, Y.; Guo, J.; Yang, H.; Su, X. Characterization and comparison of key aroma-active compounds of cocoa liquors from five different areas. *Int. J. Food Prop.* **2017**, *20*, 2396–2408. [CrossRef]
7. Magagna, F.; Guglielmetti, A.; Liberto, E.; Reichenbach, S.E.; Allegrucci, E.; Gobino, G.; Bicchi, C.; Cordero, C. Comprehensive Chemical Fingerprinting of High-Quality Cocoa at Early Stages of Processing: Effectiveness of Combined Untargeted and Targeted Approaches for Classification and Discrimination. *J. Agric. Food Chem.* **2017**, *65*, 6329–6341. [CrossRef]
8. Tran, P.D.; Van de Walle, D.; De Clercq, N.; De Winne, A.; Kadow, D.; Lieberei, R.; Messens, K.; Tran, D.N.; Dewettinck, K.; Van Durme, J. Assessing cocoa aroma quality by multiple analytical approaches. *Food Res. Int.* **2015**, *77*, 657–669. [CrossRef]
9. Utrilla-Vázquez, M.; Rodríguez-Campos, J.; Avendaño-Arazate, C.H.; Gschaedler, A.; Lugo-Cervantes, E. Analysis of volatile compounds of five varieties of Maya cocoa during fermentation and drying processes by Venn diagram and PCA. *Food Res. Int.* **2020**, *129*, 108834. [CrossRef]
10. Torres-Moreno, M.; Torrescasana, E.; Salas-Salvado, J.; Blanch, C. Nutritional composition and fatty acids profile in cocoa beans and chocolates with different geographical origin and processing conditions. *Food Chem.* **2015**, *166*, 125–132. [CrossRef]
11. Rottiers, H.; Tzompa Sosa, D.A.; De Winne, A.; Ruales, J.; De Clippeleer, J.; De Leersnyder, I.; De Wever, J.; Everaert, H.; Messens, K.; Dewettinck, K. Dynamics of volatile compounds and flavor precursors during spontaneous fermentation of fine flavor Trinitario cocoa beans. *Eur. Food Res. Technol.* **2019**, *245*, 1917–1937. [CrossRef]
12. Assi-Clair, B.J.; Koné, M.K.; Kouamé, K.; Lahon, M.C.; Berthiot, L.; Durand, N.; Lebrun, M.; Julien-Ortiz, A.; Maraval, I.; Boulanger, R.; et al. Effect of aroma potential of Saccharomyces cerevisiae fermentation on the volatile profile of raw cocoa and sensory attributes of chocolate produced thereof. *Eur. Food Res. Technol.* **2019**, *245*, 1459–1471. [CrossRef]
13. Koné, M.K.; Guéhi, S.T.; Durand, N.; Ban-Koffi, L.; Berthiot, L.; Tachon, A.F.; Brou, K.; Boulanger, R.; Montet, D. Contribution of predominant yeasts to the occurrence of aroma compounds during cocoa bean fermentation. *Food Res. Int.* **2016**, *89*, 910–917. [CrossRef]
14. Janek, K.; Niewienda, A.; Wöstemeyer, J.; Voigt, J. The cleavage specificity of the aspartic protease of cocoa beans involved in the generation of the cocoa-specific aroma precursors. *Food Chem.* **2016**, *211*, 320–328. [CrossRef]
15. Hue, C.; Gunata, Z.; Breysse, A.; Davrieux, F.; Boulanger, R.; Sauvage, F.X. Impact of fermentation on nitrogenous compounds of cocoa beans (*Theobroma cacao* L.) from various origins. *Food Chem.* **2016**, *192*, 958–964. [CrossRef]
16. Magalhães da Veiga Moreira, I.; de Figueiredo Vilela, L.; da Cruz Pedroso Miguel, M.; Santos, C.; Lima, N.; Freitas Schwan, R. Impact of a Microbial Cocktail Used as a Starter Culture on Cocoa Fermentation and Chocolate Flavor. *Molecules* **2017**, *22*, 766. [CrossRef]

17. Rodriguez-Campos, J.; Escalona-Buendía, H.B.; Contreras-Ramos, S.M.; Orozco-Avila, I.; Jaramillo-Flores, E.; Lugo-Cervantes, E. Effect of fermentation time and drying temperature on volatile compounds in cocoa. *Food Chem.* **2012**, *132*, 277–288. [CrossRef]

18. Hinneh, M.; Abotsi, E.E.; Van de Walle, D.; Tzompa-Sosa, D.A.; De Winne, A.; Simonis, J.; Messens, K.; Van Durme, J.; Afoakwa, E.O.; De Cooman, L.; et al. Pod storage with roasting: A tool to diversifying the flavor profiles of dark chocolates produced from 'bulk' cocoa beans? (part I: Aroma profiling of chocolates). *Food Res. Int.* **2019**, *119*, 84–98. [CrossRef]

19. Hinneh, M.; Van de Walle, D.; Tzompa-Sosa, D.A.; De Winne, A.; Termote, S.; Messens, K.; Van Durme, J.; Afoakwa, E.O.; De Cooman, L.; Dewettinck, K. Tuning the aroma profiles of FORASTERO cocoa liquors by varying pod storage and bean roasting temperature. *Food Res. Int.* **2019**, *125*, 108550. [CrossRef]

20. Frauendorfer, F.; Schieberle, P. Key aroma compounds in fermented Forastero cocoa beans and changes induced by roasting. *Eur. Food Res. Technol.* **2019**, *245*, 1907–1915. [CrossRef]

21. Granvogl, M.; Bugan, S.; Schieberle, P. Formation of Amines and Aldehydes from Parent Amino Acids during Thermal Processing of Cocoa and Model Systems: New Insights into Pathways of the Strecker Reaction. *J. Agric. Food Chem.* **2006**, *54*, 1730–1739. [CrossRef]

22. Frauendorfer, F.; Schieberle, P. Changes in Key Aroma Compounds of Criollo Cocoa Beans During Roasting. *J. Agric. Food Chem.* **2008**, *56*, 10244–10251. [CrossRef] [PubMed]

23. Van Durme, J.; Ingels, I.; De Winne, A. Inline roasting hyphenated with gas chromatography–mass spectrometry as an innovative approach for assessment of cocoa fermentation quality and aroma formation potential. *Food Chem.* **2016**, *205*, 66–72. [CrossRef] [PubMed]

24. Afoakwa, E.O.; Paterson, A.; Fowler, M.; Ryan, A. Flavor Formation and Character in Cocoa and Chocolate: A Critical Review. *Crit. Rev. Food Sci. Nutr.* **2008**, *48*, 840–857. [CrossRef] [PubMed]

25. Kongor, J.E.; Hinneh, M.; de Walle, D.V.; Afoakwa, E.O.; Boeckx, P.; Dewettinck, K. Factors influencing quality variation in cocoa (*Theobroma cacao*) bean flavour profile—A review. *Food Res. Int.* **2016**, *82*, 44–52. [CrossRef]

26. Aprotosoaie, A.C.; Luca, S.V.; Miron, A. Flavor Chemistry of Cocoa and Cocoa Products—An Overview. *Comp. Rev. Food Sci. Food Saf.* **2016**, *15*, 73–91. [CrossRef]

27. Counet, C.; Callemien, D.; Ouwerx, C.; Collin, S. Use of Gas Chromatography–Olfactometry To Identify Key Odorant Compounds in Dark Chocolate. Comparison of Samples before and after Conching. *J. Agric. Food Chem.* **2002**, *50*, 2385–2391. [CrossRef]

28. Liu, J.; Liu, M.; He, C.; Song, H.; Guo, J.; Wang, Y.; Yang, H.; Su, X. A comparative study of aroma-active compounds between dark and milk chocolate: Relationship to sensory perception. *J. Sci. Food Agric.* **2015**, *95*, 1362–1372. [CrossRef]

29. Seyfried, C.; Granvogl, M. Characterization of the Key Aroma Compounds in Two Commercial Dark Chocolates with High Cocoa Contents by Means of the Sensomics Approach. *J. Agric. Food Chem.* **2019**, *67*, 5827–5837. [CrossRef]

30. Acierno, V.; Liu, N.; Alewijn, M.; Stieger, M.; van Ruth, S.M. Which cocoa bean traits persist when eating chocolate? Real-time nosespace analysis by PTR-QiToF-MS. *Talanta* **2019**, *195*, 676–682. [CrossRef]

31. Acierno, V.; Yener, S.; Alewijn, M.; Biasioli, F.; van Ruth, S. Factors contributing to the variation in the volatile composition of chocolate: Botanical, and geographical origin of the cocoa beans, and brand-related formulation and processing. *Food Res. Int.* **2016**, *84*, 86–95. [CrossRef]

32. Deuscher, Z.; Andriot, I.; Sémon, E.; Repoux, M.; Preys, S.; Roger, J.-M.; Boulanger, R.; Labouré, H.; Le Quéré, J.-L. Volatile compounds profiling by using Proton Transfer Reaction—Time of Flight—Mass Spectrometry (PTR-ToF-MS). The case study of dark chocolates organoleptic differences. *J. Mass Spectrom.* **2019**, *54*, 92–119. [CrossRef]

33. Pollien, P.; Ott, A.; Montigon, F.; Baumgartner, M.; Muñoz-Box, R.; Chaintreau, A. Hyphenated headspace-gas chromatography-sniffing technique: Screening of impact odorants and quantitative aromagram comparisons. *J. Agric. Food Chem.* **1997**, *45*, 2630–2637. [CrossRef]

34. Linssen, J.P.H.; Janssens, J.L.G.M.; Roozen, J.P.; Posthumus, M.A. Combined gas chromatography and sniffing port analysis of volatile compounds of mineral water packed in laminated packages. *Food Chem.* **1993**, *46*, 367–371. [CrossRef]

35. Engel, W.; Bahr, W.; Schieberle, P. Solvent assisted flavour evaporation—A new and versatile technique for the careful and direct isolation of aroma compounds from complex food matrices. *Eur. Food Res. Technol.* **1999**, *209*, 237–241. [CrossRef]

36. Acree, T.E. Gas chromatography-olfactometry. In *Flavor Measurement*; Ho, C.T., Manley, C.H., Eds.; Dekker, M.: New York, NY, USA, 1993; pp. 77–94.

37. Grosch, W. Evaluation of the key odorants of foods by dilution experiments, aroma models and omission. *Chem. Senses* **2001**, *26*, 533–545. [CrossRef] [PubMed]

38. Abbott, N.; Etievant, P.; Issanchou, S.; Langlois, D. Critical evaluation of two commonly used techniques for the treatment of data from extract dilution sniffing analysis. *J. Agric. Food Chem.* **1993**, *41*, 1698–1703. [CrossRef]

39. Brattoli, M.; Cisternino, E.; Dambruoso, P.R.; De Gennaro, G.; Giungato, P.; Mazzone, A.; Palmisani, J.; Tutino, M. Gas Chromatography Analysis with Olfactometric Detection (GC-O) as a Useful Methodology for Chemical Characterization of Odorous Compounds. *Sensors* **2013**, *13*, 16759–16800. [CrossRef]

40. Malfondet, N.; Gourrat, K.; Brunerie, P.; Le Quéré, J.-L. Aroma characterization of freshly-distilled French brandies; their specificity and variability within a limited geographic area. *Flavour Fragr. J.* **2016**, *31*, 361–376. [CrossRef]

41. Etiévant, P.; Chaintreau, A. Les analyses olfactométriques: Synthèse critique et recommandations. In Proceedings of the XX èmes Journées Internationales Huiles Essentielles et Extraits, Dignes les Bains, France, 5–7 September 2001.

42. McDaniel, M.R.; Miranda-Lopez, R.; Watson, B.T.; Michaels, N.J.; Libbey, L.M. Pinot noir aroma: A sensory/gas chromatographic approach. In *Flavors and Off-flavors*; Charalambous, G., Ed.; Elsevier: Amsterdam, The Netherlands, 1990; pp. 23–36.

43. Serot, T.; Prost, C.; Visan, L.; Burcea, M. Identification of the main odor-active compounds in musts from French and Romanian hybrids by three olfactometric methods. *J. Agric. Food Chem.* **2001**, *49*, 1909–1914. [CrossRef]

44. Le Guen, S.; Prost, C.; Demaimay, M. Critical Comparison of Three Olfactometric Methods for the identification of the Most Potent Odorants in Cooked Mussels (*Mytilus edulis*). *J. Agric. Food Chem.* **2000**, *48*, 1307–1314. [CrossRef]

45. Van Ruth, S.M.; OConnor, C.H. Evaluation of three gas chromatography-olfactometry methods: Comparison of odour intensity-concentration relationships of eight volatile compounds with sensory headspace data. *Food Chem.* **2001**, *74*, 341–347. [CrossRef]

46. Ott, A.; Fay, L.B.; Chaintreau, A. Determination and origin of the aroma impact compounds of yogurt flavour. *J. Agric. Food Chem.* **1997**, *45*, 850–858. [CrossRef]

47. Marco, A.; Navarro, J.L.; Flores, M. Quantitation of Selected Odor-Active Constituents in Dry Fermented Sausages Prepared with Different Curing Salts. *J. Agric. Food Chem.* **2007**, *55*, 3058–3065. [CrossRef] [PubMed]

48. Wu, W.; Tao, N.-P.; Gu, S.-Q. Characterization of the key odor-active compounds in steamed meat of Coilia ectenes from Yangtze River by GC–MS–O. *Eur. Food Res. Technol.* **2014**, *238*, 237–245. [CrossRef]

49. Mastello, R.B.; Capobiango, M.; Chin, S.-T.; Monteiro, M.; Marriott, P.J. Identification of odour-active compounds of pasteurised orange juice using multidimensional gas chromatography techniques. *Food Res. Int.* **2015**, *75*, 281–288. [CrossRef]

50. Owusu, M.; Petersen, M.A.; Heimdal, H. Effect of fermentation method, roasting and conching conditions on the aroma volatiles of dark chocolate. *J. Food Process. Preserv.* **2012**, *36*, 446–456. [CrossRef]

51. Schieberle, P.; Pfnuer, P. Characterization of key odorants in chocolate. In *Flavor Chemistry: 30 Years of Progress*; Teranishi, R., Wick, E.L., Hornstein, I., Eds.; Kluwer Academic/Plenum Publishers: New York, NY, USA, 1999; pp. 147–153.

52. Owusu, M.; Petersen, M.A.; Heimdal, H. Relationship of sensory and instrumental aroma measurements of dark chocolate as influenced by fermentation method, roasting and conching conditions. *J. Food Sci. Technol.* **2013**, *50*, 909–917. [CrossRef]

53. Kunert-Kirchhoff, J.; Baltes, W. Model reactions on roast aroma formation. *Z. Lebensm. Unters. Forsch.* **1990**, *190*, 9–13. [CrossRef]

54. Nijssen, B.; van Ingen-Vissher, K.; Donders, J. *VCF Online. Vol. V 16.6.1*; BeWiDo BV: Zeist, The Netherlands, 2002.

55. Diab, J.; Hertz-Schünemann, R.; Streibel, T.; Zimmermann, R. Online measurement of volatile organic compounds released during roasting of cocoa beans. *Food Res. Int.* **2014**, *63*, 344–352. [CrossRef]

56. Moreira, I.; Vilela, L.F.; Santos, C.; Lima, N.; Schwan, R.F. Volatile compounds and protein profiles analyses of fermented cocoa beans and chocolates from different hybrids cultivated in Brazil. *Food Res. Int.* **2018**, *109*, 196–203. [CrossRef]

57. Batista, N.N.; Ramos, C.L.; Dias, D.R.; Pinheiro, A.C.; Schwan, R.F. The impact of yeast starter cultures on the microbial communities and volatile compounds in cocoa fermentation and the resulting sensory attributes of chocolate. *J. Food Sci. Technol.* **2016**, *53*, 1101–1110. [CrossRef]

58. Suzuki, D.; Sato, Y.; Nishiura, H.; Harada, R.; Kamasaka, H.; Kuriki, T.; Tamura, H. A Novel Extraction Method for Aroma Isolation from Dark Chocolate Based on the Oiling-Out Effect. *Food Anal. Methods* **2019**, *12*, 2857–2869. [CrossRef]

59. Serra Bonvehí, J. Investigation of aromatic compounds in roasted cocoa powder. *Eur. Food Res. Technol.* **2005**, *221*, 19–29. [CrossRef]

60. Crafack, M.; Keul, H.; Eskildsen, C.E.; Petersen, M.A.; Saerens, S.; Blennow, A.; Skovmand-Larsen, M.; Swiegers, J.H.; Petersen, G.B.; Heimdal, H.; et al. Impact of starter cultures and fermentation techniques on the volatile aroma and sensory profile of chocolate. *Food Res. Int.* **2014**, *63*, 306–316. [CrossRef]

61. Schnermann, P.; Schieberle, P. Evaluation of Key Odorants in Milk Chocolate and Cocoa Mass by Aroma Extract Dilution Analyses. *J. Agric. Food Chem.* **1997**, *45*, 867–872. [CrossRef]

62. Magi, E.; Bono, L.; Di Carro, M. Characterization of cocoa liquors by GC-MS and LC-MS/MS: Focus on alkylpyrazines and flavanols. *J. Mass Spectrom.* **2012**, *47*, 1191–1197. [CrossRef]

63. Afoakwa, E.O.; Paterson, A.; Fowler, M.; Ryan, A. Matrix effects on flavour volatiles release in dark chocolates varying in particle size distribution and fat content using GC–mass spectrometry and GC–olfactometry. *Food Chem.* **2009**, *113*, 208–215. [CrossRef]

64. Counet, C.; Ouwerx, C.; Rosoux, D.; Collin, S. Relationship between procyanidin and flavor contents of cocoa liquors from different origins. *J. Agric. Food Chem.* **2004**, *52*, 6243–6249. [CrossRef]

65. Braga, S.C.G.N.; Oliveira, L.F.; Hashimoto, J.C.; Gama, M.R.; Efraim, P.; Poppi, R.J.; Augusto, F. Study of volatile profile in cocoa nibs, cocoa liquor and chocolate on production process using GC × GC-QMS. *Microchem. J.* **2018**, *141*, 353–361. [CrossRef]

66. Schlüter, A.; Hühn, T.; Kneubühl, M.; Chatelain, K.; Rohn, S.; Chetschik, I. Novel Time- and Location-Independent Postharvest Treatment of Cocoa Beans: Investigations on the Aroma Formation during "Moist Incubation" of Unfermented and Dried Cocoa Nibs and Comparison to Traditional Fermentation. *J. Agric. Food Chem.* **2019**. [CrossRef]

67. Ducki, S.; Miralles-Garcia, J.; Zumbe, A.; Tornero, A.; Storey, D.M. Evaluation of solid-phase micro-extraction coupled to gas chromatography-mass spectrometry for the headspace analysis of volatile compounds in cocoa products. *Talanta* **2008**, *74*, 1166–1174. [CrossRef]

68. Waehrens, S.S.; Zhang, S.; Hedelund, P.I.; Petersen, M.A.; Byrne, D.V. Application of the fast sensory method 'Rate-All-That-Apply' in chocolate Quality Control compared with DHS-GC-MS. *Int. J. Food Sci. Technol.* **2016**, *51*, 1877–1887. [CrossRef]

69. Vitzhum, O.G.; Werkhoff, P.; Hubert, P. Volatile components of roasted cocoa: Basic fraction. *J. Food Sci.* **1975**, *40*, 911–916. [CrossRef]

70. Carlin, J.T.; Lee, K.N.; Hsieh, O.A.L.; Hwang, L.S.; Ho, C.-T.; Chang, S.S. Comparison of acidic and basic volatile compounds of cocoa butters from roasted and unroasted cocoa beans. *J. Am. Oil Chem. Soc.* **1986**, *63*, 1031–1036. [CrossRef]

71. Chetschik, I.; Kneubühl, M.; Chatelain, K.; Schlüter, A.; Bernath, K.; Hühn, T. Investigations on the Aroma of Cocoa Pulp (*Theobroma cacao* L.) and Its Influence on the Odor of Fermented Cocoa Beans. *J. Agric. Food Chem.* **2018**, *66*, 2467–2472. [CrossRef]

72. Stewart, A.; Grandison, A.S.; Ryan, A.; Festring, D.; Methven, L.; Parker, J.K. Impact of the Skim Milk Powder Manufacturing Process on the Flavor of Model White Chocolate. *J. Agric. Food Chem.* **2017**, *65*, 1186–1195. [CrossRef]

73. Ramli, N.; Hassan, O.; Said, M.; Samsudin, W.; Idris, N.A. Influence of roasting conditions on volatile flavor of roasted malaysian cocoa beans. *J. Food Process. Preserv.* **2006**, *30*, 280–298. [CrossRef]

74. Scalone, G.L.L.; Textoris-Taube, K.; De Meulenaer, B.; De Kimpe, N.; Wöstemeyer, J.; Voigt, J. Cocoa-specific flavor components and their peptide precursors. *Food Res. Int.* **2019**, *123*, 503–515. [CrossRef]

75. Ascrizzi, R.; Flamini, G.; Tessieri, C.; Pistelli, L. From the raw seed to chocolate: Volatile profile of Blanco de Criollo in different phases of the processing chain. *Microchem. J.* **2017**, *133*, 474–479. [CrossRef]

76. Mohamadi Alasti, F.; Asefi, N.; Maleki, R.; SeiiedlouHeris, S.S. Investigating the flavor compounds in the cocoa powder production process. *Food Sci. Nutr.* **2019**, *7*, 3892–3901. [CrossRef]

77. Frauendorfer, F.; Schieberle, P. Identification of the Key Aroma Compounds in Cocoa Powder Based on Molecular Sensory Correlations. *J. Agric. Food Chem.* **2006**, *54*, 5521–5529. [CrossRef]

78. Toker, O.S.; Sagdic, O.; Sener, D.; Konar, N.; Zorlucan, T.; Daglıoglu, O. The influence of particle size on some physicochemical, rheological and melting properties and volatile compound profile of compound chocolate and cocolin samples. *Eur. Food Res. Technol.* **2016**, *242*, 1253–1266. [CrossRef]

79. Buhr, K.; Pammer, C.; Schieberle, P. Influence of water on the generation of Strecker aldehydes from dry processed foods. *Eur. Food Res. Technol.* **2009**, *230*, 375. [CrossRef]

80. Albak, F.; Tekin, A.R. Variation of total aroma and polyphenol content of dark chocolate during three phase of conching. *J. Food Sci. Technol.* **2016**, *53*, 848–855. [CrossRef]

81. Johnsen, L.G.; Skou, P.B.; Khakimov, B.; Bro, R. Gas chromatography—Mass spectrometry data processing made easy. *J. Chromatogr. A* **2017**, *1503*, 57–64. [CrossRef]

82. Qin, X.-W.; Lai, J.-X.; Tan, L.-H.; Hao, C.-Y.; Li, F.-P.; He, S.-Z.; Song, Y.-H. Characterization of volatile compounds in Criollo, Forastero, and Trinitario cocoa seeds (*Theobroma cacao* L.) in China. *Int. J. Food Prop.* **2017**, *20*, 2261–2275. [CrossRef]

83. Hartmann, S.; Schieberle, P. On the Role of Amadori Rearrangement Products as Precursors of Aroma-Active Strecker Aldehydes in Cocoa. In *Browned Flavors: Analysis, Formation, and Physiology*; Granvogl, M., Peterson, D., Schieberle, P., Eds.; American Chemical Society: Washington, DC, USA, 2016; Volume 1237, pp. 1–13. [CrossRef]

84. Van den Dool, H.; Kratz, P.D. A generalization of the retention index system including linear temperature programmed gas-liquid partition chromatography. *J. Chromatogr.* **1963**, *11*, 463–471. [CrossRef]

85. Harrison, A.G. *Chemical Ionization Mass Spectrometry*, 2nd ed.; CRC Press: Boca Raton, FL, USA, 1992.

86. Sarris, J.; Etiévant, P.X.; Le Quéré, J.L.; Adda, J. The chemical ionisation mass spectra of alcohols. In *Progress in Flavour Research 1984*; Adda, J., Ed.; Elsevier: Amsterdam, The Netherlands, 1985; pp. 591–601.

**Sample Availability:** Samples of the compounds are not available from the authors.

*Article*

# HS-SPME-MS-Enose Coupled with Chemometrics as an Analytical Decision Maker to Predict In-Cup Coffee Sensory Quality in Routine Controls: Possibilities and Limits

Erica Liberto [1,*], Davide Bressanello [1], Giulia Strocchi [1], Chiara Cordero [1],
Manuela Rosanna Ruosi [2], Gloria Pellegrino [2], Carlo Bicchi [1] and Barbara Sgorbini [1]

[1] Dipartimento di Scienza e Tecnologia del Farmaco, Università degli Studi di Torino, 10125 Turin, Italy;
davide.bressanello@unito.it (D.B.); giulia.strocchi@unito.it (G.S.); chiara.cordero@unito.it (C.C.);
carlo.bicchi@unito.it (C.B.); barbara.sgorbini@unito.it (B.S.)

[2] Luigi Lavazza S.p.A, Strada Settimo 410, 10156 Turin, Italy; Manuela.Ruosi@lavazza.com (M.R.R.);
Gloria.Pellegrino@lavazza.com (G.P.)

* Correspondence: erica.liberto@unito.it; Tel.: +39-011-670-7134

Academic Editor: Eugenio Aprea
Received: 19 November 2019; Accepted: 8 December 2019; Published: 10 December 2019

**Abstract:** The quality assessment of the green coffee that you will go to buy cannot be disregarded from a sensory evaluation, although this practice is time consuming and requires a trained professional panel. This study aims to investigate both the potential and the limits of the direct headspace solid phase microextraction, mass spectrometry electronic nose technique (HS-SPME-MS or MS-EN) combined with chemometrics for use as an objective, diagnostic and high-throughput technique to be used as an analytical decision maker to predict the in-cup coffee sensory quality of incoming raw beans. The challenge of this study lies in the ability of the analytical approach to predict the sensory qualities of very different coffee types, as is usual in industry for the qualification and selection of incoming coffees. Coffees have been analysed using HS-SPME-MS and sensory analyses. The mass spectral fingerprints (MS-EN data) obtained were elaborated using: (i) unsupervised principal component analysis (PCA); (ii) supervised partial least square discriminant analysis (PLS-DA) to select the ions that are most related to the sensory notes investigated; and (iii) cross-validated partial least square regression (PLS), to predict the sensory attribute in new samples. The regression models were built with a training set of 150 coffee samples and an external test set of 34. The most reliable results were obtained with acid, bitter, spicy and aromatic intensity attributes. The mean error in the sensory-score predictions on the test set with the available data always fell within a limit of ±2. The results show that the combination of HS-SPME-MS fingerprints and chemometrics is an effective approach that can be used as a Total Analysis System (TAS) for the high-throughput definition of in-cup coffee sensory quality. Limitations in the method are found in the compromises that are accepted when applying a screening method, as opposed to human evaluation, in the sensory assessment of incoming raw material. The cost-benefit relationship of this and other screening instrumental approaches must be considered and weighed against the advantages of the potency of human response which could thus be better exploited in modulating blends for sensory experiences outside routine.

**Keywords:** HS-SPME-MS-enose; coffee; prediction of in-cup sensory quality; chemometrics

---

## 1. Introduction

Coffee is universally considered a comfort food and is widely consumed because of its particular flavour. The flavour of coffee is the result of the transformations of the harvested bean to the final roasted product. The chemical composition of coffee is variable, meaning that its sensory profile can radically change according to species, origin, year of harvest and post-harvest treatment. Roasters therefore need to constantly control the quality of their incoming beans.

Coffee is evaluated by its visual appearance; colour, bean uniformity, shape and size, number of "defective beans" and taste. However, coffee beans may have a pleasant aspect, but present an unpleasant taste because of contamination and chemical modification that may have occurred during storage, processing and transport from origin to the roaster's warehouse. Tasting, of course, plays a fundamental role in coffee quality evaluation, meaning that "cupping" is routinely used [1–3], to evaluate a lot (or a crop) for blend formulation, or "single origin" coffee and ultimately also to determine its price [4]. Nevertheless, cup tasting is time-consuming as it requires a specialised panel, who must be trained and aligned. Furthermore, the present trend in the food industry is to move panels from routine to the development of new finished products with given or peculiar flavour characteristics.

Flavour can be considered the signature of a product [5–8], and defining a relationship between chemical profile and aroma sensory impact is an important challenge for both the analytical and industrial fields, as they aim to achieve an objective and fast routine evaluation of a product with an automatic analytical procedure [9–14].

The use of rapid techniques in coffee analysis is constantly increasing. For instance, NIRS has been used to discriminate between coffee species and blends [15], to define the roasting degree of coffee beans and to quantify several bioactive coffee compounds, such as caffeine, trigonelline and chlorogenic acids and to predict sensory attributes, such as acidity, body, bitterness and the quality of espresso coffee [16,17]. Proton transfer reaction mass spectrometry (PTR-MS) and Laser Ionisation Mass Spectrometry (REMPI/TOFMS) have been used on-line, coupled to a Probat roaster, to control the roasting process, from volatile formation, and to study the kinetics of flavour development [18–23].

E-nose technology based on an array of electronic chemical sensors has also been attractive for industry. E-noses have been used in coffee research to differentiate Robusta from Arabica beans and to discriminate aromas via a fine-tuning process that involves altering the sensor materials [24]. The main advantages of these technologies are their cost-effectiveness, the fact that they can be easily integrated into a productive process, and the rapidity with which results can be obtained compared to traditional chemical and chromatographic methods. Despite these features, only few applications in industry have been described for these techniques, mainly because of the relatively low robustness, selectivity and reproducibility of the sensors, the large amount of data required to calibrate instrumentation and the resulting need for complex data analysis and algorithms [25–27].

These limits can be overcome by non-separative MS methods, better known as mass spectrometry-based electronic noses or MS-EN, which, when combined with headspace sampling, provides a representative, diagnostic and generalised mass spectrometric fingerprint of the volatile fraction of a sample, without prior chromatographic separation. With this approach, each $m/z$ ratio acts as a "sensor" whose intensity derives from the contribution of each compound that produces that fragment. It was introduced by Marsili in 1999 [28] to study off-flavours in milk and, since then, has been applied in the quality control of herbs and spices, in the authentication of food, to classify defective products, and to predict the sensory properties of food [13,29–33].

The present study applies this method to the coffee headspace, as sampled by HS-SPME, to develop an instrumental prediction model as an analytical decision maker for routine controls to define in-cup sensory quality in accepting incoming samples [10,12,14,34]. Coffee samples underwent sensory evaluation via monadic profiling, according to SCA [35] protocols, and were analysed using HS-SPME-MS in combination with multivariate statistical analysis.

## 2. Results and Discussion

### 2.1. Sensory Analysis

Sensory data show that the scores for aroma properties were spread over the full range (scale 0–10), although the highest values were poorly represented, as expected, because of the intrinsic characteristics of the samples, such as species, origin, primary processing, and the life-span of the study. The standard deviations (SD) of the attributes (Table 1) are very low considering the high numbers of both of the samples investigated (184) and judges (6). A high Coefficient of Variation (CV) was observed for some attributes, fruity, flowery and spicy, meaning that these sensory properties were rated as very high or very low by the panel. CV relates the SD to the mean values of the aroma properties and provides a more representative evaluation of the importance of SD. In addition, it is a useful measure for comparing the dispersions of two or more attributes measured on different scales.

**Table 1.** Descriptive statistics for the sensory attributes of coffee samples.

| Attributes | Mean | S.D. | Minimum | Maximum | CV |
|---|---|---|---|---|---|
| Acid | 1.79 | 1.68 | 0.10 | 7.80 | 0.93 |
| Bitter | 1.50 | 1.70 | 0.20 | 9.00 | 1.14 |
| Aromatic Intensity | 6.71 | 1.29 | 1.00 | 10.00 | 0.19 |
| Flowery | 1.08 | 1.76 | 0.00 | 9.00 | 1.62 |
| Fruity | 0.69 | 1.49 | 0.00 | 10.00 | 2.16 |
| Nutty | 1.41 | 2.06 | 0.00 | 9.00 | 1.46 |
| Woody | 1.36 | 2.02 | 0.00 | 8.00 | 1.49 |
| Spicy | 0.76 | 1.57 | 0.00 | 8.00 | 2.07 |
| Overall Quality | 6.63 | 1.46 | 0.60 | 10.00 | 0.22 |

ANOVA analyses and a post-hoc Tukey's test provided information on the judges' ability to evaluate the sensory attributes. Figure 1 shows that judge 3 does not perform similarly to the others for the aromatic intensity coffee property, and, together with judge 2, for acid notes, while judge 2 has a different evaluation for bitter attribute compared to the others. All judges are, however, aligned in rating the scores of the other attributes. These two judges were therefore not taken into consideration for the attributes in which their variance was not comparable to the others. The score averages were used as the "main scores" for the nine attributes in the following elaborations.

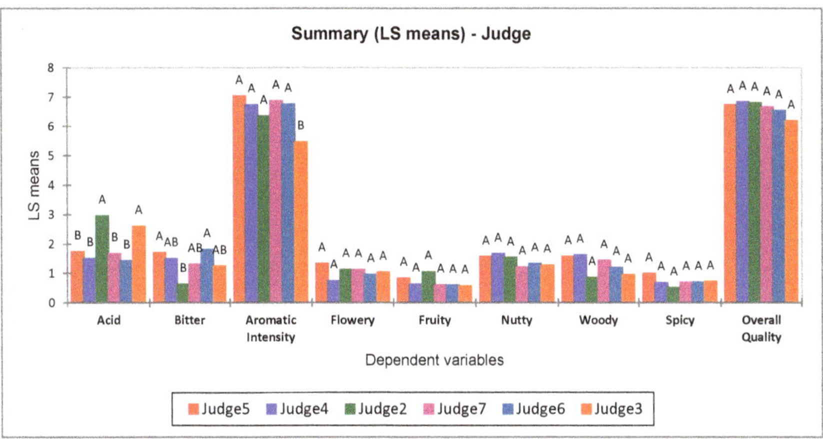


**Figure 1.** ANOVA and post-hoc Tukey's test results on the ability of the judges to rate the different attributes. The same letter means that the judges involved rate the attributes in the same way at a confidence level of 95%.

### 2.2. How a TAS System Based on the MS-Enose Works

The platform adopted allowed experiments to be run with an high throughput Total Analysis System (TAS) [36], which consisted of an autosampler for fully automated HS-SPME sample preparation on-line, and was directly combined to a mass spectrometer (MS), through a void column that was thermostatted in a GC oven, whose output signal (data) was elaborated on-line and then processed using chemometric software. The HS-SPME-MS TIC (Total ion Current) pattern is a single peak whose mass spectrum is representative of the fingerprints of the whole coffee volatile fraction (Figure 2).

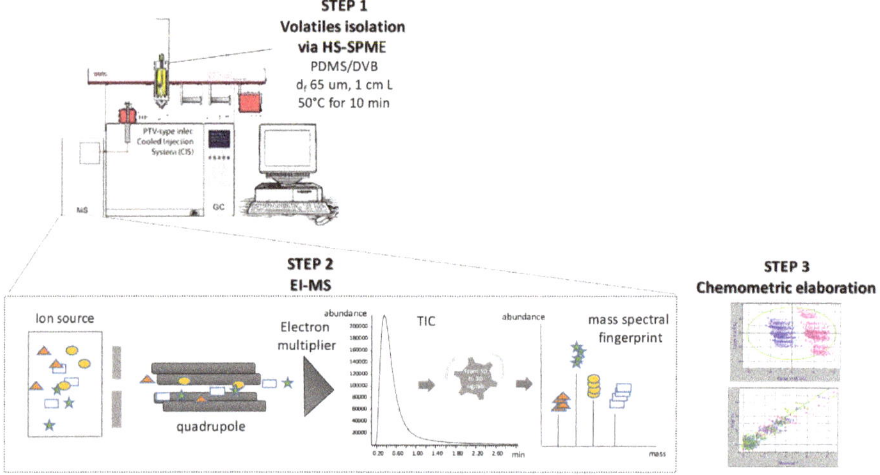


**Figure 2.** Schematic representation of the Total Analysis System (TAS) system used.

The corresponding mass spectral fingerprint is highly reproducible and ideally suited for further chemometric elaboration as it only consists of whole masses [37,38].

Compared to conventional GC-MS, mass spectral fingerprints provide total information about each sample and may even be more helpful and meaningful, in routine control screening, than the characterisation of each individual component in that sample. The reliability of the mass spectral fingerprint is demonstrated by the comparison of the high degree of overlap between the average mass profile of an Arabica volatile fraction obtained using the non-separative techniques (HS-SPME-MS-enose) and the average total spectrum over the total analysis time from a conventional HS-SPME-GC-MS analysis, as reported in Figure 3.

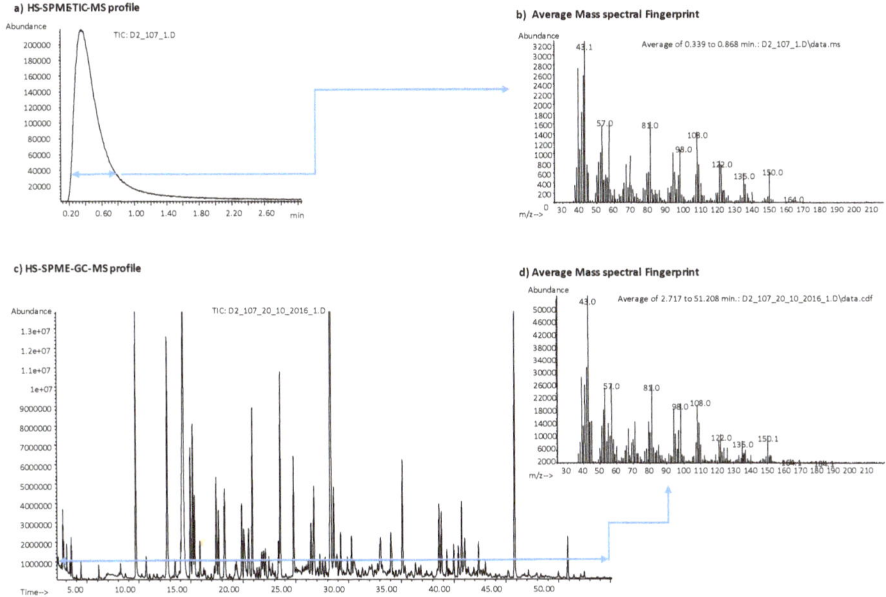

**Figure 3.** Analytical output signals of an Arabica roasted coffee sample from: (**a**) HS-SPME-MS-enose profile; (**b**) average HS-SPME-MS-enose mass spectral fingerprint that corresponds to the TIC data from MS-enose; (**c**) HS-SPME-GC-MS chromatogram; (**d**) average HS-SPME-GC-MS mass spectral fingerprint of the whole chromatographic profile.

### 2.3. Signal Processing and Chemometric Workflow

The mass spectral fingerprint encompasses all the chemical information on the volatile fraction of an analysed sample, while diagnostic and informative fragments can be correlated to a compound or a class of compounds. The mass fingerprint is displayed on a plot that reports the mass fragments ($m/z$) within the selected mass range on the $X$-axis, and the ion abundances for the mass fragments on the $Y$-axis (Figure 3).

The use of chemometrics to extract the significant and useful information from the complex data matrix, however, requires the profile to be precise, in particular when data monitoring is carried out over a long period and when a mathematical model for classification or correlation has to be generated. The chemometric tools adopted in this study were, in sequence: (i) Principal Component Analysis (PCA) to identify outliers; (ii) Partial Least Square Discriminant Analysis (PLS-DA) carried out on the sensory scaled samples (low-high score range) to identify the fragment ions that are most closely related to each sensory attribute; and (iii) Partial Least Square Regression (PLS) to correlate chemicals to sensory attributes, and to evaluate the ability of extracted chemical variables to predict sensory scores. The data processing work-flow is reported in Figure 4.

The consistency of the SPME fibres over time was ensured by testing six fibres of the same lot with a text mixture, and selecting those whose responses could not be distinguished using ANOVA; their performance was periodically monitored using the same test mixture. Before chemometric processing, the matrix was cleaned of ground fragments that may have interfered with data elaboration, e.g., the fragments at $m/z = 44$ ($CO_2$), $m/z = 73, 133, 147$ and $177$ (system bleeding), and $m/z = 149$ (derived from phthalates). The resulting data matrix was than subjected to internal normalisation vs. the most abundant ion ($m/z$ 43) and standardisation using Pareto scaling [39,40].

The HS-SPME-MS pattern is informative because the intensity of each ion ($m/z$) derives from the contribution of all components that present that fragment in their ionisation pattern. Chemometric elaboration consists of a series of steps to "extract" significant information from the MS fingerprint for further sensory score prediction.

The data matrix from the Pareto scaling was first submitted to an unsupervised exploratory investigation using principal component analysis (PCA) to detect sample outliers. The samples were found to be homogeneously distributed along the first three PCs with a cumulative explained variance of 83.15%, and the data indicated two populations of samples on the first PC that were related to the coffee species (Figure S1 in Supplementary Material). The fragment ions that derived from the volatile fraction therefore provided information on the chemical diversity of the investigated set of samples.

A supervised Partial Least Square Discriminant Analysis (PLS-DA) was then applied to the reprocessed MS spectral fingerprints on selections of samples that had the highest and the lowest scores for each sensory attribute in order to extract ions that had a high impact on sample discrimination (low vs. high scores). A cross-validation (CV = 5) was set to run the PLS-DA. The variable importance for projection (VIPs) scores estimate the importance of each variable in the projection that was used in a PLS-DA model, and is often used to select variables. VIPs higher than 1 and with a low standard deviation were considered for the extraction of the relevant ions that described each sensory attribute.

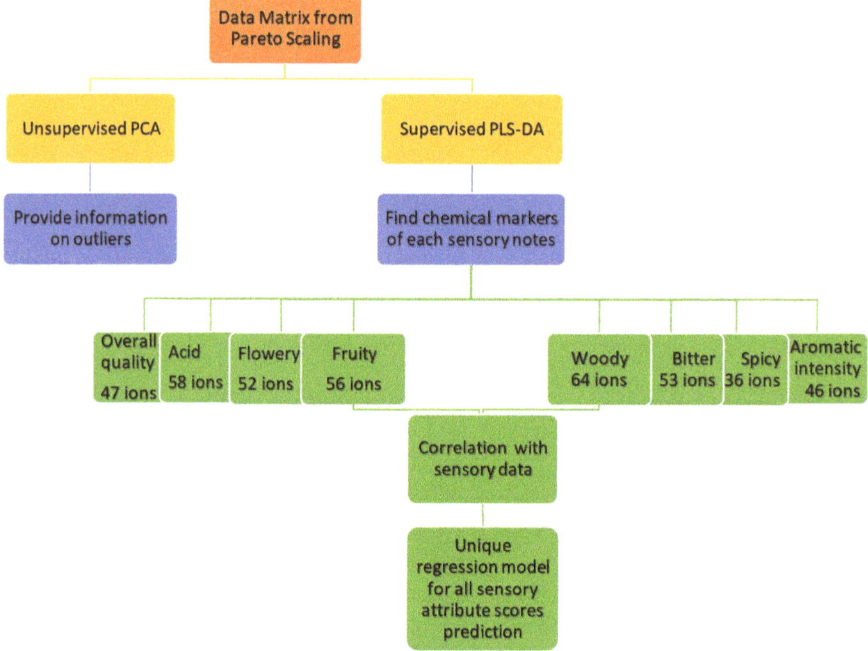


**Figure 4.** Workflow of the chemical data processing used to obtain the regression model.

Table 2 shows the significant ions that were selected for each sensory attribute under investigation, together with their VIP value and standard deviation. Results show that the total number of relevant ions is different for each considered sensory attribute and, in particular, that sensory qualities share several ions Table 2 and Figure 5.

**Table 2.** Significant ions selected from the partial least square discriminant analysis (PLS-DA) by variable importance for projection (VIP) for each sensory attribute.

| Flowery | | | Fruity | | | Acid | | | Bitter | | | Nutty | | | Spicy | | | Woody | | | Aromatic Intensity | | | Overall Quality | | |
| --- | --- | --- | --- | --- | --- | --- | --- | --- | --- | --- | --- | --- | --- | --- | --- | --- | --- | --- | --- | --- | --- | --- | --- | --- | --- | --- |
| $m/z$ | VIP | SD | $m/z$ | VIP | SD | $m/z$ | VIP | SD | $m/z$ | VIP | SD | $m/z$ | VIP | SD | $m/z$ | VIP | SD | $m/z$ | VIP | SD | $m/z$ | VIP | SD | $m/z$ | VIP | SD |
| 37 | 2.012 | 0.201 | 42 | 2.060 | 0.197 | 36 | 1.822 | 0.208 | 37 | 1.751 | 0.171 | 42 | 2.668 | 0.795 | 96 | 1.839 | 0.175 | 110 | 1.626 | 0.085 | 100 | 1.819 | 0.262 | 79 | 2.060 | 0.316 |
| 42 | 1.967 | 0.184 | 45 | 1.999 | 0.176 | 37 | 1.814 | 0.201 | 42 | 1.722 | 0.170 | 61 | 2.266 | 0.557 | 36 | 1.830 | 0.182 | 36 | 1.612 | 0.114 | 36 | 1.717 | 0.096 | 37 | 2.054 | 0.265 |
| 45 | 1.960 | 0.213 | 46 | 1.887 | 0.188 | 38 | 1.775 | 0.168 | 45 | 1.714 | 0.180 | 64 | 2.218 | 0.542 | 37 | 1.824 | 0.189 | 37 | 1.603 | 0.108 | 37 | 1.717 | 0.118 | 38 | 1.925 | 0.220 |
| 56 | 1.953 | 0.196 | 52 | 1.884 | 0.183 | 45 | 1.769 | 0.212 | 56 | 1.682 | 0.140 | 66 | 2.003 | 0.751 | 38 | 1.819 | 0.168 | 42 | 1.602 | 0.109 | 38 | 1.697 | 0.106 | 39 | 1.858 | 0.296 |
| 57 | 1.821 | 0.157 | 54 | 1.877 | 0.304 | 46 | 1.752 | 0.190 | 57 | 1.667 | 0.079 | 67 | 1.955 | 0.408 | 45 | 1.800 | 0.144 | 45 | 1.564 | 0.103 | 39 | 1.666 | 0.113 | 40 | 1.816 | 0.140 |
| 59 | 1.795 | 0.349 | 56 | 1.847 | 0.138 | 56 | 1.703 | 0.211 | 60 | 1.666 | 0.074 | 72 | 1.954 | 0.452 | 46 | 1.699 | 0.123 | 56 | 1.549 | 0.109 | 40 | 1.664 | 0.198 | 41 | 1.781 | 0.258 |
| 60 | 1.757 | 0.133 | 57 | 1.826 | 0.172 | 57 | 1.675 | 0.176 | 62 | 1.659 | 0.103 | 80 | 1.890 | 0.478 | 50 | 1.681 | 0.224 | 57 | 1.546 | 0.036 | 41 | 1.664 | 0.269 | 42 | 1.729 | 0.079 |
| 61 | 1.755 | 0.139 | 58 | 1.805 | 0.264 | 60 | 1.655 | 0.165 | 63 | 1.642 | 0.056 | 93 | 1.868 | 0.480 | 53 | 1.645 | 0.054 | 60 | 1.529 | 0.049 | 45 | 1.651 | 0.208 | 44 | 1.696 | 0.255 |
| 62 | 1.748 | 0.127 | 60 | 1.787 | 0.197 | 62 | 1.621 | 0.153 | 64 | 1.632 | 0.088 | 94 | 1.810 | 0.396 | 55 | 1.609 | 0.074 | 62 | 1.514 | 0.038 | 48 | 1.634 | 0.164 | 50 | 1.671 | 0.409 |
| 63 | 1.734 | 0.113 | 61 | 1.775 | 0.252 | 63 | 1.610 | 0.141 | 65 | 1.594 | 0.050 | 95 | 1.693 | 0.527 | 56 | 1.609 | 0.203 | 63 | 1.512 | 0.074 | 50 | 1.592 | 0.265 | 51 | 1.655 | 0.370 |
| 64 | 1.732 | 0.097 | 62 | 1.754 | 0.184 | 64 | 1.609 | 0.114 | 66 | 1.591 | 0.185 | 96 | 1.648 | 0.667 | 57 | 1.575 | 0.180 | 64 | 1.501 | 0.047 | 51 | 1.584 | 0.130 | 52 | 1.644 | 0.346 |
| 65 | 1.716 | 0.169 | 63 | 1.750 | 0.213 | 65 | 1.608 | 0.178 | 67 | 1.589 | 0.071 | 100 | 1.646 | 0.674 | 60 | 1.572 | 0.126 | 65 | 1.497 | 0.053 | 52 | 1.580 | 0.197 | 53 | 1.641 | 0.286 |
| 66 | 1.709 | 0.119 | 64 | 1.731 | 0.129 | 66 | 1.608 | 0.154 | 74 | 1.580 | 0.045 | 106 | 1.638 | 0.764 | 61 | 1.532 | 0.166 | 66 | 1.491 | 0.056 | 53 | 1.548 | 0.145 | 54 | 1.607 | 0.388 |
| 67 | 1.688 | 0.261 | 65 | 1.723 | 0.121 | 67 | 1.597 | 0.132 | 75 | 1.579 | 0.096 | 107 | 1.629 | 0.609 | 63 | 1.531 | 0.098 | 67 | 1.489 | 0.047 | 55 | 1.544 | 0.104 | 55 | 1.596 | 0.279 |
| 69 | 1.687 | 0.174 | 66 | 1.697 | 0.315 | 72 | 1.565 | 0.067 | 76 | 1.564 | 0.206 | 108 | 1.612 | 0.471 | 64 | 1.530 | 0.067 | 69 | 1.480 | 0.095 | 57 | 1.525 | 0.152 | 58 | 1.571 | 0.383 |
| 72 | 1.673 | 0.173 | 67 | 1.654 | 0.294 | 74 | 1.557 | 0.198 | 77 | 1.561 | 0.030 | 109 | 1.567 | 0.493 | 65 | 1.503 | 0.057 | 72 | 1.480 | 0.075 | 59 | 1.522 | 0.274 | 59 | 1.562 | 0.092 |
| 74 | 1.643 | 0.232 | 74 | 1.647 | 0.279 | 76 | 1.524 | 0.129 | 78 | 1.534 | 0.107 | 110 | 1.438 | 0.586 | 66 | 1.497 | 0.160 | 74 | 1.477 | 0.097 | 60 | 1.517 | 0.194 | 61 | 1.559 | 0.388 |
| 75 | 1.612 | 0.291 | 75 | 1.621 | 0.144 | 77 | 1.511 | 0.097 | 80 | 1.504 | 0.039 | 114 | 1.429 | 0.719 | 68 | 1.497 | 0.180 | 75 | 1.471 | 0.068 | 61 | 1.503 | 0.129 | 67 | 1.551 | 0.223 |
| 77 | 1.585 | 0.222 | 76 | 1.618 | 0.153 | 78 | 1.492 | 0.129 | 89 | 1.477 | 0.076 | 121 | 1.418 | 0.711 | 69 | 1.493 | 0.075 | 76 | 1.461 | 0.068 | 68 | 1.492 | 0.197 | 68 | 1.548 | 0.327 |
| 78 | 1.578 | 0.259 | 77 | 1.609 | 0.203 | 79 | 1.492 | 0.101 | 92 | 1.451 | 0.093 | 122 | 1.382 | 0.613 | 70 | 1.488 | 0.219 | 77 | 1.451 | 0.077 | 69 | 1.481 | 0.178 | 69 | 1.533 | 0.253 |
| 80 | 1.559 | 0.349 | 78 | 1.585 | 0.146 | 80 | 1.470 | 0.153 | 93 | 1.423 | 0.061 | 135 | 1.379 | 0.693 | 72 | 1.478 | 0.119 | 78 | 1.446 | 0.105 | 70 | 1.468 | 0.230 | 70 | 1.497 | 0.190 |
| 89 | 1.512 | 0.415 | 79 | 1.580 | 0.199 | 91 | 1.464 | 0.143 | 94 | 1.417 | 0.063 | 136 | 1.357 | 0.494 | 74 | 1.464 | 0.203 | 80 | 1.435 | 0.101 | 72 | 1.463 | 0.220 | 71 | 1.478 | 0.128 |
| 92 | 1.495 | 0.475 | 80 | 1.565 | 0.344 | 92 | 1.462 | 0.125 | 95 | 1.417 | 0.049 | 159 | 1.356 | 0.614 | 75 | 1.445 | 0.250 | 89 | 1.430 | 0.118 | 73 | 1.462 | 0.190 | 72 | 1.477 | 0.520 |
| 93 | 1.487 | 0.285 | 92 | 1.554 | 0.344 | 93 | 1.435 | 0.096 | 96 | 1.412 | 0.230 | 160 | 1.327 | 0.645 | 77 | 1.444 | 0.130 | 91 | 1.425 | 0.110 | 74 | 1.457 | 0.310 | 75 | 1.450 | 0.107 |
| 94 | 1.480 | 0.186 | 93 | 1.487 | 0.365 | 94 | 1.399 | 0.157 | 100 | 1.399 | 0.074 | | | | 78 | 1.440 | 0.113 | 92 | 1.419 | 0.115 | 79 | 1.450 | 0.112 | 78 | 1.416 | 0.363 |
| 95 | 1.468 | 0.172 | 94 | 1.435 | 0.252 | 95 | 1.392 | 0.140 | 104 | 1.394 | 0.063 | | | | 80 | 1.436 | 0.165 | 93 | 1.418 | 0.122 | 81 | 1.435 | 0.190 | 80 | 1.409 | 0.433 |
| 96 | 1.445 | 0.439 | 95 | 1.372 | 0.383 | 96 | 1.388 | 0.195 | 105 | 1.392 | 0.045 | | | | 89 | 1.425 | 0.181 | 94 | 1.417 | 0.045 | 82 | 1.421 | 0.359 | 81 | 1.386 | 0.436 |
| 100 | 1.423 | 0.137 | 96 | 1.366 | 0.237 | 100 | 1.376 | 0.127 | 106 | 1.374 | 0.075 | | | | 91 | 1.422 | 0.277 | 95 | 1.405 | 0.108 | 83 | 1.421 | 0.246 | 82 | 1.352 | 0.198 |
| 106 | 1.419 | 0.286 | 103 | 1.354 | 0.330 | 104 | 1.374 | 0.300 | 107 | 1.366 | 0.116 | | | | 92 | 1.403 | 0.138 | 96 | 1.401 | 0.157 | 86 | 1.415 | 0.136 | 83 | 1.340 | 0.291 |
| 107 | 1.412 | 0.302 | 104 | 1.316 | 0.205 | 105 | 1.374 | 0.247 | 108 | 1.362 | 0.162 | | | | 93 | 1.384 | 0.190 | 100 | 1.396 | 0.059 | 87 | 1.401 | 0.222 | 86 | 1.338 | 0.405 |
| 108 | 1.404 | 0.242 | 105 | 1.287 | 0.170 | 106 | 1.368 | 0.086 | 109 | 1.350 | 0.225 | | | | 94 | 1.344 | 0.146 | 104 | 1.392 | 0.126 | 91 | 1.375 | 0.265 | 87 | 1.317 | 0.297 |
| 109 | 1.389 | 0.275 | 106 | 1.285 | 0.182 | 107 | 1.367 | 0.125 | 110 | 1.347 | 0.114 | | | | 95 | 1.338 | 0.367 | 105 | 1.378 | 0.072 | 92 | 1.343 | 0.258 | 88 | 1.295 | 0.280 |
| 110 | 1.328 | 0.158 | 107 | 1.254 | 0.158 | 108 | 1.366 | 0.241 | 117 | 1.340 | 0.129 | | | | 97 | 1.320 | 0.351 | 106 | 1.369 | 0.066 | 95 | 1.310 | 0.216 | 94 | 1.277 | 0.272 |
| 112 | 1.307 | 0.322 | 108 | 1.253 | 0.197 | 109 | 1.365 | 0.163 | 118 | 1.339 | 0.204 | | | | 98 | 1.312 | 0.151 | 107 | 1.351 | 0.103 | 96 | 1.280 | 0.333 | 97 | 1.272 | 0.338 |
| 118 | 1.274 | 0.259 | 109 | 1.235 | 0.438 | 110 | 1.345 | 0.101 | 119 | 1.324 | 0.070 | | | | 100 | 1.298 | 0.185 | 108 | 1.329 | 0.166 | 97 | 1.267 | 0.319 | 98 | 1.271 | 0.265 |

Table 2. *Cont.*

| Flowery | | | Fruity | | | Acid | | | Bitter | | | Nutty | | | Spicy | | | Woody | | | Aromatic Intensity | | | Overall Quality | | |
|---|---|---|---|---|---|---|---|---|---|---|---|---|---|---|---|---|---|---|---|---|---|---|---|---|---|---|
| *m/z* | VIP | SD | *m/z* | VIP | SD | *m/z* | VIP | SD | *m/z* | VIP | SD | *m/z* | VIP | SD | *m/z* | VIP | SD | *m/z* | VIP | SD | *m/z* | VIP | SD | *m/z* | VIP | SD |
| 119 | 1.239 | 0.301 | 110 | 1.228 | 0.315 | 115 | 1.335 | 0.174 | 120 | 1.309 | 0.120 | | | | 103 | 1.267 | 0.209 | 109 | 1.327 | 0.085 | 98 | 1.255 | 0.134 | 99 | 1.251 | 0.228 |
| 120 | 1.211 | 0.231 | 113 | 1.223 | 0.272 | 117 | 1.329 | 0.109 | 121 | 1.291 | 0.165 | | | | 104 | 1.260 | 0.425 | 112 | 1.303 | 0.099 | 99 | 1.214 | 0.393 | 111 | 1.213 | 0.446 |
| 121 | 1.202 | 0.215 | 117 | 1.221 | 0.308 | 118 | 1.318 | 0.150 | 122 | 1.284 | 0.120 | | | | 105 | 1.254 | 0.371 | 113 | 1.285 | 0.088 | 109 | 1.210 | 0.424 | 112 | 1.206 | 0.226 |
| 122 | 1.186 | 0.459 | 118 | 1.220 | 0.228 | 119 | 1.309 | 0.263 | 123 | 1.283 | 0.120 | | | | 106 | 1.219 | 0.136 | 115 | 1.285 | 0.175 | 110 | 1.185 | 0.510 | 113 | 1.190 | 0.333 |
| 123 | 1.176 | 0.525 | 119 | 1.206 | 0.375 | 120 | 1.292 | 0.207 | 124 | 1.276 | 0.125 | | | | 107 | 1.209 | 0.231 | 116 | 1.258 | 0.144 | 111 | 1.133 | 0.324 | 123 | 1.187 | 0.382 |
| 124 | 1.150 | 0.193 | 120 | 1.187 | 0.329 | 121 | 1.288 | 0.164 | 125 | 1.274 | 0.081 | | | | 108 | 1.180 | 0.122 | 117 | 1.255 | 0.153 | 112 | 1.131 | 0.483 | 125 | 1.143 | 0.411 |
| 126 | 1.112 | 0.242 | 121 | 1.180 | 0.249 | 122 | 1.268 | 0.196 | 131 | 1.265 | 0.137 | | | | 109 | 1.173 | 0.196 | 118 | 1.244 | 0.121 | 116 | 1.109 | 0.305 | 126 | 1.120 | 0.353 |
| 134 | 1.100 | 0.479 | 122 | 1.167 | 0.391 | 123 | 1.265 | 0.186 | 132 | 1.263 | 0.195 | | | | 110 | 1.168 | 0.209 | 119 | 1.230 | 0.179 | 126 | 1.055 | 0.445 | 138 | 1.096 | 0.196 |
| 135 | 1.093 | 0.381 | 123 | 1.156 | 0.277 | 124 | 1.225 | 0.167 | 134 | 1.259 | 0.141 | | | | 112 | 1.148 | 0.203 | 120 | 1.224 | 0.128 | 140 | 1.036 | 0.459 | 139 | 1.068 | 0.285 |
| 136 | 1.085 | 0.304 | 124 | 1.146 | 0.348 | 125 | 1.215 | 0.186 | 135 | 1.231 | 0.118 | | | | 115 | 1.146 | 0.138 | 121 | 1.223 | 0.125 | 141 | 1.025 | 0.370 | 140 | 1.063 | 0.585 |
| 137 | 1.072 | 0.387 | 125 | 1.145 | 0.401 | 129 | 1.195 | 0.207 | 136 | 1.175 | 0.145 | | | | 117 | 1.142 | 0.142 | 122 | 1.214 | 0.232 | 166 | 1.014 | 0.527 | 141 | 1.060 | 0.184 |
| 139 | 1.053 | 0.336 | 132 | 1.132 | 0.187 | 134 | 1.173 | 0.166 | 137 | 1.174 | 0.160 | | | | 118 | 1.131 | 0.188 | 123 | 1.206 | 0.130 | | | | 161 | 1.028 | 0.269 |
| 146 | 1.044 | 0.371 | 134 | 1.102 | 0.327 | 135 | 1.139 | 0.155 | 145 | 1.160 | 0.137 | | | | 119 | 1.124 | 0.224 | 124 | 1.205 | 0.111 | | | | | | |
| 147 | 1.040 | 0.379 | 135 | 1.101 | 0.322 | 136 | 1.138 | 0.241 | 146 | 1.157 | 0.176 | | | | 120 | 1.108 | 0.210 | 125 | 1.188 | 0.156 | | | | | | |
| 150 | 1.021 | 0.471 | 136 | 1.099 | 0.220 | 137 | 1.125 | 0.411 | 150 | 1.138 | 0.122 | | | | 121 | 1.102 | 0.257 | 127 | 1.177 | 0.157 | | | | | | |
| 160 | 1.018 | 0.521 | 137 | 1.097 | 0.306 | 145 | 1.108 | 0.348 | 151 | 1.137 | 0.271 | | | | 122 | 1.098 | 0.253 | 131 | 1.176 | 0.106 | | | | | | |
| 164 | 1.002 | 0.467 | 145 | 1.092 | 0.248 | 146 | 1.096 | 0.246 | 152 | 1.126 | 0.146 | | | | 123 | 1.085 | 0.199 | 132 | 1.168 | 0.190 | | | | | | |
| | | | 146 | 1.074 | 0.213 | 148 | 1.092 | 0.371 | 164 | 1.112 | 0.214 | | | | 124 | 1.084 | 0.296 | 134 | 1.145 | 0.166 | | | | | | |
| | | | 150 | 1.039 | 0.274 | 150 | 1.077 | 0.204 | | | | | | | 126 | 1.056 | 0.303 | 135 | 1.131 | 0.146 | | | | | | |
| | | | 152 | 1.006 | 0.280 | 151 | 1.046 | 0.225 | | | | | | | 132 | 1.037 | 0.488 | 136 | 1.119 | 0.226 | | | | | | |
| | | | 164 | 1.001 | 0.272 | 152 | 1.031 | 0.405 | | | | | | | 134 | 1.036 | 0.420 | 137 | 1.114 | 0.208 | | | | | | |
| | | | | | | 160 | 1.028 | 0.231 | | | | | | | 135 | 1.034 | 0.318 | 139 | 1.107 | 0.152 | | | | | | |
| | | | | | | 164 | 1.011 | 0.142 | | | | | | | 136 | 1.023 | 0.331 | 146 | 1.092 | 0.169 | | | | | | |
| | | | | | | | | | | | | | | | 137 | 1.016 | 0.487 | 148 | 1.089 | 0.250 | | | | | | |
| | | | | | | | | | | | | | | | 145 | 1.015 | 0.500 | 150 | 1.086 | 0.218 | | | | | | |
| | | | | | | | | | | | | | | | 150 | 1.015 | 0.383 | 151 | 1.082 | 0.300 | | | | | | |
| | | | | | | | | | | | | | | | 151 | 1.010 | 0.214 | 152 | 1.017 | 0.183 | | | | | | |
| | | | | | | | | | | | | | | | 164 | 1.001 | 0.384 | 160 | 1.014 | 0.263 | | | | | | |
| | | | | | | | | | | | | | | | | | | 164 | 1.003 | 0.129 | | | | | | |
| **Total number** | | | | | | | | | | | | | | | | | | | | | | | | | | |
| 52 | | | 56 | | | 58 | | | 53 | | | 24 | | | 63 | | | 64 | | | 46 | | | 47 | | |

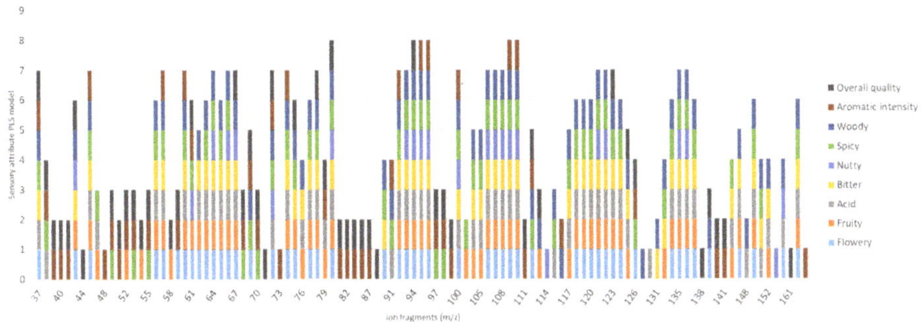

**Figure 5.** The importance and occurrence of each selected mass fragment ($m/z$) in the partial least square (PLS) regression model of every sensory attribute.

A PLS model was built for each single sensory attribute to verify the relationship between the chemical ion fragments ($m/z$) (i.e., the chemical components of the volatile fraction) and the sensory profile, and to predict the sensory scores. Figure 5 shows that the ions selected from the PLS-DA and used to design the corresponding regression model for each sensory feature are involved in more than one feature.

For example, the typical base fragment of organic acids $m/z$ 60 in coffee (the most abundant of them being acetic acid and 3-methyl butanoic acid), is depicted in 7/9 regression models, while $m/z$ 150/152, predominantly related to methoxy phenols, are only present in 4/9 regression models. This result is in agreement with results reported in Ribeiro et al., who underlined the importance of the co-participation of several volatiles in describing various sensory features in fifty three Arabica coffees [41].

Moreover, the mass spectral fingerprint provides some information on the volatiles that characterise the samples. For example, $m/z$ 108 is mainly related to several alkylated pyrazines and pyrrole derivatives, $m/z$ 95/96 are associated to furfuryl products, and 135/137/150/152 are primarily related to methyl-ethyl pyrazine and methoxy phenols.

The ability of the VIP-selected ions to describe the sensory characteristics of samples can be visualised for the most fruity and woody samples in the heat-map in Figure 6, which shows a clear discrimination between the samples with these two sensory profiles thanks to mass fingerprinting. The rows indicate the $m/z$ ions, and the columns the investigated samples. The colour scale varies from blue (low abundance) to red (high abundance). A hierarchical cluster analysis (HCA) of both the rows and columns shows that volatile distributions differ according to their normalised response across samples. Figure 6 highlights that the ion-intensity ratios across samples in place of the different quality volatiles is very effective to discriminate the two cluster of samples linked to their sensory peculiarity [3,42,43].

The ions selected using PLS-DA were then used as independent variables to evaluate the relationship with sensory data and the ability to predict scores (dependent variables) of each sensory attribute by developing a specific and optimised regression model for each feature. All sensory notes have been modelled through a PLS algorithm; the evolution of each sensory note over the sample sets has been related a different number of variables (Table 2). The flexibility of the prediction model for each sensory attribute was evaluated in samples that covered a range of seasonality, origins and crops and then tested the models with an external test set (Table 3). Acid, Bitter, Spicy and (to a lesser extent) aromatic intensity and flowery PLS prediction models show good performance. The R2 values indicate that nearly more than 50% of the variance in the measured scores is explained by the models (i.e., the selected mass spectral fragments used to describe the model). This is quite a good result in consideration of the high variability of the training set. The goodness of the predictive capability is confirmed by the acceptable values of the root mean squared errors (RMSECV and RMSEP) reported for these attributes. The limit of acceptability for predicted values has been strictly fixed by the sensory panel, in ±1 score points. All RMSECV values are within or close to this interval while prediction on new samples show RMSEP slightly higher in particular for the overall quality and to a lesser extent for woody and nutty.

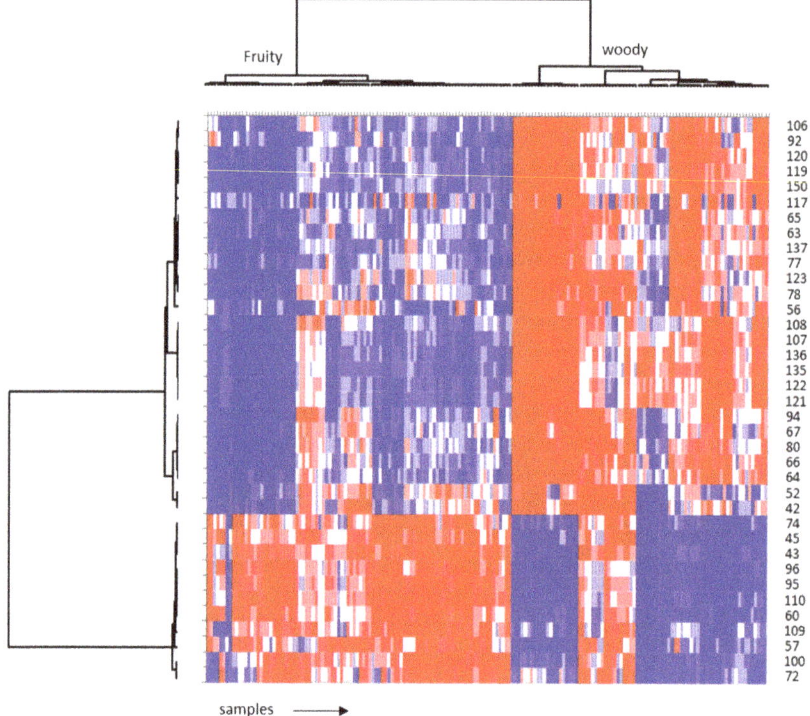


**Figure 6.** Heat-map of a group of the selected samples that present woody and fruity features.

The models so far developed, however, still require nine different data elaborations. This is a serious limit for routine HS-SPME-MS-enose applications that can only be overcome with a unique multi-note sensory-score prediction model. The variable selection for the multi-note model was carried out by combining the matrices (ions fingerprints) used for the single-note prediction models. Fragment ions from the single-model note, without repetitions, were selected, the x matrix was then simplified and the number of variables reduced according to VIP values. This variable reduction was carried out to reduce the statistical noise, and maximise the information provided by each single note chemical fingerprint. According to the VIP values (VIP > 0.8), 78 ions out of 104 in the volatile fraction were retained in building the model, thus allowing the multidimensional structure of the prediction model to be simplified with negligible loss in performance.

The regression model was built with a training set of 150 objects and an external test set of 34. The leave-p-out cross-validation method ($n = 20$) was used to select a suitable number of components from the PLS regression and to reduce the errors when the calibration model was used for the feature predictions of unknown samples. The results of the developed multi-note regression model for the prediction of sensory-attribute scores show that acid, bitter and spicy notes were the most reliable as they present a lower root mean square error in prediction (RMSEP). However, the mean error in the sensory-score prediction RMSEP in the external test set with these data fall within the range of ±2 (Table 3). The other attributes show discrete to good fitting between chemical and sensory data from the R2val values, and a better ability to predict the scores of the training and internal evaluation set, but fail to meet the expectation limit for acceptability given by the panellist (±1) when used to estimate new samples Table 3 and Figure 7. The high errors in prediction, in particular for samples with high sensory scores, probably occurred because of the unbalanced samples (i.e., the number of high scored samples was lower than that of the low scored ones for some attributes such as fruity, nutty, woody), making this part of the score range less well represented in the sample sets.

**Table 3.** Multi-note model performance summary.

| | Single-Note Model Performance | | | | Multi-Note Model Performance | | | |
|---|---|---|---|---|---|---|---|---|
| *Sensory Note* | *Model Factors* | *R2val* | *RMSEV* | *RMSEP* | *Model Factors* | *R2val* | *RMSEV* | *RMSEP* |
| Acid | 3 | 0.663 | 1.129 | 0.946 | 3 | 0.856 | 0.726 | 1.192 |
| Bitter | 4 | 0.817 | 1.142 | 1.063 | 4 | 0.936 | 0.626 | 1.315 |
| Woody | 4 | 0.669 | 1.570 | 1.725 | 4 | 0.884 | 1.003 | 2.306 |
| Flowery | 4 | 0.746 | 1.038 | 1.345 | 4 | 0.907 | 0.651 | 1.964 |
| Fruity | 4 | 0.661 | 1.026 | 1.499 | 2 | 0.790 | 0.785 | 1.598 |
| Spicy | 1 | 0.792 | 0.963 | 1.209 | 3 | 0.784 | 0.977 | 1.194 |
| Nutty | 6 | 0.544 | 1.506 | 1.661 | 4 | 0.893 | 0.864 | 1.891 |
| Aroma intensity | 1 | 0.557 | 0.936 | 1.296 | 4 | 0.764 | 0.627 | 1.642 |
| Overall quality | 4 | 0.556 | 0.936 | 2.120 | 4 | 0.756 | 0.726 | 2.239 |

The low predictive ability for new set of samples may be due to: (a) the different species, origin seasonality and post-harvest treatments compared to other published work [23,41]; (b) the high noise caused by an unbalanced pool of samples for some attributes, such as nutty or flowery; (c) the difficulties linked to an excessively general lexicon to define the notes; and (d) compromises in the abilities of modelling for each sensory characteristic must be considered when building a multi-note model.

**Figure 7.** Comparison of the measured sensory profiles (from the panel) and predicted sensory profiles, from the developed model, of a selection of external test set samples. Sensory and chemical data were pre-processed using pareto scaling.

## 3. Materials and Methods

### 3.1. Samples

One hundred and eighty four coffee samples, with distinctive sensory notes, originating from a number of different countries were analysed. Coffee samples were kindly supplied by Lavazza S.p.A. (Turin, Italy). The roasting degree of each sample was carefully measured by ground bean light reflectance, with a single-beam Color Test 2 instrument Neuhaus Neotec (Genderkesee, Germany) at a wavelength of 900 nm, on 25–30 g of ground coffee. The roasting degree was set at 55° Nh, in order to be close to the international standardisation protocol for cupping [35]. Samples were roasted no more than 24 h before cupping and left at least 8 h to stabilise.

### 3.2. Descriptive Sensory Analysis of Coffee Aroma

The samples were submitted to sensory evaluation by a panel of six coffee experts. Aroma quality was assessed for a set of nine attributes, namely flowery, fruity, woody, nutty, spicy, acidity, bitterness, aroma intensity and overall quality. The quality and intensity of each attribute were simultaneously evaluated, on a scale from 0 to 10. ANOVA analysis with a post-hoc test were run to verify panel alignment on each attribute. Average scores from experts whose evaluations were similar were used as the "main scores" for the investigated attributes.

### 3.3. Head Space Solid Phase Micro Extraction Sampling

Volatiles were sampled using HS-SPME and an MPS-2 multipurpose sampler (Gerstel, Mulheim a/d Ruhr, Germany) which was integrated online with an Agilent 7890 GC coupled to a 5975 MS detector (Agilent, Little Falls, DE, USA). One point five grams of ground roasted coffee in a 20 mL vial were directly sampled by HS-SPME for 10 min at 50 °C at a stirring speed of 350 rpm. The SPME fibre was a PDMS/DVB df 65 μm, and 1 cm long (Supelco, Bellefonte, PA, USA). After sampling, the recovered analytes were thermally desorbed, by heating the fibre for 5 min at 250 °C, into the GC injector body, from where they were transferred on-line to the gas-chromatographic column. All samples were analysed in duplicate.

### 3.4. MS-eNose Instrument Set-Up

The GC oven and injector were maintained at 250 °C; injection mode, split; split ratio, 1/10; carrier gas, helium; flow rate, 0.4 mL/min; fibre desorption time and reconditioning, 3 min. The transfer column was uncoated deactivated fused silica tubing (dc = 0.10 mm, length = 6.70 m) from MEGA (Legnano, Italy).

MSD Conditions: ionisation, EI mode at 70 eV; temperatures: ion source: 230 °C, transfer line: 280 °C. Standard tuning was used and the scan range was set at $m/z$ 35–350 with a scanning rate of 1.000 amu/s.

### 3.5. Data Acquisition and Elaboration

Data were acquired and processed using an Agilent MSD Chem Station ver. E.02.01.1177 (Agilent, Little Falls, DE, USA). Raw data were transformed using RapidDataInterpretation software by Gerstel (Gerstel, Mulheim a/d Ruhr, Germany). This is a post-run macro that expands the scope of the function of the Agilent ChemStation software, which allows the 3-dimensional raw data supplied by the mass spectrometry (retention time, $m/z$ fragmentation and intensities) to be reduced to 2-dimensional data that can then be properly used by statistical software for further elaboration. The intensities of a sample are added as a function of the masses. The result is a data matrix of 91,980 data in which the rows report the samples and the columns report the intensity assigned to each mass.

Chemometric analyses were carried out using Pirouette software ver. 4.5 (Infometrix, Inc., Bothell, WA, USA). Principal component analysis (PCA), Partial Least Square Discriminant Analysis (PLS-DA) and Partial Least Square (PLS) regression were used. Heat map visualisation, One-way ANOVA and t-tests were performed on the sensorial results using XLSTAT (Addinsoft, New York, NY, USA).

## 4. Conclusions

The results show that the combination of HS-SPME-MS fingerprints and chemometrics is a promising technique for use as a TAS system working as a high throughput solution for the prediction of the in-cup coffee sensory quality of incoming coffee beans. Sensory quality control and evaluation is crucial if the coffee industry is to satisfy the ever-increasing demand for coffee with specific sensory attributes. The described methods would allow trained panellists to be exempted (at least partially) from routine tasting and focus their activity on new products and sensory attributes. The ambitious challenge of this study was based on the exploration of the ability of this analytical approach to predict in-cup coffee quality, including representative coffee samples of different origins, species and postharvest treatments, as occurs in industry quality control upon the acceptance of incoming beans. The study has demonstrated that this approach and the use of a multi-note model to predict global coffee sensory profiles requires a number of compromises, in terms of model robustness and acceptance of the errors in prediction. The high errors in prediction, in particular for samples with high sensory scores, probably occurred because the number of high scored samples was lower than that of the low scored samples, making this part of the score range less well represented in the sample sets. A second explanation may involve the sensory scores measured by the panel; high scores are more difficult to define and require a precise alignment.

As a general consideration, the main limit of this study is the number of coffee samples, which is only a snapshot of the number of coffees that may be processed in a plant, and is therefore not sufficient to obtain fully reliable and robust models. Automatic screening to predict the cup-quality of the raw material requires a huge repository of sensory and instrumental data. Furthermore, this approach operates, in terms of chemometric data processing, within an order of magnitude of hundreds of samples with similar qualities, as shown by Ribeiro on fifty three Arabica coffees and Lindinger on 18 espresso coffees (Ristretto and Lungo types). For higher numbers of samples, other data mining approaches should be considered in the development of the prediction tool, e.g., artificial neural networks and deep learning algorithms [23,41,43].

**Supplementary Materials:** The following is present in supplementary material available online, Figure S1: PCA scores plots of coffee samples.

**Author Contributions:** Conceptualization, E.L. and C.B.; methodology, E.L., B.S. and C.B.; software, D.B. and E.L.; validation, G.P., M.R.R. and C.C.; formal analysis, D.B. and G.S.; investigation, D.B. and G.S.; resources, G.P. and M.R.R.; data curation, D.B. and G.S.; writing—original draft preparation, E.L. and D.B.; writing—review and editing, E.L., C.B., G.P., M.R.R.; visualization, B.S. and C.C.; supervision, E.L. and C.B.; project administration, E.L.

**Funding:** The research is part of a PhD project financially supported by Luigi Lavazza S.p.A, Torino.

**Conflicts of Interest:** The authors declare that they have no known competing financial interests or personal relationships that could have appeared to influence the work reported in this paper.

## References

1. Flament, I. *Coffee Flavor Chemistry*; Wiley: Chichester, UK, 2002.
2. Folmer, B. *The Craft and Science of Coffee*; Academic Press: London, UK, 2017.
3. Folmer, B. How Can Science Help to Create New Value in Coffee? *Food Res. Int.* **2014**, *63*, 477–482. [CrossRef]
4. Lingle, T.R.; Menon, S.N. Cupping and grading—Discovering character and quality. In *The Craft and Science of Coffee*; Folmer, B., Ed.; Academic Press: London, UK, 2017; pp. 181–203.
5. Sunarharum, W.B.; Williams, D.J.; Smyth, H.E. Complexity of Coffee Flavor: A Compositional and Sensory Perspective. *Food Res. Int.* **2014**, *62*, 315–325. [CrossRef]
6. Ruosi, M.R.; Cordero, C.; Cagliero, C.; Rubiolo, P.; Bicchi, C.; Sgorbini, B.; Liberto, E. A Further Tool to Monitor the Coffee Roasting Process: Aroma Composition and Chemical Indices. *J. Agric. Food Chem.* **2012**, 11283–11291. [CrossRef] [PubMed]

7.   Liberto, E.; Ruosi, M.R.; Cordero, C.; Rubiolo, P.; Bicchi, C.; Sgorbini, B. Non-Separative Headspace Solid Phase Microextraction-Mass Spectrometry Profile as a Marker to Monitor Coffee Roasting Degree. *J. Agric. Food Chem.* **2013**, 1652–1660. [CrossRef] [PubMed]

8.   Bressanello, D.; Liberto, E.; Cordero, C.; Sgorbini, B.; Rubiolo, P.; Pellegrino, G.; Ruosi, M.R.; Bicchi, C. Chemometric Modelling of Coffee Sensory Notes through Their Chemical Signatures: Potential and Limits in Defining an Analytical Tool for Quality Control. *J. Agric. Food Chem.* **2018**, 7096–7109. [CrossRef] [PubMed]

9.   Dirinck, I.; Van Leuven, I.; Dirinck, P. *ChemSensor Classification of Red Wines*; Elsevier: Amsterdam, The Netherlands, 2006; Volume 43. [CrossRef]

10.   Dirinck, I.; Van Leuven, I.; Dirinck, P. Hyphenated Electronic Nose Technique for Aroma Analysis of Foods and Beverages. *LC GC Eur.* **2009**, *22*, 525–531.

11.   Liang, Y.-Z.; Xie, P.; Chan, K. Quality Control of Herbal Medicines. *J. Chromatogr. B Anal. Technol. Biomed. Life Sci.* **2004**, *812*, 53–70. [CrossRef]

12.   Pérès, C.; Begnaud, F.; Eveleigh, L.; Berdagué, J.-L. Fast Characterization of Foodstuff by Headspace Mass Spectrometry (HS-MS). *TrAC - Trends Anal. Chem.* **2003**, *22*, 858–866. [CrossRef]

13.   Lesiak, A.D.; Musah, R.A. Rapid High-Throughput Species Identification of Botanical Material Using Direct Analysis in Real Time High Resolution Mass Spectrometry. *J. Vis. Exp.* **2016**, *116*, 1–11. [CrossRef]

14.   Goodner, K.; Rouseff, R. *Practical Analysis of Flavor and Fragrance Materials*; Wiley: Chirchester, UK, 2011. [CrossRef]

15.   Pizarro, C.; Esteban-Díez, I.; González-Sáiz, J.M. Mixture Resolution According to the Percentage of Robusta Variety in Order to Detect Adulteration in Roasted Coffee by near Infrared Spectroscopy. *Anal. Chim. Acta* **2007**, *585*, 266–276. [CrossRef]

16.   Ribeiro, J.S.; Ferreira, M.M.C.; Salva, T.J.G. Chemometric Models for the Quantitative Descriptive Sensory Analysis of Arabica Coffee Beverages Using near Infrared Spectroscopy. *Talanta* **2011**, *83*, 1352–1358. [CrossRef] [PubMed]

17.   Barbin, D.F.; Felicio, A.L.d.S.M.; Sun, D.W.; Nixdorf, S.L.; Hirooka, E.Y. Application of Infrared Spectral Techniques on Quality and Compositional Attributes of Coffee: An Overview. *Food Res. Int.* **2014**, *61*, 23–32. [CrossRef]

18.   Sánchez-López, J.A.; Zimmermann, R.; Yeretzian, C. Insight into the Time-Resolved Extraction of Aroma Compounds during Espresso Coffee Preparation: Online Monitoring by PTR-ToF-MS. *Anal. Chem.* **2014**, *86*, 11696–11704. [CrossRef] [PubMed]

19.   Wieland, F.; Gloess, A.N.; Keller, M.; Wetzel, A.; Schenker, S.; Yeretzian, C. Online Monitoring of Coffee Roasting by Proton Transfer Reaction Time-of-Flight Mass Spectrometry (PTR-ToF-MS): Towards a Real-Time Process Control for a Consistent Roast Profile. *Anal. Bioanal. Chem.* **2012**, *402*, 2531–2543. [CrossRef]

20.   Romano, A.; Gaysinsky, S.; Czepa, A.; Del Pulgar, J.S.; Cappellin, L.; Biasioli, F. Static and Dynamic Headspace Analysis of Instant Coffee Blends by Proton-Transfer-Reaction Mass Spectrometry. *J. Mass Spectrom.* **2015**, *50*, 1057–1062. [CrossRef]

21.   Charles, M.; Romano, A.; Yener, S.; Barnabà, M.; Navarini, L.; Märk, T.D.; Biasoli, F.; Gasperi, F. Understanding Flavour Perception of Espresso Coffee by the Combination of a Dynamic Sensory Method and In-Vivo Nosespace Analysis. *Food Res. Int.* **2015**, *69*, 9–20. [CrossRef]

22.   Zimmermann, R.; Heger, H.J.; Yeretzian, C.; Nagel, H.; Boesl, U. Application of Laser Ionization Mass Spectrometry for On-Line Monitoring of Volatiles in the Headspace of Food Products: Roasting and Brewing of Coffee. *Rapid Commun. Mass Spectrom.* **1996**, *10*, 1975–1979. [CrossRef]

23.   Lindinger, C.; Labbe, D.; Pollien, P.; Rytz, A.; Juillerat, M.A.; Yeretzian, C.; Blank, I. When Machine Tastes Coffee: Instrumental Approach To Predict the Sensory Profile of Espresso Coffee. *Anal. Chem.* **2008**, *80*, 1574–1581. [CrossRef]

24.   Wang, G.; He, X.; Zhou, F.; Li, Z.; Fang, B.; Zhang, X.; Wang, L. Application of Gold Nanoparticles/TiO$_2$ modified Electrode for the Electrooxidative Determination of Catechol in Tea Samples. *Food Chem.* **2012**, *135*, 446–451. [CrossRef]

25.   Loutfi, A.; Coradeschi, S.; Mani, G.K.; Shankar, P.; Rayappan, J.B.B. Electronic Noses for Food Quality: A Review. *J. Food Eng.* **2015**, *144*, 103–111. [CrossRef]

26.   Roberts, J.; Power, A.; Chapman, J.; Chandra, S.; Cozzolino, D. A Short Update on the Advantages, Applications and Limitations of Hyperspectral and Chemical Imaging in Food Authentication. *Appl. Sci.* **2018**, *8*, 505. [CrossRef]

27. Grassi, S.; Alamprese, C. Advances in NIR Spectroscopy Applied to Process Analytical Technology in Food Industries. *Curr. Opin. Food Sci.* **2018**, *22*, 17–21. [CrossRef]

28. Marsili, R.T. SPME–MS–MVA as an Electronic Nose for the Study of Off-Flavors in Milk. *J. Agric. Food Chem.* **1999**, *47*, 648–665. [CrossRef] [PubMed]

29. Cozzolino, D.; Smyth, H.E.; Cynkar, W.; Janik, L.; Dambergs, R.G.; Gishen, M. Use of Direct Headspace-Mass Spectrometry Coupled with Chemometrics to Predict Aroma Properties in Australian Riesling Wine. *Anal. Chim. Acta* **2008**, *621*, 2–7. [CrossRef] [PubMed]

30. Majchrzak, T.; Wojnowski, W.; Dymerski, T.; Gębicki, J.; Namieśnik, J. Electronic Noses in Classification and Quality Control of Edible Oils: A Review. *Food Chem.* **2018**, *246*, 192–201. [CrossRef] [PubMed]

31. Majcher, M.A.; Kaczmarek, A.; Klensporf-Pawlik, D.; Pikul, J.; Jeleń, H.H. SPME-MS-Based Electronic Nose as a Tool for Determination of Authenticity of PDO Cheese, Oscypek. *Food Anal. Methods* **2015**, *8*, 2211–2217. [CrossRef]

32. Gliszczyńska-Świgło, A.; Chmielewski, J. Electronic Nose as a Tool for Monitoring the Authenticity of Food. A Review. *Food Anal. Methods* **2017**, *10*, 1800–1816. [CrossRef]

33. Sgorbini, B.; Bicchi, C.; Cagliero, C.; Cordero, C.; Liberto, E.; Rubiolo, P. Herbs and Spices: Characterization and Quantitation of Biologically-Active Markers for Routine Quality Control by Multiple Headspace Solid-Phase Microextraction Combined with Separative or Non-Separative Analysis. *J. Chromatogr. A* **2015**, *1376*, 9–17. [CrossRef]

34. Sandra, P.; David, F.; Tienpont, B. From CGC-MS to MS-Based Analytical Decision Makers. *Chromatographia* **2004**, *60*. [CrossRef]

35. SCAA. SCAA Protocols Cupping Specialty Coffee. Available online: https://www.scaa.org/PDF/resources/cupping-protocols.pdf (accessed on 26 July 2017).

36. Dittrich, P.S.; Tachikawa, K.; Manz, A. Micro Total Analysis Systems. Latest Advancements and Trends. *Anal. Chem.* **2006**, *78*, 3887–3908. [CrossRef]

37. Heiden, A.C.; Gil, C.; Ramos, L.S. Comparison of Different Approaches to Rapid Screening of Headspace Samples: Pros and Cons of Using MS-Based Electronic Noses versus Fast Chromatography. Available online: https://www.gerstel.com/pdf/p-cs-an-2002-08.pdf (accessed on 26 April 2018).

38. Kinton, V.; Pfannkoch, E.; Whitecavage, J. Discrimination of Soft Drinks Using a Chemical Sensor and Principal Component Analysis. Available online: https://www.scaa.org/PDF/resources/cupping-protocols.pdf (accessed on 26 April 2018).

39. Pérez Pavón, J.L.; del Nogal Sánchez, M.; Pinto, C.G.; Fernández Laespada, M.E.; Cordero, B.M.; Peña, A.G. Strategies for Qualitative and Quantitative Analyses with Mass Spectrometry-Based Electronic Noses. *TrAC Trends Anal. Chem.* **2006**, *25*, 257–266. [CrossRef]

40. Hou, W.; Tian, Y.; Liao, T.; Huang, Y.; Tang, Z.; Wu, Y.; Duan, Y. Development of the Mass Spectral Fingerprint by Headspace-Solid-Phase Microextraction-Mass Spectrometry and Chemometric Methods for Rapid Quality Control of Flavoring Essence. *Microchem. J.* **2016**, *128*, 75–83. [CrossRef]

41. Ribeiro, J.S.; Augusto, F.; Salva, T.J.G.; Ferreira, M.M.C. Prediction Models for Arabica Coffee Beverage Quality Based on Aroma Analyses and Chemometrics. *Talanta* **2012**, *101*, 253–260. [CrossRef] [PubMed]

42. Bressanello, D.; Liberto, E.; Cordero, C.; Rubiolo, P.; Pellegrino, G.; Ruosi, M.R.; Bicchi, C. Coffee Aroma: Chemometric Comparison of the Chemical Information Provided by Three Different Samplings Combined with GC–MS to Describe the Sensory Properties in Cup. *Food Chem.* **2017**, 218–226. [CrossRef] [PubMed]

43. Esposito, A.; Bassis, S.; Morabito, F.C.; Pasero, E. Some Notes on Computational and Theoretical Issues in Artificial Intelligence and Machine Learning. In *Smart Innovation, Systems and Technologies*; Springer Science and Business Media Deutschland GmbH: Canberra, Australia, 2016; Volume 54, pp. 3–12. [CrossRef]

**Sample Availability:** Samples of the several key odor compounds of coffee are available from the authors.

 *molecules*

*Article*

# Biophenolic Compounds Influence the In-Mouth Perceived Intensity of Virgin Olive Oil Flavours and Off-Flavours

**Alessandro Genovese [1],*, Ferdinando Mondola [1], Antonello Paduano [2] and Raffaele Sacchi [1]**

[1]  Department of Agricultural Sciences, University of Naples Federico II, Via Università 100, 80055 Portici, Italy; fe.mondola@libero.it (F.M.); sacchi@unina.it (R.S.)

[2]  Department of Agricultural and Environmental Science, University of Bari, Via Amendola 165/A, 70126 Bari, Italy; antonello.paduano@uniba.it

*  Correspondence: alessandro.genovese@unina.it; Tel.: +39-081-2539352

Academic Editor: Eugenio Aprea
Received: 5 April 2020; Accepted: 20 April 2020; Published: 23 April 2020

**Abstract:** In this study, the influence of phenolic compounds on the sensory scores attributed to extra virgin olive oil (EVOO) by panel test was investigated. Two model olive oils (MOOs) with identical concentrations of volatile compounds, differing only in the amount of biophenols (297 vs. 511 mg kg$^{-1}$), were analysed by two official panels and by SPME-GC/MS. Six other MOOs set up by the two previous models were also tested and analysed. They were formulated separately with the addition of three off-flavours ('rancid', 'winey–vinegary' and 'fusty–muddy'). While high levels of EVOO phenolic compounds did not produce any effect on the headspace concentration of volatile compounds, they did affect the scores of both positive and negative sensory attributes of EVOO, due to the well-known in-mouth interactions between EVOO phenols, saliva and volatile compounds. In particular, a decrease of about 39% in the positive fruity score was found in the presence of a higher concentration of phenols. Regarding EVOO off-flavours, the higher level of phenolic compounds decreased by about 23% the score of 'fusty–muddy' defect and increased the score of 'winey–vinegary' defect about 733%. No important effect of EVOO phenolics on the perceived intensity of the 'rancid' defect was found. These findings could be helpful in explaining some discrepancies of panel test responses observed during extra virgin olive oil shelf life.

**Keywords:** extra virgin olive oil; sensory analysis; phenolic compounds; virgin olive oil off-flavours; panel test; volatile compounds; SPME-GC/MS

---

## 1. Introduction

The quality of extra virgin olive oil (EVOO) is defined by several chemical indices (acidity, peroxide value, UV and alkyl esters), but the sensory perception of its flavour is the ultimate determinant [1–3]. Sensory assessment identifies mainly positive attributes and defects in the oil, and it is critical for the oil's quality classification according to the International Olive Council [4] and the EU legislation [1]. In fact, to be classified as "extra virgin" (highest quality), olive oil must have the presence of 'fruity' notes in its flavour and the absence of any unpleasant sensations, also defined as defects or off-flavours. The four main frequent off-flavours of virgin olive oil are 'musty', 'winey', 'fusty–muddy' and 'rancid'.

The positive attributes 'bitterness' and 'pungency' in-mouth are also evaluated in EVOO by sensory assessors and, although they are not considered important in the quality classification of olive oil, they are very desirable [3,5,6].

The bitterness and pungency are mainly related to the quali-quantitative presence of phenolic compounds in EVOO [7,8], characterised by different health functions [6]. Olive oil phenolic compounds comprise simple phenols, lignans ((+)-1-acetoxypinoresinol and (+)-1-pinoresinol), flavonoids (luteolin and apigenin) and hydroxyl-isochromans, but the most abundant are the secoiridoid derivatives of the glycosides oleuropein and ligstroside. Among these, secoiridoid derivatives of oleuropein such as the dialdehydic form of elenoic acid linked to hydroxytyrosol (3,4-DHPEA-EDA or 'oleacein',) and the aldehydic form of elenoic acid linked to hydroxytyrosol (3,4-DHPEA-EA or 'oleuropeine aglycon') have been reported to be as the main contributors to EVOO bitterness. In contrast, the dialdehydic form of elenoic acid linked to tyrosol (or 'oleocanthal', p-HPEA-EDA) has been reported to be as the main compound responsible for the pungency of EVOO, perceived typically at the back of the tongue [7,8]. Commonly, four categories of EVOO can be pointed to in relation to the phenolic content. A level of phenolics equal to or lower than 220 mg kg$^{-1}$ corresponds to non-bitter oils or oils with almost imperceptible bitterness; slight bitterness corresponds to 220–340 mg kg$^{-1}$; bitter oils have phenol contents ranging from 340 to 410 mg kg$^{-1}$; and a phenol content higher than 410 mg kg$^{-1}$ corresponds to quite bitter or very bitter oils [9]. The concentration range of phenolic compounds is very wide in EVOOs because these substances are mainly affected by the agronomic and technological conditions of EVOO production such as cultivar, ripening stage, geographic origin of olives, crushing, malaxation, etc. [9].

It has been reported that EVOO phenolic compound–aroma interactions can affect the release of EVOO aroma compounds in the presence of human saliva [10,11]. It has been hypothesised that the complex formed from the interaction between EVOO phenolics and proline-rich proteins could bind aroma compounds and consequently decrease their level during head-space analysis and during organoleptic assessment of olive oil. In particular, an in vivo study showed that 1-penten-3-one, *trans*-2-hexenal and esters had a lower release in the presence of higher levels of biophenols (about 600 mg kg$^{-1}$) after swallowing 3.5 mL olive oil. In contrast, linalool and 1-hexanol had a longer persistence in the breath than other compounds [10]. Another in vitro study of EVOO with a low–medium level of phenolic compounds (about 300 mg kg$^{-1}$) showed the lowest headspace release for some ethyl esters, acetates, alcohols and ketones [11]. Consequently, the sensory assessment of EVOO could be affected by the presence and content of phenolic compounds. This effect could influence the score given by panellists for EVOOs e.g., slight 'fusty–muddy' notes, mainly related to some ester compounds [12], would not be perceived in very bitter–pungent EVOOs because of a physicochemical trapping effect by phenolics and saliva on esters [10,11].

Therefore, the wide-ranging level of 'bitter' and 'pungent' taste in EVOOs makes the use of sensory assessment to classify EVOOs into categories a tool not exempt from risk.

Although the volatile and phenolic compounds in EVOO have been widely studied [7,9,13–15], to date, no study has aimed to verify the effect of biophenols on the sensory assessment of olive oil flavour and off-flavour.

Therefore, the aim of this work was to investigate the effect of EVOO phenolic compounds on the sensory scores of positive and negative attributes assessed by a panel test. For this purpose, two model olive oils (MOOs) with identical concentrations of volatile compounds, differing only in the amount of biophenols, were used. This allowed us to study the positive sensory attributes ('fruity' quality) in oils differing in phenolic compound content only, without major differences in EVOO flavour composition. Six other MOOs, created by the two previous models, were formulated with the addition of the 'rancid', 'winey–vinegary' and 'fusty–muddy' off-flavour olive oil references (supplied by IOOC) in order to verify the possible masking or salting out effects of EVOO biophenols on the perception of common defects. The sensory assessment of MOOs was made by two different official panels according to the EU legislation [1].

## 2. Results and Discussion

### 2.1. Quality Indices and Phenolic Compounds

Table 1 reports the free acidity, peroxide value, ultraviolet indices ($K_{232}$, $K_{270}$, $\Delta K$) and phenolic compounds in the MOO samples. The free acidity, peroxide value, $K_{232}$, $K_{270}$ and $\Delta K$ of the two samples remained within the legal limits of the category of extra virgin olive oil [1], showing no difference between the two MOOs. Total phenolic compounds, determined by colorimetric measurement using Folin–Ciocalteau reagent, were 297.5 and 510.8 mg kg$^{-1}$, respectively, in the MOO control sample with lowest concentration of phenolic compounds (MOO+P) and in the MOO sample with highest level of phenolic compounds (MOO++P). The addition of the phenolic extract increased the level of total phenolic compounds of about 71%. This increase varied for each phenolic compound e.g., the phenolic compound that increased the least was hydroxytyrosol (26%); in contrast, the ligstroside *p*-HPEA-EA showed the greatest increase (77%). These findings confirmed that the extractability of the phenolic compounds is linked to the different hydrophobicities of these compounds [16]. However, the average level of 3,4-DHPEA-EDA and 3,4-DHPEA-EA, responsible for EVOO bitterness and the level of p-HPEA-EDA responsible for pungency feeling [7,8], increased in the same way by about 60% and 62%, respectively.

**Table 1.** Legal quality indices and phenolic compounds of MOO samples at two different level of EVOO phenolic compounds.

|  | MOO+P | MOO++P | Legal Limits |
|---|---|---|---|
| Quality indices |  |  |  |
| Acidity | 0.38 ± 0.05a | 0.37 ± 0.02a | ≤0.80 |
| Peroxide value | 6.4 ± 0.1a | 6.3 ± 0.1a | ≤20 |
| K$_{232}$ | 1.821 ± 0.050a | 1.837 ± 0.012a | ≤2.50 |
| K$_{270}$ | 0.120 ± 0.003a | 0.128 ± 0.003a | ≤0.22 |
| ΔK | 0.004 ± 0.001a | 0.004 ± 0.001a | ≤0.01 |
| Phenolic compound |  |  |  |
| Hydroxytyrosol | 10.0 ± 0.4a | 12.6 ± 0.2b | - |
| Tyrosol | 7.4 ± 0.1a | 11.6 ± 0.6b | - |
| 3,4-DHPEA-EDA | 48.3 ± 2.1a | 72.1 ± 0.8b | - |
| *p*-HPEA-EDA | 44.7 ± 2.0a | 72.2 ± 0.6b | - |
| Lignans | 29.2 ± 1.8a | 46.2 ± 0.5b | - |
| 3,4-DHPEA-EA | 40.2 ± 0.8a | 68.5 ± 0.3b | - |
| *p*-HPEA-EA | 13.1 ± 0.9a | 23.2 ± 0.0b | - |
| Total phenolics (HPLC) | 192.8 ± 4.3a | 306.4 ± 3.0b | - |
| Total phenolics (Folin–Ciocalteau) | 297.5 ± 8.7a | 510.8 ± 9.3b | - |

Acidity is expressed as oleic acid equivalent. Peroxide value is expressed as meq O$_2$ kg$^{-1}$ oil. Phenolic compounds obtained by HPLC analysis are expressed as mg tyrosol kg$^{-1}$ of oil. Total phenolics obtained by Folin–Ciocalteau essay are expressed as mg caffeic acid kg$^{-1}$ of oil. 3,4-DHPEA-EDA: dialdehydic form of elenoic acid linked to hydroxytyrosol; *p*-HPEA-EDA: dialdehydic form of elenoic acid linked to tyrosol; 3,4-DHPEA-EA: aldehydic form of elenoic acid linked to hydroxytyrosol; *p*-HPEA-EA: aldehydic form of elenoic acid linked to tyrosol; lignans: sum of pinoresinol and acetoxypinoresinol. Values are the average of three replicates of analysis. Values followed by different letters are significantly different ($p < 0.05$).

### 2.2. Volatile Compounds

Table 2 shows the headspace concentration of volatile compounds in the MOO samples. As expected, the two MOO samples without the added defects had a lower number of volatile compounds than MOOs with the added defects. The volatile compounds that characterised MOOs without the addition of defects were 1-penten-3-one, 3-pentanone, hexanal, 1-penten-3-ol, *trans*-2-hexenal, hexyl acetate, cis-3-hexen-1-ol acetate, 1-hexanol, *cis*-3-hexen-1-ol and *trans*-2-hexen-1-ol. These compounds are typically found in EVOO and are generated by endogenous olive enzymes in the LOX pathway, starting from linoleic and linolenic acids [17]. Other volatile compounds present in MOO samples without the addition of the defects were ethanol and ethyl acetate. The presence of ethanol is a residual due to the use of ethanol for the recovery from EVOO of the phenolic compounds subsequently added to MOO samples.

**Table 2.** Volatile compounds, expressed as mg kg$^{-1}$ ± standard deviation, in MOO samples without and with the addition of EVOO off-flavours at two different concentrations of EVOO phenolic compounds.

| Compound | MOO | | MOO with the Addition of VOO Off-Flavour | | | | | |
| --- | --- | --- | --- | --- | --- | --- | --- | --- |
| | | | Rancid | | Winey–Vinegary | | Fusty–Muddy | |
| | MOO+P | MOO++P | MOO+P | MOO++P | MOO+P | MOO++P | MOO+P | MOO++P |
| octane | nf A | nf A | 0.36 ± 0.02 aC | 0.30 ± 0.01 aC | 0.07 ± 0.02 aB | 0.09 ± 0.00 aB | 0.09 ± 0.00 aB | 0.08 ± 0.00 aB |
| ethyl acetate | 0.18 ± 0.07 aAB | 0.22 ± 0.00 aA | 0.25 ± 0.01 aB | 0.45 ± 0.05 bC | 0.12 ± 0.00 aAB | 0.33 ± 0.01 bAB | 0.10 ± 0.00 aA | 0.35 ± 0.01 bBC |
| ethanol | 30.08 ± 2.43 aA | 31.89 ± 1.17 aA | 32.6 ± 2.51 aA | 34.97 ± 0.13 aA | 43.31 ± 0.89 aB | 44.97 ± 0.69 aB | 34.28 ± 2.96 aAB | 31.43 ± 1.82 aA |
| ethyl propanoate | nf A | nf A | nf A | nf A | nf A | nf A | 0.01 ± 0.00 aB | 0.01 ± 0.00 aB |
| 3-pentanone | 0.32 ± 0.00 aA | 0.32 ± 0.03 aAB | 0.40 ± 0.01 aB | 0.36 ± 0.02 aB | 0.29 ± 0.00 aA | 0.28 ± 0.01 aA | 0.31 ± 0.02 aA | 0.29 ± 0.00 aAB |
| 1-penten-3-one | 0.24 ± 0.01 aA | 0.23 ± 0.01 aA | 0.26 ± 0.02 aA | 0.23 ± 0.02 aA | 0.28 ± 0.03 aA | 0.25 ± 0.00 aA | 0.24 ± 0.03 aA | 0.24 ± 0.00 aA |
| ethyl butanoate | nf A | nf A | nf A | nf A | nf A | nf A | 0.04 ± 0.00 aB | 0.04 ± 0.00 aB |
| hexanal | 2.23 ± 0.15 aA | 2.24 ± 0.06 aA | 3.12 ± 0.11 aB | 2.87 ± 0.23 aB | 2.27 ± 0.10 aA | 2.25 ± 0.12 aA | 2.38 ± 0.27 aA | 2.23 ± 0.04 aA |
| *trans*-2-pentenal | nf A | nf A | 0.09 ± 0.01 aB | 0.09 ± 0.00 aC | 0.11 ± 0.02 aB | 0.07 ± 0.01 aBC | 0.10 ± 0.01 aB | 0.07 ± 0.01 aB |
| 1-penten-3-ol | 0.25 ± 0.03 aA | 0.26 ± 0.03 aA | 0.21 ± 0.02 aA | 0.18 ± 0.01 aA | 0.25 ± 0.01 aA | 0.24 ± 0.01 aA | 0.23 ± 0.00 aA | 0.23 ± 0.03 aA |
| 2-heptanone | nf A | nf A | 0.38 ± 0.04 aC | 0.38 ± 0.00 aC | 0.11 ± 0.01 aB | 0.15 ± 0.00 aB | nf A | nf A |
| heptanal | nf A | nf A | 0.26 ± 0.01 aB | 0.21 ± 0.02 aB | nf A | nf A | nf A | nf A |
| 3-methyl-1-butanol | nf A | nf A | nf A | nf A | 0.08 ± 0.00 aB | 0.09 ± 0.01 aB | 0.07 ± 0.01 aB | 0.08 ± 0.00 aB |
| *trans*-2-hexenal | 29.03 ± 1.84 aA | 26.91 ± 2.17 aA | 32.4 ± 0.79 aA | 27.9 ± 1.98 aA | 28.55 ± 0.81 aA | 26.13 ± 3.63 aA | 31.22 ± 2.44 aA | 26.43 ± 1.50 aA |
| 1-butanol | nf A | nf A | 0.04 ± 0.00 aB | 0.04 ± 0.00 aB | nf A | nf A | nf A | nf A |
| hexyl acetate | 0.31 ± 0.02 aAB | 0.28 ± 0.03 aA | 0.34 ± 0.00 aB | 0.30 ± 0.02 aA | 0.26 ± 0.00 aA | 0.25 ± 0.03 aA | 0.31 ± 0.02 aAB | 0.26 ± 0.00 aA |
| 2-octanone | nf A | nf A | 0.10 ± 0.00 aB | 0.08 ± 0.01 aB | nf A | nf A | nf A | nf A |
| octanal | nf A | nf A | 0.84 ± 0.04 aC | 0.69 ± 0.05 aB | 0.05 ± 0.01 aBC | 0.05 ± 0.00 aA | 0.1 ± 0.01 aB | 0.11 ± 0.01 aA |
| *trans, trans*-2,6,10-dodecatrienal | nf A | nf A | nf A | nf A | nf A | nf A | 0.12 ± 0.02 aB | 0.13 ± 0.00 aB |
| *cis*-3-hexen-1-ol acetate | 1.89 ± 0.17 aA | 1.71 ± 0.20 aA | 2.03 ± 0.05 aA | 1.74 ± 0.18 aA | 1.70 ± 0.11 aA | 1.58 ± 0.25 aA | 1.97 ± 0.21 aA | 1.62 ± 0.07 aA |
| *trans*-2-heptenal | nf A | nf A | 0.20 ± 0.01 aB | 0.30 ± 0.03 aC | nf A | nf A | 0.2 ± 0.01 aB | 0.19 ± 0.01 aB |
| 1-hexanol | 2.58 ± 0.21 aA | 2.36 ± 0.15 aA | 2.89 ± 0.05 aA | 2.39 ± 0.16 aA | 2.66 ± 0.16 aA | 2.42 ± 0.44 aA | 2.84 ± 0.19 aA | 2.34 ± 0.13 aA |
| *cis*-3-hexen-1-ol | 3.04 ± 0.21 aA | 2.81 ± 0.27 aA | 3.39 ± 0.07 aA | 2.82 ± 0.24 aA | 3.20 ± 0.17 aA | 2.92 ± 0.44 aA | 3.37 ± 0.02 aA | 2.76 ± 0.21 aA |
| nonanal | nf A | nf A | 0.60 ± 0.03 aC | 0.52 ± 0.05 aC | 0.16 ± 0.01 aB | 0.19 ± 0.03 aB | 0.2 ± 0.02 aB | 0.17 ± 0.00 aB |
| *trans*-2-hexen-1-ol | 4.46 ± 0.32 aA | 4.12 ± 0.34 aA | 5.06 ± 0.16 aA | 4.16 ± 0.36 aA | 4.70 ± 0.26 aA | 4.24 ± 0.69 aA | 4.97 ± 0.31 aA | 4.06 ± 0.26 aA |
| decanal | nf A | nf A | 0.05 ± 0.00 aB | 0.03 ± 0.00 aB | nf A | nf A | nf A | nf A |
| *trans*-2-octenal | nf A | nf A | 0.06 ± 0.01 aB | 0.06 ± 0.00 aB | nf A | nf A | nf A | nf A |
| *trans, trans*-2,4-heptadienal | nf A | nf A | 0.08 ± 0.00 aC | 0.08 ± 0.01 aC | 0.06 ± 0.00 aB | 0.05 ± 0.00 aB | nf A | nf A |
| acetic acid | nf A | nf A | nf A | nf A | 1.22 ± 0.11 aB | 0.83 ± 0.09 aB | nf A | nf A |
| butanoic acid | nf A | nf A | nf A | nf A | nf A | nf A | 0.34 ± 0.01 aB | 0.33 ± 0.01 aB |
| pentanoic acid | nf A | nf A | 0.50 ± 0.02 aB | 0.37 ± 0.05 aB | nf A | nf A | nf A | nf A |
| hexanoic acid | nf A | nf A | 2.7 ± 0.11 aB | 2.3 ± 0.09 aC | 0.2 ± 0.07 aA | 0.22 ± 0.04 aB | 0.25 ± 0.07 aA | 0.23 ± 0.03 aB |

Values with different letters are significantly different ($p < 0.05$). Small letters indicate significant differences between MOO+P and MOO++P samples. Capital letters indicate significant differences among the samples without and with the addition of the defects. nf = Not found.

However, its sensory contribution was very modest since it was found at levels very close to its odour threshold of 30 mg kg$^{-1}$ [12]. Ethyl acetate was found in all the samples at concentration levels below its odour threshold (0.94 mg kg$^{-1}$ [12]). Therefore, it did not contribute to the aroma of the MOOs. Ethanol and ethyl acetate are generated through sugar fermentation by microorganisms found on olive fruits. When LOX plays a major role during EVOO production, the odour of these two volatile compounds is not expected to be defective. On the other hand, ethanol and ethyl acetate become defective when they are present at higher levels [17]. In all MOO samples, the volatile compounds derived from the lipoxygenase pathway had the same concentration level, except for hexanal and 3-pentanone, which showed elevated levels in MOO samples with the rancid defect. In fact, these two volatile compounds derive from the auto-oxidation of fatty acids and are typically found in rancid oils. As expected, the MOO samples with the addition of defects had a greater number of compounds that originated from excessive fermentations, amino acid conversion, mould activity or lipid oxidation phenomena than MOOs without defects [17]. Most of these off-flavours were also common among the defects [12]. It is important to mention that the aim of this work was to obtain slightly defective MOO samples, therefore, it is obvious that the number and the quantity of off-flavour volatile compounds were not the same as has been reported in literature for very defective olive oils [12].

In order to statistically explain the differences between defective MOOs in relation to the quality and quantity of off-flavour volatile compounds, a Principal Component Analysis (PCA) was also performed (Figure 1). For volatile compounds, those that did not present a significant difference among the MOO samples according to Tukey's test (i.e., volatile compounds generated in the LOX pathway) were not considered (Table 2). PCA explained the 91.43% of variance using the first two principal components (PCs). The first PC clearly discriminated rancid MOO samples from fusty and winey MOO samples, while the second PC differentiated fusty MOO samples from winey MOO samples.

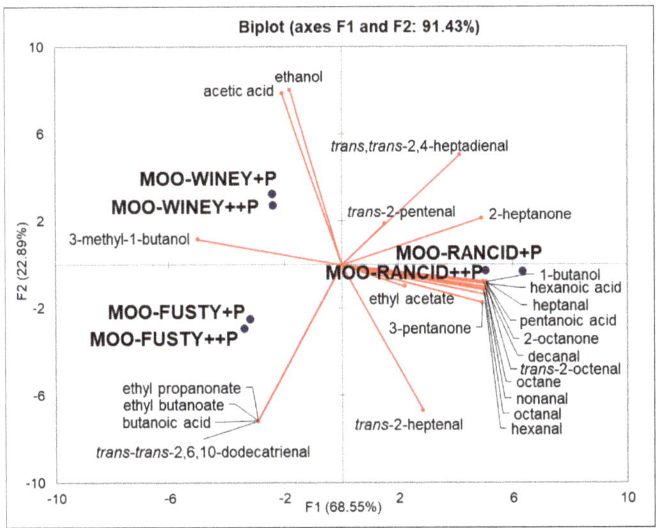


**Figure 1.** PCA based on SPME-GC/MS analysis of volatile compounds characterising MOO samples with the presence of sensory defects of 'rancid', 'winey–vinegary' and 'fusty–muddy'. MOO+P code indicates the lowest level of phenolic compounds in model olive oil while MOO++P indicates the highest level.

The volatile compounds predominant in 'rancid' MOOs, mainly produced by auto-oxidation of fatty acids, were hexanal, 3-pentanone, octane, *trans*-2-pentenal, 2-heptanone, heptanal, 1-butanol, 2-octanone, octanal, *trans*-2-heptenal, nonanal, decanal, *trans*-2-octenal, *trans,trans*-2-4-heptadienal, pentanoic acid and hexanoic acid. All these volatile compounds are usually linked to the rancidity of olive oil [15]. In fact, they are mainly characterised by rancid, oily and fatty sensory notes. Generally, aldehydes are the first compounds produced by the oxidation of unsaturated fatty acids, while acids appear at the end of the oxidative process due to the oxidation of the aldehydes previously formed [12].

The volatile compounds that characterised 'winey' and 'fusty' MOOs were acetic acid and ethanol for the first defect and *trans, trans*-2,6,10-dodecatrienal, ethyl propanoate, ethyl butanoate and butanoic acid for the second, while 3-methyl-1-butanol was significant for both defects. Both defects are due to microbial contamination of the olives. The winey defect is related to fermentation of *Lactobacillus* and acetic acid bacteria, while the fusty defect is linked to the Enterobacteriaceae genera *Aerobacter* and *Escherichia* at the beginning of olive storage, and the genera *Pseudomonas*, *Clostridium* and *Serratia* after extended olive storage [12].

It is evident that all volatile compounds analysed by SPME-GC/MS showed no differences between MOO+P and MOO++P samples (Table 2). These results were in agreement with previous studies [10,11] in which it was found that biophenolic compounds alone do not influence the headspace concentration of some $C_5$ and $C_6$ aroma compounds of EVOO. On the contrary, the studies demonstrated that phenolic compounds influenced the release of volatile compounds only in the presence of saliva. The first study was conducted in vitro with a retronasal aroma simulator (RAS) device, simulating mouth conditions with human saliva, while the second study was conducted in vivo by atmospheric pressure chemical ionisation–mass spectrometry (APCI/MS).

Therefore, it is correct to hypothesise that during olive oil tasting, i.e., when EVOO phenolic compounds enter into contact with saliva in the mouth, the sensory perception of EVOO is influenced by a different content of phenolic compounds. In order to verify this hypothesis, a sensorial analysis was performed to better understand whether and to what extent EVOO phenolic compounds affect the sensory perception of EVOO.

### 2.3. Sensory Analysis

Official sensory analysis includes a limited number of sensory attributes, namely four groups of defects and the positive attributes of olive 'fruitiness', 'bitterness' and 'pungency', of which only the first positive attribute classifies the category [4]. In addition, it should be highlighted that in the IOOC methodology panellists do not rate aroma and flavour attributes separately, but evaluate each perception as the whole olfactory–gustatory–tactile sensation. In fact, it is reported in the literature that the sensory description of EVOO is not impacted by the separate assessment of all sensations that form the flavour [18,19].

Figure 2 shows the median value of the scores of positive and negative attributes in MOO samples without and with the addition of defects. As expected, the increase of phenolic compounds from 298 to 511 mg kg$^{-1}$ (increase of 72%), previously verified by the Folin–Ciocalteu method (Table 1), determined an increase of the 'bitterness' and 'pungency' sensory attributes in all MOO++P samples. In particular, in the sample without the addition of defects (Figure 2a), the median values assigned by panels to MOO++P samples for these two sensory attributes were 3.5 and 5.3 for bitterness (for Panels 1 and 2, respectively) and 2.8 and 2.3 for pungency (for Panels 1 and 2, respectively). In contrast, the median values of MOO+P samples were 0.8 and 2 for 'bitterness' (for Panels 1 and 2, respectively) and 1.2 and 1 for 'pungency' sensory attributes (for Panels 1 and 2, respectively).

In MOO samples with the addition of defects, the 'bitterness' descriptor ranged from 2.7 to 5 in MOO++P samples and from 1 to 2.5 in the MOO+P samples. The 'pungency' descriptor showed a lower intensity than 'bitterness', and varied from 1.9 to 3.7 in MOO++P samples and from 1 to 2.4 in the MOO+P samples (Figure 2b–d).

In MOO samples without the addition of the defects, the highest level of biophenols (MOO++P) reduced the score of the sensory 'fruity' attribute by about 50% (1.9 vs. 4 and 1.5 vs. 2.8 for Panels 1 and 2, respectively) compared to MOO+P samples (Figure 2a), although the headspace concentration of the volatile compounds responsible for the 'fruity' note (1-penten-3-one, hexanal, *trans*-2-pentenal,

1-penten-3-ol, *trans*-2-hexenal, hexyl acetate, *cis*-3-hexen-1-ol acetate, 1-hexanol, *cis*-3-hexen-1-ol and *trans*-2-hexen-1-ol) showed no difference between MOO++P and MOO+P samples (Table 2).

These results were in agreement with the previous instrumental works on olive oil aroma [10,11].

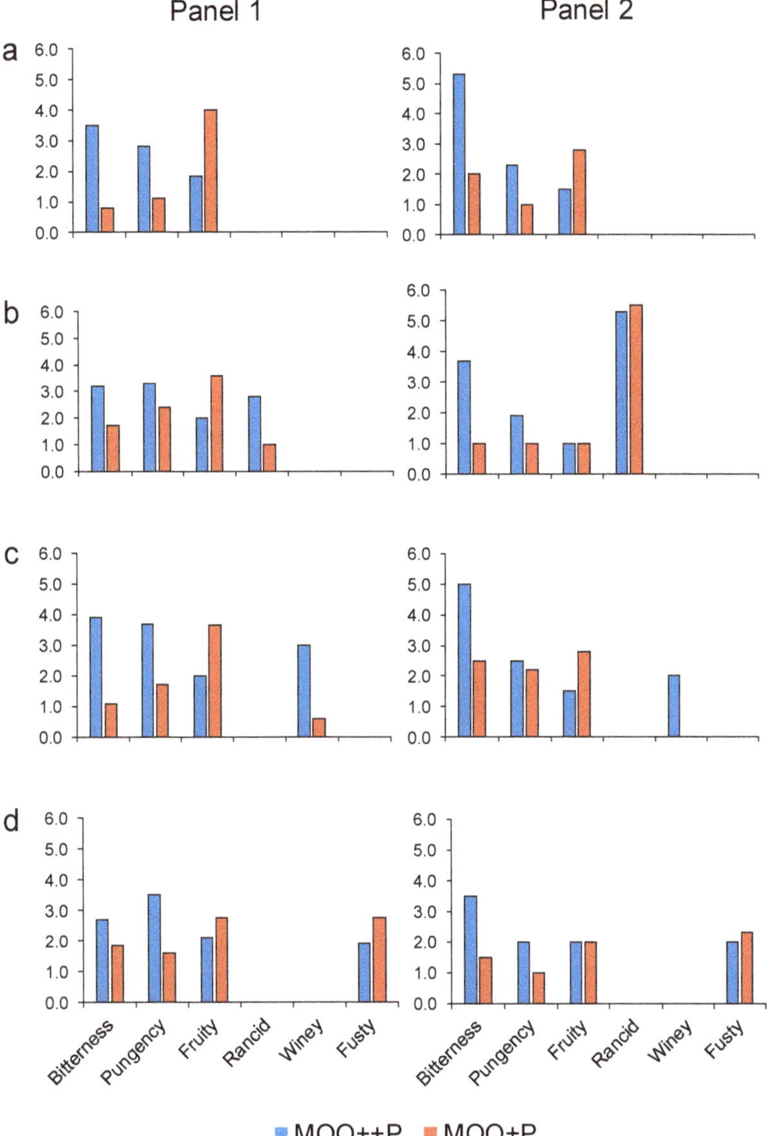


**Figure 2.** Sensory profiles of two panel tests obtained by analysing MOO+P in comparison to MOO++P samples without off-flavours (**a**) and with 'rancid' (**b**), 'winey–vinegary' (**c**) and 'fusty–muddy' (**d**) defects. MOO+P code indicates the lowest level of phenolic compounds in model olive oil while MOO++P is the highest level. Sensory attributes are expressed as median on an unstructured 0–10 scale.

The first work mentioned reported that EVOO phenolic compounds reduce the headspace concentrations of hexyl acetate, *cis*-3-hexen-1-ol acetate, 1-hexanol, *cis*-3-hexen-1-ol, *trans*-2-hexenal and 1-penten-3-one only in the presence of the saliva. The above compounds are responsible for the fruity descriptor as has been reported in literature for EVOO [13]. The second work studied the effect of phenolic compounds on the release of olive oil aroma compounds under in vivo conditions. Among eight monitored volatile compounds, 1-penten-3-one, *trans*-2-hexenal, *cis*-3-hexen-1-ol acetate, hexanal and 1-hexanol were included. The first three compounds had a lower release in the presence of higher levels of biophenols, while hexanol had a greater release. For hexanal, no important differences were found [10]. However, among the $C_5$ and $C_6$ volatile compounds that contribute to the fruity sensory note of EVOO, 1-penten-3-one and *trans*-2-hexenal are the main active odour volatile compounds. The former is important because of its low odour threshold, and the latter because of its usual high concentration in EVOO [20,21]. Therefore, it is possible to state that the score of positive 'fruity' descriptor was lower in MOO++P than MOO+P samples because the headspace concentration level mainly decreased for 1-penten-3-one and *trans*-2-hexenal in the presence of phenolic compounds and saliva. As a possible explanation, it has been reported that the formation of EVOO phenolic compound–proline-rich protein complexes could retain volatile compounds in the hydrophobic cavities and consequently decrease the concentration level in the headspace [10]. Saliva, increasing the stickiness of phenolic compounds to the oral surface, prolongs their retention even for long periods in the oral cavity despite a constant saliva flow [22], thus, easing the possible polyphenol–salivary proteins–aroma interactions. The effect of saliva–phenol interactions on aroma release has also been shown for red wines with both instrumental and sensory approaches [23–25]. In particular, sensory approaches showed how the intensities of 'fruity' and 'floral' aromas seemed to decrease when the level of polyphenols increased [26].

In the rancid MOO samples (Figure 2b), the scores assigned by panels to MOO++P samples for 'fruity' attribute were 2 and 1, respectively, for Panels 1 and 2. In MOO+P samples, the median values were 3.6 and 1, respectively, for Panels 1 and 2. The rancid descriptor showed for Panel 1 a higher value in the MOO++P sample (2.8) than the MOO+P sample (1), while Panel 2 showed higher intensity than Panel 1 but very few differences in the rancid attribute between the two MOO samples (5.3 vs. 5.5). Among the volatile compounds that impacted on 'rancidity' in MOO samples (hexanal, 3-pentanone, octane, *trans*-2-pentenal, 2-heptanone, heptanal, 1-butanol, 2-octanone, octanal, *trans*-2-heptenal, nonanal, decanal, *trans*-2-octenal, *trans,trans*-2-4-heptadienal, pentanoic acid and hexanoic acid) (Table 2, Figure 1), it has been reported in the literature that EVOO phenolics have no effect on hexanal and *trans*-2-pentenal [10,11]. Therefore, it is possible to hypothesise a similar behaviour to that of hexanal and *trans*-2-pentenal for the other aldehydes found in the 'rancid' MOO samples, such as heptanal, octanal, nonanal, *trans*-2-heptenal and *trans*-2-octenal.

In the 'winey–vinegary' MOO samples (Figure 2c), the higher level of phenolic compounds determined a lower intensity of the 'fruity' descriptor and a higher intensity of the 'winey–vinegary' defect. The median values assigned by panels to MOO++P samples for fruity attribute were 2 and 1.5, respectively, for Panels 1 and 2. In the MOO+P samples, the median values were 3.7 and 2.8, respectively, for Panels 1 and 2. The 'winey–vinegary' off-flavour was reported by Panel 1 to have a higher value in the MOO++P sample (3) than the MOO+P sample (0.6), while Panel 2 reported the 'winey–vinegary' attribute only in the MOO++P sample (2). The most involved volatile compounds for this defect are acetic acid and ethanol, followed by 3-methylbutanol (Table 2, Figure 1). Unfortunately, no published data were found regarding a possible influence of phenolic compounds on the aroma release of these compounds in the mouth or in in vitro systems in the presence of saliva.

Figure 2d shows the sensory results of MOO+P and MOO++P samples with the addition of the 'fusty–muddy' defect. In contrast to the previous defects, the fusty–muddy defect showed higher median values in MOO+P than MOO++P samples (2.8 vs. 1.9, Panel 1; 2.3 vs. 2.0 Panel 2). The most involved volatile compounds for this defect were ethyl propanoate, ethyl butanoate, butanoic acid, trans, trans-2,6,10-dodecatrienal and 3-methyl-1-butanol (Table 2, Figure 1). In the case of esters, it has been reported that phenolic compounds interacting with saliva are able to trap ethyl butyrate, ethyl isobutyrate and ethyl-2-methyl butyrate, reducing their in-mouth release [10,11]. Therefore, in the presence of a high level of phenolics, the impact of esters on the 'fusty–muddy' defect could decrease, making it less evident. Here again, no published data were found regarding the possible influence of olive oil phenolic compounds on the aroma release of *trans, trans*-2,6,10-dodecatrienal and 3-methyl-1-butanol.

The higher level of phenolic compounds also determined a lower intensity of the 'fruity' attribute descriptor, even if it was less than the other two off-flavours. In particular, the scores assigned by the panels to MOO++P samples for the 'fruity' attribute were 2.1 and 2, respectively, for Panels 1 and 2. In MOO+P samples, the median values were 2.8 and 2, respectively, for Panels 1 and 2.

Finally, comparing the average of all MOO++P with all MOO+P samples, it is possible to state that 511 mg kg$^{-1}$ of phenolic compounds in MOO++P samples increased the score of 'bitterness' and 'pungency' attributes by 144% and 87%, respectively, in comparison to the MOO+P samples, in which the phenolics were 297 mg kg$^{-1}$. In contrast, the high level of phenolic compounds decreased the score of fruity attribute of 39% (Figure 3).

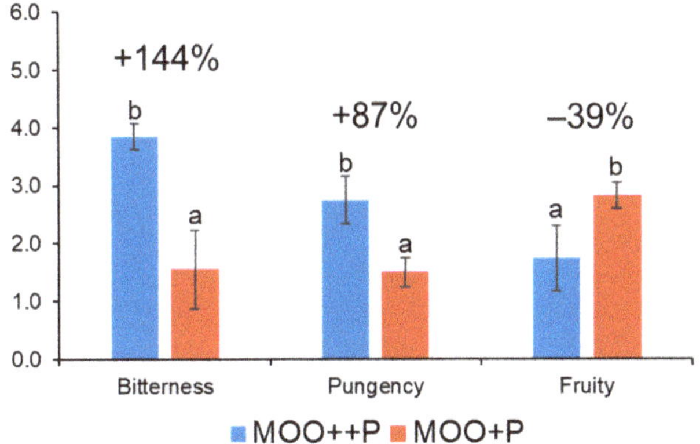


**Figure 3.** Sensory scores of 'bitterness', 'pungency' and 'fruity' attributes for all MOO samples obtained from the two panels. MOO+P code indicates the lowest level of phenolic compounds in model olive oil, while MOO++P is the highest level. Sensory attributes are expressed as the average of the median values obtained by using an unstructured 0–10 scale.

To statistically explain the differences between MOO++P and MOO+P samples in relation to positive and negative sensory attributes, a PCA was conducted on the sensory scores (Figure 4). The two main principal components explained 63.8% of the variance, with Principal Component 1 (F1) discriminating MOO++P and MOO+P samples. All MOO+P samples were characterised by 'fruity' and 'fusty' attributes. In contrast, the attributes 'bitterness', 'pungency' and 'winey' characterised all MOO++P samples. The 'rancid' off-flavour had little to no influence by the polyphenol content in the MOO. Therefore, a high level of phenolic compounds in EVOO could mask a 'fusty–muddy' off-flavour and enhance a 'winey–vinegary' off-flavour.

In conclusion, the sensory findings discussed here confirmed the hypothesis (one previously formulated in other studies, conducted by in vivo and in vitro instrumental approaches) that phenolic compounds, affecting the release of EVOO's aroma compounds during its consumption, can influence the scores of perceived sensory attributes.

These results could be helpful in explaining some discrepancies of panel test responses commonly observed during extra virgin olive oil shelf life. In particular, in a 1 year old extra virgin olive oil, the in-mouth appearance of a 'fusty–muddy' defect could be due to the biophenol natural decrease over time (autoxidation). Therefore, this effect may change over time the original sensory classification made by the panel test from "extra virgin" to "virgin" olive oil.

Consequently, a high biophenol content (over 500–600 mg/kg) in bottled oils, apart from the possible use of EFSA nutritional claims on the label to inform consumers [5], could also ensure a sensory profile stability of bottled EVOO, avoiding the risk of slight sensory defects appearing during storage and preserving the EVOO legal classification.

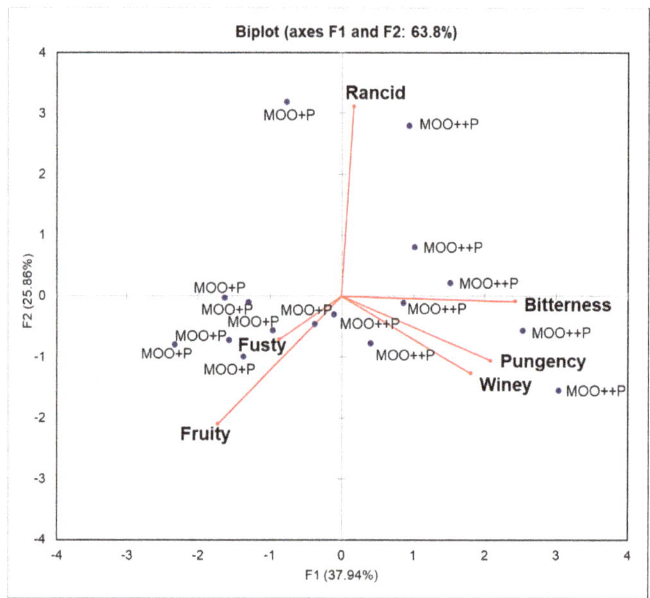


**Figure 4.** PCA based on sensory scores of the positive attributes of 'fruity', 'bitterness' and 'pungency' and the negative attributes of 'rancid', 'winey–vinegary' and 'fusty–muddy'. MOO+P code indicates the lowest level of phenolic compounds in model olive oil, while MOO++P is the highest level. The sensory attributes were assessed by two different panels.

## 3. Materials and Methods

### 3.1. Samples, Standards and Reagents

EVOO, named "Teti" and obtained from Rotondella (60%), Carpellese (30%) and Frantoio (10%) cultivars, was supplied by Torretta srl (Battipaglia, Salerno, Italy). For the production of Teti EVOO, the olives were harvested between 20 October and 30 November. Olives were washed and crushed using a disk crusher. The olive paste was then centrifuged using a three-phase low-volume water decanter. EVOO was placed into green glass bottles (500 mL) and stored in the dark at 19 °C to prevent oxidation until the moment of the chemical analysis, which was carried out at the fourth month of storage.

EVOO samples were characterised by the following quality indices: acidity 0.39 (±0.05), PV 6.40 (±0.1), $K_{232}$ 1.775 (±0.018), $K_{270}$ 0.121 (±0.001) and $\Delta K$ 0.005 (±0.001). The median of the fruity attribute (MF) was 5.2, while the median of each defect (MD) was equal to 0.

The composition of the biophenols was as follows: hydroxytyrosol 9.1 ± 0.1 mg kg$^{-1}$, tyrosol 7.4 ± 0.2 mg kg$^{-1}$, dialdehydic form of elenoic acid linked to hydroxytyrosol (3,4-DHPEA-EDA) 50.9 ± 0.6 mg kg$^{-1}$, dialdehydic form of elenoic acid linked to tyrosol (*p*-HPEA-EDA) 47.9 ± 0.5 mg kg$^{-1}$, pinoresinol and acetoxypinoresol (lignans) 30.1 ± 1.1 mg kg$^{-1}$, aldehydic form of elenoic acid linked to hydroxytyrosol (3,4-DHPEA-EA) 45.7 ± 1.0 mg kg$^{-1}$, aldehydic form of elenoic acid linked to tyrosol (*p*-HPEA-EA) 14.7 ± 0.4 mg kg$^{-1}$, total phenolic compounds by HPLC 205.6 ± 1.0 mg kg$^{-1}$ and by Folin–Ciocalteau method 348.4 ± 8.2 mg kg$^{-1}$.

The refined olive oil (ROO) was supplied by Dorella Oleificio Candela srl (Castellamare di Stabia, Napoli, Italy). The International Olive Oil Council (IOOC, Madrid, Spain) supplied samples for each of the three defected reference oils, usually used in the process of training assessors to detect sensory defects. Each standard oil was characterised by one of the following off-flavours with its median of the defect (MD): 'rancid' (MD = 9.6), 'winey–vinegary' (MD = 5.6) and 'fusty–muddy' (MD = 6.9).

Hexanal (97%), *cis*-3-hexenylacetate (98%), ethyl acetate (99%), *trans*-2-hexenal (95%), 1-hexanol (98%), 1-penten-3-one (95%), octane (99.7%), 3-pentanone (99%), *trans*-2-pentenal (95%), 1-penten-3-ol (99%), 2-heptanone (99%), heptanal (95%), 3-methyl-1-butanol (99%), hexyl acetate (99%), 2-octanone (99.9%), octanal (99%), *cis*-3-hexen-1-ol (98%), nonanal (95%), *trans*-2-hexen-1-ol (96%), decanal (99%), *trans*-2-octenal (94%) *trans*, *trans*-2,4-heptadienal (97%) and hexanoic acid (99%) were supplied by Sigma–Aldrich (St. Louis, MO, USA). Acetic acid (99%) and *trans*-2-heptenal 98% were supplied by Fluka (Buchs, Switzerland). The reagents glacial acetic acid, diethyl ether and distilled water were used for the analysis and were supplied by Romil (Cambridge, UK). HPLC-grade methanol (>99.9% purity), hexane (>95%), Folin–Ciocalteu reagent, anhydrous sodium carbonate (>99.5%), caffeic acid (97%), ammonium acetate (0.4%) and sodium hydroxide were bought from Sigma–Aldrich (St. Louis, MO, USA). Potassium iodide was provided by AppliChem (Darmstadt, Germany). Phenolphthalein and starch were provided by Titolchimica s.p.a. (Rovigo, Italy). Food-grade ethyl alcohol (96%) was supplied by Selex S.p.A. (Trezzano sul Naviglio, Milano, Italy).

### 3.2. Sample Preparation

To study the effect of phenolic compounds on the sensory scores of the panel test, two MOOs were set up with identical volatile compound concentrations but with different concentrations of polyphenols. Teti EVOO was used as the control sample with the lowest concentration of phenolic compounds (encoded MOO+P). The MOO with the highest level of phenolic compounds (encoded MOO++P) was obtained by adding to EVOO Teti ROO phenolic compounds extracted from the same Teti EVOO in order to obtain a MOO sample with about a double concentration of phenolic compounds compared to MOO+P. A ROO without phenolics was also added to the control sample (MOO+P), which was then subjected to the same protocol as MOO++P. In order to understand the possible effect of phenolic compounds on EVOO off-flavours, six other MOOs were prepared by adding a separate aliquot of each off-flavour IOOC olive oil reference ('rancid', 'winey–vinegary' and 'fusty–muddy') to the previous MOOs (MOO+P and MOO++P). Finally, a total of four MOO samples with the lowest concentration of phenolic compounds (MOO+P) were built, one without off-flavour and the remaining three with one off-flavour each: one 'rancid', another one with the off-flavour 'winey–vinegary' and the last one with off-flavour 'fusty–muddy'. The other four MOOs were built at highest concentration of phenolic compounds (MOO++P) without and with the same three off-flavour additions, for a total of eight samples. MOOs were stored before sensory and SPME-GC/MS analyses in room conditions (19 °C), avoiding light exposure and high temperatures in order to prevent oxidation, and were used within one month of their preparation.

### 3.2.1. Preparation of MOO Samples with the Addition of Virgin Olive Oil Phenolic Compounds

MOOs were prepared according to Genovese et al. [10,11]. The phenolic extract was obtained from Teti EVOO. An aliquot of the oil sample (50 g) was dissolved in hexane (100 mL). A subsequent extraction was carried out using a water/methanol mixture (40/60 v/v) in a separating funnel (500 mL), after having shaken it vigorously for 15 min in a 500 mL bottle. This step was repeated twice using a total of 140 mL solvent. Subsequently, the hydro-alcoholic extract that was obtained was washed with hexane to remove any oil contamination and was centrifuged for 5 min at 3500 rpm (ALC International srl, PK-120, Milan, Italy). The organic phase was removed from the sample, and the hydro-alcoholic phase was collected in the flask and evaporated under a vacuum in a rotary evaporator at 35 °C (Heidolph VV 2000, Germany). The phenolic compounds were suspended using 10 mL ethyl alcohol (food-grade). A total of 1.6 kg of Teti EVOO was used to extract phenolics. This phase was repeated several times in order to obtain a total of 280 mL of biophenol extract in ethyl alcohol. The phenolic extract was subsequently concentrated up to a final volume of 40 mL using a rotary evaporator at 35 °C (Heidolph, VV 2000). The 40 mL phenolic extract was added to a flask with 170 g of ROO. The oil mixture was stirred and treated in an ultrasonic bath for 5 min. Ethanol was then evaporated in a vacuum evaporator (Heidolph VV 200) at 35 °C.

ROO with phenolic compounds was diluted in 1680 g of Teti EVOO in order to obtain MOO++P with a level of total phenolic compounds of $511.0 \pm 9.3$ mg kg$^{-1}$ in order to reproduce a very bitter oil. Indeed, Beltrán et al. [9] reported that above mentioned phenolic compounds level is to be considered the level of a very bitter EVOO. This level of phenolic compounds was chosen because in in vivo experiments it was shown to affect the release of aromatic compounds of EVOO [10]. In the control sample (MOO+P), phenolic extract was not added, but the sample was subjected to the same protocol previously described for the addition of the phenolic compounds. Therefore, 40 mL ethanol food grade was added to a flask with 170 g of ROO. The mixture was then stirred and treated in an ultrasonic bath, followed by the evaporation of ethanol in a vacuum evaporator. Finally, ROO was diluted in 1680 g of Teti EVOO in order to obtain a MOO+P sample with a level of total phenolic compounds of $297.5 \pm 8.7$ mg kg$^{-1}$. This level has been reported as the typical level found in bitter oil [9].

### 3.2.2. Preparation of MOO+P and MOO++P Samples with the Addition of the Defected Reference Oils

The amount of 'standard defect oil' to be added was established by preliminary sensory tests in laboratory. The aim was to achieve a slight defect in each MOO. In order to build MOO samples with the 'rancid', 'winey–vinegary' and 'fusty–muddy' off-flavours, 1.71 g, 1.64 g and 0.74 g of 'rancid', 'winey–vinegary' and 'fusty–muddy' reference oils, respectively, were weighed and diluted in 410 g of each of the two MOO samples (MOO+P and MOO++P).

### 3.3. EVOO and MOO Analysis

### 3.3.1. Free Acidity, Peroxide Value and Specific Ultraviolet Absorbance K232 and K270

The Teti EVOO and MOO samples were analysed to determine their acidity levels, peroxide value (PV), $K_{232}$, $K_{270}$ and $\Delta K$, according to the EU legislation [1]. Acidity was expressed as oleic acid percentage (%); PV was expressed as meq $O_2$ kg$^{-1}$ oil. For the analysis of spectrophotometric indices, an ultraviolet–visible UV-1601 spectrophotometer (Shimadzu, Kyoto, Japan) was used. All the analyses were performed in triplicate.

### 3.3.2. Extraction and Analysis of Phenolic Compounds

The extraction and analysis of phenolic compounds were carried out according to Sacchi et al. [27]. The quantification of phenolic compounds was carried out using the Folin–Ciocalteau and HPLC methods. The phenolic compounds were analysed in both Teti EVOO and MOO samples to confirm the quantity and composition of the added phenolics in the MOOs. The analyses were performed in triplicate for each extraction.

### 3.3.3. Extraction and Analysis of Volatile Compounds

Dynamic headspace (DH) SPME-GC/MS was used for the analysis of volatile compounds in Teti EVOO and MOO samples according to Romero et al., [28]. The SPME device (Supelco Co., Bellefonte, CA, USA) was equipped with a 50/30 µm thickness divinylbenzene/carboxen/polydimethylsiloxane (DVB/CAR/PDMS) fibre coated with a 1 cm length stationary phase. A 2 g aliquot of EVOO was added to a 20 mL vial with 10 µL of isobutyl acetate (Sigma–Aldrich, St. Louis, USA; 560.1 mg kg$^{-1}$ in refined olive oil), which was used as the internal standard (IS). The vial was then closed with a polytetrafluoroethylene (PTFE) septum. The fibre was exposed for 40 min at 40 °C after 10 min at 40 °C for equilibration [21]. Volatile compounds were analysed by GC/MS according to Morales et al. [12]. A Shimadzu QP5050A GC/MS instrument (Kyoto, Japan) was used with a Supelcowax–10 capillary column (60 m, 0.32 mm i.d., 0.5 µm thickness; Supelco Co., Bellefonte, PA, USA). Thermal desorption of volatile compounds was carried out by putting the SPME fibre in the injector for 10 min. The temperature was set at 40 °C for 6 min, followed by an increase of 2 °C min$^{-1}$ up to 200 °C, and was then held for 10 min. The injector was kept at 230 °C. Helium was used as a carrier gas (0.9 mL min$^{-1}$). The peak areas were calculated using Lab solutions acquisition system (GCMS solutions version 1.20; Shimadzu) and were normalised with respect to the area of the internal standard peak. Compound identification was performed by comparing retention times and mass spectra obtained by analysing pure reference compounds in the same conditions. The identification was further confirmed by comparing mass spectra with those of the National Institute of Standards and Technology (NIST) database. Mass spectra were recorded at 70 eV. The source temperature was 200 °C and the interface temperature was 230 °C. Before use, the fibre was conditioned at 270 °C for 1 h for the analysis. A blank test was carried out before every run to prevent the release of undesirable compounds. All analyses were performed in triplicate.

### 3.3.4. Sensory Analysis

EVOO assessment was performed by two different panels according to the method described in the European Commission Regulation and further amendments [1], one at the Laboratorio Chimico Merceologico of CCIAA (Camera di Commercio Industria Artigianato e Agricoltura) of Naples (Italy) and the other at the CCIAA of Salerno (Italy), each of them made up of ten trained panellists. They scored the descriptors on a normalised unstructured sheet (from 0 to 10) according to the IOOC/UE official method. Panel tests were made in 'blind' conditions, and only after they had been performed the identities of the MOO samples were revealed to panel leaders.

### 3.4. Statistical Analysis of Data

Significant differences among the different model systems were determined for each compound by one-way ANOVA statistical analysis. Tukey's test was used to discriminate among the means of the variables. Differences with $p < 0.05$ were considered significant. Principal component analysis (PCA) was chosen as an exploratory technique to investigate the separation among groups of observations. Data elaboration was carried out using XLSTAT (version 2014.5.03), an add-in software package for Microsoft Excel (Addinsoft Corp., Paris, France).

**Author Contributions:** Conceptualization, A.G.; methodology, A.G.; formal analysis, A.G.; investigation, F.M. and A.P.; resources, R.S.; data curation, A.G.; writing—original draft preparation, A.G.; writing—review and editing, R.S.; visualization, A.G.; supervision, R.S.; project administration, A.G. and R.S.; funding acquisition, R.S. All authors have read and agreed to the published version of the manuscript.

**Funding:** This research was funded by AGER 2 Project, grant n. 2016-0174, COMPETITIVE (Claims of olive oil to improve the market value of the product).

**Acknowledgments:** We are grateful to official sensory panels of CCIAAs of Naples and Salerno for performing sensory tests, and to panel leaders Maria Luisa Ambrosino and Giovanni Pipolo for their collaboration and fruitful discussions. Marco Genovese is also acknowledged for proofreading the English manuscript.

**Conflicts of Interest:** The authors declare no conflict of interest.

## References

1. The European Commission. *Commission Regulation (EEC) No. 2568/91 of 11 July 1991 on the Characteristics of Olive Oil and Olive Residue Oil and on the Relevant Methods of Analysis*; The European Commission: Brussels, Belgium, 1991; pp. 1–83.
2. Tena, N.; Wang, S.C.; Aparicio-Ruiz, R.; Garcia-Gonzalez, D.L.; Aparicio, R. In-depth assessment of analytical methods for olive oil purity, safety, and quality characterization. *J. Agr. Food Chem.* **2015**, *63*, 4509–4526. [CrossRef]
3. De Santis, D.; Frangipane, M.T. Sensory perceptions of virgin olive oil: New panel evaluation method and the chemical compounds responsible. *Nat. Sci.* **2015**, *7*, 132–142. [CrossRef]

4. International Olive Council. *Sensory Analysis of Olive Oil Method for the Organoleptic Assessment of Virgin Olive Oil*; International Olive Council: Madrid, Spain, 2015.

5. EFSA. Scientific Opinion on the substantiation of health claims related to polyphenols in olive and protection of LDL particles from oxidative damage (ID ,1696, 2865), "maintenance of normal blood HDL-cholesterol concentrations" (ID 1639), "maintenance of normal blood pressure" (ID 3781), "anti-inflammatory" (ID 1882), "contributes to the upper respiratory tract health" (ID 3468), "can help to maintain a function of gastrointestinal tract" (3779), and "contributes to body defences against external agents" (ID 3467) pursuant to Article 13 (1) of Regulation (EC) No 1924/2006. *EFSA J.* **2011**, *9*, 2033–2058.

6. Vitaglione, P.; Savarese, M.; Paduano, A.; Scalfi, L.; Fogliano, V.; Sacchi, R. Healthy virgin olive oil: A matter of bitterness. *Crit. Rev. Food Sci.* **2015**, *55*, 1808–1818. [CrossRef]

7. Servili, M.; Esposto, S.; Fabiani, R.; Urbani, S.; Taticchi, A.; Mariucci, F.; Montedoro, G.F. Phenolic compounds in olive oil: Antioxidant, health and organoleptic activities according to their chemical structure. *Inflammopharmacology* **2009**, *17*, 76–84. [CrossRef]

8. Taticchi, A.; Esposto, S.; Servili, M. The basis of the sensory properties of virgin olive oil. In *Olive Oil Sensory Science*; Monteleone, E., Langstaff, S., Eds.; John Wiley & Sons: New York, NY, USA, 2014; pp. 33–54.

9. Beltrán, G.; Ruano, M.T.; Jiménez, A.; Uceda, M.; Aguilera, M.P. Evaluation of virgin olive oil bitterness by total phenol content analysis. *Eur. J. Lipid Sci. Tech.* **2007**, *108*, 193–197. [CrossRef]

10. Genovese, A.; Yang, N.; Linforth, R.; Sacchi, R.; Fisk, I. The role of phenolic compounds on olive oil aroma release. *Food Res. Int.* **2018**, *112*, 319–327. [CrossRef]

11. Genovese, A.; Caporaso, N.; Villani, V.; Paduano, A.; Sacchi, R. Olive oil phenolic compounds affect the release of aroma compounds. *Food Chem.* **2015**, *181*, 284–294. [CrossRef]

12. Morales, M.T.; Luna, G.; Aparicio, R. Comparative study of virgin olive oil sensory defects. *Food Chem.* **2005**, *91*, 293–301. [CrossRef]

13. Angerosa, F.; Servili, M.; Selvaggini, R.; Taticchi, A.; Esposto, S.; Montedoro, G. Volatile compounds in virgin olive oil: Occurrence and their relationship with the quality. *J. Chromatogr. A* **2004**, *1054*, 17–31. [CrossRef]

14. García-González, D.L.; Aparicio, R. Research in olive oil: Challenges for the near future. *J. Agr. Food Chem.* **2010**, *58*, 12569–12577. [CrossRef]

15. Aparicio, R.; Morales, M.T.; García-González, D.L. Towards new analyses of aroma and volatiles to understand sensory perception of olive oil. *Eur. J. Lipid Sci. Tech.* **2012**, *114*, 1114–1125. [CrossRef]

16. Paiva-Martins, F.; Gordon, M.H.; Gameiro, P. Activity and location of olive oil phenolic antioxidants in liposomes. *Chem. Phys. Lipids*, **2003**, *124*, 23–36. [CrossRef]

17. Angerosa, F. Influence of volatile compounds on virgin olive oil quality evaluated by analytical approaches and sensor panels. *Eur. J. Lipid Sci. Tech.* **2002**, *104*, 639–660. [CrossRef]

18. Angerosa, F.; Campestre, C. Sensory Quality: Methodologies and Applications. In *Handbook of Olive Oil-Analysis and Properties*, 2nd ed.; Aparicio, R., Harwood, J., Eds.; Springer: Massachusetts, MA, USA, 2013; pp. 523–560.

19. Langstaff, S. Sensory Quality Control. In *Olive Oil Sensory Science*; Monteleone, E., Langstaff, S., Eds.; John Wiley & Sons: New York, NY, USA, 2014; pp. 81–108.

20. Reiners, J.; Grosch, W. Odorants of virgin olive oils with different flavor profiles. *J. Agr. Food Chem.* **1998**, *46*, 2754–2763. [CrossRef]

21. Genovese, A.; Caporaso, N.; Leone, T.; Paduano, A.; Mena, C.; Perez-Jimenez, M.A.; Sacchi, R. Use of odorant series for extra virgin olive oil aroma characterisation. *J. Sci. Food Agric.* **2019**, *99*, 1215–1224. [CrossRef]

22. Ginsburg, I.; Koren, E.; Shalish, M.; Kanner, J.; Kohen, R. Saliva increases the availability of lipophilic polyphenols as antioxidants and enhances their retention in the oral cavity. *Arch. Oral Biol.* **2012**, *57*, 1327–1334. [CrossRef]

23. Muñoz-González, C.; Feron, G.; Guichard, E.; Rodríguez-Bencomo, J.J.; Martin-Alvarez, P.J.; Moreno-Arribas, M.V.; Pozo-Bayón, M.A. Understanding the role of saliva in aroma release from wine by using static and dynamic headspace conditions. *J. Agr. Food Chem.* **2014**, *62*, 8274–8288. [CrossRef]

24. Esteban-Fernández, A.; Muñoz-González, C.; Jiménez-Girón, A.; Pérez-Jiménez, M.; Pozo-Bayón, M.Á. Aroma release in the oral cavity after wine intake is influenced by wine matrix composition. *Food Chem.* **2018**, *243*, 125–133. [CrossRef]

25. Perez-Jiménez, M.; Chaya, C.; Pozo-Bayón, M.Á. Individual differences and effect of phenolic compounds in the immediate and prolonged in-mouth aroma release and retronasal aroma intensity during wine tasting. *Food Chem.* **2019**, *285*, 147–155. [CrossRef]

26. Goldner, M.C.; di Leo Lira, P.; van Baren, C.; Bandoni, A. Influence of polyphenol levels on the perception of aroma in Vitis vinifera cv. Malbec wine. *S. Afr. J. Enol. Vitic.* **2016**, *32*, 21–27. [CrossRef]

27. Sacchi, R.; Caporaso, N.; Paduano, A.; Genovese, A. Industrial-scale filtration affects volatile compounds in extra virgin olive oil cv. Ravece. *Eur. J. Lipid Sci. Tech.* **2015**, *117*, 2007–2014. [CrossRef]

28. Romero, I.; García-González, D.L.; Aparicio-Ruiz, R.; Morales, M.T. Validation of SPME–GCMS method for the analysis of virgin olive oil volatiles responsible for sensory defects. *Talanta* **2015**, *134*, 394–401. [CrossRef] [PubMed]

**Sample Availability:** Samples are not available from the authors.

 *molecules*

Article

# From Extra Virgin Olive Oil to Refined Products: Intensity and Balance Shifts of the Volatile Compounds versus Odor

Jing Yan [1,2], Martin Alewijn [2] and Saskia M. van Ruth [1,2,*]

[1] Food Quality and Design Group, Wageningen University and Research, 6700AA Wageningen, The Netherlands; jing.yan@wur.nl

[2] Wageningen Food Safety Research, Wageningen University and Research, 6700AE Wageningen, The Netherlands; martin.alewijn@wur.nl

* Correspondence: saskia.vanruth@wur.nl; Tel.: +31-317-480250

Academic Editor: Eugenio Aprea

Received: 29 April 2020; Accepted: 24 May 2020; Published: 26 May 2020

**Abstract:** To explore relationships between the volatile organic compounds (VOCs) of different grades of olive oils (OOs) (extra virgin olive oil (EVOO), refined olive oil (ROO), and pomace olive oil (POO)) and odor quality, VOCs were measured in the headspace of the oils by proton transfer reaction quadrupole ion guide time-of-flight mass spectrometry. The concentrations of most VOCs differed significantly between the grades (EVOO > ROO > POO), whereas the abundance of $m/z$ 47.012 (formic acid), $m/z$ 49.016 (fragments), $m/z$ 49.027 (fragments), and $m/z$ 115.111 (heptanal/heptanone) increased in that order. Although the refined oils had considerably lower VOC abundance, the extent of the decline varied with the VOCs. This results in differences in VOCs proportions. The high VOC abundance in the EVOO headspace in comparison to ROO and POO results in a richer and more complex odor. The identified C5–C6 compounds are expected to contribute mainly to the green odor notes, while the identified C1–C4 and C7–C15 are mainly responsible for odor defects of OOs. Current results reveal that processing strongly affects both the quantitative and relative abundance of the VOCs and, therefore, the odor quality of the various grades of OOs.

**Keywords:** extra virgin olive oil; odor quality; processing grades; quantitation; VOCs proportion

---

## 1. Introduction

Olive oils (OOs) are very popular with customers due to their pleasant flavor and odor, as well as their health benefits. The concentrations of the volatile compounds (odor profile) in OOs are affected by many factors, including the cultivar [1], environmental factors [2], olive fruit maturity [3], the technical processing [4], as well as storage of the fruit (long time storage may be responsible for odor defects) [5], or storage of OOs (oxidative degradation) [6]. Therefore, OOs come with great variation in odor quality. Among those factors, processing methods (e.g., cold pressing and refining steps) cause dramatic effects on the concentrations of the volatile organic compounds (VOCs) in different grades of OOs [7], such as extra virgin olive oil (EVOO), refined olive oil (ROO), and pomace olive oil (POO).

The VOC molecules comprising 5 and 6 carbon atoms (called C5 and C6 compounds), which are mainly produced through enzymatic reactions leading to degradation of polyunsaturated fatty acids during processing [8], are considered the most important VOCs for the green odor notes of EVOO [9,10]. These compounds are more likely to be formed under cold-pressed conditions. Some oils are subjected to refining and most of C5 and C6, as well as many other VOCs are removed during the deodorization step in the refining process [11,12]. VOCs of OOs have been studied before [13–15], but so far, the differences in headspace concentrations of the VOCs of different OO grades has not yet been studied in depth. Therefore, it may be interesting to understand the relationship between the variation of the VOCs concentrations and the processing methods, as well as the relationship between the variation of the VOCs concentrations and the odor quality of the oils.

Proton transfer reaction time-of-flight mass spectrometry (PTR-ToF-MS) is based on soft chemical ionization by proton transfer from hydronium ions [16,17]. It is free of complex sample pre-processing and can provide VOCs fingerprints within seconds. Thus, it has been considered as an alternative method for the real-time and rapid analysis of OOs. PTR-ToF-MS has been used to study for example olive fruits [18], coffee [19], honey [20], peppers [17], ham [21], milk [16,22], and chocolate [23]. Although PTR-ToF-MS has been carried out for analyzing the VOCs of OOs [7,24,25] too, the influence of the OO grades on VOCs has not been compared extensively so far. Taiti and Marone [7] investigated the capability of PTR-ToF-MS in grading OOs, but the effect of the processing on the quantitative and relative abundance of the VOCs was not considered. Furthermore, PTR-ToF-MS coupled with a quadrupole ion guide (PTR-QiToF-MS), which has an improved transmission efficiency of ions and thus an increased sensitivity of the ion detection, has not been applied to measure the VOCs in OOs of different processing grades.

The present work is designed to elucidate the quantitative and relative differences of the VOCs in different grades of OOs (EVOO, ROO, and POO) by PTR-QiToF-MS and to explore the odor quality of the corresponding oils.

## 2. Results and Discussion

### 2.1. PTR-QiToF-MS Spectral Profile

Two hundred OOs were subjected to PTR-QiToF-MS analysis, and 295 mass peaks in the range of *m/z* 18.033 to *m/z* 207.204 were obtained for each sample. The $^{10}$log transformed mass spectra for three grades, i.e., EVOO (*n* = 140), ROO (*n* = 45), and POO (*n* = 15), are presented in Figure 1. For each average spectral profile in Figure 1, the summed observed concentration of the mass peaks ranging from *m/z* 18.033 to *m/z* 207.204 in the headspace of the samples were calculated. The value for EVOO (782 parts per million by volume (ppmv)) was about 3 times higher than the summed value for ROO (276 ppmv) and 5 times higher than that value for POO (127 ppmv). This indicates that EVOO is richer in VOCs than the other lower grades of OOs. Furthermore, the concentration of most mass peaks measured in the headspace of the EVOO samples is considerably higher than the other two grades. It further supports the idea that large amounts of the VOCs are removed during the refining process [11,26].

### 2.2. Concentration Differences of the VOCs

To explore the differences between EVOO and the lower grades of OOs, the 295 mass peaks were subjected to non-parametric Kruskal–Wallis tests ($p < 0.05$). Subsequently, 291 out of 295 mass peaks were present in significant differences across the three types of OO. Fifty-four mass peaks were tentatively identified based on their accurate molar masses and likely chemical formulas [16], as well as based on information from literature [9]. The headspace concentrations, odor characteristics, and odor thresholds (OTs) of the 54 tentatively identified VOCs of the three types of OO (EVOO, ROO, and POO) are listed in Table 1. The carbon numbers of the 54 tentatively identified VOCs varied from 1 to 15 (C1–C15). Furthermore, 20 out of 54 mass peaks were tentatively identified as several possible isomeric compounds.

**Table 1.** Tentatively identified volatile organic compounds of extra virgin olive oil (EVOO, n = 140), refined olive oil (ROO, n = 45), and pomace olive oil (POO, n = 15), their odor notes, odor thresholds (OT, ppbv) in air, average headspace concentrations (ppbv), standard deviation (SD), and statistical comparisons (Kruskal–Wallis tests and Mann–Whitney U-tests, $p < 0.05$). ppbv: parts per billion by volume.

| Measured Protonated Mass $m/z$ | Protonated Chemical Formula | Tentative Identification | Type | Reference | Odor Notes | Reference | OT | Average ± SD (ppbv) | | |
|---|---|---|---|---|---|---|---|---|---|---|
| | | | | | | | | EVOO | ROO | POO |
| **C1–C4 [A]** | | | | | | | | | | |
| 33.033 | $CH_5O^+$ | Methanol | Alcohols | [27,28] | Sweet | [29] | $100 \times 10^3$ | $(348 \pm 98) \times 10^{3\,a}$ | $(106 \pm 93) \times 10^{3\,b}$ | $(18 \pm 37) \times 10^{3\,b}$ |
| 43.018 | $C_2H_3O^+$ | Esters | Esters | [30] | Ester, green, pungent, sweet, fruity | [30] | | $(44 \pm 40) \times 10^3$ | $(16 \pm 18) \times 10^{3\,b}$ | $(6 \pm 4) \times 10^{3\,b}$ |
| 45.034 | $C_2H_5O^+$ | Acetaldehyde | Aldehydes | [9] | Pungent, sweet | [9] | 210 | $(64 \pm 39) \times 10^{3\,a}$ | $(41 \pm 28) \times 10^{3\,b}$ | $(39 \pm 34) \times 10^{3\,ab}$ |
| 47.012 | $CH_3O_2^+$ | Formic acid | Carboxylic acids | [31,32] | Pungent, penetrating | [33] | $28 \times 10^3$ | $(3 \pm 3) \times 10^{3\,b}$ | $(12 \pm 11) \times 10^{3\,a}$ | $(17 \pm 16) \times 10^{3\,a}$ |
| 47.049 | $C_2H_7O^+$ | Ethanol | Alcohols | [1,9] | Apple, sweet, alcohol | [1,9] | $10 \times 10^3$ | $(45 \pm 55) \times 10^{3\,a}$ | $(18 \pm 25) \times 10^{3\,b}$ | $(4 \pm 8) \times 10^{3\,b}$ |
| 57.033 | $C_3H_5O^+$ | 2-Propenal | Aldehydes | [32] | Unpleasant odor, irritating | [34,35] | 210 | $(39 \pm 33) \times 10^{3\,a}$ | $(11 \pm 12) \times 10^{3\,b}$ | $(4 \pm 5) \times 10^{3\,b}$ |
| 59.049 | $C_3H_7O^+$ | Propanal | Aldehydes | [28,36] | Pungent, sweet | [36] | 419 | $(16 \pm 12) \times 10^{3\,a}$ | $(12 \pm 9) \times 10^{3\,b}$ | $(10 \pm 7) \times 10^{3\,b}$ |
| | | Acetone | Ketones | [28] | Sweet, pungent | [29] | $100 \times 10^3$ | | | |
| 61.028 | $C_2H_5O_2^+$ | Acetic acid | Carboxylic acids | [9,37] | Sour, vinegary | [9] | 162 | $(33 \pm 33) \times 10^{3\,a}$ | $(20 \pm 24) \times 10^{3\,ab}$ | $(7 \pm 5) \times 10^{3\,b}$ |
| 63.026 | $C_2H_7S^+$ | Dimethyl sulfide | Others | [30,38] | Wet earth, organic, beetroot, sulfury | [30,38] | 1 | $(1 \pm 2) \times 10^{3\,a}$ | $45 \pm 54^{\,b}$ | $16 \pm 11^{\,b}$ |
| 73.064 | $C_4H_9O^+$ | Butan-2-one | Ketones | [9] | Ethereal, fruity | [9] | $10 \times 10^3$ | $(4 \pm 3) \times 10^{3\,a}$ | $(3 \pm 2) \times 10^{3\,a}$ | $(3 \pm 2) \times 10^{3\,b}$ |
| 75.044 | $C_3H_7O_2^+$ | Propanoic acid | Carboxylic acids | [9,32,39] | Pungent, sour, mold | [9,39] | 33 | $(22 \pm 21) \times 10^{3\,a}$ | $(3 \pm 3) \times 10^{3\,b}$ | $391 \pm 471^{\,b}$ |
| | | Methyl acetate | Esters | [1] | Ethereal, sweet | [1] | $561 \times 10^3$ | | | |
| 79.021 | $C_2H_7OS^+$ | Dimethyl sulfoxide | Others | [37,40] | Unpleasant | [37] | 1 | $711 \pm 810^{\,a}$ | $197 \pm 183^{\,b}$ | $68 \pm 51^{\,b}$ |
| 89.059 | $C_4H_9O_2^+$ | Ethyl acetate | Esters | [1,9] | Sticky, sweet, ethereal | [1,9] | $1 \times 10^3$ | $(4 \pm 4) \times 10^{3\,a}$ | $(1 \pm 1) \times 10^{3\,b}$ | $143 \pm 236^{\,b}$ |
| | | Butanoic acid | Carboxylic acids | [9,39] | Rancid, cheese | [9,39] | 1 | | | |
| **C5** | | | | | | | | | | |
| 85.064 | $C_5H_9O^+$ | trans-2-Pentenal; trans-2-methyl-2-butenal | Aldehydes | [1,9] | Green, apple, bitter almond; green fruity, aromatic | [1,9] | $437^{\,B}$ | $(7 \pm 6) \times 10^{3\,a}$ | $284 \pm 204^{\,b}$ | $105 \pm 109^{\,b}$ |
| | | 1-Penten-3-one | Ketones | [1,9,37] | Green, pungent, mustard | [1,9] | | | | |
| 87.081 | $C_5H_{11}O^+$ | 3-Penten-2-ol; 2-methyl-3-buten-2-ol | Alcohols | [1,9] | Perfumery, woody; grassy, earth, oily | [1,9] | | $(13 \pm 8) \times 10^{3\,a}$ | $(1 \pm 1) \times 10^{3\,b}$ | $319 \pm 208^{\,b}$ |
| | | 2-Methylbutanal; 3-methylbutanal; pentanal | Aldehydes | [9,37] | Malty; malty; woody, bitter, oily | [9] | $11^{\,C}$ | | | |
| | | 3-Pentanone | Ketones | [30,40] | Sweet, green | [30] | | | | |
| 103.075 | $C_5H_{11}O_2^+$ | Ethyl propionate | Esters | [1,37] | Strawberry, apple, fruity, sweet | [1] | $2 \times 10^3$ | $208 \pm 98^{\,a}$ | $69 \pm 77^{\,b}$ | $91 \pm 252^{\,b}$ |
| | | 3-Methylbutanoic acid; pentanoic acid | Carboxylic acids | [9,32,37,39] | Sweaty; unpleasant, pungent | [9,39] | 19; 9 | | | |

**Table 1.** *Cont.*

| Measured Protonated Mass *m/z* | Protonated Chemical Formula | Tentative Identification | Type | Reference | Odor Notes | Reference | OT | Average ± SD (ppbv) | | |
|---|---|---|---|---|---|---|---|---|---|---|
| | | | | | | | | EVOO | ROO | POO |
| **C6** | | | | | | | | | | |
| 81.070 | $C_6H_9{}^+$ | Terpene fragment or fragments from cis-/trans-hexenal | Others | [41,42] | - | - | | $(12 \pm 12) \times 10^3$ [a] | $337 \pm 280$ [b] | $74 \pm 68$ [b] |
| 83.085 | $C_6H_{11}{}^+$ | Terpene fragment | Others | [41] | - | - | | $(8 \pm 3) \times 10^3$ [a] | $(1 \pm 1) \times 10^3$ [b] | $(1 \pm 1) \times 10^3$ [b] |
| 85.101 | $C_6H_{13}{}^+$ | Cyclohexane | Others | [43] | Sweet, aromatic | [33] | $728 \times 10^3$ | $(3 \pm 2) \times 10^3$ [a] | $259 \pm 220$ [b] | $110 \pm 89$ [b] |
| 97.064 | $C_6H_9O^+$ | 2,4-Hexadienal | Aldehydes | [1] | Fresh, green, floral, citric | [1] | | $277 \pm 159$ [a] | $25 \pm 14$ [b] | $20 \pm 17$ [b] |
| | | Ethyl furan | Others | [1,43] | Sweet, ethereal | [1,43] | | | | |
| 99.081 | $C_6H_{11}O^+$ | trans-2-Hexenal; cis-3-hexenal | Aldehydes | [1,9,37] | Green, apple, bitter almonds, astringent; green, leaf-like | [1,9] | 1 [D] | $(13 \pm 11) \times 10^3$ [a] | $243 \pm 240$ [b] | $66 \pm 125$ [b] |
| 101.095 | $C_6H_{13}O^+$ | cis-3-Hexen-1-ol; trans-3-hexen-1-ol; cis-2-hexen-1-ol; trans-2-hexen-1-ol | Alcohols | [1,9,10,37] | Green; green grassy, sweet; almond, grassy, astringent; green grassy, leaves, fruity, astringent, bitter | [1,9] | | $154 \pm 78$ [a] | $57 \pm 35$ [b] | $35 \pm 31$ [b] |
| | | Hexanal; 3-methyl pentanal | Aldehydes | [1,9,37] | Green-sweet, green-apple, grassy; | [9] | 0.27 [E] | | | |
| | | 4-Methylpentan-2-one | Ketones | [1,43] | Strawberry, fruity, sweet, ethereal | [1,43] | 470 | | | |
| 113.059 | $C_6H_9O_2{}^+$ | Sorbic acid | Carboxylic acids | [18] | - | - | | $171 \pm 308$ [a] | $10 \pm 7$ [b] | $5 \pm 4$ [b] |
| | | 5-Ethyl-2-(5H)-furanone | Ketones | [40] | - | - | | | | |
| 117.091 | $C_6H_{13}O_2{}^+$ | Butyl acetate; ethyl butyrate; ethyl isobutyrate | Esters | [1,9,37] | Green, fruity, pungent; sweet, fruity, cheesy; fruity | [1,9] | 29; 8; 5 | $657 \pm 488$ [a] | $66 \pm 69$ [b] | $110 \pm 358$ [b] |
| | | Hexanoic acid | Carboxylic acids | [9,39] | Pungent, rancid, sour, sharp | [9,39] | 127 | | | |
| **C7–C15** | | | | | | | | | | |
| 93.070 | $C_7H_9{}^+$ | Toluene | Others | [30,37,40] | Gasoline vapors | [30,37] | $2 \times 10^3$ | $254 \pm 223$ [a] | $33 \pm 51$ [b] | $7 \pm 12$ [b] |
| 105.090 | $C_8H_9{}^+$ | Ethenyl benzene | Others | [1,30] | Gasoline vapors | [30] | 47 | $324 \pm 247$ [a] | $48 \pm 49$ [b] | $18 \pm 13$ [b] |
| 107.086 | $C_8H_{11}{}^+$ | Ethyl benzene | Others | [1] | Strong | [1] | $39 \times 10^3$ | $338 \pm 275$ [a] | $75 \pm 131$ [b] | $17 \pm 27$ [b] |
| 109.101 | $C_8H_{13}{}^+$ | Methyl norbornene | Others | [18] | - | - | | $50 \pm 24$ [a] | $29 \pm 24$ [b] | $33 \pm 20$ [ab] |
| 111.080 | $C_7H_{11}O^+$ | 2,4-Heptadienal | Aldehydes | [9,39] | Fatty, rancid, nutty | [9,39] | 8 | $86 \pm 62$ [a] | $38 \pm 32$ [b] | $27 \pm 34$ [b] |
| 113.096 | $C_7H_{13}O^+$ | trans-2-Heptenal | Aldehydes | [9] | Oxidized, tallowy, pungent | [9] | 19 | $131 \pm 125$ [a] | $30 \pm 22$ [b] | $26 \pm 30$ [b] |
| 115.111 | $C_7H_{15}O^+$ | Heptanal | Aldehydes | [9,10,39] | Oily, fatty, woody, rancid | [9,39] | 0.19 | $51 \pm 31$ [b] | $81 \pm 65$ [a] | $77 \pm 63$ [ab] |
| | | Heptan-2-one | Ketones | [1,9] | Sweet, fruity, cinnamon | [1,9] | 1 | | | |
| 121.099 | $C_9H_{13}{}^+$ | 1,2,4-Trimethylbenzene | Others | [1,10] | - | - | $25 \times 10^3$ | $78 \pm 59$ [a] | $7 \pm 13$ [b] | $2 \pm 5$ [b] |
| | $C_8H_9O^+$ | Acetophenone | Ketones | [40] | - | - | | | | |
| 123.080 | $C_8H_{11}O^+$ | 2-Phenylethanol | Alcohols | [10,40] | - | - | | $4 \pm 2$ [a] | $1 \pm 2$ [b] | $1 \pm 1$ [b] |

**Table 1.** *Cont.*

| Measured Protonated Mass *m/z* | Protonated Chemical Formula | Tentative Identification | Type | Reference | Odor Notes | Reference | OT | Average ± SD (ppbv) | | |
|---|---|---|---|---|---|---|---|---|---|---|
| | | | | | | | | EVOO | ROO | POO |
| 125.096 | $C_8H_{13}O^+$ | cis-1,5-Octadien-3-one; octan-2-one | Ketones | [1,9,37] | Geranium-like; mold, overripe | [1,9] | | 34 ± 26 [a] | 13 ± 11 [b] | 5 ± 5 [b] |
| | | trans,trans-2,4-Octadienal | Aldehydes | [10] | - | - | | | | |
| 127.111 | $C_8H_{15}O^+$ | trans-2-Octenal | Aldehydes | [1,9] | Herbaceous, spicy | [1,9] | 1 | 37 ± 17 [a] | 17 ± 12 [b] | 19 ± 13 [b] |
| | | 6-Methyl-5-hepten-2-one; 1-octen-3-one; 3-octen-2-one | Ketones | [1,9,10,37] | Pungent, green fruity, grassy; mushroom, mold, pungent; rose | [1,9,44] | 1 [F] | | | |
| 129.091 | $C_7H_{13}O_2^+$ | 3-Methyl-2-butenyl acetate | Esters | [1] | Pungent | [1] | | 16 ± 8 [a] | 5 ± 4 [b] | 5 ± 5 [b] |
| 129.127 | $C_8H_{17}O^+$ | Octanal | Aldehydes | [9,37,39] | Fatty, sharp, citrus-like, rancid | [9,39] | 0.08 | 27 ± 15 | 41 ± 36 | 59 ± 81 |
| | | 6-Methyl-5-hepten-3-ol; 1-octen-3-ol | Alcohols | [9,39] | Perfumery, nutty, perfumery, nutty; mold, earthy | [9,39] | 1 [G] | | | |
| | | Octan-2-one | Ketones | [9] | Mold, green | [9] | 27 | | | |
| 131.106 | $C_7H_{15}O_2^+$ | Propyl butyrate; ethyl 2-methylbutyrate; ethyl 3-methylbutyrate | Esters | [1,9,37] | Pineapple, sharp; fruity; fruity, green, banana | [1,9] | $2 \times 10^3$ [H] | 72 ± 52 [a] | 21 ± 28 [b] | 6 ± 6 [b] |
| | | Heptanoic acid | Carboxylic acids | [9,39] | Rancid, fatty | [9,39] | | | | |
| 137.132 | $C_{10}H_{17}^+$ | Terpene fragments (α-pinene; β-pinene; limonene; tricyclene; camphene; sabinene; myrcene; β-ocimene) | Others | [10,32] | - | - | $3 \times 10^3$; $6 \times 10^3$; 130 [I] | 89 ± 68 [a] | 37 ± 70 [b] | 12 ± 18 [b] |
| 139.112 | $C_9H_{15}O^+$ | trans,trans-2,4-Nonadienal | Aldehydes | [9,37] | Soapy, penetrating, deep-fried | [9] | 0.04 | 19 ± 12 [a] | 3 ± 1 [b] | 5 ± 3 [b] |
| 139.147 | $C_{10}H_{19}^+$ | 3-Ethyl-1,5-octadiene | Others | [40] | - | - | | 82 ± 50 [a] | 6 ± 5 [b] | 4 ± 3 [b] |
| 141.127 | $C_9H_{17}O^+$ | tran-2-Nonenal; cis-2-nonenal | Aldehydes | [9,36,37] | Paper-like, fatty; green, fatty | [9,36] | 0.02 [J] | 8 ± 8 [a] | 3 ± 3 [b] | 5 ± 4 [a] |
| 143.107 | $C_8H_{15}O_2^+$ | cis-3-Hexenyl acetate; trans-3-hexenyl acetate; 3-methyl-4-penten-1-ol-acetate | Esters | [1,9,10,37] | Green, banana-like; green, banana, green leaves, fruity | [1,9] | | 49 ± 25 [a] | 8 ± 7 [b] | 3 ± 3 [b] |
| 143.142 | $C_9H_{19}O^+$ | Nonanal | Aldehydes | [9,10,37] | Fatty, waxy, pungent | [9] | 0.45 | 44 ± 23 [a] | 17 ± 11 [b] | 14 ± 12 [b] |
| | | Nonan-2-one | Ketones | [1] | Fruity, floral | [1] | 5 | | | |
| 145.122 | $C_8H_{17}O_2^+$ | Hexyl acetate; 2-methylpropyl butanoate; ethyl hexanoate | Esters | [1,9,10] | Green, fruity, sweet, apple; unpleasant, winey, fusty | [1,9] | 307 [K] | 58 ± 36 [a] | 8 ± 9 [b] | 4 ± 3 [b] |
| | | Octanoic acid | Carboxylic acids | [9,39] | Oily, fatty | [9,39] | | | | |
| 153.125 | $C_{10}H_{17}O^+$ | 2,4-Decadienal; trans, trans-2,4-decadienal; trans, cis-2,4-decadienal | Aldehydes | [9,37] | Strong, fatty; deep-fried; deep-fried | [9] | 0.37 [L] | 8 ± 3 [a] | 3 ± 3 [b] | 3 ± 3 [b] |
| | | 2,3-Dehydro-1,8-cineole | Others | [10] | - | - | | | | |

**Table 1.** *Cont.*

| Measured Protonated Mass *m/z* | Protonated Chemical Formula | Tentative Identification | Type | Reference | Odor Notes | Reference | OT | Average ± SD (ppbv) | | |
|---|---|---|---|---|---|---|---|---|---|---|
| | | | | | | | | EVOO | ROO | POO |
| 155.141 | $C_{10}H_{19}O^+$ | trans-2-Decenal | Aldehydes | [9] | Painty, fishy, fatty | [9] | 0.43 | 6 ± 3 [a] | 3 ± 3 [b] | 2 ± 4 [b] |
| | | 3,7-Dimethylocta-1,6-dien-3-ol; cis-p-menth-2-en-1-ol | Alcohols | [10] | - | - | | | | |
| 157.124 | $C_9H_{17}O_2^+$ | Ethyl cyclohexanecarboxylate | Esters | [9,37] | Aromatic, fruity | [9] | | 11 ± 6 [a] | 1 ± 1 [b] | 1 ± 1 [b] |
| 157.158 | $C_{10}H_{21}O^+$ | Decanal | Aldehydes | [9,10,43] | Penetrating, sweet, waxy, painty | [9,43] | 0.41 | 5 ± 3 [a] | 3 ± 3 [b] | 4 ± 6 [ab] |
| 169.123 | $C_{10}H_{17}O_2^+$ | trans-4,5-Epoxy-trans-2-decenal | Aldehydes | [9,10,37] | Metallic | [9] | | 2 ± 2 [a] | 0.5 ± 0.4 [b] | 0.3 ± 0.3 [b] |
| 171.174 | $C_{11}H_{23}O^+$ | Undecanal | Aldehydes | [43] | Fatty, tallowy | [43] | | 3 ± 1 [a] | 0.4 ± 0.4 [b] | 0.3 ± 0.4 [b] |
| 173.154 | $C_{10}H_{21}O_2^+$ | Ethyl octanoate; methyl nonanoate | Esters | [32,36,37] | Green, fruity; - | [36] | | 11 ± 24 [a] | 1 ± 1 [b] | 0.8 ± 0.9 [b] |
| 183.082 | $C_{13}H_{11}O^+$ | Benzophenone | Others | [40] | - | - | | 2 ± 2 [a] | 0.8 ± 0.4 [b] | 0.6 ± 0.3 [b] |
| 205.194 | $C_{15}H_{25}^+$ | β-Caryophyllene; copaene; β-selinene; α-farnesene; eremophilene | Others | [10,32,40] | - | - | | 23 ± 22 [a] | 4 ± 7 [b] | 0.4 ± 0.7 [b] |

Superscript letters a and b in a row indicate significant differences ($p < 0.05$). [A] C1–C7 indicate that the VOCs with 1–7 carbon atoms in the molecule, respectively. [B] The OT of trans-2-pentenal. [C] The OT of pentanal. [D] The OT of trans-2-hexenal. [E] The OT of hexanal. [F] The OT of 3-octen-2-one. [G] The OT of 1-octen-3-ol. [H] The OT of propyl butyrate. [I] 2981, 5465, 119 refer to the OTs of α-pinene, β-pinene, limonene, respectively. [J] The OT of tran-2-nonenal. [K] The OT of hexyl acetate. [L] 2,4-Decadienal. Green indicates that the odor activity value (OAV, the average concentration of the volatile compound of the oils divided by its OT) is more than 2. Yellow indicates that the OAV is between 1 and 2. Red indicates that the OAV is less than 1. Rows without color indicate that no OT was found for this compound.

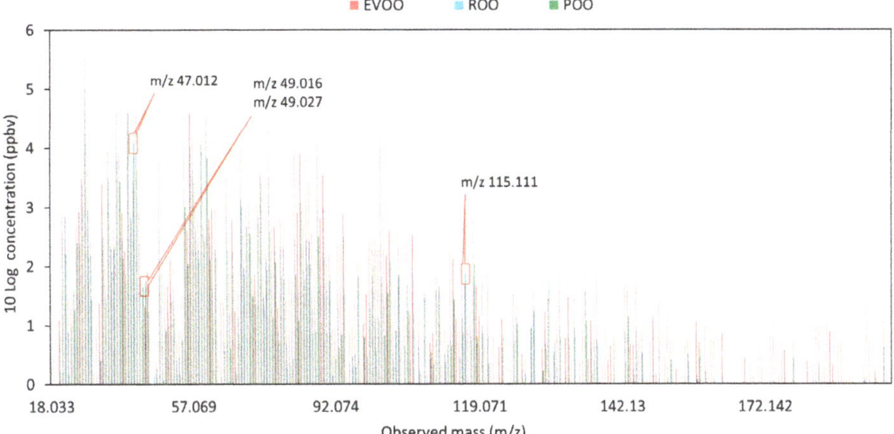



**Figure 1.** [10]Log averaged spectral profiles of volatile organic compounds of three olive oil grades (extra virgin olive oil, EVOO; refined olive oil, ROO; pomace olive oil, POO). The four highlighted mass peaks (*m/z* 47.012, *m/z* 49.016, *m/z* 49.027, and *m/z* 115.111) were present in significantly higher abundance in the headspace of ROO and POO compared to EVOO.

### 2.2.1. VOCs with Higher Concentrations in the EVOO Headspace

Among the 295 mass peaks, 287 mass peaks were present in significantly higher concentrations in the headspace of the EVOO samples than for the other OOs. It confirms that most of the VOCs are removed during the refining process [11]. Similarly, for 51 out of the 54 tentatively identified VOCs, significantly higher concentrations were observed for EVOO than for the other oils (Table 1).

Except for the mass peak *m/z* 47.012 (formic acid), significantly higher headspace concentrations were determined for all identified C1–C4 compounds in EVOO compared to ROO/POO. Mass peak *m/z* 33.033 (methanol) was the most abundant compound in the EVOO headspace, followed by mass peaks *m/z* 45.034 (acetaldehyde), *m/z* 47.049 (ethanol), *m/z* 43.018 (esters), *m/z* 57.033 (2-propenal), and *m/z* 61.028 (acetic acid). However, the detection of methanol is rarely reported, and/or methanol is often found to be present at low concentrations using gas chromatography-mass spectroscopy-based techniques. There are several possible reasons for the detection of high concentrations of the VOCs with low molecular mass in the headspace of OOs using PTR-QiToF-MS. One of the possible reasons is that this mass may be a mixture of a small amount of methanol and a large amount of fragments from higher molecular masses. It may also be due to differences in the set-up between the current method and gas chromatography approaches. For instance, the way of sampling and injection differ considerably.

Significantly higher concentrations were determined for all identified C5 and C6 compounds in the EVOO headspace compared to the lower grades of OOs. Regarding C5 compounds, the headspace concentration in EVOO of mass peak *m/z* 87.081 (13 ± 8 ppmv, 3-penten-2-ol/2-methyl-3-buten-2-ol/2-methylbutanal/3-methylbutanal/pentanal/3-pentanone) was higher than *m/z* 85.064 (7 ± 6 ppmv, trans-2-pentenal/1-penten-3-one/trans-2-methyl-2-butenal) and *m/z* 103.075 (208 ± 98 parts per billion by volume (ppbv), ethyl propionate/3-methylbutanoic acid/pentanoic acid). This may be due to the malaxation step, which involves temperatures that favor amino acid conversion. This conversion results in an elevated production of 2-methylbutanal and 3-methylbutanal [8]. Among the eight identified C6 compounds, mass peak *m/z* 99.081 (13 ± 11 ppmv, trans-2-hexenal/cis-3-hexenal) were present in highest abundance in the EVOO headspace. Two isomers were tentatively identified in mass peak *m/z* 99.081, which are trans-2-hexenal and cis-3-hexenal.

Furthermore, significantly higher concentrations of 28 out of 30 identified C7–C15 compounds were determined in the headspace of EVOO than for the other OO counterparts. Among these 28 identified compounds, the concentrations of mass peaks *m/z* 93.070 (254 ± 223 ppbv, toluene,

C1-benzene), *m/z* 105.09 (324 ± 247 ppbv, ethenyl benzene, C2-benzene) and *m/z* 107.086 (338 ± 275 ppbv, ethyl benzene, C2-benzene) found in the headspace of EVOO were higher than the other compounds. It is reported that oils can be easily contaminated by these potentially harmful VOCs because of their lipophilic nature and widely distribution [45]. The presence of these compounds in OOs is likely due to the contamination by gasoline vapors in the oil mill [46,47]. These compounds could also originate from the packaging materials [48]. The contamination of these compounds in OOs deserves a special consideration in the future due to their potential harm to human body [49]. Furthermore, significantly higher concentrations of those three compounds in the EVOO headspace compared to the other OOs is likely due to removal during the refining process [48].

2.2.2. VOCs with Higher Concentrations in the ROO/POO Headspace

It is interesting to note that four mass peaks (*m/z* 47.012, *m/z* 49.012, *m/z* 49.027, and *m/z* 115.111) were present in significantly higher headspace concentrations in ROO and POO compared to EVOO (Figure 1 and Table 1).

Mass peak *m/z* 47.012 was tentatively identified as formic acid [31,32]. The headspace concentration of formic acid in EVOO (3 ± 3 ppmv) was significantly lower than in ROO (12 ± 11 ppmv) and POO (17 ± 16 ppmv). One of the possible pathways contributing to the formation of formic acid (*m/z* 47.012) in the lower grades of OOs is oxidation during storage [31], such as the decomposition of unstable volatiles (2,4-(E-E)-decadienal) [31,50]. Another possible pathway is microbial metabolism during storage [51]. The other two mass peaks *m/z* 49.012 and *m/z* 49.027 (non-identified) were also present in significantly lower concentrations for EVOO compared to ROO and POO. They are most likely fragments of higher molecular masses. In addition, the concentration of mass peak *m/z* 115.111 (51 ± 31 ppbv, heptanal/heptan-2-one) for EVOO was significantly lower than that for ROO (81 ± 65 ppbv) and POO (77 ± 63 ppbv). This mass peak was tentatively identified as heptanal or heptan-2-one (Table 1) [9,39], which originates from the decomposition of linoleic acid [43]. The formation of this compound in ROO and POO most likely occurs during storage, because the steam deodorization step before storage would have removed such organic compounds [12,52].

Taken together, the concentrations of most VOCs were significantly lower in the headspace of OOs that have been subjected to a refining step, whereas the concentrations of four mass peaks, i.e. *m/z* 47.012 (formic acid), *m/z* 49.016 (non-identified), *m/z* 49.027 (non-identified) and *m/z* 115.111 (heptanal/heptan-2-one), presented a reversed trend.

*2.3. Odor Implications*

Some groups of scientists relate the odor contribution of a certain VOC to the human perceivable odor not only to be related to its concentration, but also to its OT [8,9,36,39]. This approach allows some ranking of the VOCs in terms of their relevance to the odor. When the odor activity value (OAV, the average concentration of the volatile compound of the oils divided by its OT) of the volatile compound is greater than one, the odor of this compound is expected to contribute to the odor of the oils according to this theory [53]. In this study, we looked into the odor relevance of compounds using the OTs.

Considering the average concentrations of the identified C1–C4 compounds in OOs and their OTs in Table 1, the odor of those compounds (acetaldehyde, OAV = 185; 2-propenal, OAV = 19; propanal, OAV = 24; acetic acid, OAV = 43; dimethyl sulfide, OAV = 16; propanoic acid, OAV = 12; dimethyl sulfoxide, OAV = 68; butanoic acid, OAV = 143. OAV is calculated based on the average concentration for the OO grade with lowest intensity and its OT in Table 1) are considered to contribute strongly to the odor of the oils due to their high OAVs. Surprisingly, most of those compounds are associated with odor defects.

Regarding the identified C5 compounds, trans-2-pentenal (*m/z* 85.064), associated with green-fruity odor note [1,9], was present with an OAV of 16 for EVOO. Regarding the identified C6 compounds, trans-2-hexenal (*m/z* 99.081), associated with a green-fruity odor note [1,9], was present with an OAV of 13000 for EVOO. Hexanal (*m/z* 101.095), associated with a green-sweet odor note [9], was present with an OAV of 570 for EVOO. Butyl acetate, ethyl butyrate, and ethyl isobutyrate (*m/z* 117.091), associated with a green-sweet-fruity note [1,9], were present with OAVs of 23, 82, and 131 for EVOO, respectively. Therefore, trans-2-pentenal, trans-2-hexenal, hexanal, butyl acetate, ethyl butyrate, and ethyl isobutyrate might be the relevant contributors to the green odor notes of EVOO. This is in agreement with previous studies [9,10] reported that the C5-C6 compounds were described as the most important VOCs in terms of the contribution to the green odor notes for EVOO. Moreover, trans-2-hexenal and hexanal are most likely the most important contributors to the green odor notes of the EVOO due to the highest OAV (13000 and 570, respectively) compared to the other compounds. These results agree with those in a previous study, which reported that the identified C6 aldehydes (especially trans-2-hexenal and hexanal) contribute to the green odor notes in European EVOO [37]. In addition, the OAVs of those compounds mentioned above for EVOO were higher than its lower grade counterparts. Furthermore, it is reported that a great amount of the VOCs associated with the green odor notes have been found in high-quality/grade OO (EVOO) [3,54]. Therefore, the odor of those compounds most likely contributes to the differences in perception of the green odor notes between the premium grade EVOO and the lower grades of OOs.

Although the identified C7–C15 compounds have relatively low concentrations in OOs compared to the identified C1–C6 compounds, they were also components of the volatile odor fraction in OOs, especially the identified C7–C10 compounds [10]. The OAVs of some of these compounds were over one in OOs (2,4 heptadienal, OAV = 3; trans-2-heptenal, OAV = 1.37; heptanal, OAV = 268; heptan-2-one, OAV = 51; trans-2-octenal, OAV = 17; 3-octen-2-one, OAV = 17; octanal, OAV = 338; 1-octen-3-ol, OAV = 27; octan-2-one, OAV = 1.00; trans,trans-2,4-nonadienal, OAV = 75; trans-2-nonenal, OAV = 150; nonanal, OAV = 31; nonan-2-one, OAV = 3; trans,trans-2,4-decadienal, OAV = 8; trans-2-decenal, OAV = 5; decanal, OAV = 7). However, most of those compounds are associated with odor defects. Hexyl acetate (*m/z* 145.122), associated with a green-fruity note, was present with an OAV less than one in OOs, which support previous research [55]. This indicates that this compound might not be a relevant contributor to the green odor notes of OOs. Thus, those minor compounds are more likely related to odor defects of OOs due to their high OAV value and related odor notes.

Summarizing, the identified C5–C6 compounds mainly possess the green odor notes, while the identified C1–C4 and C7–C15 compounds are mainly associated with odor defects. EVOO has 31 volatile compounds exceeding an OAV of one, which is more than ROO (30 volatile compounds) and POO (26 volatile compounds). EVOO is also present with higher OAV values for 29 out of these 31 compounds compared to ROO and POO. Thus, most likely, these VOCs contribute to the richer and more complex odor of EVOO compared to ROO and POO. This is similar to the result in Section 2.2 that EVOO were present with significantly higher headspace concentrations of the VOCs in comparison to ROO and POO.

Consumers' preference in OOs is mainly related to the odor descriptors qualified with the 'green' note [56]. Therefore, the green notes are fairly important sensory traits. In Table 1, trans-2-hexenal (*m/z* 99.081), hexanal (*m/z* 101.095), butyl acetate, ethyl butyrate, and ethyl isobutyrate (*m/z* 117.091) are expected to contribute to the green odor notes of OOs, since their OAVs are greater than one. In order to compare the full sets of samples, the scatter plots of the $^{10}$log transformed concentrations of *m/z* 99.081, *m/z* 101.095, and *m/z* 117.091 are presented for all samples in Figure 2. The plots show distinct clustering of the three grades of OOs. EVOO (located in the upper right corner in Figure 2) grouped separately from the lower grades of OOs (widely spread in the lower left corner). This indicates that EVOO was present with consistently higher concentrations of these compounds with green notes, and with OAV values >1, in the headspace of EVOO, which is in agreement with previous studies [3,54].

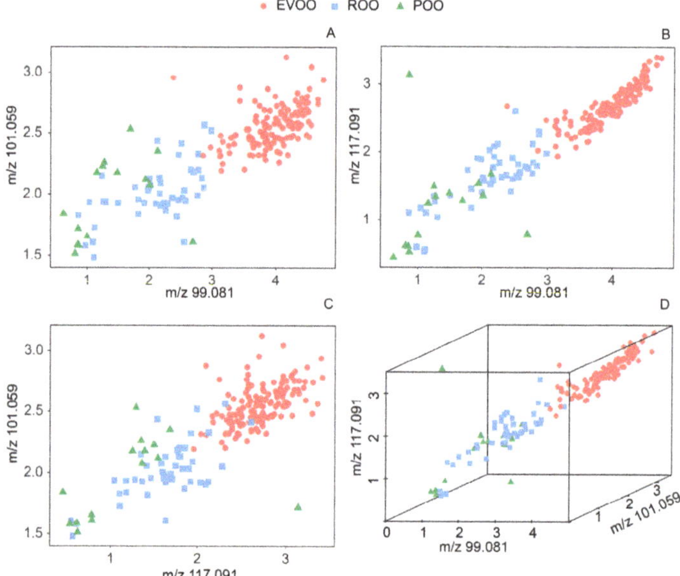


**Figure 2.** Scatter plots of the $^{10}$log transformed concentrations (ppbv) of (**A**) *m/z* 99.081 versus *m/z* 101.095; (**B**) *m/z* 99.081 versus *m/z* 117.091; (**C**) *m/z* 101.095 versus *m/z* 117.091; (**D**) *m/z* 99.081, *m/z* 101.095 and *m/z* 143.107 of three olive oil grades, including extra virgin olive oil (EVOO, *n* = 140), refined olive oil (ROO, *n* = 45), and pomace olive oil (POO, *n* = 15).

## 2.4. Relative Concentration Differences of the VOCs

To explore the VOCs proportions of OO grades, the relative average concentrations and standard deviation of 54 tentatively identified VOCs in the headspace of each grade (EVOO, ROO, and POO) are shown in Table 2.

**Table 2.** Tentative identification volatile organic compounds (VOCs), average relative concentrations, standard deviation (SD), and statistical comparisons (Kruskal–Wallis tests and Mann–Whitney U-tests, $p < 0.05$) of the VOCs in the headspace of extra virgin olive oil (EVOO), refined olive oil (ROO), and pomace olive oil (POO). The relative concentration of each mass per sample was expressed as the ratio (%, *w/w*) of the single mass peak intensity per sample to the total mass intensity per sample.

| Measured Protonated Mass *m/z* | | Tentative Identification | Average ± SD (%) | | |
|---|---|---|---|---|---|
| | | | EVOO (*n* = 140) | ROO (*n* = 45) | POO (*n* = 15) |
| C1–C4 [A] | | | | | |
| | 33.033 | Methanol | $46 \pm 9$ [a] | $33 \pm 14$ [b] | $10 \pm 10$ [c] |
| | 43.018 | Esters | $5 \pm 3$ [b] | $6 \pm 3$ [a] | $5 \pm 1$ [ab] |
| | 45.034 | Acetaldehyde | $8 \pm 3$ [b] | $16 \pm 6$ [a] | $30 \pm 17$ [a] |
| | 47.012 | Formic acid | $(4 \pm 5) \times 10^{-1}$ [b] | $5 \pm 4$ [a] | $14 \pm 7$ [a] |
| | 47.049 | Ethanol | $5 \pm 5$ | $5 \pm 4$ | $3 \pm 2$ |
| | 57.033 | 2-Propenal | $6 \pm 5$ | $6 \pm 8$ | $3 \pm 3$ |
| | 59.049 | Propanal; acetone | $2 \pm 2$ [b] | $5 \pm 3$ [a] | $9 \pm 3$ [a] |
| | 61.028 | Acetic acid | $4 \pm 3$ [b] | $7 \pm 3$ [a] | $6 \pm 3$ [a] |
| | 63.026 | Dimethyl sulfide | $(1 \pm 2) \times 10^{-1}$ [a] | $(2 \pm 2) \times 10^{-2}$ [b] | $(1 \pm 0) \times 10^{-2}$ [b] |
| | 73.064 | Butan-2-one | $1 \pm 1$ [b] | $2 \pm 1$ [a] | $3 \pm 2$ [a] |
| | 75.044 | Propanoic acid; methyl acetate | $3 \pm 2$ [a] | $1 \pm 0$ [b] | $(3 \pm 3) \times 10^{-1}$ [c] |
| | 79.021 | Dimethyl sulfoxide | $(8 \pm 7) \times 10^{-2}$ | $(1 \pm 1) \times 10^{-1}$ | $(6 \pm 2) \times 10^{-2}$ |
| | 89.059 | Ethyl acetate; butanoic acid | $(4 \pm 3) \times 10^{-1}$ [a] | $(2 \pm 2) \times 10^{-1}$ [b] | $(2 \pm 5) \times 10^{-1}$ [c] |

**Table 2.** *Cont.*

| Measured Protonated Mass *m/z* | | Tentative Identification | Average ± SD (%) | | |
|---|---|---|---|---|---|
| | | | EVOO (*n* = 140) | ROO (*n* = 45) | POO (*n* = 15) |
| **C5–C6** | | | | | |
| | 81.070 | Terpene fragment or fragments from cis-/trans-hexenal | $2 \pm 2$ [a] | $(2 \pm 1) \times 10^{-1}$ [b] | $(7 \pm 5) \times 10^{-2}$ [b] |
| | 83.085 | Terpene fragment | $1 \pm 0$ [a] | $1 \pm 0$ [b] | $1 \pm 0$ [b] |
| | 85.064 | trans-2-Pentenal; trans-2-methyl-2-butenal; 1-penten-3-one | $1 \pm 1$ [a] | $(1 \pm 1) \times 10^{-1}$ [b] | $(1 \pm 0) \times 10^{-1}$ [b] |
| | 85.099 | Cyclohexane | $(4 \pm 3) \times 10^{-1}$ [a] | $(9 \pm 4) \times 10^{-2}$ [b] | $(9 \pm 4) \times 10^{-2}$ [b] |
| | 87.081 | 3-Penten-2-ol; 2-methyl-3-buten-2-ol; 2-methylbutanal; 3-methylbutanal; pentanal; 3-pentanone | $2 \pm 1$ [a] | $(4 \pm 2) \times 10^{-1}$ [b] | $(3 \pm 1) \times 10^{-1}$ [b] |
| | 97.064 | 2,4-Hexadienal; ethyl furan | $(4 \pm 2) \times 10^{-2}$ [a] | $(1 \pm 1) \times 10^{-2}$ [b] | $(2 \pm 1) \times 10^{-2}$ [b] |
| | 99.081 | trans-2-Hexenal; cis-3-hexenal | $2 \pm 2$ [a] | $(1 \pm 1) \times 10^{-1}$ [b] | $(1 \pm 2) \times 10^{-1}$ [b] |
| | 101.095 | cis-3-Hexen-1-ol; trans-3-hexen-1-ol; cis-2-hexen-1-ol; trans-2-hexen-1-ol; hexanal; 3-methyl pentanal; 4-methylpentan-2-one | $(5 \pm 2) \times 10^{-2}$ [b] | $(6 \pm 7) \times 10^{-2}$ [b] | $(1 \pm 1) \times 10^{-1}$ [a] |
| | 103.075 | Ethyl propionate; 3-methylbutanoic acid; pentanoic acid | $(3 \pm 1) \times 10^{-2}$ [b] | $(2 \pm 2) \times 10^{-2}$ [b] | $(2 \pm 6) \times 10^{-1}$ [a] |
| | 113.059 | Sorbic acid; 5-ethyl-2-(5H)-furanone | $(2 \pm 4) \times 10^{-2}$ [a] | $(4 \pm 3) \times 10^{-3}$ [b] | $(4 \pm 5) \times 10^{-3}$ [b] |
| | 117.091 | Butyl acetate; ethyl butyrate; ethyl isobutyrate; hexanoic acid | $(9 \pm 7) \times 10^{-2}$ [a] | $(2 \pm 1) \times 10^{-2}$ [b] | $(2 \pm 9) \times 10^{-1}$ [b] |
| **C7–C15** | | | | | |
| | 93.070 | Toluene | $(3 \pm 4) \times 10^{-2}$ [a] | $(1 \pm 1) \times 10^{-2}$ [b] | $(1 \pm 1) \times 10^{-2}$ [b] |
| | 105.090 | Ethenyl benzene | $(4 \pm 2) \times 10^{-2}$ [a] | $(2 \pm 2) \times 10^{-2}$ [b] | $(2 \pm 1) \times 10^{-2}$ [b] |
| | 107.086 | Ethyl benzene | $(4 \pm 3) \times 10^{-2}$ [a] | $(2 \pm 2) \times 10^{-2}$ [b] | $(2 \pm 5) \times 10^{-2}$ [b] |
| | 109.101 | Methyl norbornene | $(1 \pm 0) \times 10^{-2}$ [c] | $(1 \pm 1) \times 10^{-2}$ [b] | $(3 \pm 2) \times 10^{-2}$ [a] |
| | 111.080 | 2,4-Heptadienal | $(1 \pm 1) \times 10^{-2}$ [b] | $(2 \pm 1) \times 10^{-2}$ [a] | $(2 \pm 1) \times 10^{-2}$ [ab] |
| | 113.096 | trans-2-Heptenal | $(2 \pm 1) \times 10^{-2}$ [ab] | $(1 \pm 1) \times 10^{-2}$ [b] | $(2 \pm 1) \times 10^{-2}$ [a] |
| | 115.111 | Heptanal; heptan-2-one | $(1 \pm 0) \times 10^{-2}$ [b] | $(6 \pm 1) \times 10^{-2}$ [a] | $(1 \pm 1) \times 10^{-1}$ [a] |
| | 121.099 | 1,2,4-Trimethylbenzene; acetophenone | $(9 \pm 6) \times 10^{-3}$ [a] | $(2 \pm 3) \times 10^{-3}$ [b] | $(3 \pm 11) \times 10^{-3}$ [b] |
| | 123.080 | 2-Phenylethanol | $(6 \pm 2) \times 10^{-4}$ [a] | $(5 \pm 6) \times 10^{-4}$ [b] | $(9 \pm 8) \times 10^{-4}$ [a] |
| | 125.096 | cis-1,5-Octadien-3-one; octan-2-one; trans, trans-2,4-octadienal | $(5 \pm 3) \times 10^{-3}$ | $(6 \pm 5) \times 10^{-3}$ | $(4 \pm 2) \times 10^{-3}$ |
| | 127.111 | trans-2-Octenal; 6-methyl-5-hepten-2-one; 1-octen-3-one; 3-octen-2-one | $(5 \pm 3) \times 10^{-3}$ [c] | $(9 \pm 7) \times 10^{-3}$ [b] | $(2 \pm 1) \times 10^{-2}$ [a] |
| | 129.091 | 3-Methyl-2-butenyl acetate | $(2 \pm 1) \times 10^{-3}$ [ab] | $(3 \pm 5) \times 10^{-3}$ [b] | $(6 \pm 9) \times 10^{-3}$ [a] |
| | 129.127 | Octanal; 6-methyl-5-hepten-3-ol; 1-octen-3-ol; octan-2-one | $(4 \pm 2) \times 10^{-3}$ [b] | $(3 \pm 6) \times 10^{-2}$ [a] | $(9 \pm 15) \times 10^{-2}$ [a] |
| | 131.106 | Propyl butyrate; ethyl 2-methylbutyrate; ethyl 3-methylbutyrate; heptanoic acid | $(9 \pm 5) \times 10^{-3}$ [a] | $(7 \pm 4) \times 10^{-3}$ [b] | $(7 \pm 12) \times 10^{-3}$ [b] |
| | 137.132 | Terpene fragments (α-pinene; β-pinene; limonene; tricyclene; camphene; sabinene; myrcene; β-ocimene) | $(1 \pm 1) \times 10^{-2}$ [b] | $(2 \pm 8) \times 10^{-2}$ [a] | $(1 \pm 2) \times 10^{-2}$ [ab] |
| | 139.112 | trans,trans-2,4-Nonadienal | $(3 \pm 2) \times 10^{-3}$ [b] | $(1 \pm 2) \times 10^{-3}$ [c] | $(6 \pm 4) \times 10^{-3}$ [a] |
| | 139.147 | 3-Ethyl-1,5-octadiene | $(1 \pm 1) \times 10^{-2}$ [a] | $(2 \pm 1) \times 10^{-3}$ [b] | $(5 \pm 3) \times 10^{-3}$ [b] |
| | 141.127 | tran-2-Nonenal; cis-2-nonenal; cis-3-nonenal | $(1 \pm 1) \times 10^{-3}$ [b] | $(1 \pm 1) \times 10^{-3}$ [b] | $(5 \pm 2) \times 10^{-3}$ [a] |
| | 143.107 | cis-3-Hexenyl acetate; trans-3-hexenyl acetate; 3-methyl-4-penten-1-ol-acetate | $(7 \pm 5) \times 10^{-3}$ [a] | $(3 \pm 3) \times 10^{-3}$ [b] | $(2 \pm 1) \times 10^{-3}$ [b] |
| | 143.142 | Nonanal; nonan-2-one | $(6 \pm 4) \times 10^{-3}$ [b] | $(1 \pm 1) \times 10^{-2}$ [ab] | $(1 \pm 1) \times 10^{-2}$ [a] |
| | 145.122 | Hexyl acetate; 2-methylpropyl butanoate; ethyl hexanoate; octanoic acid | $(8 \pm 4) \times 10^{-3}$ [a] | $(3 \pm 1) \times 10^{-3}$ [b] | $(4 \pm 5) \times 10^{-3}$ [b] |
| | 153.125 | 2,4-Decadienal; trans, trans-2,4-decadienal; trans, cis-2,4-decadienal; 2,3-dehydro-1,8-cineole | $(1 \pm 0) \times 10^{-3}$ [ab] | $(1 \pm 1) \times 10^{-3}$ [b] | $(2 \pm 3) \times 10^{-3}$ [a] |
| | 155.141 | trans-2-Decenal; 3,7-dimethylocta-1,6- dien-3-ol; cis-p-menth-2-en-1-ol | $(1 \pm 0) \times 10^{-3}$ [ab] | $(1 \pm 1) \times 10^{-3}$ [b] | $(2 \pm 2) \times 10^{-3}$ [a] |
| | 157.124 | Ethyl cyclohexanecarboxylate | $(2 \pm 1) \times 10^{-3}$ [a] | $(7 \pm 8) \times 10^{-4}$ [b] | $(1 \pm 1) \times 10^{-3}$ [a] |
| | 157.158 | Decanal | $(7 \pm 5) \times 10^{-4}$ [b] | $(2 \pm 4) \times 10^{-3}$ [b] | $(6 \pm 10) \times 10^{-3}$ [a] |
| | 169.123 | trans-4,5-Epoxy-trans-2-decenal | $(3 \pm 2) \times 10^{-4}$ [a] | $(2 \pm 4) \times 10^{-4}$ [b] | $(3 \pm 3) \times 10^{-4}$ [ab] |
| | 171.174 | Undecanal | $(4 \pm 3) \times 10^{-4}$ [a] | $(1 \pm 2) \times 10^{-4}$ [b] | $(4 \pm 6) \times 10^{-4}$ [b] |
| | 173.154 | Ethyl octanoate; methyl nonanoate | $(1 \pm 3) \times 10^{-3}$ [a] | $(4 \pm 3) \times 10^{-4}$ [b] | $(6 \pm 5) \times 10^{-4}$ [b] |
| | 183.082 | Benzophenone | $(4 \pm 7) \times 10^{-4}$ [b] | $(4 \pm 3) \times 10^{-4}$ [a] | $(7 \pm 6) \times 10^{-4}$ [a] |
| | 205.194 | β-Caryophyllene; copaene; β-selinene; α-farnesene; eremophilene | $(3 \pm 3) \times 10^{-3}$ [a] | $(1 \pm 1) \times 10^{-3}$ [b] | $(3 \pm 3) \times 10^{-4}$ [c] |

Superscript letters a, b, and c indicate the significant differences ($p < 0.05$). [A] C1–C7 are the VOCs with 1–7 carbon atoms in the molecule.

Significant differences in the relative concentrations (proportions) in OOs headspace were observed in 50 out of 54 tentatively identified VOCs in Table 2. The major constituents of the VOCs obtained from the headspace of EVOO were identified as methanol (*m/z* 33.033), acetaldehyde (*m/z* 45.034), 2-propenal (*m/z* 57.033), esters (*m/z* 43.018), and ethanol (*m/z* 47.049) in descending order. The VOCs proportions were different in the headspace of the lower grades of OOs. The proportions of the top five major compounds in the ROO headspace were the same as those in the EVOO headspace, except that ethanol changes to acetic acid. However, the top five major compounds in the POO headspace in descending order are acetaldehyde (*m/z* 45.034), formic acid (*m/z* 47.012), methanol (*m/z* 33.033), propanal/acetone (*m/z* 59.049), and acetic acid (*m/z* 61.028).

The relative concentrations of five identified C1–C4 compounds (esters, acetaldehyde, ethanol, acetone, and butan-2-one) in the headspace of ROO/POO were significantly higher than for EVOO (Table 2). Whereas the relative concentrations of the other compounds (2-propenal, acetic acid, dimethyl sulfide, propanoic acid, dimethyl sulfoxide, ethyl acetate, and butanoic acid) in the headspace of ROO/POO were lower than for EVOO (Table 2). Furthermore, 9 out of 11 identified C5–C6 compounds were present in significantly higher relative concentrations for EVOO than for ROO/POO. However, significantly lower proportions were presented in two mass peaks *m/z* 101.095 and *m/z* 103.075 (non-identified) for POO compared to EVOO and ROO. Regarding the identified C7–C15 compounds, 29 out of 30 compounds were present in significant differences in OOs in Table 2. Only 11 out of 29 compounds were present in significantly higher proportions in the EVOO headspace than for the ROO/POO headspace. The relative concentrations of five mass peaks *m/z* 109.101 (methyl norbornene), *m/z* 115.111 (heptanal/heptan-2-one), *m/z* 127.111 (trans-2-octenal/6-methyl-5-hepten-2-one/1-octen-3-one/3-octen-2-one), *m/z* 129.127 (octanal/6-methyl-5-hepten-3-ol/1-octen-3-ol/octan-2-one), and *m/z* 183.082 (benzophenone) in the headspace of ROO and POO were significantly higher than for EVOO. This is different from the concentration results observed in Section 2.2.2 that the other four mass peaks (*m/z* 47.012, *m/z* 49.012, *m/z* 49.027 and *m/z* 115.111) were present in significantly higher concentrations in the headspace of ROO/POO compared to EVOO.

In total, the VOCs proportions were different in the headspace of different OO grades, which is most likely due to the different processing methods. It is known from the literature that absolute concentrations of the VOCs are important for the odor traits of products, but the balance of the VOCs is just as important [57].

## 3. Materials and Methods

### 3.1. Samples Preparation

For this study, 240 OOs were gathered from producers, traders, and retailers across Europe in 2016 and 2017. The authenticity of the 180 EVOO samples was verified by fatty acid fingerprints combined with chemometrics [58], ultraviolet-visible spectra analysis [59], and evaluated by 2/3-monochloropropane-1,2-diol and glycidyl esters analysis [60]. A total of 40 EVOO samples did not meet the requirements for one or more of these methods and were therefore removed from the sample set. In total, 200 OOs were used in this study, which consisted of 140 EVOO, 45 ROO, and 15 POO. Prior to analysis, all the oils were sealed and stored in the dry and dark environment at 18 ± 1 °C. Sampling was completed within 6 months. To avoid long-term storage, the analyses were carried out within two weeks after sampling of each sample.

The sample preparation method was similar to our previous study [25] with minor modifications. Firstly, 5 mL of oil was transferred into a 250 mL flask. Then, the closed flask was kept in a water bath at 30 °C for 30 min to equilibrate the headspace before instrumental analysis.

### 3.2. PTR-QiToF-MS Analysis

PTR-QiToF-MS (Ionicon Analytik GmbH, Innsbruck, Austria) was operated with a drift voltage of 999 V, a drift temperature of 61 ± 1 °C, a drift pressure of 3.803 mbar, and an E/N value of 134 ± 1 Townsend. The laboratory air was measured for the first 10 s as a blank before each sample. Then, the VOCs in the headspace of the flask were transported to the PTR-QiToF-MS through a peek inlet tube with a temperature of 60 ± 0.5 °C. The flow rate of the air in the tube was 61 ± 2 mL min$^{-1}$.

The measurement time was 30 s. The acquisition rate was 1 spectrum per second. On each of two different days, one independent replicates per sample was measured. Samples were analyzed in a random sequence to avoid any order bias. Results were stored in the system automatically.

### 3.3. VOCs Data Pre-Processing

All the raw data obtained from the PTR-QiToF-MS machine were integrated by PTRwid software (Utrecht University, Utrecht, the Netherlands; http://www.staff.science.uu.nl/~{}holzi101/ptrwid/) [61]. The unified mass list with the ion count per second (cps) of each sample were provided after the autonomous mass scale calibration, as described by Holzinger [61]. The average of the 30 sample scans and the average of the 10 blank scans were calculated separately. The VOC concentrations (molecules per $cm^3$) were calculated from cps according to Equation (1) [62]:

$$[VOC] = \frac{1}{kt} \times \frac{[VOC \cdot H^+]_{measured}}{[H_3O^+]_{measured} \times 487} \times \frac{\sqrt{(m/z)_{H_3O^+}}}{\sqrt{(m/z)_{VOC \cdot H^+}}} \tag{1}$$

where t is the residence time of the primary ions in the drift tube, k is the coefficient of the reaction rate with a value of $2 \times 10^{-9}$ $cm^3/s$, $[VOC \cdot H^+]_{measured}$ is the ion count rate of the protonated VOC, $[H_3O^+]_{measured}$ is the ion count rate of the protonated water at $m/z$ 21.022, 487 is the intensity ratio of the protonated water at $m/z$ 19.018 (100%) to the protonated water at $m/z$ 21.022 (0.2055%) [63], $(m/z)_{H3O+}$ and $(m/z)_{VOC \cdot H+}$ are the molecular weight of protonated water and protonated VOC.

Subsequently, the unit of molecules per $cm^3$ was converted to ppbv, on the basis of ideal gas using Equation (2) [64]:

$$PV = nRT \tag{2}$$

where P (Pa) is the pressure, V ($cm^3$) is the volume, n (mol) is the number of moles, R (J $K^{-1}$/mol) is the gas constant, and T (K) is the temperature.

After unit conversion, the average of each sample's 10 blank cycles were subtracted from the sample's averaged scan. The replicates of each sample were checked using autocorrelation [65], and the sample will be removed when the correlation value was below 0.9. In this study, the correlation values of all the samples were over 0.9. Finally, sample averages were calculated from the data of two replicate measurements.

Sample independent ions, such as $N_2^+$, $NO^+$, $O_2^+$, $H_2O^+$, $H_3{}^{18}O^+$, $(H_2O)_2 \cdot H^+$, $H_2O \cdot H_2{}^{18}O \cdot H^+$ and $(H_2O)_3 \cdot H^+]$ signals at mass peaks $m/z$ 28.005, $m/z$ 29.997, $m/z$ 31.989, $m/z$ 18.010, $m/z$ 21.022, $m/z$ 37.028, $m/z$ 39.032, and $m/z$ 55.039, respectively, were removed. After the data pre-processing, 295 mass peaks in the range of $m/z$ 18.033 to 207.204 remained.

### 3.4. Relative Concentration

The relative concentration (C, %) of each mass of each sample was calculated by the intensity of the single mass ($I_s$, ppbv) per sample and the total mass intensity ($I_t$, ppbv) per sample using Equation (3):

$$C = \frac{I_s}{I_t} \times 100\%. \tag{3}$$

### 3.5. Data Analysis

Significant differences of the (relative) concentrations of tentatively identified VOCs for different grades of OOs were assessed using non-parametric Kruskal–Wallis tests ($p < 0.05$). Mann–Whitney U-tests were used to perform pairwise comparisons between OO grades. These analyses methods were performed using SPSS (version 23, IBM, Chicago, IL, USA).

### 3.6. Odor Threshold in Air

The OTs of the identified compounds were collected from several publications [29,39,66–69]. Firstly, the OT in oil were converted into the OT in air using Equation (4):

$$C_{a1} = KC_o\rho \tag{4}$$

where $C_{a1}$ (µg/L) is the OT in air, $C_o$ (µg/kg) is the OT in oil, K is the air/liquid partition coefficient of compound [67], and $\rho$ is the density of olive oil (0.916 kg/L).

Then, the unit of the OT in air was converted using Equation (5) [64]:

$$C_{a2} = V_m \frac{C_{a1}}{M} \times 1000 \tag{5}$$

where $C_{a2}$ (ppbv) is the OT in air, $V_m$ is the standard molar volume of ideal gas at 1 bar and 298 K with a value of 24.77 L/mol [70], and M (g/mol) is the molecular weight of the compound.

Subsequently, the lowest OT of each compound was used in this study. The calculated OTs in air from various literature sources are provided in the Supplementary Material (Table S1). The OAV was calculated using the average concentration of the volatile compound for each of the type of OO divided by the corresponding OT.

## 4. Conclusions

Significant differences in VOC headspace concentrations were determined for the different grades of OOs. Most of the VOCs were present in significantly higher concentrations for EVOO than for ROO/POO. However, significantly higher concentrations of mass peaks *m/z* 47.012 (formic acid), *m/z* 49.016, *m/z* 49.027, and *m/z* 115.111 (heptanal/heptanone) were found for the lower grades of OOs (ROO/POO) compared to EVOO. Furthermore, significant differences of the VOCs proportions were observed indicating a distinct change in the balance of the VOCs across OO grades. Thus, EVOO and ROO/POO not only differ quantitatively (concentrations of compounds) but also qualitatively (proportions of compounds). Comparison with OAVs of the compounds revealed the expected change in contribution to the odor of the OOs. Our results underpin the well-known richer and more complex odor of EVOO by the elevated contribution of the VOCs with green notes exceeding the minimal OAV. Furthermore, the consistent differences in VOCs concentrations between EVOO and other grades of OO may provide potential for verification of the identity of OOs.

**Supplementary Materials:** The following are available online, Table S1: List of the air/liquid partition coefficients (K) and the calculated odor thresholds (OTs) in air (ppbv) of volatile organic compounds.

**Author Contributions:** Conceptualization, S.M.v.R.; methodology, J.Y.; formal analysis, J.Y.; data curation, J.Y. and M.A.; writing—original draft, J.Y.; writing—review and editing, J.Y., M.A. and S.M.v.R.; supervision, S.M.v.R. All authors have read and agreed to the published version of the manuscript.

**Funding:** The first author received financially support for her PhD project from the China Scholarship Council (201506910048).

**Acknowledgments:** We thank Alex Koot and Rita Boerrigter-Eenling of WFSR Wageningen University and Research for their technical support with the PTR-QiToF-MS analysis. The authors are also grateful to the producers, traders and retailers across Europe for their supply of the large number of samples.

**Conflicts of Interest:** The authors declare no conflict of interest.

## References

1. Luna, G.; Morales, M.T.; Aparicio, R. Characterisation of 39 varietal virgin olive oils by their volatile compositions. *Food Chem.* **2006**, *98*, 243–252. [CrossRef]
2. Ranalli, A.; De Mattia, G.; Patumi, M.; Proietti, P. Quality of virgin olive oil as influenced by origin area. *Grasas Aceites* **1999**, *50*, 249–259. [CrossRef]

3. Aparicio, R.; Morales, M.T. Characterization of olive ripeness by green aroma compounds of virgin olive oil. *J. Agric. Food Chem.* **1998**, *46*, 1116–1122. [CrossRef]

4. Tura, D.; Prenzler, P.D.; Bedgood, D.R.; Antolovich, M.; Robards, K. Varietal and processing effects on the volatile profile of Australian olive oils. *Food Chem.* **2004**, *84*, 341–349. [CrossRef]

5. Koprivnjak, O.; Conte, L.; Totis, N. Influence of olive fruit storage in bags on oil quality and composition of volatile compounds. *Food Technol. Biotech.* **2002**, *40*, 129–133.

6. Vichi, S.; Pizzale, L.; Conte, L.S.; Buxaderas, S.; Lopez-Tamames, E. Solid-phase microextraction in the analysis of virgin olive oil volatile fraction: Modifications induced by oxidation and suitable markers of oxidative status. *J. Agric. Food Chem.* **2003**, *51*, 6564–6571. [CrossRef] [PubMed]

7. Taiti, C.; Marone, E. EVOO or not EVOO? A new precise and simple analytical tool to discriminate virgin olive oils. *Adv. Hortic. Sci.* **2017**, *31*, 329–337.

8. Angerosa, F.; Servili, M.; Selvaggini, R.; Taticchi, A.; Esposto, S.; Montedoro, G. Volatile compounds in virgin olive oil: Occurrence and their relationship with the quality. *J. Chromatogr. A* **2004**, *1054*, 17–31. [CrossRef]

9. Kalua, C.M.; Allen, M.S.; Bedgood, D.R.; Bishop, A.G.; Prenzler, P.D.; Robards, K. Olive oil volatile compounds, flavour development and quality: A critical review. *Food Chem.* **2007**, *100*, 273–286. [CrossRef]

10. Baccouri, B.; Ben Temime, S.; Campeol, E.; Cioni, P.L.; Daoud, D.; Zarrouk, M. Application of solid-phase microextraction to the analysis of volatile compounds in virgin olive oils from five new cultivars. *Food Chem.* **2007**, *102*, 850–856. [CrossRef]

11. Morales, M.T.; Angerosa, F.; Aparicio, R. Effect of the extraction conditions of virgin olive oil on the lipoxygenase cascade: Chemical and sensory implications. *Grasas Aceites* **1999**, *50*, 114–121. [CrossRef]

12. Antonopoulos, K.; Valet, N.; Spiratos, D.; Siragakis, G. Olive oil and pomace olive oil processing. *Grasas Aceites* **2006**, *57*, 56–67.

13. Morales, M.T.; Aparicio, R. Effect of extraction conditions on sensory quality of virgin olive oil. *J. Am. Oil Chem. Soc.* **1999**, *76*, 295–300. [CrossRef]

14. Romero, I.; Garcia-Gonzalez, D.L.; Aparicio-Ruiz, R.; Morales, M.T. Validation of SPME-GCMS method for the analysis of virgin olive oil volatiles responsible for sensory defects. *Talanta* **2015**, *134*, 394–401. [CrossRef] [PubMed]

15. Ben Brahim, S.; Amanpour, A.; Chtourou, F.; Kelebek, H.; Selli, S.; Bouaziz, M. Gas chromatography-mass spectrometry-olfactometry to control the aroma fingerprint of extra virgin olive oil from three tunisian cultivars at three harvest times. *J. Agric. Food Chem.* **2018**, *66*, 2851–2861. [CrossRef] [PubMed]

16. Fabris, A.; Biasioli, F.; Granitto, P.M.; Aprea, E.; Cappellin, L.; Schuhfried, E.; Soukoulis, C.; Mark, T.D.; Gasperi, F.; Endrizzi, I. PTR-TOF-MS and data-mining methods for rapid characterisation of agro-industrial samples: Influence of milk storage conditions on the volatile compounds profile of Trentingrana cheese. *J. Mass Spectrom.* **2010**, *45*, 1065–1074. [CrossRef]

17. Taiti, C.; Costa, C.; Menesatti, P.; Comparini, D.; Bazihizina, N.; Azzarello, E.; Masi, E.; Mancuso, S. Class-modeling approach to PTR-TOFMS data: A peppers case study. *J. Sci. Food Agric.* **2015**, *95*, 1757–1763. [CrossRef]

18. Masi, E.; Romani, A.; Pandolfi, C.; Heimler, D.; Mancuso, S. PTR-TOF-MS analysis of volatile compounds in olive fruits. *J. Sci. Food Agric.* **2015**, *95*, 1428–1434. [CrossRef]

19. Yener, S.; Romano, A.; Cappellin, L.; Mark, T.D.; del Pulgar, J.S.; Gasperi, F.; Navarini, L.; Biasioli, F. PTR-ToF-MS characterisation of roasted coffees (*C-arabica*) from different geographic origins. *J. Mass Spectrom.* **2014**, *49*, 929–935. [CrossRef]

20. Schuhfried, E.; del Pulgar, J.S.; Bobba, M.; Piro, R.; Cappellin, L.; Mark, T.D.; Biasioli, F. Classification of 7 monofloral honey varieties by PTR-ToF-MS direct headspace analysis and chemometrics. *Talanta* **2016**, *147*, 213–219. [CrossRef]

21. del Pulgar, J.S.; Soukoulis, C.; Biasioli, F.; Cappellin, L.; Garcia, C.; Gasperi, F.; Granitto, P.; Mark, T.D.; Piasentier, E.; Schuhfried, E. Rapid characterization of dry cured ham produced following different PDOs by proton transfer reaction time of flight mass spectrometry (PTR-ToF-MS). *Talanta* **2011**, *85*, 386–393. [CrossRef] [PubMed]

22. Liu, N.J.; Koot, A.; Hettinga, K.; de Jong, J.; van Ruth, S.M. Portraying and tracing the impact of different production systems on the volatile organic compound composition of milk by PTR-(Quad)MS and PTR-(ToF)MS. *Food Chem.* **2018**, *239*, 201–207. [CrossRef]

23. Acierno, V.; Yener, S.; Alewijn, M.; Biasioli, F.; van Ruth, S. Factors contributing to the variation in the volatile composition of chocolate: Botanical and geographical origins of the cocoa beans, and brand-related formulation and processing. *Food Res. Int.* **2016**, *84*, 86–95. [CrossRef]

24. Marone, E.; Masi, E.; Taiti, C.; Pandolfi, C.; Bazihizina, N.; Azzarello, E.; Fiorino, P.; Mancuso, S. Sensory, spectrometric (PTR–ToF–MS) and chemometric analyses to distinguish extra virgin from virgin olive oils. *J. Food Sci. Tech.* **2017**, *54*, 1368–1376. [CrossRef] [PubMed]

25. Araghipour, N.; Colineau, J.; Koot, A.; Akkermans, W.; Rojas, J.M.M.; Beauchamp, J.; Wisthaler, A.; Mark, T.D.; Downey, G.; Guillou, C.; et al. Geographical origin classification of olive oils by PTR-MS. *Food Chem.* **2008**, *108*, 374–383. [CrossRef]

26. Aparicio-Ruiz, R.; Romero, I.; García-González, D.L.; Oliver-Pozo, C.; Aparicio, R. Soft-deodorization of virgin olive oil: Study of the changes of quality and chemical composition. *Food Chem.* **2017**, *220*, 42–50. [CrossRef] [PubMed]

27. Angerosa, F.; d'Alessandro, N.; Basti, C.; Vito, R. Biogeneration of volatile compounds in virgin olive oil: Their evolution in relation to malaxation time. *J. Agric. Food Chem.* **1998**, *46*, 2940–2944. [CrossRef]

28. Davis, B.M.; Senthilmohan, S.T.; Wilson, P.F.; McEwan, M.J. Major volatile compounds in head-space above olive oil analysed by selected ion flow tube mass spectrometry. *Rapid Commun. Mass Sp.* **2005**, *19*, 2272–2278. [CrossRef] [PubMed]

29. Leonardos, G.; Kendall, D.; Barnard, N. Odor threshold determination of 53 odorant chemicals. *J. Air Pollut. Control Assoc.* **1969**, *19*, 91–95. [CrossRef]

30. Vezzaro, A.; Boschetti, A.; Dell'Anna, R.; Canteri, R.; Dimauro, M.; Ramina, A.; Ferasin, M.; Giulivo, C.; Ruperti, B. Influence of olive (cv *Grignano*) fruit ripening and oil extraction under different nitrogen regimes on volatile organic compound emissions studied by PTR-MS technique. *Anal. Bioanal. Chem.* **2011**, *399*, 2571–2582. [CrossRef]

31. Paradiso, V.M.; Pasqualone, A.; Summo, C.; Caponio, F. An "omics" approach for lipid oxidation in foods: The case of free fatty acids in bulk purified olive oil. *Eur. J. Lipid Sci. Tech.* **2018**, *120*. [CrossRef]

32. Vichi, S.; Castellote, A.I.; Pizzale, L.; Conte, L.S.; Buxaderas, S.; Lopez-Tamames, E. Analysis of virgin olive oil volatile compounds by headspace solid-phase microextraction coupled to gas chromatography with mass spectrometric and flame ionization detection. *J. Chromatogr. A* **2003**, *983*, 19–33. [CrossRef]

33. Ruth, J.H. Odor thresholds and irritation levels of several chemical substances: A review. *Am. Ind. Hyg. Assoc. J.* **1986**, *47*, A142–A151. [CrossRef] [PubMed]

34. Jo, C.; Ahn, D.U. Production of volatile compounds from irradiated oil emulsion containing amino acids or proteins. *J. Food Sci.* **2000**, *65*, 612–616. [CrossRef]

35. Gomez-Cortes, P.; Sacks, G.L.; Brenna, J.T. Quantitative analysis of volatiles in edible oils following accelerated oxidation using broad spectrum isotope standards. *Food Chem.* **2015**, *174*, 310–318. [CrossRef] [PubMed]

36. Reiners, J.; Grosch, W. Odorants of virgin olive oils with different flavor profiles. *J. Agric. Food Chem.* **1998**, *46*, 2754–2763. [CrossRef]

37. Iraqi, R.; Vermeulen, C.; Benzekri, A.; Bouseta, A.; Collin, S. Screening for key odorants in Moroccan green olives by gas chromatography-olfactometry/aroma extract dilution analysis. *J. Agric. Food Chem.* **2005**, *53*, 1179–1184. [CrossRef]

38. Gracka, A.; Jelen, H.H.; Majcher, M.; Siger, A.; Kaczmarek, A. Flavoromics approach in monitoring changes in volatile compounds of virgin rapeseed oil caused by seed roasting. *J. Chromatogr. A* **2016**, *1428*, 292–304. [CrossRef]

39. Morales, M.T.; Luna, G.; Aparicio, R. Comparative study of virgin olive oil sensory defects. *Food Chem.* **2005**, *91*, 293–301. [CrossRef]

40. Cappellin, L.; Aprea, E.; Granitto, P.; Wehrens, R.; Soukoulis, C.; Viola, R.; Mark, T.D.; Gasperi, F.; Biasioli, F. Linking GC-MS and PTR-TOF-MS fingerprints of food samples. *Chemometr. Intell. Lab.* **2012**, *118*, 301–307. [CrossRef]

41. Yener, S.; Sanchez-Lopez, J.A.; Granitto, P.M.; Cappellin, L.; Mark, T.D.; Zimmermann, R.; Bonn, G.K.; Yeretzian, C.; Biasioli, F. Rapid and direct volatile compound profiling of black and green teas (*Camellia sinensis*) from different countries with PTR-ToF-MS. *Talanta* **2016**, *152*, 45–53. [CrossRef] [PubMed]

42. Taiti, C.; Costa, C.; Menesatti, P.; Caparrotta, S.; Bazihizina, N.; Azzarello, E.; Petrucci, W.A.; Masi, E.; Giordani, E. Use of volatile organic compounds and physicochemical parameters for monitoring the post-harvest ripening of imported tropical fruits. *Eur. Food Res. Technol.* **2015**, *241*, 91–102. [CrossRef]

43. Morales, M.T.; Rios, J.J.; Aparicio, R. Changes in the volatile composition of virgin olive oil during oxidation: Flavors and off-flavors. *J. Agric. Food Chem.* **1997**, *45*, 2666–2673. [CrossRef]

44. Yang, D.S.; Shewfelt, R.L.; Lee, K.-S.; Kays, S.J. Comparison of odor-active compounds from six distinctly different rice flavor types. *J. Agric. Food Chem.* **2008**, *56*, 2780–2787. [CrossRef] [PubMed]

45. Vichi, S.; Pizzale, L.; Conte, L.S.; Buxaderas, S.; Lopez-Tamames, E. The occurrence of volatile and semi-volatile aromatic hydrocarbons in virgin olive oils from north-eastern Italy. *Food Control* **2007**, *18*, 1204–1210. [CrossRef]

46. Biedermann, M.; Grob, K.; Morchio, G. On the origin of benzene, toluene, ethylbenzene and xylene in extra virgin olive oil. *Z. Lebensm. Unters. For.* **1995**, *200*, 266–272. [CrossRef]

47. Biedermann, M.; Grob, K.; Morchio, G. On the origin of benzene, toluene, ethylbenzene, and the xylenes in virgin olive oil—Further results. *Z. Lebensm. Unters. For.* **1996**, *203*, 224–229. [CrossRef]

48. Carrillo-Carrión, C.; Lucena, R.; Cárdenas, S.; Valcárcel, M. Liquid–liquid extraction/headspace/gas chromatographic/mass spectrometric determination of benzene, toluene, ethylbenzene,(o-, m- and p-) xylene and styrene in olive oil using surfactant-coated carbon nanotubes as extractant. *J. Chromatogr. A* **2007**, *1171*, 1–7. [CrossRef]

49. Irwin, R.J.; Mouwerik, M.V.; Stevens, L.; Seese, M.D.; Basham, W. *Environmental Contaminants Encyclopedia, Naphthalene Entry*; National Park Service: Washington, WA, USA, 1997; pp. 1–80.

50. Spiteller, P.; Kern, W.; Reiner, J.; Spiteller, G. Aldehydic lipid peroxidation products derived from linoleic acid. *Bba-Mol. Cell Biol. L.* **2001**, *1531*, 188–208. [CrossRef]

51. Piperidou, C.I.; Chaidou, C.I.; Stalikas, C.D.; Soulti, K.; Pilidis, G.A.; Balis, C. Bioremediation of olive oil mill wastewater: Chemical alterations induced by *Azotobacter vinelandii. J. Agric. Food Chem.* **2000**, *48*, 1941–1948. [CrossRef]

52. Holser, R. Odor development in refined meadowfoam (*Limnanthes alba*) oil. *Ind. Crop. Prod.* **2002**, *16*, 129–132. [CrossRef]

53. Pérez, A.G.; de la Rosa, R.; Pascual, M.; Sánchez-Ortiz, A.; Romero-Segura, C.; León, L.; Sanz, C. Assessment of volatile compound profiles and the deduced sensory significance of virgin olive oils from the progeny of Picual × Arbequina cultivars. *J. Chromatogr. A* **2016**, *1428*, 305–315. [CrossRef] [PubMed]

54. Gómez-Rico, A.; Salvador, M.D.; La Greca, M.; Fregapane, G. Phenolic and volatile compounds of extra virgin olive oil (*Olea europaea* L. Cv. Cornicabra) with regard to fruit ripening and irrigation management. *J. Agric. Food Chem.* **2006**, *54*, 7130–7136.

55. Morales, M.T.; Aparicio, R.; Calvente, J.J. Influence of olive ripeness on the concentration of green aroma compounds in virgin olive oil. *Flavour Frag. J.* **1996**, *11*, 171–178. [CrossRef]

56. Mcewan, J.A. Consumer attitudes and olive oil acceptance: The potential consumer. *Grasas Aceites* **1994**, *45*, 9–15. [CrossRef]

57. Tateo, F.; Brunelli, N.; Cucurachi, S.; Ferrillo, A. New trends in the study of the merits and short comings of olive oil in organoleptic terms, in correlation with the GC/MS analysis of the aromas. *Dev. Food Sci.* **2013**, *32*, 301–311.

58. Yang, Y.; Ferro, M.D.; Cavaco, I.; Liang, Y.Z. Detection and identification of extra virgin olive oil adulteration by GC-MS combined with chemometrics. *J. Agric. Food Chem.* **2013**, *61*, 3693–3702. [CrossRef] [PubMed]

59. International Olive Council, Spectrophotometric investigation in the ultraviolet. COI/T.20/Doc. No 19/Rev. 4. Available online: https://www.internationaloliveoil.org/wp-content/uploads/2019/11/Method-COI-T.20-Doc.-No-19-Rev.-5-2019-2.pdf (accessed on 21 November 2019).

60. Yan, J.; Oey, S.B.; van Leeuwen, S.P.J.; van Ruth, S.M. Discrimination of processing grades of olive oil and other vegetable oils by monochloropropanediol esters and glycidyl esters. *Food Chem.* **2018**, *248*, 93–100. [CrossRef]

61. Holzinger, R. PTRwid: A new widget tool for processing PTR-TOF-MS data. *Atmos. Meas. Tech.* **2015**, *8*, 3903–3922. [CrossRef]

62. Cappellin, L.; Karl, T.; Probst, M.; Ismailova, O.; Winkler, P.M.; Soukoulis, C.; Aprea, E.; Mark, T.D.; Gasperi, F.; Biasioli, F. On quantitative determination of volatile organic compound concentrations using proton transfer reaction time-of-flight mass spectrometry. *Environ. Sci. Technol.* **2012**, *46*, 2283–2290. [CrossRef]

63. Malásková, M.; Henderson, B.; Chellayah, P.D.; Ruzsanyi, V.; Mochalski, P.; Cristescu, S.M.; Mayhew, C.A. Proton transfer reaction time-of-flight mass spectrometric measurements of volatile compounds contained in peppermint oil capsules of relevance to real-time pharmacokinetic breath studies. *J. Breath Res.* **2019**, *13*. [CrossRef] [PubMed]

64. Chemguide. Ideal Gases and the Ideal Gas Law. Available online: https://www.chemguide.co.uk/physical/kt/idealgases.html (accessed on 15 August 2017).

65. Box, G.E.; Jenkins, G.M. *Time Series Analysis: Forecasting and Control, Revised Edition*; Holden-Day: San Francisco, CA, USA, 1976.

66. Xu, L.; Yu, X.; Li, M.; Chen, J.; Wang, X. Monitoring oxidative stability and changes in key volatile compounds in edible oils during ambient storage through HS-SPME/GC–MS. *Int. J. Food Prop.* **2017**, *20*, S2926–S2938. [CrossRef]

67. van Ruth, S.M.; Grossmann, I.; Geary, M.; Delahunty, C.M. Interactions between artificial saliva and 20 aroma compounds in water and oil model systems. *J. Agric. Food Chem.* **2001**, *49*, 2409–2413. [CrossRef] [PubMed]

68. Nagata, Y.; Takeuchi, N. Measurement of odor threshold by triangle odor bag method. *Odor Meas. Rev.* **2003**, *118*, 118–127.

69. Nielsen, G.D.; Hansen, L.F.; Andersen, B.; Poulsen, N.; Melchior, O. Indoor air guideline levels for formic, acetic, propionic and butyric acid. *Indoor Air* **1998**, *8*, 8–24. [CrossRef]

70. Shubert, D.; Leyba, J. *Chemistry and Physics for Nurse Anesthesia: A Student-Centered Approach*; Springer Publishing Company: New York, NY, USA, 2013.

**Sample Availability:** Samples of the compounds are not available from the authors.

*Article*

# Impact of Drying Method on the Evaluation of Fatty Acids and Their Derived Volatile Compounds in 'Thompson Seedless' Raisins

Dong Wang [1,2,3,4,†], Hafiz Umer Javed [1,4,5,†], Ying Shi [1,4], Safina Naz [6], Sajid Ali [6] and Chang-Qing Duan [1,4,*]

1   Center for Viticulture & Enology, College of Food Science and Nutritional Engineering, China Agricultural University, Beijing 100083, China; wangdong@cau.edu.cn (D.W.); qaziumerjaved@cau.edu.cn (H.U.J.); shiy@cau.edu.cn (Y.S.)
2   Beijing Food Industrial Research Institute, Beijing 100075, China
3   Beijing Industrial Technology Research Institute, Beijing 101111, China
4   Key Laboratory of Viticulture and Enology, Ministry of Agriculture, Beijing 100083, China
5   Department of Plant Sciences, School of Agriculture and Biology, Shanghai Jiao Tong University, 800 Dongchuan Road, Minhang District, Shanghai 200240, China
6   Department of Horticulture, Faculty of Agricultural Sciences and Technology, Bahauddin Zakariya University, Multan 60800, Pakistan; safinz bzu@yahoo.com (S.N.); ch.sajid15@yahoo.com (S.A.)
*   Correspondence: chqduan@cau.edu.cn; Tel.: +86-10-62738537
†   These authors contributed equally to this work.

Academic Editor: Eugenio Aprea
Received: 27 December 2019; Accepted: 19 January 2020; Published: 30 January 2020

**Abstract:** Air- and sun-dried raisins from Thompson Seedless (TS) grapes were analyzed under GC/MS to evaluate fatty acids (FAs) and their derived volatile compounds, coming from unsaturated fatty acids oxidation. A total of 16 FAs were identified in TS raisins, including 10 saturated fatty acids (SFAs) and 6 unsaturated fatty acids (USFAs). The contents of C18:0, C15:0, and C16:0 among SFAs and C18:3, C18:2 and C18:1 in USFAs were significantly higher. Furthermore, USFAs such as C16:1 and C20:1 were only identified in air-dried raisins. The principal component analysis showed the increased content of FAs and FA-derived compounds were in air-dried and sun-dried raisins, respectively. Among FA-derived compounds, 2-pentyl furan, 3-octen-2-one, 1-hexanol and heptanoic acid were more potent. This study shows that air-drying is more favorable for the production of fatty acids (SFAs and USFAs), whereas sun-drying is more advantageous in terms of fatty acid-derived volatiles.

**Keywords:** fatty acids; UFAO-derived compounds; air- and sun-drying; raisins; GC/MS

---

## 1. Introduction

Raisin is a dried grape blessed with a particularly appealing sweet taste, and nutritional and energy value. A wide variety of raisins are being used around the world either as fruits, brewing or in cultural cuisines as dessert. According to the United States Department of Agriculture [1], China is a major producer of raisins (1.90 million tons per year), and raisins have an important role in the economy. The global raisin production is about 1.25 million metric tons, with China ranked third following America and Turkey in the year 2016–17 [2]. About 20 kinds of grape varieties are cultivated for raisin production. Among these, TS is a leading grape variety and is consumed as fresh grapes, raisins, and wines. The 'TS' is a versatile variety because it can be used as a green raisin (sun-dried) or golden raisin (commercial dehydrator). The relatively high sugar content is the dominant feature of 'TS' variety, which accounts for 85–90% of the acreage in the populated raisin-growing region 'Turpan', in China [2].

Traditionally, raisins have been produced from fully ripened grapes by air- or sun-drying methods, or by using prevailing technologies such as microwave, freezing, and oven drying [3]. Various drying methods have been introduced for raisin production, but air-drying has gained more popularity than others in Turpan (China). The ultimate purpose of any drying technique is to produce top-quality raisins at a low price. The air-drying method may significantly affect fatty acids and their volatile compounds in comparison with the sun-drying process. The direct comparison of both drying methods with respect to fatty acid composition and their derived compounds has generally been overlooked so far.

The aroma is the most striking feature of consumer preferences. Raisin flavor generally increased and changed, when they are being used as an ingredient of processed food. In recent years, a large number of volatiles affecting the aroma of raisins have been identified. It has already been reported that raisin volatile compounds are produced due to fresh grapes, glycosidic binding, Maillard reaction, unsaturated fatty acids, and carotenoids [4]. FA-derived volatile compounds were the major contributor to raisin aroma [4–6], which mainly consists of six-carbon ($C_6$) and nine-carbon ($C_9$) compounds of aldehydes, alcohols, and esters. These volatiles have also been known as a major source of aroma in grapes and wine [7], and are produced by the oxidative breakdown of USFAs. This process takes place through the autoxidation of USFAs [8] or the lipoxygenase (LOX) hydroperoxide lyase (HPL) pathway, in which the major enzyme such as alcohol dehydrogenases (ADHs), LOX and HPL play an important role in volatile production [9,10].

The oxidative degradation of USFAs produces the volatile compounds, and is also a key concern of food scientists and consumers because it significantly influences sensory quality, nutritional value, and the shelf life of the product. The production of a large number of compounds, such as (*E*)-2-nonenal, (*E*,*E*)-2,4-nonadienal, (*E*)-2-octenal, heptanal, (*E*)-2-heptenal and (*E*)-2-hexenal, has already been reported by the oxidation of linoleic acid whereas decanal and octanal are produced by the oxidation of oleic acid. Pentanoic acid may originate from the degradation of SFAs [11,12]. In addition, the compounds derived from USFAs include ester, alcohol, and aldehydes, and these compounds play a pivotal role in the generation of fruity, floral and green leafy aroma in raisins.

Storage causes a deterioration in the quality, freshness, and aroma of foodstuffs, which can easily be recognized by the consumer [13]. During storage, the autoxidation of fatty acids, in addition to the formation of volatile compounds, also causes rancidity. Packaging materials fulfill several purposes, including reduction of rancidity [14], protection against adulteration, and maintenance of the anticipated quality, aroma and flavor of the product [15]. Likewise, for consumer acceptance, the aroma and flavor should be maintained during raisin storage. The concentration and composition of FAs and their derived volatile compounds produced in the air-drying raisins may differ from those of sun-drying. Therefore, the aim of our research work was to investigate the generation and change the regularity of FA and volatile compounds in TS raisin during storage.

## 2. Results

FAs are an essential component of our body that we usually take from our foods. Raisin is one of the main sources of FAs. Furthermore, their derived compounds produce different kinds of aromas (fatty, roasted, fruity, etc.). FAs and their derived compounds can be released or generated under the influence of air- and sun-drying methods during storage.

### 2.1. Fatty Acid Composition

The compositions of 16 FAs were obtained from TS raisins (air- and sun-dried) during storage as depicted in Table 1. The USFAs mainly composed of erucic acid (C22:1), paullinic acid (C20:1), linolenic acid (C18:3), linoleic acid (C18:2), oleic acid (C18:1), and palmitoleic acid (C16:1). Among USFAs, linoleic acid (C18:2) was the most dominant compound. Furthermore, the SFAs consisted of 10 major compounds, including; lignoceric acid (C24:0), tricosylic acid (C23:0), behenic acid (C22:0), arachidic acid (C20:0), stearic acid (C18:0), margaric acid (C17:0), palmitic acid (C16:0), pentadecylic acid (C15:0), myristic acid (C14:0) and lauric acid (C12:0). The C16:0 and C18:0 were the most highly significant compounds of the SFAs. Overall, the concentrations of C18:3, C18:2, C18:1, and C16:0 were higher in raisin. Both drying methods had a different effect on the FA profile during storage. There were 14 FAs that were present in both air-dried and sun-dried raisins. In addition to this, two FAs (C16:1; C20:1) were only identified in air-dried raisins.

The total concentration of SFAs, USFAs, and total FAs was comparatively higher in those raisins dried by the air as compared to those dried by the sun (Figure 1). The most striking result emerging from the data is that the concentrations of all FAs (SFAs and USFAs) were significantly higher in air-dried raisins than sun-dried raisins (Table 1). The results showed that the concentrations of C16:0, C18:0, and C18:2 were higher in air-dried raisins at the start of storage and then decreased with the time duration, while the C15:0 content increased with storage duration. In contrast, the concentrations of other FAs did not change in storage with respect to drying methods.

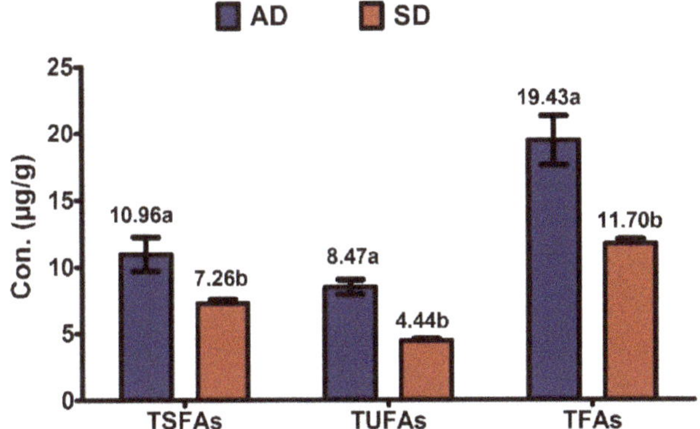


**Figure 1.** Effect of drying method on total saturated fatty acids (TSFAs), total unsaturated fatty acids (TUSFAs) and total fatty acids (TFAs) of raisins. Different lettering indicates a significance level $p < 0.005$; $n = 3$.

**Table 1.** Identification of different fatty acids (saturated and unsaturated) in air-dried and sun-dried raisins during storage.

| S/No | RT | Fatty Acids | Common Name | ID | Linear Equation | R | Air-Dried | | | | Sun-Dried | | | |
|---|---|---|---|---|---|---|---|---|---|---|---|---|---|---|
| | | | Saturated Fatty Acid | | | | 0-M | 4-M | 8-M | 12-M | 0-M | 4-M | 8-M | 12-M |
| 1 | 10.4 | C12:0 | Lauric acid | 1 | y = 0.1282x + 0.4898 | 0.999 | 2.39a | 0.56b | 0.56b | 0.56b | 0.51b | 0.49b | 0.50b | 0.50b |
| 2 | 11.8 | C14:0 | Myristic acid | 1 | y = 0.1344x + 0.919 | 0.992 | 1.16a | 1.07b | 1.09b | 1.09b | 0.99c | 0.95d | 0.95d | 0.94d |
| 3 | 12.6 | C15:0 | Pentadecylic acid | 1 | y = 0.1395x + 1.8572 | 0.991 | 1.89b | 1.90b | 1.91a | 1.91a | 1.88c | 1.86d | 1.87d | 1.86d |
| 4 | 13.4 | C16:0 | Palmitic acid | 1 | y = 0.0148x + 0.8343 | 0.990 | 1.68a | 1.37b | 1.26c | 1.27c | 0.97d | 0.91e | 0.91e | 0.92e |
| 5 | 14.4 | C17:0 | Margaric acid | 1 | y = 0.1428x + 0.5216 | 0.993 | 0.63a | 0.62a | 0.62a | 0.62a | 0.55b | 0.53b | 0.54b | 0.53b |
| 6 | 15.4 | C18:0 | Stearic acid | 1 | y = 0.1428x + 0.5216 | 0.993 | 4.41a | 3.75b | 3.45c | 3.38c | 1.14d | 1.10d | 1.04d | 1.10d |
| 7 | 18.2 | C20:0 | Arachidic acid | 1 | y = 0.2098x + 0.6079 | 0.991 | 0.99a | 0.96a | 0.95a | 0.96a | 0.69b | 0.65b | 0.67b | 0.66b |
| 8 | 21.9 | C22:0 | Behenic acid | 1 | y = 0.1511x + 0.2175 | 0.995 | 0.70a | 0.68a | 0.69a | 0.68a | 0.29b | 0.27b | 0.28b | 0.28b |
| 9 | 24.4 | C23:0 | Tricosylic acid | 1 | y = 0.1259x + 0.2172 | 0.995 | 0.28ab | 0.28ab | 0.29a | 0.29a | 0.23c | 0.23c | 0.23c | 0.23c |
| 10 | 27.6 | C24:0 | Lignoceric acid | 1 | y = 0.0953x + 0.1182 | 0.997 | 0.26a | 0.24a | 0.24a | 0.24a | 0.15b | 0.14b | 0.14b | 0.14b |
| | | | Unsaturated Fatty Acid | | | | 0-M | 4-M | 8-M | 12-M | 0-M | 4-M | 8-M | 12-M |
| 11 | 13.8 | C16:1 | Palmitoleic acid | 1 | y = 0.1042x + 0.7286 | 0.991 | 0.73a | 0.74a | 0.74a | 0.74a | NF | NF | NF | NF |
| 12 | 15.8 | C18:1 | Oleic acid | 1 | y = 0.1283x + 1.5322 | 0.993 | 2.03a | 1.51b | 1.58b | 1.62b | 0.70c | 0.66d | 0.64d | 0.66d |
| 13 | 16.4 | C18:2 | Linoleic acid | 1 | y = 0.0883x + 1.3784 | 0.993 | 3.52a | 2.78b | 2.66bc | 2.53c | 1.61d | 1.47d | 1.50d | 1.52d |
| 14 | 17.3 | C18:3 | Linolenic acid | 1 | y = 0.1006x + 0.5955 | 0.997 | 1.96a | 1.93ab | 1.90b | 1.92ab | 1.59c | 1.57c | 1.57c | 1.59c |
| 15 | 18.6 | C20:1 | Paullinic acid | 1 | y = 0.053x + 0.2264 | 0.994 | 0.24a | 0.24a | 0.24a | 0.24a | NF | NF | NF | NF |
| 16 | 22.4 | C22:1 | Erucic acid | 2 | | | 1.54a | 1.10b | 1.04b | 1.10b | 0.68c | 0.63c | 0.67c | 0.66c |

RT: retention time. ID (identification method): 1, identified, mass spectrum and RI were in accordance with standards; 2, tentatively identified, mass spectrum matched in the standard NIST 2008 library and RI matched with NIST Standard Reference Database (NIST Chemistry WebBook). Linear equation: concentration in mg/L; x, peak area ratio of a compound into the internal standard (4-methyl-2-pentanol). R: Regression coefficient.

## 2.2. Fatty Acid-Derived Volatile Compounds

Twenty-six aroma compounds, including 7 aldehydes, 9 alcohols, 4 esters, 4 acids, 1 ketone, and 1 furan, were identified in raisins during storage. Acids were noticed to have the highest content in raisins dried by both methods, followed by alcohol, aldehyde, ketone, furan, and ester. Overall, higher contents were observed in sun-dried raisins (Figure 2).

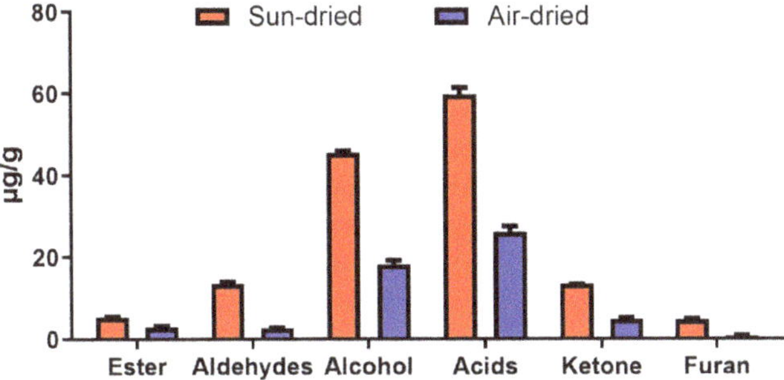


**Figure 2.** Effect of sun- and air-drying methods on the different classes of fatty acid-derived compounds.

During storage, in sun-dried raisins, the aldehydes, alcohol and furan compound contents decreased as time went by, but the ester and ketone contents increased. The concentrations of main class (acids) were less at the fresh stage compared to at 4, 8 and 12 months of storage, but similar results were found during storage (Figure 3). Furthermore, during storage, as time passed in the air-dried raisins, the concentration of ester and aldehydes increased, but the alcohol and furan concentration decreased. The compounds belonging to the acid and the ketone groups exhibited asymmetrical effects during storage (Figure 4).

A total number of 26 FA-derived volatile compounds in raisins were listed, among which 20 compounds were identified in both drying methods. Of the remainder, 5 compounds were only detected in sun-dried raisins and 1 compound was only identified in air-dried raisins (Supplementary Table S1). Ethyl octanoate, methyl octanoate, 2-ethyl-1-hexanol, (*E*)-2-heptenal, and (*E*)-2-octen-1-ol were not found in air-dried raisins, while 2-nonanol was not recognized in those raisins that were dried under the sun. A few volatile compounds such as (*Z*)-3-hexen-1-ol and methyl octanoate in sun-dried raisins, and (*E,E*)-2,4-heptadienal and decanal in air-dried raisins, were not found in the fresh sample, but were quantified during storage. Furthermore, compounds such as decanal, 1-hexanol, 3-octen-2-one and 2-pentyl furan were the leading compounds on the basis of concentration, and their contents decreased with the passage of storage. On the other hand, the concentrations of heptanoic acid and octanoic acid were directly proportional to storage intervals (Supplementary Table S1).

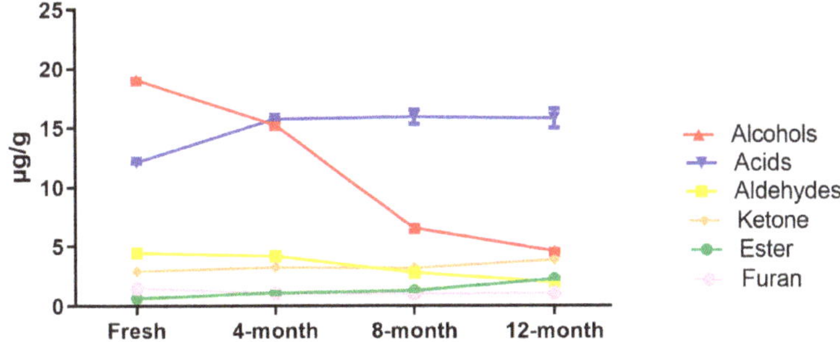


**Figure 3.** Post-storage changes of different classes of fatty acid-derived compounds in sun-dried raisins.

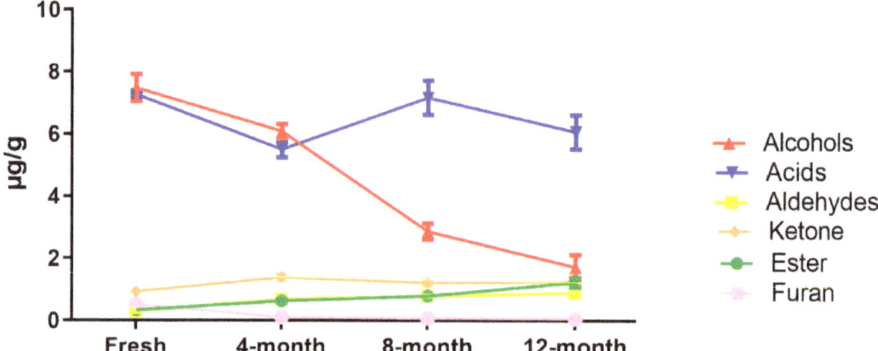

**Figure 4.** Post-storage changes of different classes of fatty acid-derived compounds in air-dried raisins.

*2.3. Effect of Drying Method on Fatty Acids and UFAO-Derived Volatile Compounds*

During storage, the FA composition and intensity of UFAO-derived volatiles of TS raisins were analyzed using principal component and K-means analysis to determine the effect of the sun- and air-drying methods on raisin FAs and their volatiles. With respect to the drying method (air and sun), the first two principal components (PCs) represented 88.36% of the total variance (Figure 5A). The sun-dried raisins were well separated from the air-dried raisins by PC1. The PC1, which accounted for 77.09% of the total variance, was characterized by all SFAs and USFAs, as well as three VOCs, including pentanol, 2-nonanol and (Z)-3-hexen-1-ol (Figure 5B). The remaining UFAO-derived compounds were considered to be the part of belonging to sun-drying methods.

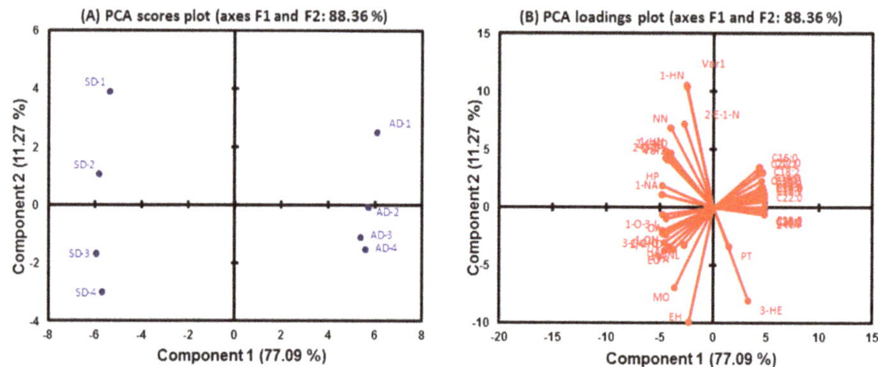

**Figure 5.** PCA (**A**) score plot for samples and (**B**) loadings plot based on fatty acids and their derived volatile compounds during storage.

*2.4. Hierarchical Cluster Analysis*

With respect to air- and sun-drying methods, Figure 6 shows the accumulation pattern of different FA-derived volatile compounds in TS raisins. The cluster of volatile compounds was drawn using hierarchical cluster analysis. The volatile compounds with a similar effect were mainly classified into two main clusters on the basis of the calculated distance. In the air-drying and sun-drying method, higher amounts of volatile compounds were noted in clusters 1 and 2, respectively. Due to the changing trend of volatile compounds during storage, cluster 2 was further divided into two sub-clusters, 3 and 4. Cluster 3 represents those compounds whose concentration decreased during storage, while increasing trends are shown in 4.

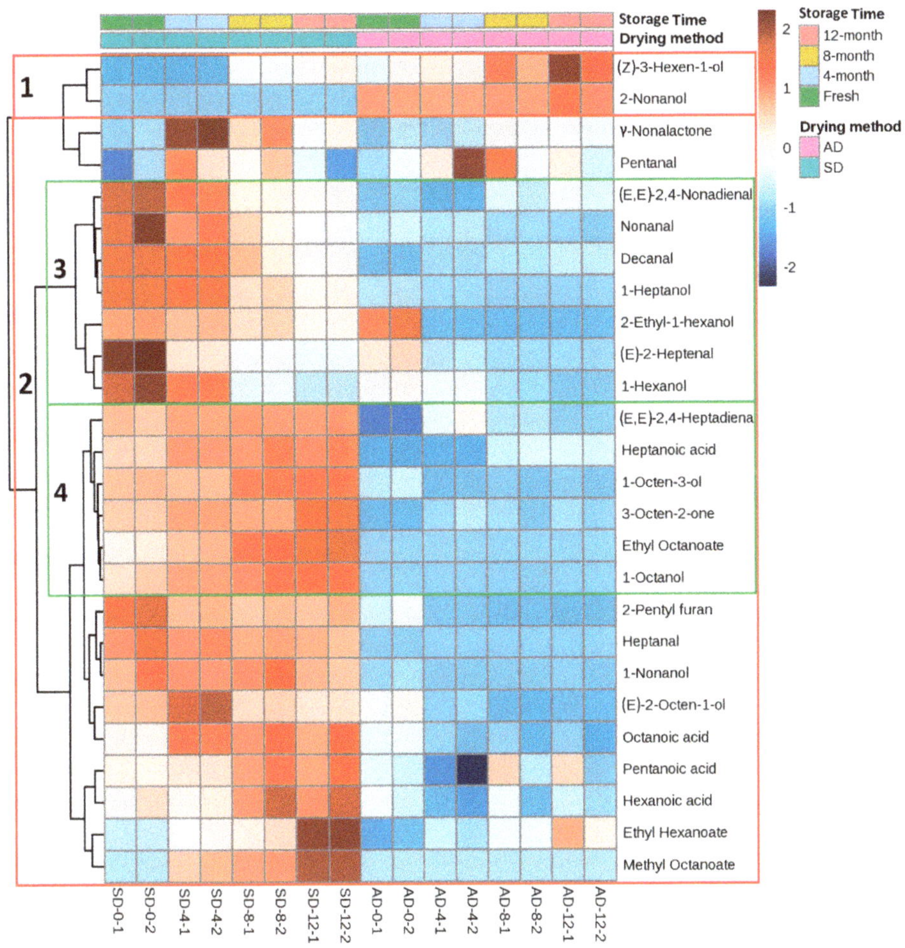



**Figure 6.** Heatmap visualization of fatty acid-derived volatile compounds of Thompson Seedless raisins.

## 3. Discussion

From a nutritional point of view, FAs play a key role in human life. In particular, USFAs have a significant importance for human health, and are responsible for decreasing the cholesterol level, as well as preventing different diseases such as cancer, heart disease, atherosclerosis, and diabetes [16]. Among USFAs, linoleic acid (C18:2) was the most dominant compound, as mentioned in the previous studies with respect to different raisins varieties [17]. As per the results, the number of USFAs, as well as the concentrations of SFAs and USFAs, was significantly higher in the air-dried raisins. The temperature and heat are significantly higher during the sun-drying of raisins as compared to air-drying, and this has been found to be more favorable for the autoxidation reaction [8], and is responsible for the conversion of FAs into their volatile compounds. Regardless of C16:0, C18:0, and C18:2, the concentration of other FAs compounds did not change during the storage of raisins.

With respect to the aroma, FAs are key precursors for the formation of flavor in different fruits [18]. Autoxidation plays a dynamic role in catalyzing USFAs into C6 and C9 volatiles (aldehydes and alcohol), which develop a distinct aroma [19]. Overall, 26 FA-derived compounds were quantified in this study that had already been described in our previous work [4,6,20]. The compounds generated by sun-drying had higher concentration compared to those generated by air-drying. This is due to higher temperature and heat, which can be considered to be favorable for producing a higher concentration of FA-derived volatile compounds [6].

The descriptions, precursors of identified compounds [11,12,21–24], and aroma descriptors [20,25–27] are presented in Table 2. The fruity, green and sweet aromas [25,26] producing compounds such as decanal, 1-hexanol, 3-octen-2-one and 2-pentyl furan were the leading compounds in terms of concentration, and their contents decreased with storage time. In addition, the fatty and cheesy aroma [26]-producing compounds (heptanoic acid and octanoic acid) were considered, and these had a direct relationship with storage. Overall the concentration of fruity, floral and herbaceous aroma-producing compounds was higher throughout the study. Only the amount of (Z)-3-hexen-1-ol was higher in air-dried raisins and increased gradually as the storage period progressed, and their precursor had not been reported in the literature. The compounds (E)-2-heptenal, (E,E)-2,4-nonadienal and nonanal were derived from linoleic acids [11,12], while decanal and 1-heptanol came from oleic acids [22,23], and showed higher concentration at fresh stage, with volatility decreasing with the extension of storage duration, which might be due to the loss of their high proportion of volatilization by evaporation [28] in storage. The derivates of oleic acid (1-octanol), linoleic acids (1-octen-3-ol, heptanoic acid, and ethyl octanoate), linolenic acid (E,E)-2,4-heptadienal) and arachidonic acid (3-octen-2-one) showed lower content in fresh raisins, with their concentration being increased as the storage time increased. In sun-dried raisin, all acidic compounds octanoic acid, heptanoic acid, pentanoic acid, and hexanoic acid showed higher concentrations when specifically stored for 8 and 12 months, but in air-dried raisin, these showed irregular trends during storage. The acid compounds were derived from the autoxidation of methyl linoleic acid [24]. Some volatile compounds are specific to the drying method, such as 2-nonanol and methyl octanoate, which were only present in air-dried and sun-dried raisins, respectively. These showed higher concentration at 12 months of storage. The only furan compound, '2-pentyl furan', came from linoleic acid [11,23] and had a higher concentration in fresh raisin as compared to storage for both drying methods. These results coincided with our previous report on changes of volatile compounds in pre-treated raisins during storage [4].

**Table 2.** The description of identified UFAO-derived compounds in Thompson seedless raisins.

| S/No | RI | Volatile Compounds | Source | ID-M | Ion ($m/z$) | Formula | Precursor | Aroma Descriptor |
|---|---|---|---|---|---|---|---|---|
| 1 | 1227 | Ethyl hexanoate | Ester | 1 | 88 | $C_8H_{16}O_2$ | Linoleic acid [F] | Fruity, apple-like [e] |
| 2 | 1432 | Ethyl octanoate | Ester | 2 | 74 | $C_{10}H_{20}O_2$ | Linoleic acid, Linolenic acid [F] | Apple, fruity, sweet [e] |
| 3 | 1378 | Methyl octanoate | Ester | 2 | 88 | $C_9H_{18}O_2$ | Oleic acid [A] | Fruity, citrus like [e] |
| 4 | 2035 | γ-Nonalactone | Ester | 2 | 85 | $C_9H_{16}O_2$ | Linoleic acid [F] | Coconut, peach [e] |
| 5 | 975 | Pentanal | Aldehyde | 2 | 44 | $C_5H_{10}O$ | Linoleic acid, Arachidonic acid [E] | Fat, Green [e] |
| 6 | 1178 | Heptanal | Aldehyde | 2 | 44 | $C_7H_{14}O$ | Linoleic acid [D,E], Oleic acid [E] | Dry fish, solvent, smoky [e] |
| 7 | 1325 | (E)-2-Heptenal | Aldehyde | 2 | 41 | $C_7H_{12}O$ | Linoleic acid [D,E] | Fatty, soapy, tallow [e] |
| 8 | 1393 | Nonanal | Aldehyde | 1 | 57 | $C_9H_{18}O$ | Linoleic acid [B] | Green, Fruity [e] |
| 9 | 1501 | Decanal | Aldehyde | 1 | 43 | $C_{10}H_{20}O$ | Oleic acid [D,E] | Sweet, citrus, green [c] |
| 10 | 1497 | (E,E)-2,4-Heptadienal | Aldehyde | 2 | 81 | $C_7H_{10}O$ | Linolenic acid [E] | Fatty, hay [a] |
| 11 | 1705 | (E,E)-2,4-Nonadienal | Aldehyde | 2 | 81 | $C_9H_{14}O$ | Linoleic acid [D,E] | Fatty, oily [a] |
| 12 | 1349 | 1-Hexanol | Alcohol | 2 | 56 | $C_6H_{14}O$ | NF | green [a] |
| 13 | 1395 | (Z)-3-Hexen-1-ol | Alcohol | 2 | 67 | $C_6H_{12}O$ | NF | Fruity, green [a] |
| 14 | 1449 | 1-Octen-3-ol | Alcohol | 1 | 57 | $C_8H_{16}O$ | Arachidonic acid [D], Linoleic acid [A] | Mushroom, fruity [e] |
| 15 | 1453 | 1-Heptanol | Alcohol | 2 | 70 | $C_7H_{16}O$ | Oleic acid [A] | Grape, sweet [b] |
| 16 | 1487 | 2-Ethyl-1-hexanol | Alcohol | 1 | 57 | $C_8H_{18}O$ | NF | Floral, sweet fruity [b] |
| 17 | 1555 | 1-Octanol | Alcohol | 1 | 56 | $C_8H_{18}O$ | Methyl Oleate [B] | Citrus, rose [c] |
| 18 | 1614 | (E)-2-Octen-1-ol | Alcohol | 2 | 57 | $C_8H_{16}O$ | Oleic acid [A] | Fatty, rancid [a] |
| 19 | 1657 | 1-Nonanol | Alcohol | 1 | 56 | $C_9H_{20}O$ | NF | Floral [a] |
| 20 | 1488 | 2-Nonanol | Alcohol | 2 | 45 | $C_9H_{20}O$ | NF | NF |
| 21 | 1740 | Pentanoic acid | Acid | 2 | 60 | $C_5H_{10}O_2$ | Methyl Linoleic acid [C] | Sweet [a] |
| 22 | 1847 | Hexanoic acid | Acid | 1 | 60 | $C_6H_{12}O_2$ | Methyl Linoleic acid [C] | Rancid, Cheese, Fatty [e] |
| 23 | 1953 | Heptanoic acid | Acid | 1 | 60 | $C_7H_{14}O_2$ | Methyl Linoleic acid [C] | Sweety, cheesy [d] |
| 24 | 2060 | Octanoic acid | Acid | 1 | 60 | $C_8H_{16}O_2$ | Methyl Linoleic acid [C] | Rancid, Cheese, Fatty [e] |
| 25 | 1416 | 3-Octen-2-one | Ketone | 2 | 55 | $C_8H_{14}O$ | Arachidonic acid [E] | Green, fruity [a] |
| 26 | 1224 | 2-Pentyl furan | Furan | 2 | 81 | $C_9H_{14}O$ | Linoleic acid [B,E] | Fruity, green, sweet [e] |

Reported the precursor (fatty acids) of their respected volatile compounds ([A] Frankel 1980; [B] Frankel, Neff, and Edward S. 1981; [C] Horvat et al. 1968; [D] Meeting, Ho, and Hartman 1994; [E] Whitfield and Mottram 1992; [F] Belitz, Garcia and Wayne, 2004); NF = not found; Retention Indices (RI): Kovats RI was calculated based on the $n$-alkane series (C6–C24) on the poly (ethylene glycol) (PEG) column under the same chromatographic conditions. Aroma descriptors were obtained from "Flavornet and human odor space" the LRI and odor database ([a] http://www.odour.org.uk/odour/index.html) and from reported literature ([b] Jiang & Zhang, 2010; [c] Wang et al., 2017; [d] Welke et al., 2014; [e] Wu et al., 2016). Identification method (ID-M): 1, identified, mass spectrum and RI were in accordance with standards; 2, tentatively identified, mass spectrum matched in the standard NIST 2008 library and RI matched with NIST Standard Reference Database (NIST Chemistry WebBook). Ion ($m/z$): The characteristic ion ($m/z$) was used for choosing the corresponding compound and evaluating the peak areas of them in order to avoid possible interference by other compounds.

## 4. Materials and Methods

TS grapevine was planted in a commercial orchard located in Turpan (at northern hemisphere with latitude 42.948 and longitude 89.179 coordinates) Xinjiang province, China. The specific air-drying method reported by our lab [6] and common sun-drying methods were used for dehydrating the fully ripe grapes (TSS > 20 °Brix), which required 42 and 28 days, respectively. The drying process continued until the weight of TS raisins remained unchangeable, up to a maximum of 3 days, and the moisture contents were less than 15%. Then, 1 kg raisins samples from each drying method (sun and air) were packed into plastic bags and preserved at room temperature (24 ± 2 °C) for 12 months (October 2014 to September 2015). The post-storage changes of TSS (°Brix), pH, and moisture content (%) of the sun- and air-dried raisins are shown in Supplementary Figure S1. The FAs and their derived volatile compounds were detected and identified after 4-month intervals of storage (i.e., at 0, 4, 8 and 12 months). All of the samples were kept in the freezer (at −40 °C) until use, and each one was analyzed in triplicate.

### 4.1. Analysis of Fatty Acids

#### 4.1.1. Sample Preparation

The extraction of FA was conducted as in our earlier study [29], with minor modifications. Firstly, raisins were deep-frozen using liquid nitrogen with added quartz sand (1:1), and then immediately pulverized by an electronic shaker.

#### 4.1.2. Extraction and Methylation

A two-gram raisin sample was added to a 50 mL glass flask with 25 mL *n*-hexane and 8 mL methanol as the extraction solvent. Prior to extraction, the solvent was shaken by an electronic oscillator for 30 min at 28 °C and the residue was collected in a trice using the same procedure. The collected organic layers were immediately put into a vacuum rotary evaporator (30 °C) for drying. After evaporation of the solvent, 5 mL of 1% $H_2SO_4$/methanol (*w/v*) solution was added to methylate the lipid extracts for 2 h at 65 °C and then cooled down on room temperature. Afterward, *n*-hexane (3 mL) and distilled water (3 mL) were added to separate the methyl ester of FA from this two-phase mixture, and this was repeated in triplicate. The hexane layer was collected and dried under a gentle stream of nitrogen. Finally, 990 µL *n*-hexane and 10 µL methyl nonadecanoate (0.04 mg/mL) were added, and filtered through the organic phase microfiltration membrane (0.22 µm) and then immediately analyzed by gas chromatography (GC) coupled with a 5975 mass spectrum (MS).

#### 4.1.3. GC-MS Condition

The FA methyl esters were detected by an Agilent GC (7890, (J&W Scientific, Folsom, CA, USA)) equipped with MS (5975) (J&W Scientific, Folsom, CA, USA). A capillary column 0.25 µm in thickness (60 m × 0.25 mm id HP-INNOWAX) (J&W Scientific, Folsom, CA, USA) was used to determine the FA. In this experiment, the temperature condition of GC-MS was quite different from our previous work [29]. Initially, the flow rate of helium (carrier gas) was 1 mL/min, and the oven temperature was 40 °C for 1 min, elevated to 220 °C at 25 °C/min, and then changed to 250 °C at 5 °C/min. In the end, the temperature was held for 20 min at 250 °C. The temperatures of the injector and transfer line were maintained at 250 °C and 280 °C, respectively.

### 4.2. Fatty Acid-Derived Volatile Compounds

#### 4.2.1. Sample Preparation

The raisin samples for the extraction of FA-derived volatile compounds were prepared according to precedent [6]. The slurries were prepared by soaking frozen raisins in an equal weight of distilled water and stored overnight at 4 °C. On the very next day, the samples were thoroughly homogenized and macerated (4 h). The pulp was immediately centrifuged (at 8000 rpm; 15 min; 4 °C) and the whole supernatant was obtained thereafter. The clear supernatant was used for the identification of free-form VOCs.

### 4.2.2. SPME Condition

For extraction, the clear supernatant (5 mL) and 4-methyl-2-pentanol (10 µL; 1.0018 mg/L; as an internal standard) were blended with a magnetic stirrer in a vial (15 mL). NaCl (1.3 g) was mixed in, and the vial was tightly capped with a stopper (PTFE-silicon) (CNW Technologies, Dusseldorf, Germany). The sample-containing vial was then equilibrated (40 min; 60 °C) on a heating stand. The extraction was executed by inserting solid-phase micro-extraction fiber (CAR/PDMS/DVB) (Supleco, Darmstadt, Germany) into the headspace over a period of 40 min, with constant heating and agitation. At the end of the extraction of volatiles, the fiber was instantly desorbed for 8 min into the GC injection port.

### 4.2.3. GC-MS Condition

The detection of FA-derived compounds was accomplished on the same GC-MS (mentioned earlier), but the oven temperature was changed, and was automated as follows: temperature and time: 50 °C for 1 min, and increased to 220 °C at a rate of 3 °C/min and then maintained at 220 °C for 5 min. Mass spectra were accomplished using the electron impact (EI) approach by obtaining the ionization energy (IE, 70 eV) and source temperature (230 °C). The full-scan approach was applied along with mass range ($m/z$; 20–450), and then further action took place in a selective ion manner under auto-tune conditions.

ChemStation software (Agilent Technologies, Santa Clara, CA, USA) was used for managing and investigating the data. The identification of compounds was carried out using retention indices (RI) of reference standards, while the mass spectra were matched with the NIST 08 (National Institute of Standards and Technologies, Gaithersburg, MD, USA) library. Wherever the reference standard was not available, tentative identification was performed by comparing the mass spectrum of the NIST library and RI that was cited in the literature.

### 4.2.4. Quantification Method

The quantification method was standardized in our previous work [6], with minor modifications. A simulated solution was prepared by taking the average concentration of the sugar and acids in the raisin supernatant. Then, the solution was prepared in distilled water containing glucose (400 g/L) and tartaric acid (5 g/L). The pH of the solution was acclimatized to 4.2 with 1 M solution of NaOH. Now, the known concentrations of standard compounds were mixed in ethanol (HPLC-grade) and diluted up to the fifteenth level using the simulated raisin solution. Each level was extracted and analyzed in a similar manner as that of the raisin supernatant. Furthermore, the volatile compounds that did not fulfill chemical standards were juxtaposed with those standards that had identical functional groups/similar numbers of C-atoms. In the end, the content of identified volatile compounds was computed by the characteristic ion peak area with regards to the IS.

### 4.3. Statistical Analysis

Heatmap cluster analysis (HCA) and principal component analysis (PCA) were performed based on the concentration of identified volatile compounds and FAs by the Metabo-Analyst 3.0 software (http://www.metaboanalyst.ca/) through time series analysis. Auto-scaling (mean-centered and divided by the standard deviation of each variable) was used to normalize the data. A two-way statistical analysis of variance was utilized to calculate the influence of drying techniques during the storage time on different classes of volatile compounds employing a significance level $p < 0.005$.

## 5. Conclusions

Raisins were prepared from TS grapes using air- and sun-drying techniques. The dried raisins were stored for 0, 4, 8 and 12 months at room temperature and the FAs were then identified and quantified along with their derived compounds under GC/MS. A total of 16 and 14 FAs were identified in air-dried and sun-dried raisins, respectively. Paullinic acid (C20:1) and palmitoleic acid (C16:1) were only identified in air-dried raisins. Overall, a higher content was seen in air-dried raisins than sun-dried raisins, specifically at the fresh stage. Generally, non-significant changing effects of FAs were seen during storage. The concentrations of FA-derived raisin compounds were significantly higher in sun-dried raisins. The compounds (*E*)-2-heptenal, (*E*,*E*)-2,4-nonadienal, nonanal, 1-hexanol, 2-ethyl-1-hexanol, decanal, 1-heptanol were higher in fresh raisins, while ester compounds (ethyl hexanoate, ethyl octanoate, and methyl octanoate) and ketone compound (3-octen-2-one) were higher after a storage period of 12 months. The outcomes of this research indicate that the air-drying method improves the nutritional quality in terms of FA, and the sun-dried method enhances the flavor of TS raisins.

**Supplementary Materials:** The following are available online: Figure S1: Effect of drying method on TSS, pH and moisture content of stored raisins, Table S1: Effect of drying methods on UFAO-derived compounds of Thompson Seedless raisins during storage.

**Author Contributions:** C.-Q.D. and Y.S. conceived the idea and designed the experiment. H.U.J. and D.W. conducted the experiment. H.U.J. wrote an initial draft of the manuscript. S.A. and S.N. edited, revised and significantly improved the whole manuscript. All authors have read and agreed to the published version of the manuscript.

**Funding:** This research was funded by the China Agriculture Research System to C.-Q. Duan. The grant number is CARS-31.

**Conflicts of Interest:** The authors declare no conflict of interest.

## Abbreviations

| | |
|---|---|
| USFAs | Unsaturated Fatty Acids |
| SFAs | Saturated Fatty Acids |
| FAs | Fatty Acids |
| TS | Thompson Seedless |
| GC/MS | Gas Chromatography/Mass Spectrometry |

## References

1. *Raisins: World Markets and Trade*; USDA: Washington, DC, USA, 2017. Available online: https://apps.fas.usda.gov/psdonline/circulars/raisins.pdf (accessed on 20 January 2020).

2. Inouye, A. *China—Peoples Republic of Raisin Annual*; Global Agricultural Information Network (GAIN): Turpan, China, 2017.

3. Christensen, L.P. *Raisin Production Manual*; UCANR Publication: Oakland, CA, USA, 2000; ISBN 978-1879906440.

4. Javed, H.U.; Wang, D.; Shi, Y.; Wu, G.F.; Xie, H.; Pan, Y.Q.; Duan, C.Q. Changes of free-form volatile compounds in pre-treated raisins with different packaging materials during storage. *Food Res. Int.* **2018**, *107*, 649–659. [CrossRef]

5. Javed, H.U.; Wang, D.; Wu, G.; Muhammad, Q.; Duana, C.-Q.; Shi, Y. Post-storage changes of volatile compounds in air- and sun-dried raisins with different packaging materials using HS-SPME with GC/MS. *Food Res. Int.* **2019**, *119*, 23–33. [CrossRef] [PubMed]

6. Wang, D.; Cai, J.; Zhu, B.Q.; Wu, G.F.; Duan, C.Q.; Chen, G.; Shi, Y. Study of free and glycosidically bound volatile compounds in air-dried raisins from three seedless grape varieties using HS-SPME with GC-MS. *Food Chem.* **2015**, *177*, 346–353. [CrossRef] [PubMed]

7. Buttery, R.G.; Turnbaugh, J.G.; Ling, L.C. Contribution of volatiles to rice aroma. *J. Agric. Food Chem.* **1988**, *36*, 1006–1009. [CrossRef]

8. Jelen, H.; Wa, E. Lipid-derived flavor compounds. In *Food Flavors: Chemical, Sensory, and Technological Properties*; Jelen, H., Ed.; CRC Press Taylor & Francis Group: Boca Raton, FL, USA, 2011; pp. 65–89.

9. Feussner, I.; Wasternack, C. Te lipoxygenase pathway. *Annu. Rev. Plant Biol.* **2002**, *53*, 275–297. [CrossRef]

10. Zhu, B.Q.; Xu, X.Q.; Wu, Y.W.; Duan, C.Q.; Pan, Q.H. Isolation and characterization of two hydroperoxide lyase genes from grape berries HPL isogenes in Vitis vinifera grapes. *Mol. Biol. Rep.* **2012**, *39*, 7443–7455. [CrossRef]

11. Whitfield, F.B.; Mottram, D.S. Volatiles from interactions of Maillard reactions and lipids. *Crit. Rev. Food Sci. Nutr.* **1992**, *31*, 1–58. [CrossRef]

12. Meeting, A.C.S.; Ho, C.-T.; Hartman, T.G. *Lipids in Food Flavors*; ACS Publications; American Chemical Society: Washington, DC, USA, 1994.

13. Rizzo, V.; Muratore, G. Effects of packaging on shelf life of fresh celery. *J. Food Eng.* **2009**, *90*, 124–128. [CrossRef]

14. Sebranek, J.; Neel, S. Rancidity and antioxidants. *WFLO Commod. Storage Man* **2008**, 1–3.

15. Quezada-Gallo, J.A.; Debeaufor, F. Mechanism of aroma transfer through edible and plastic packaging. In *Food Packaging Testing Method and Applications*; Risch, S.J., Ed.; American Chemical Society: Washington, DC, USA, 2000; pp. 125–140.

16. Khiari, R.; Zemni, H.; Mihoubi, D. Raisin processing: Physicochemical, nutritional and microbiological quality characteristics as affected by drying process. *Food Rev. Int.* **2018**, *35*, 246–298. [CrossRef]

17. Aydin, A. Determination of fatty acid compositions of some raisin cultivars in Turkey. *Asian J. Chem* **2011**, *23*, 1819–1821.

18. Núñez, V.; Monagas, M.; Gomez-Cordovés, M.C.; Bartolomé, B. Vitis vinifera L. cv. Graciano grapes characterized by its anthocyanin profile. *Postharvest Biol. Technol.* **2004**, *31*, 69–79. [CrossRef]

19. Han, Y.; Barringer, S. Formation of volatiles in the lipoxygenase pathway as affected by fruit type and temperature. *J. Exp. Food Chem.* **2015**, *1*, 1–7. [CrossRef]

20. Wang, D.; Duan, C.-Q.; Shi, Y.; Zhu, B.-Q.; Javed, H.U.; Wang, J. Free and glycosidically bound volatile compounds in sun-dried raisins made from different fragrance intensities grape varieties using a validated HS-SPME with GC–MS method. *Food Chem.* **2017**, *228*, 125–135. [CrossRef]

21. Belitz, H.-D.; Grosch, W.; Schieberle, P. *Food Chemistry*, 5th ed.; Springer: Berlin, Germany, 2004; ISBN 9783540408185.

22. Frankel, E.N. Lipid oxidation. *Prog. Lipid Res.* **1980**, *19*, 1–22. [CrossRef]

23. Frankel, E.N.; Neff, W.E.; Edward, S. Analysis of autoxidized fats by gas chromatography-mass spectrometry: VII. volatile thermal decomposition products of pure hydroperoxides from autoxidized and photosensitized oxidized methyl oleate, linoleate and linolenate. *Lipids* **1981**, *16*, 279–285. [CrossRef]

24. Horvat, R.I.; Mcfadden, W.H.; Ng, H.; Lane, W.G.; Lee, A.; Lundin, R.E.; Scherer, R. Identification of some acids from autoxidation of methyl linoleate. *J. Am. Oil Chem. Soc.* **1968**, *46*, 94–96. [CrossRef]

25. Welke, J.E.; Zanus, M.; Lazzarotto, M.; Alcaraz Zini, C. Quantitative analysis of headspace volatile compounds using comprehensive two-dimensional gas chromatography and their contribution to the aroma of Chardonnay wine. *Food Res. Int.* **2014**, *59*, 85–99. [CrossRef]

26. Wu, Y.; Duan, S.; Zhao, L.; Gao, Z.; Luo, M.; Song, S.; Xu, W.; Zhang, C.; Ma, C.; Wang, S. Aroma characterization based on aromatic series analysis in table grapes. *Sci. Rep.* **2016**, *6*, 1–16. [CrossRef]

27. Jiang, B.; Zhang, Z. Volatile compounds of young wines from cabernet sauvignon, cabernet gernischet and chardonnay varieties grown in the loess plateau region of China. *Molecules* **2010**, *15*, 9184–9196. [CrossRef]

28. Arancibia, M.Y.; López-Caballero, M.E.; Gómez-Guillén, M.C.; Montero, P. Release of volatile compounds and biodegradability of active soy protein lignin blend films with added citronella essential oil. *Food Control* **2014**, *44*, 7–15. [CrossRef]

29. Xu, X.Q.; Cheng, G.; Duan, L.L.; Jiang, R.; Pan, Q.H.; Duan, C.Q.; Wang, J. Effect of training systems on fatty acids and their derived volatiles in Cabernet Sauvignon grapes and wines of the north foot of Mt. Tianshan. *Food Chem.* **2015**, *181*, 198–206. [CrossRef]

**Sample Availability:** Samples of the compounds are not available from the authors.

 *molecules*

Article

# Effect of Different Clarification Treatments on the Volatile Composition and Aromatic Attributes of 'Italian Riesling' Icewine

Teng-Zhen Ma [1], Peng-Fei Gong [1], Rong-Rong Lu [1], Bo Zhang [1], Antonio Morata [2] and Shun-Yu Han [1,*]

[1] Gansu Key Laboratory of Viticulture and Enology, College of Food Science and Engineering, Gansu Agricultural University, Lanzhou 730070, China; matz@gsau.edu.cn (T.-Z.M.); qpxhgpf@163.com (P.-F.G.); lurongrong0416@163.com (R.-R.L.); zhangbo@gsau.edu.cn (B.Z.)

[2] Food Technology Department, Technical College of Agricultural Engineers, Technical University of Madrid, Avenida Complutense S/N, 28040 Madrid, Spain; antonio.morata@upm.es

* Correspondence: lzhansy@126.com; Tel.: +86-0931-7632-968

Academic Editor: Eugenio Aprea
Received: 31 March 2020; Accepted: 3 June 2020; Published: 8 June 2020

**Abstract:** The aim of this study was to evaluate the influence of clarification treatments on volatile composition and aromatic attributes of wine samples. 'Italian Riesling' icewines from the Hexi Corridor Region of China were clarified by fining agents (bentonite (BT) and soybean protein (SP)), membrane filtration (MF), and centrifugation (CF) methods. The clarity, physicochemical indexes, volatile components, and aromatic attributes of treated wines were investigated. Both the fining agents and mechanical clarification treatments increased the transmittance and decreased the color intensity of icewine samples. Bentonite fining significantly influenced the total sugar content, total acidity and volatile acidity. Total acidity decreased 2–3.5% and volatile acidity 2–12%. MF showed the greatest influence on total phenol content, decreasing the initial content by 12%, while other treatments by less than 8%. Volatile analysis indicated that both the categories and contents of volatile compounds of wine samples decreased. MF treatment showed the most significant influence, while SP fining showed much lower impact. Odor activity values indicated the compound with the highest odor activity in Italian Riesling icewines was β-damascenone. For this compound, BT and SP did not show significant differences, however, in MF and CF it decreased by 20% and 63%, respectively. Furthermore, with high impact on aroma were: ethyl hexanoate which reduced by 20–80% especially in MF; rose oxide which extremely reduced in MF and undetected in BT, SP, and CF; isoamyl acetate which reduced by 3–33% and linalool decreased by 10–20% and undetected for BT. Principle component analysis indicated that icewine clarified by different methods could be distinguished and positively correlated with odor-active compounds. Floral and fruity were the dominant aroma series in icewine samples followed by fatty, earthy, spicy, vegetative and pungent flavor. The total odor active value of these series significantly ($p < 0.5$) decreased in different clarification treatments. Sensory evaluation showed similar results, but the SP and CF wine samples achieved better sensory quality. This study provides information that could help to optimize the clarification of ice wines.

**Keywords:** fining; filtration; icewine; volatile compounds; aroma series; sensory analysis

## 1. Introduction

Clarity is an essential quality attribute of wine and is recognized and valued by consumers together with the taste and aroma properties. Nowadays, wine stabilization and limpidity can be enhanced by centrifugation, filtration, fining or other clarification technologies [1]. The aim of this process is to decrease turbidity by removing suspended and colloidal particles in wine, meanwhile, macromolecules such as unstable proteins, which can later be denatured or aggregated and cause stability problems can also be removed [2,3].

The simplest and most economical way of clarification is natural settling or spontaneous sedimentation, where particles causing the turbidity settles to the bottom of wine by gravity. However, this method is time consuming and there are some particles with poor settling characteristics that cannot settle [3]. The aim of centrifugation is to accelerate the settling process by rotating it very fast around an axis. It is a rapid method for removing sediments and obtaining clean, stable and ready-to-drink wines, however, this system is restricted by the volume capacity or the enormous investment cost [3]. Filtration could eliminate solids or particles by passing wines through a filter medium, however, this separation technique was sometimes restricted by the clogging of filter surfaces, the throughput, the efficiency, the cost, and the practicability [3,4]. Clarifying agent addition (fining) is a process used to modulate and protect the organoleptic properties of the wines and to ensure their physicochemical stability by preventing the formation of hazes and deposits [5–7]. Unlike filtration and sedimentation that remove mainly particulates, fining could also remove some soluble substances including aroma components, coloring matter, and polymerized tannins in wines [8].

Bentonite, polyvinylpolypyrrolidone (PVPP), and protein-based fining agents are widely used in wine clarification. Among them, bentonite was the most efficient fining agent to obtain wine protein stability, however, its non-selectiveness may reduce both wine quality and quantity [9]. Horvat et al. found that bentonite added during fermentation positively affected wine quality by enhancing the preservation of key fermentation volatiles in relation to the control and exhibited positive sensory effects [9]. Other fining agents also influence wine aroma compounds, as Gil et al. found that thiol compounds in rose wine fining were PVPP dose-dependent, and a possible explanation was that PVPP would adsorb glutathione-S-conjugates aroma precursors, thus reducing the aroma content of the finished wine [7]. Proteins used for fining could come from animal origin (including egg albumin, casein, serum albumin, gelatin and isinglass) or plant origin (obtained from wheat, rice, pea, lupin and maize). However, the issues involved with animal diseases and their possible transmission to human beings have led the restriction on the use of animal products in wine fining, which increased the viable possibility use of plant-derived proteins [8,10,11]. As liquid chromatography-mass spectrometry analysis has detected the residual egg and wheat protein in wines, the use of these proteins was restricted by their potential allergenic features [12,13]. As a consequence, the plant-based products with lower allergenic potential such as soybean protein, were in response to winemakers' interest [8,10,12]. Recently, vegetable protein fining was studied by numerous researchers in wine clarification. Granato et al. found that lentil proteins and gluten fining showed a significant influence on the total content of fermentative aroma compounds (esters and alcohols), while soybean protein showed less impact on wine aroma components [8]. Although the influence of clarification on wine volatile compounds have been identified, further study should be considered because only a limited number of volatile fractions can be perceived and contribute to wine aroma.

Icewine is a kind of sweet wine produced by naturally frozen grapes [14]. The contents of sugar, acids, and aroma compounds are concentrated in the ice-grape must which results in icewine having a rich, balanced, and intense flavor profile [15]. To ensure that top quality in icewines, aroma profiling tends to be an important priority [16]. Quality and odor profile of icewines were affected by many aspects from grape growing, winemaking, and aging technology. Among them, clarification tends to be an important technology because the high soluble solids levels of icewines make the clarification more difficult than dry wines, while a strength clarification treatment could significantly impact the final flavor and sensory quality of bottled wines. In this study, the effect of different clarification methods: membrane filtration (MF), centrifugation (CF), bentonite (BT) and soybean protein (SP) on clarity, physicochemical indexes, aroma quality and sensory characteristics of 'Italian Riesling' icewine were investigated, thus to provide useful insight and technical support in icewine clarification.

## 2. Results and Discussion

### 2.1. Chemical Composition and Transmittance of Wines

The first requirement for wine fining is to decrease turbidity or increase transmittance, thus, to get a crystal clear and translucent wine with good stability, whiles minimizing the influence on other parameters of the wine [2]. As can be seen from Figure 1A, both fining agents and mechanical clarification caused a significant increase in wine transmittance. All the treated wine samples showed a transmittance higher than 90%. Among them, MF treatment showed higher efficiency in wine clarification, which exhibited a significantly higher transmittance value (97.98%) compared to the other treatments. On the contrary, color intensity (CI) showed a significant decrease in treated wines (Figure 1B). The MF treatment showed the lowest CI value with a 35.44% decrease while SP fining only decreased by 9.54%. In conclusion, the effectiveness of the clarification would depend on the different treatments, both mechanical clarification and fining agent could increase wine transmittance, but decrease the wine color, among them, SP fining could clarify the wine and minimize the decrease in wine color intensity. This is in agreement with what other studies have found [8].

The chemical composition of the wines after clarification are shown in Table 1. Most parameters of the wine showed a decrease after clarification treatments, but the major differences revealed at total acidity, volatile acidity, and total phenolic content. BT fining showed the most significant decrease in total sugar and total acidity content, which may change the flavor, sweetness, and acidity of the wine, and was confirmed in sensory analysis. Total acidity was decreased by 2–3.5% and volatile acidity by 2–12%. BT and CF treatments decreased the protein content while wine treated by SP fining showed an increase. However, the change was not significant. Other researchers have reported that plant protein fining did not influence wine protein stability [8], and it could be settled together with wine suspended solids during the fining process [16]. Total phenol content showed a significant decrease in BT, SP, and MF treatments, which may delay or reduce the browning of the wine as Cosme and Laborde et al. reported [17,18]. MF showed the greatest influence on total phenol content decreasing the initial content by 12%, while other treatments by less than 8%.

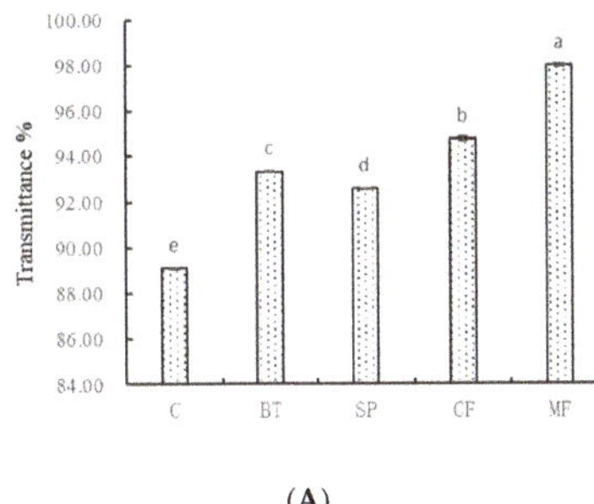

(A)

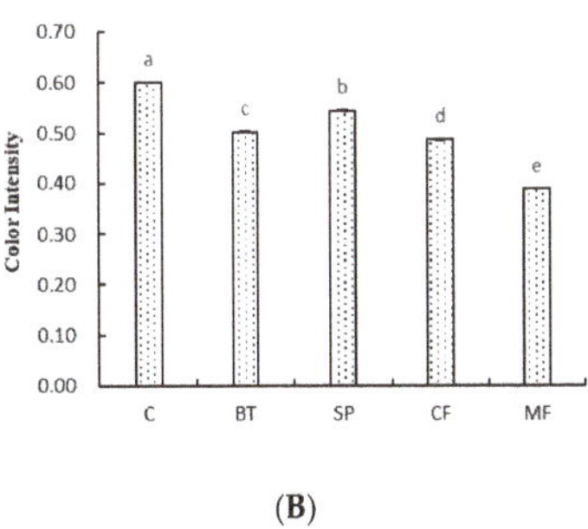

(B)

**Figure 1.** Change of (**A**) light transmittance and (**B**) color intensity in wine samples with different treatments. Different letters represent significant differences at a significant level of 0.05.

**Table 1.** Basic physicochemical indexes of wine samples processed under different clarification process.

| Treatment | Total Sugar (g/L) | Total Acidity (g/L) | Volatile Acidity (g/L) | Protein (g/L) | Total Phenol (g/L) |
|---|---|---|---|---|---|
| C | 190.00 ± 1.00 [a] | 9.74 ± 0.01 [a] | 0.51 ± 0.02 [a] | 0.115 ± 0.001 [a] | 0.477 ± 0.000 [a] |
| BT | 171.33 ± 3.21 [b] | 9.41 ± 0.05 [d] | 0.45 ± 0.02 [c] | 0.114 ± 0.001 [ab] | 0.454 ± 0.011 [b] |
| SP | 188.67 ± 0.58 [a] | 9.50 ± 0.03 [bc] | 0.50 ± 0.01 [ab] | 0.116 ± 0.000 [a] | 0.441 ± 0.006 [b] |
| CF | 189.33 ± 1.53 [a] | 9.45 ± 0.05 [cd] | 0.48 ± 0.01 [b] | 0.113 ± 0.001 [ab] | 0.470 ± 0.006 [a] |
| MF | 190.00 ± 1.00 [a] | 9.55 ± 0.03 [b] | 0.49 ± 0.01 [ab] | 0.115 ± 0.001 [a] | 0.392 ± 0.011 [c] |

Data are means ± SD ($n$ = 3). Different letters represent significant differences at a significant level of 0.05.

## 2.2. Volatile Compounds Analysis

It was reported that over 1000 volatile compounds were identified in wine. The chemical classes and character of volatile compounds, the contents, and sensory impacts influenced wine flavor complexity and was pursued by the consumers [18–20]. The interaction between clarification treatments on aroma compounds depended on chemical features of the target compounds, the membrane materials, the physical-chemical characteristics of the fining agent, and the possible interactions between volatiles and other macromolecules previously linked to the fining agent [1,8,21,22]. In order to evaluate the impact of the fining agent or mechanical clarification on wine aroma components, solid-phase microextraction (SPME) coupled with a gas-chromatography-mass spectrometer (GC/MS) were used for the identification of volatile compounds.

As can be seen from Table 2, 57 volatile compounds were identified, including 20 esters, 14 alcohols, 9 fatty acids, 9 terpenes and norisoprenoids, 3 carbonyls, and 2 volatile phenol compounds. Compared to the control wine, the categories, and contents of volatile compounds changed significantly. However, SP fining showed lower influence compared to the other treatments, which has been found to be in agreement with Granato [8,12].

**Table 2.** Volatile compounds and concentration (mg/L) with standard deviation (SD) of 'Italian Riesling' icewine samples.

| Compounds | RI DB-WAX | C mg/L ± SD | BT mg/L ± SD | SP mg/L ± SD | CF mg/L ± SD | MF mg/L ± SD |
|---|---|---|---|---|---|---|
| Esters | | | | | | |
| Ethyl acetate | 883 | 4.498 ± 0.065 [a] | 3.453 ± 0.144 [c] | 4.465 ± 0.115 [a] | 4.052 ± 0.116 [b] | 3.141 ± 0.064 [d] |
| Isobutyl acetate | 1006 | 0.002 ± 0.000 [a] | 0.001 ± 0.000 [d] | 0.002 ± 0.000 [b] | 0.001 ± 0.000 [c] | ND |
| Ethyl butyrate | 1044 | 0.049 ± 0.002 [a] | 0.027 ± 0.001 [e] | 0.042 ± 0.001 [c] | 0.046 ± 0.001 [b] | 0.034 ± 0.002 [d] |
| Isoamyl acetate | 1120 | 0.493 ± 0.012 [a] | 0.353 ± 0.006 [c] | 0.463 ± 0.007 [b] | 0.449 ± 0.028 [b] | 0.330 ± 0.007 [c] |
| Ethyl hexanoate | 1232 | 2.697 ± 0.085 [a] | 1.356 ± 0.074 [c] | 2.142 ± 0.120 [b] | 2.158 ± 0.061 [b] | 0.568 ± 0.023 [d] |
| Ethyl (*E*)-hex-2-enoate | 1245 | 0.040 ± 0.001 [a] | 0.021 ± 0.001 [c] | 0.038 ± 0.001 [b] | 0.036 ± 0.004 [b] | ND |
| Ethyl heptanoate | 1332 | 0.045 ± 0.001 [a] | 0.016 ± 0.001 [d] | 0.034 ± 0.001 [b] | 0.029 ± 0.002 [c] | 0.003 ± 0.001 [e] |
| Ethyl octanoate | 1434 | 2.870 ± 0.080 [a] | 0.768 ± 0.022 [c] | 1.706 ± 0.086 [b] | 1.790 ± 0.140 [b] | 0.574 ± 0.027 [d] |
| Ethyl nonanoate | 1530 | 0.005 ± 0.000 [a] | ND | 0.002 ± 0.000 [b] | ND | ND |
| Ethyl decanoate | 1638 | 1.737 ± 0.026 [a] | 0.758 ± 0.013 [d] | 1.249 ± 0.039 [b] | 1.023 ± 0.05 [c] | 0.27 ± 0.034 [e] |
| Ethyl benzoate | 1644 | 0.028 ± 0.001 [a] | 0.021 ± 0.001 [c] | 0.028 ± 0.002 [ab] | 0.027 ± 0.001 [b] | ND |
| Diethyl succinate | 1687 | 0.773 ± 0.009 [a] | 0.653 ± 0.023 [c] | 0.726 ± 0.017 [b] | 0.717 ± 0.030 [b] | 0.695 ± 0.036 [bc] |
| Ethyl phenylacetate | 1776 | 0.019 ± 0.002 [a] | 0.014 ± 0.001 [b] | 0.015 ± 0.001 [b] | 0.018 ± 0.001 [a] | ND |
| Phenethyl acetate | 1825 | 0.092 ± 0.001 [a] | 0.090 ± 0.003 [a] | 0.089 ± 0.004 [a] | 0.088 ± 0.003 [a] | 0.034 ± 0.001 [b] |
| Ethyl dodecanoate | 1847 | 0.727 ± 0.026 [a] | 0.444 ± 0.023 [c] | 0.655 ± 0.033 [b] | 0.377 ± 0.017 [d] | 0.075 ± 0.007 [e] |
| Ethyl 3-phenylpropionate | 1914 | 0.010 ± 0.001 [a] | 0.007 ± 0.001 [b] | 0.010 ± 0.001 [a] | 0.010 ± 0.002 [a] | ND |
| Ethyl myristate | 2043 | 0.216 ± 0.018 [a] | 0.126 ± 0.015 [b] | 0.205 ± 0.007 [a] | 0.142 ± 0.012 [b] | 0.032 ± 0.002 [c] |
| Diethyl malate | 2060 | 0.029 ± 0.001 [a] | ND | 0.028 ± 0.001 [ab] | 0.026 ± 0.003 [bc] | 0.024 ± 0.001 [c] |
| Ethyl hexadecanoate | 2243 | 0.134 ± 0.005 [a] | 0.13 ± 0.009 [ab] | 0.123 ± 0.001 [b] | 0.109 ± 0.006 [c] | 0.040 ± 0.001 [d] |
| δ-Dodecalactone | 2438 | 0.044 ± 0.001 [a] | 0.039 ± 0.002 [bc] | 0.037 ± 0.001 [c] | 0.042 ± 0.003 [ab] | ND |
| Subtotal | | 14.508 ± 0.162 [a] | 8.274 ± 0.176 [d] | 12.058 ± 0.235 [b] | 11.141 ± 0.420 [c] | 5.821 ± 0.162 [e] |
| Alcohols | | | | | | |
| Isobutanol | 1077 | 0.109 ± 0.001 [a] | 0.091 ± 0.002 [b] | 0.109 ± 0.003 [a] | 0.104 ± 0.004 [a] | 0.084 ± 0.005 [c] |
| n-Butanol | 1150 | 0.018 ± 0.001 [a] | 0.010 ± 0.001 [bc] | 0.011 ± 0.001 [b] | 0.009 ± 0.001 [c] | 0.010 ± 0.001 [bc] |
| 1-Pentanol | 1255 | 5.989 ± 0.008 [a] | 5.558 ± 0.181 [c] | 5.980 ± 0.105 [a] | 5.778 ± 0.023 [b] | 4.910 ± 0.124 [d] |
| 3-Methyl-1-pentanol | 1318 | 0.006 ± 0.001 [a] | 0.005 ± 0.000 [b] | 0.005 ± 0.000 [b] | 0.006 ± 0.000 [a] | 0.004 ± 0.000 [c] |
| 1-Hexanol | 1353 | 1.098 ± 0.011 [a] | 1.117 ± 0.038 [a] | 1.081 ± 0.035 [a] | 1.064 ± 0.033 [a] | 0.819 ± 0.018 [b] |
| cis-3-Hexen-1-ol | 1366 | 0.067 ± 0.001 [a] | 0.059 ± 0.002 [b] | 0.066 ± 0.001 [a] | 0.065 ± 0.005 [a] | 0.035 ± 0.002 [c] |
| 1-Octen-3-ol | 1447 | 0.147 ± 0.012 [a] | 0.141 ± 0.012 [a] | ND | 0.146 ± 0.006 [a] | 0.067 ± 0.007 [b] |
| Heptanol | 1449 | 0.064 ± 0.001 [a] | 0.053 ± 0.003 [c] | ND | 0.064 ± 0.002 [a] | 0.058 ± 0.003 [b] |
| 1-Octanol | 1554 | 0.052 ± 0.001 [a] | 0.052 ± 0.004 [a] | 0.047 ± 0.001 [a] | 0.050 ± 0.006 [a] | 0.045 ± 0.003 [a] |
| 2,3-Butanediol | 1556 | 0.047 ± 0.002 [a] | 0.042 ± 0.002 [b] | ND | 0.046 ± 0.004 [a] | ND |
| Nonanol | 1666 | 0.019 ± 0.001 [a] | 0.014 ± 0.001 [b] | 0.015 ± 0.001 [b] | 0.019 ± 0.001 [a] | ND |
| Decanol | 1769 | 0.022 ± 0.001 [a] | ND | 0.021 ± 0.001 [a] | 0.021 ± 0.001 [a] | 0.020 ± 0.001 [b] |
| Phenethyl alcohol | 1912 | 2.740 ± 0.010 [a] | 2.719 ± 0.099 [a] | 2.685 ± 0.111 [a] | 2.569 ± 0.219 [a] | 2.312 ± 0.085 [b] |
| Dodecanol | 1970 | 0.016 ± 0.001 [a] | 0.016 ± 0.001 [a] | 0.014 ± 0.001 [b] | 0.016 ± 0.001 [a] | 0.015 ± 0.001 [a] |
| Subtotal | | 10.393 ± 0.031 [a] | 9.876 ± 0.312 [b] | 10.033 ± 0.211 [ab] | 9.956 ± 0.235 [b] | 8.379 ± 0.240 [c] |

**Table 2.** *Cont.*

| Compounds | RI DB-WAX | C mg/L ± SD | BT mg/L ± SD | SP mg/L ± SD | CF mg/L ± SD | MF mg/L ± SD |
|---|---|---|---|---|---|---|
| Acids | | | | | | |
| Acetic acid * | 1452 | 0.621 ± 0.018 [a] | 0.593 ± 0.013 [a] | 0.606 ± 0.016 [a] | 0.603 ± 0.078 [a] | 0.451 ± 0.022 [b] |
| Isobutanoic acid | 1581 | 0.012 ± 0.001 [a] | ND | 0.012 ± 0.001 [a] | 0.011 ± 0.001 [ab] | 0.010 ± 0.000 [b] |
| 2-Methyl butanoic acid | 1655 | 0.056 ± 0.001 [a] | 0.054 ± 0.001 [ab] | 0.052 ± 0.001 [b] | 0.056 ± 0.002 [a] | 0.052 ± 0.002 [b] |
| Hexanoic acid | 1851 | 0.652 ± 0.024 [a] | 0.617 ± 0.024 [abc] | 0.598 ± 0.029 [bc] | 0.640 ± 0.024 [ab] | 0.591 ± 0.017 [c] |
| Heptoic acid | 1960 | 0.029 ± 0.002 [a] | 0.027 ± 0.001 [bc] | 0.026 ± 0.001 [bc] | 0.028 ± 0.002 [ab] | 0.025 ± 0.001 [c] |
| Octanoic acid | 2050 | 1.237 ± 0.037 [a] | 1.091 ± 0.086 [b] | 1.060 ± 0.052 [b] | 1.122 ± 0.034 [b] | 0.797 ± 0.021 [c] |
| Nonanoic acid | 2169 | 0.036 ± 0.001 [a] | 0.016 ± 0.001 [c] | ND | 0.036 ± 0.002 [a] | 0.025 ± 0.001 [b] |
| Decanoic acid | 2279 | 0.757 ± 0.011 [a] | 0.596 ± 0.016 [b] | 0.506 ± 0.019 [c] | 0.745 ± 0.013 [a] | 0.109 ± 0.006 [d] |
| Dodecanoic acid | 2502 | 0.072 ± 0.001 [a] | ND | 0.035 ± 0.003 [c] | 0.069 ± 0.002 [a] | 0.064 ± 0.001 [b] |
| Subtotal | | 3.399 ± 0.049 [a] | 2.994 ± 0.107 [c] | 2.860 ± 0.068 [c] | 3.241 ± 0.074 [b] | 2.061 ± 0.054 [d] |
| Terpenes | | | | | | |
| Rose oxide | 1337 | 0.020 ± 0.001 [a] | ND | ND | ND | 0.005 ± 0.001 [b] |
| Linalool | 1552 | 0.501 ± 0.009 [a] | ND | 0.489 ± 0.003 [ab] | 0.402 ± 0.013 [c] | 0.477 ± 0.021 [b] |
| α-Terpineol | 1680 | 0.049 ± 0.002 [a] | 0.049 ± 0.001 [ab] | 0.045 ± 0.002 [ab] | 0.045 ± 0.003 [ab] | 0.045 ± 0.003 [b] |
| Citronellol | 1750 | 0.401 ± 0.009 [a] | 0.378 ± 0.025 [a] | 0.372 ± 0.015 [a] | 0.387 ± 0.013 [a] | 0.147 ± 0.009 [b] |
| Nerol | 1806 | 0.018 ± 0.000 [a] | 0.013 ± 0.001 [b] | 0.013 ± 0.000 [b] | 0.017 ± 0.001 [a] | 0.004 ± 0.001 [c] |
| Hydroxycitronellol | 1822 | 0.008 ± 0.001 [a] | ND | 0.008 ± 0.001 [ab] | 0.007 ± 0.001 [bc] | 0.007 ± 0.001 [c] |
| β-Damascenone | 1831 | 0.125 ± 0.006 [a] | 0.119 ± 0.009 [a] | 0.112 ± 0.005 [a] | 0.080 ± 0.009 [b] | 0.046 ± 0.004 [c] |
| Geranyl acetone | 1865 | 0.100 ± 0.002 [a] | 0.018 ± 0.000 [c] | 0.038 ± 0.007 [b] | 0.098 ± 0.019 [a] | 0.091 ± 0.002 [a] |
| Geranic acid | 2334 | 0.007 ± 0.001 [a] | 0.003 ± 0.000 [c] | 0.006 ± 0.001 [b] | 0.006 ± 0.001 [b] | ND |
| Subtotal | | 1.227 ± 0.022 [a] | 0.582 ± 0.023 [d] | 1.083 ± 0.022 [b] | 1.041 ± 0.038 [b] | 0.823 ± 0.022 [c] |
| carbonyls | | | | | | |
| 6-Methyl-5-hepten-2-one | 1341 | 0.023 ± 0.001 [a] | ND | ND | 0.021 ± 0.001 [b] | 0.021 ± 0.001 [b] |
| Oct-2-enal | 1434 | 0.025 ± 0.001 [b] | 0.025 ± 0.001 [b] | 0.027 ± 0.001 [a] | 0.017 ± 0.001 [c] | ND |
| Furfural | 1468 | 0.031 ± 0.001 [ab] | 0.033 ± 0.001 [a] | 0.029 ± 0.001 [b] | 0.008 ± 0.002 [c] | ND |
| Subtotal | | 0.078 ± 0.001 [a] | 0.058 ± 0.001 [b] | 0.056 ± 0.001 [b] | 0.046 ± 0.002 [c] | 0.021 ± 0.001 [d] |
| Volatile phenols | | | | | | |
| Eugenol | 2141 | 0.022 ± 0.001 [a] | ND | ND | 0.015 ± 0.004 [b] | ND |
| Guaiacol | 2203 | 0.043 ± 0.001 [a] | 0.031 ± 0.001 [d] | 0.038 ± 0.002 [c] | 0.041 ± 0.001 [b] | 0.011 ± 0.001 [e] |
| Subtotal | | 0.065 ± 0.001 [a] | 0.031 ± 0.001 [d] | 0.038 ± 0.001 [c] | 0.056 ± 0.004 [b] | 0.011 ± 0.001 [e] |

Data are means ± SD ($n = 3$). Retention indices (RI) were reported in the NIST standard reference database. Different letters represent significant differences at a significant level of 0.05. "ND" means that the aroma component is not detected or in trace amount. * The concentration of Acetic acid was presented with g/L.

### 2.2.1. Esters

Ester is one of the most important volatile species in wine aroma. It could be processed through the chemical interactions between alcohols and acids or released by yeast strains during alcohol fermentation [19,20,23]. The esters could mainly be categorized as acetate esters, ethyl esters, and other esters, which could contribute to wine fruity aroma [20,23]. A total of 20 esters were found in the experiment wine samples (Table 2) and among them, the control wine showed the highest level in both the quantities and contents of volatile compounds. MF treatment showed the most significant influence in ester compounds, however, 9 of them were not detected and the total contents reduced by 59.87%, followed by BT fining with a 47.74% decrease, while SP fining wine showed only a 16.97% decrease in total ester content and without the absence in quantities. This is in agreement with what Vincenzi et al. found. Bentonite treatment resulted in significant removal of ethyl esters and fatty acids [24]. Ethyl hexanoate, ethyl octanoate, ethyl decanoate and ethyl dodecanoate were the dominant ethyl esters present in wine in appreciable concentrations, while ethyl acetate and isoamyl acetate were the most abundant acetate esters. Most of the esters followed the same decrease tendency in treated wines. MF wine showed the lowest concentration, whereas the contents of ethyl butyrate, diethyl succinate and diethyl malate were lower in BT fining wine.

### 2.2.2. Alcohols

Alcohol is another important volatile compound in wine which is mainly produced through yeast metabolism during wine fermentation and could contribute pungent scents to wine aroma. The aromatic contribution of alcohol depends on its concentration in wines. They could provide a pleasant aroma when their concentration bellows at 300 mg/L, on the contrary, a pungent aroma would be perceived [23,25]. A total of 14 alcohols were found in the experimental wine samples (Table 2).

There was no significant difference between control wine and SP fining wine samples in total alcohol content, however, other clarification treatments showed a significant decrease, especially the wine treated by the MF. There was no obvious influence between BT and CF treated wines. Pentanol, 1-hexanol, and phenethyl alcohol showed relatively higher concentration among alcohols, however, unlike pentanol, the concentration of 1-hexanol and phenethyl alcohol only showed a significant decrease in MF treated wines.

### 2.2.3. Acids

Acids were formed enzymatically during fermentation through yeast metabolism and were reported to contribute fruity, cheese, fatty, and rancid notes to wine aroma complexity at low concentrations. However, a negative flavor would occur if the concentration is too high [23,25]. A total of 9 acids were detected (Table 2). Among them, acetic acid, hexanoic acid, octanoic acid, and decanoic acid were recognized as the dominant acids. Compared to the control wine, clarification treatments reduced the acid concentration significantly, especially wine treated with the MF method, reduced by 38.81%, while CF wine only showed a 4.65% decrease. However, there was no obvious difference in the concentration of acid compounds between the control and the CF wines except for octanoic acid.

### 2.2.4. Terpenes and C13 Nor-Isoprenoids

Terpenes and C13 nor-isoprenoids which are derived mainly from the grapes could be from the free aromas or the glycosidic aroma precursors and are known to be varietal aroma compounds [23,26]. As can be seen from Table 2, a total of 9 compounds were identified in this experiment. Compared to the control wine, clarification treatments significantly influenced the total concentration of terpenes and C13 nor-isoprenoids compounds, and among them, BT fining showed the greatest influence with a 32.36% decrease. This finding was in contrast with Vincenzi et al. [24], who found bentonite showed a low effect on the loss of terpenes, but the doses of bentonite were much lower than this experiment. On the contrary, wines treated by SP and CF showed a lower decrease with 10.34% and 13.36%, respectively. linalool, α-terpineol, and citronellol were the dominant terpenes identified. The concentration of linalool was significantly influenced by CF and BT treatments, while the concentration of α-terpineol and citronellol only showed significant decrease in MF wine. However, there was no observed difference in other treatments. β-damascenone and geranyl acetone belonged to C13 nor-isoprenoids, surprisingly, the influence of clarification treatments on these compounds was in controversy. Fining agents, both BT and SP treatment influenced geranyl acetone significantly, while mechanical clarification CF and MF treatments showed a significant impact on β-damascenone concentration.

### 2.2.5. Carbonyls and Phenols

Three carbonyl compounds (two aldehydes and 1 ketone) were detected in wine samples (Table 2). Although the concentration of carbonyl compounds was relatively low in wine, they could contribute to the overall aroma through a synergetic effect [20,23]. 6-Methyl-5-hepten-2-one was not detected in wine samples treated with both fining agents, while oct-2-enal and furfural were not identified in MF wine samples. The total content of carbonyl compounds significantly decreased in MF wine while BT and SP wines showed much less impact. Volatile phenols, which formed by the action of hydroxycinnamate carboxylase, originated by the hydrolysis of precursors known as phenolic acids or hydroxycinnamic acids in grapes [27,28]. Clarification treatments decreased the concentration of volatile phenols significantly, especially the MF treatment, while CF treatment showed the least influence.

As has been discussed, MF treatment showed the most significant decrease in the total content of different volatile compounds, except for terpenes, which showed the maximum decrease in BT fining wine samples. Vincenzi et al. [24] suggested that the mechanism of aroma losses after bentonite fining may be due to the direct adsorption of the clay. In this experiment, we also confirmed this. On the contrary, SP fining showed the least influence on esters, alcohols, and terpene compounds, while CF and BT wines showed the least influence on fatty acids, phenol, and carbonyl compounds respectively.

There was no significant difference between SP and CF wine in total terpenes. Additionally, the total carbonyls in BT and SP wines also showed no observed difference. The influence of membrane filtration on wine aroma quality could be influenced by the types of filtration, different filter media, filtration parameters and the membrane itself [3,4]. Recently, researchers have found that cross-flow filtration, a relatively new technique, produced the highest decreases in both total phenol index (TPI) and color intensity (CI) values and removed highly polymerized phenols that are related to astringency in red wines, therefore, could be used in the same way as fining [1,29]. In this study, MF treatment showed the most significant difference in volatile compounds of 'Italian Riesling' Icewine, but the mechanism still remains to be further studied.

*2.3. Odor Activity Values*

The odor activity value (OAV) was frequently used to indicate the contribution of the volatile compounds to the overall aroma in the wine [25,26]. Generally, a volatile with OAV (>1) was suggested to exhibit contributions to the overall aroma, however, it is also reported that a compound with $0.1 < OAV < 1$ should also be considered in agreement with the acknowledged theory that sub-threshold compounds may also contribute to wine aroma through additive effects of compounds with similar structure or odor [30]. According to OAV, the volatile compounds that could significantly contribute their flavor notes to the overall aroma of icewines are listed in Table 3. A total of 16 compounds with an OAV above 1 were identified, among them, β-damascenone showed the highest OAV value, followed by ethyl hexanoate. For β-damascenone, BT and SP did not show significant differences, however, in MF and CF it decreased by 20% and 63%, respectively. Furthermore, ethyl hexanoate reduced by 20–80% especially in MF. Other compounds with high OAVs were isoamyl acetate (reduced by 3–33%) and linalool (decreased by 10–20%), but both with OAVs >10. Rose oxide also showed relatively high OAV but was only detected in the control and MF wine samples.

**Table 3.** Odor activity values (OAV > 0.1) and odor description of the would-be impact odorants of different clarified ice-wine samples.

| Compounds | Odor Description | Odor Threshold (µg/L) | Reference | Odor Activity Value | | | | | Aroma Classes |
|---|---|---|---|---|---|---|---|---|---|
| | | | | C | BT | SP | CF | MF | |
| Ethyl acetate | Pineapple, fruity, balsamic | 7500 | c | 0.60 | 0.46 | 0.60 | 0.54 | 0.42 | 2,7 |
| Isobutyl acetate | Fruity, apple | 20 | a | 0.12 | 0.07 | 0.11 | 0.07 | 0.00 | 2 |
| Ethyl butyrate | Floral, fruity | 20 | a | 2.47 | 1.34 | 2.08 | 2.30 | 1.72 | 1,2 |
| Isoamyl acetate | Banana | 30 | a | 16.45 | 11.76 | 15.44 | 14.97 | 11.01 | 2 |
| Ethyl hexanoate | Fruity, green apple | 14 | a | 192.62 | 96.85 | 153.02 | 154.18 | 40.59 | 2 |
| Ethyl heptanoate | Pineapple, fruity | 220 | c | 0.21 | <0.1 | 0.16 | 0.13 | <0.1 | 2 |
| Ethyl octanoate | Ripe fruits, pear, sweety | 240 | a | 11.96 | 3.20 | 7.11 | 7.46 | 2.39 | 2 |
| Ethyl decanoate | Pleasant fruity | 200 | b | 8.68 | 3.79 | 6.25 | 5.11 | 1.35 | 2 |
| Diethyl succinate | Fruity, cheese | 6000 | c | 0.13 | 0.11 | 0.12 | 0.12 | 0.12 | 2,3 |
| Ethyl phenylacetate | Floral, honey | 73 | c | 0.25 | 0.19 | 0.20 | 0.24 | 0.00 | 1 |
| Phenethyl acetate | Floral | 250 | a | 0.37 | 0.36 | 0.36 | 0.35 | 0.14 | 1 |
| Ethyl dodecanoate | Oily, fatty, fruity | 1500 | b | 0.48 | 0.30 | 0.44 | 0.25 | 0.05 | 2,3 |
| δ-Dodecalactone | Coconut fruity | 500 | c | 0.09 | 0.08 | 0.07 | 0.08 | 0.00 | 2 |
| Ethyl myristate | Mild waxy, soapy | 500 | d | 0.43 | 0.25 | 0.41 | 0.28 | 0.06 | 3 |
| Ethyl hexadecanoate | Mild waxy | 1000 | d | 0.13 | 0.13 | 0.12 | 0.11 | 0.04 | 3 |
| 1-Pentanol | Balsamic, bitter almond | 64,000 | b | 0.09 | 0.09 | 0.09 | 0.09 | 0.08 | 7 |
| 1-Hexanol | Herbaceous, grass, woody | 1100 | b | 1.00 | 1.02 | 0.98 | 0.97 | 0.74 | 6 |
| cis-3-Hexen-1-ol | Plant, fruity, aromatic | 400 | a | 0.17 | 0.15 | 0.16 | 0.16 | 0.09 | 6,2 |
| 1-Octen-3-ol | Mushroom | 20 | d | 7.37 | 7.05 | ND | 7.30 | 3.34 | 4 |
| Heptanol | Oily | 200 | b | 0.32 | 0.27 | ND | 0.32 | 0.29 | 3 |
| 1-Octanol | Jasmine, lemon | 800 | b | 0.06 | 0.06 | 0.06 | 0.06 | 0.06 | 3,1 |
| Decanol | Waxy, fatty | 400 | b | 0.05 | 0.00 | 0.05 | 0.05 | 0.05 | 3 |
| Phenethyl alcohol | Flowery, rose, Honey | 10,000 | a | 0.27 | 0.27 | 0.27 | 0.26 | 0.23 | 1 |
| 2-Methyl butanoic acid | Fatty, rancid, cheesy | 250 | a | 0.22 | 0.22 | 0.21 | 0.22 | 0.21 | 3 |
| Hexanoic acid | Cheese, fatty | 420 | b | 1.55 | 1.47 | 1.42 | 1.52 | 1.41 | 3 |
| Heptoic acid | Fatty | 300 | d | 0.10 | 0.09 | 0.09 | 0.09 | 0.08 | 3 |
| Octanoic acid | Rancid, cheese, fattyacid | 500 | b | 2.47 | 2.18 | 2.12 | 2.24 | 1.59 | 3 |
| Nonanoic acid | Fatty acid, dry, woody | 1400 | a | 0.07 | 0.03 | 0.00 | 0.07 | 0.05 | 3 |
| Decanoic acid | Fatty acid, dry, woody | 1400 | a | 0.54 | 0.43 | 0.36 | 0.53 | 0.08 | 3 |
| Dodecanoic acid | Laurel oil | 1000 | c | 0.07 | 0.00 | 0.03 | 0.07 | 0.06 | 3 |
| Rose oxide | Lychee | 0.2 | b | 97.89 | ND | ND | ND | 26.41 | 1 |
| Linalool | Flowery, Muscat | 25 | b | 20.03 | ND | 20.35 | 16.07 | 19.08 | 1 |
| α-Terpineol | Floral | 250 | a | 0.19 | 0.20 | 0.18 | 0.18 | 0.18 | 1 |
| Citronellol | Rose | 100 | b | 4.01 | 3.79 | 3.52 | 3.87 | 1.47 | 1 |
| β-Damascenone | Rose, sweet, flowers | 0.14 | b | 889.98 | 852.97 | 802.89 | 569.67 | 330.58 | 1 |
| Geranyl acetone | Floral | 60 | c | 1.67 | 0.30 | 0.63 | 1.63 | 1.52 | 1 |
| Geranic acid | Citric, geranium | 20 | b | 0.33 | 0.17 | 0.29 | 0.30 | 0.00 | 1 |
| Eugenol | Clove | 6 | c | 3.64 | ND | ND | 2.50 | ND | 5 |
| Guaiacol | Clove, curry | 20 | d | 2.14 | 1.53 | 1.90 | 2.05 | 0.54 | 5 |

Data are mean values. ND means that the aroma component is not detected, or it is in trace amount. Odor description (ODE) and odor threshold (OTH) as reported in the literature references a-[30]; b-[31]; c-[32]; d-[33]. Odor activity value (OAV): defined as the ratio between odor concentration and OTH. Aroma classes: each compound was attributed to one or more classes depending on odor descriptors: 1, floral; 2, fruity; 3, fatty; 4, earthy; 5, spicy; 6, vegetative; 7, pungent.

## 2.4. Principal Component Analysis

In order to explore possible differentiation among wine samples clarified by different technologies, principal component analysis (PCA) was applied to the data matrix containing the volatile compounds with OAV > 1 [20,34]. Although the interaction between compounds may be partially lost, these volatiles are the major compounds that contributed to the overall aroma of the wine samples. As principal components 1 and 2 (PC1 and PC2) represented 53.51% and 25.1% of the total variance respectively, approximately 80% of the total variance has been represented (Supplement Figure S1). The relationships/correlations between wine samples and compounds could be observed from Figure 2, where the control wine was positioned on the positive side of PC1, while SP and BT wine samples were located on the negative side. Nevertheless, the CF and MF wines were located on the positive and negative sides of PC2, respectively. In addition, rose oxide and eugenol were located on the positive side of PC1 while β-damascenone, citronellol, guaiacol, ethyl hexanoate, ethyl decanoate, octanoic acid, 1-hexnol, isoamyl acetate, ethyl butyrate, and hexanoic acid were located on the negative side (Figure 2 and Supplement Figure S2). In addition, these compounds played an important role in aromatic feature which distinguished the control wine from SP and BT wines. As for PC2, ethyl decanoate, ethyl hexanoate, eugenol, and guaiacol were located on the positive side, while octanoic acid, 1-hexnol, isoamyl acetate, ethyl butyrate, hexanoic acid, rose oxide, and geranyl acetone were located on the negative side (Figure 2 and Supplement Figure S2) which indicated that CF and MF wines could be separated by these compounds. Crandles et al. also found that most odor-active compounds were positively correlated with icewine cultivar and vintage combinations by PCA analysis [34].

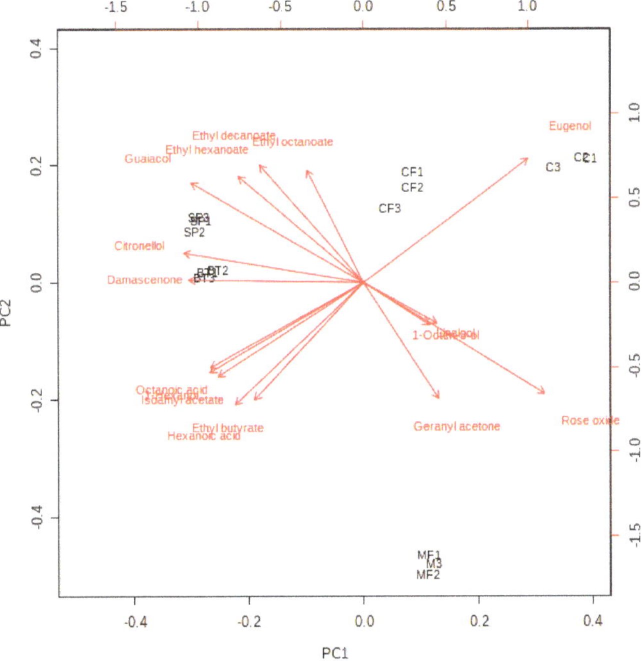


**Figure 2.** Principal component analysis of volatile compounds (OAV > 1) in wine samples with different clarification treatments.

## 2.5. Aroma Series

According to Simonetta et al. [30], wine aroma could be categorized into different series by the aroma wheel. Volatile compounds with OAV values higher than 0.1 were used to evaluate each aroma series in wine samples. In this experiment, icewine aroma was found to consist of floral, fruity, fatty, earthy, spicy, vegetative, and pungent flavor. The floral and fruity aromas were the dominant aromatic categories of 'Italian Riesling' icewine, which were mainly contributed by some terpenes (including rose oxide, linalool, citronellol, β-damascenone, and geranyl acetone) or esters (including ethyl butyrate, isoamyl acetate, ethyl hexanoate, ethyl octanoate, and ethyl decanoate compounds) (Table 3). Another study also showed that the compounds with the highest odor activity for icewines were β-damascenone, 1-octen-3-ol, ethyl octanoate, cis-rose oxide and ethyl hexanoate [35], which is similar to the results of this study. Although all tested wine samples exhibited a similar aroma series order, the content of aroma series was different which indicated that clarification treatments played an important role in altering the aromatic quality of the tested wines.

As can be seen from Figure 3, the floral was mainly influenced by mechanical clarification, where MF showed the greatest decrease followed by CF treatment. There were no observed differences between BT and SP fining wines. Fruity was the second largest category of icewine samples, in which MF treatment showed the largest decrease, followed by BT fining, while there was no significant difference between SP and CF wine samples. Fatty, earthy, spicy, vegetative, and pungent flavors could also improve wine complexity. These flavors may be contributed by fatty acids, 1-Octen-3-ol, volatile phenols, C6, and alcohol compounds, however, high levels of these flavors could also negatively influence wine aroma features [33,35,36]. A significant difference in these flavors was found between the control and clarified wines with the exception of earthy (not found in SP wine). MF wine showed the greatest decrease in these flavors, while CF treatment showed the most fatty, earthy, spicy and pungent flavors. Another interesting phenomenon is that clarification treatments showed the least influence in vegetative flavor, where only MF treatment showed a significant difference, but the mechanism still remains to be further studied.

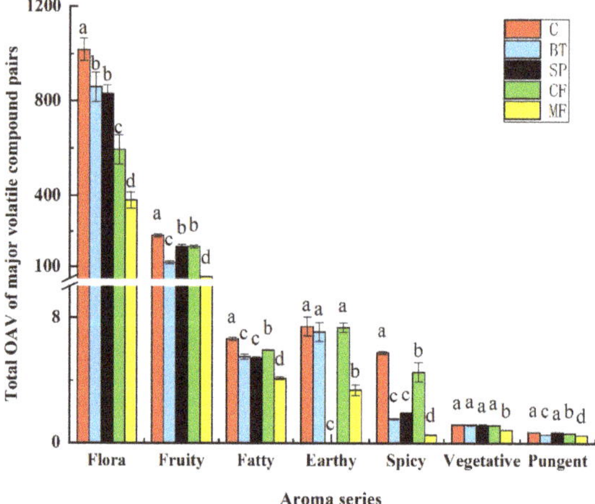


**Figure 3.** Total OAV of the aroma series after different clarification treatments. Data are means ± SD (*n* = 3). Different letters represent significant differences at a significant level of 0.05.

## 2.6. Sensory Evaluation

The icewines treated with different clarification treatments was assessed by sensory analysis to evaluate whether there were any perceivable differences among clarification treatments. As can be seen from Figure 4, clarification treatments could increase limpidity and decrease the color intensity of wine samples. Some researchers suggested that phenol compounds could be absorbed by fining or filtration, thus influence the color intensity of wine samples [17,37], which was in agreement with this study. For balance and persistence, there was no observed differences in the wine samples. Among all treated wines, BT showed the most significant influence on acidity, sweetness and complicacy.

Moreover, the aroma intensity, fruity and floral for BT and MF wines were lower than other treatments, indicated that these treatments may have a negative effect on the aroma quality of the wine samples. However, there was no observed difference between BT and MF (Figure 4). In contrast, SP and CF treatments showed little impacts on wine aroma quality (the influence was not significant) which could be alternative method in icewine clarification.

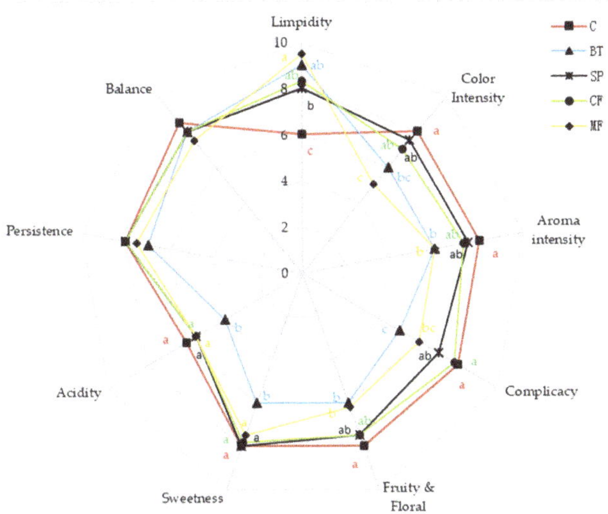


**Figure 4.** Radar map of sensory analysis. Different letters represent significant differences at a significant level of 0.05.

## 3. Materials and Methods

### 3.1. Icewine Samples

The Italian Riesling vines were planted in Gansu Qilian Winery Co., Ltd, Gaotai, Gansu Province, China (latitude 39°18′ N, altitude 99°40′ E). The vines were 22 years old and with a plant density of 1 m × 3 m and the yield was around 4500 kg/ha. Italian Riesling grapes were harvested in November 3–6, 2016 after the first snow and the temperature reduced to −7~−8 °C during the night. The grapes were picked up in the morning when they were frozen and transported to the winery immediately. After bunch sorting, the stems were removed and the grape berries were pressed in a pneumatic press (PADOVAN, Conegliano, Italy). The pressing was done in 3 stages. At the first stage, the pressure was raised to 400 mbar (low pressure) in 10 min and kept for 3 min; at the second stage, the pressure was raised to 800 mbar (middle pressure) in 10 min and kept for another 3 min; finally, the pressure was raised to 1200 mbar (high pressure) and kept for 4 min. Eighty milligram per Liter (80 mg/L) of sulfur dioxide and 35 mg/L of pectinolytic enzyme (Expresser, DSM, Cassano Spinola, Italy) were added during pressing. In this experiment, the free run must and the first pressed must were collected and transferred to a 20 m$^3$ stainless steel tank and allowed to settle for 14 h before clarification. Then clean must was obtained by removal of particles and cooled at 0 °C in an insulated tank after which 750 mg/L of bentonite was added and mixed thoroughly. After 8 days of settling, the clean must was further separated/clarified. The chemical variables of the final must were as follows: soluble solids 37.6 °Brix, total sugar content 385.12 g/L, titratable acidity 8.72 g/L, and pH 3.5. A commercial *Saccharomyces cerevisiae* strain (Aroma White, Enartis, San Martino Trecate NO, Italy) was inoculated to induce alcoholic fermentation with a 30 g/hL doses. The fermentation temperature was controlled at 10–12 °C. When the alcohol degree reached 11%/vol, 70 mg/L of sulfur dioxide was added, and the wine was cooled at a temperature of 0 °C to stop alcoholic fermentation. After natural settling, the dead yeast cells were separated from the wine and 40 L of clarified wine was transported to the laboratory and kept at 0–4 °C.

### 3.2. Chemicals and Standards

The standards of volatile compounds were purchased from Sigma-Aldrich (St. Louis, MO, USA), including isobutanol, 3-methyl-1-butanol, *cis*-3-hexen-1-ol, 1-hexanol, heptanol, octanol, phenethyl alcohol, 2,6-nonadien-1-ol, dodecanol, ethyl acetate, isoamyl acetate, ethyl lactate, ethyl hexanoate, ethyl octanoate, diethyl succinate, phenethyl acetate, ethyl decanoate, butanoic acid, hexanoic acid, octanoic acid, dodecanoic acid, octanal, (2e,6z)-2,6-nonadienal, benzaldehyde, furfural, citronellol, linalool, rose oxide, geraniol, nerol, β-damascenone, β-ionone, nerolidol, hexanal, guaiacol, 4-Ethylphenol, and 2-octanol which were used as the internal standard.

Folin–Ciocalteu reagent and sodium chloride were purchased from Beijing Chemical Works (Beijing, China). Potassium hydrogen tartrate were purchased at Kemiou Chemical Reagent Co. (Tianjin, China). Bovine serum albumin was obtained from Asahi Kasei Corporation (Tokyo, Japan). Deionized water (<18 MW resistance) was purified by using a Milli-Q purification system (Molecular, Chongqing, China). Bentonite, soybean protein and potassium metabisulfite were purchased from Lallemand Company (Lallemand, Toulouse, France).

### 3.3. Vinifications and Samples

The different clarification processes were applied as follows: (The diagram of experiment can be found in Supplement Figure S3)

(i)    8 L of wine without any treatment was kept at 0–4 °C and employed as control © wine;

(ii)   8 L of wine was clarified with bentonite (BT) and another 8 L of wine was treated with soybean protein (SP) respectively. Different amounts of fining agents were used in preliminary experiment and the stability of wine was tested by heat test. Finally, 1000 mg/L of bentonite and 500 mg/L of soybean protein were used, and the wine was recognized as protein stable. In the fining process, the wine was thoroughly mixed with fining agent solution and held at 0–4 °C for 14 days until the sediments were separated;

(iii)  Another 8 L of wine was subjected to a plate filter with GS-100 cellulose filter sheets (FLOM-men-0015, Fumei, Xiamen, China) and with a membrane diameter of 0.2 μm. When the filtration was finished, the wine was kept at 0–4 °C in a refrigerator;

(iv)   Another 8 L of wine was clarified by using a high-speed refrigerated centrifuge (H2050R, Xiangqi, Changsha, China) at 8000 r/min with controlled temperature at 5 °C. The sediments were separated, and the treated wine was kept at 0–4 °C for parameters analysis.

In the clarification processes, a total of 40 L of wine samples were distributed into 2.5 L brown glass tanks, each treatment was performed in triplicate. Samples for analysis were taken from control w©(C), wines clarified with bentonite (BT) and soybean protein (SP), as well as wines submitted to membrane filtration (MF) and centrifugation (CF). Wine samples from each treatment were kept at 0–4 °C in a refrigerator and physical-chemical parameters were measured immediately when the clarification was finished.

### 3.4. Oenological Parameters Analysis

Physico-chemical parameters of the wine samples such as the content of total sugar, total acidity, and volatile acid were measured according to the analytical methods of wine and fruit wine (National Standard of the People's Republic of China) [38]. The total sugar was measured by Fehling regent titration method whiles the total acidity and volatile acid were measured by acid–alkali titration method, and phenolphthalein was used as the indicator. The transmittance was measured by a

Genesis 10S ultraviolet-visible spectrophotometer (Thermo Fisher Scientific, Waltham, MA, USA) at 680 nm. The color intensity and total phenol content were measured according to the Compendium of International Methods of Wine and Must Analysis methods by measuring absorbance at 420 nm and 750 nm (10 mm cell) respectively using the UV-VIS spectrophotometer [39]. The content of protein was measured by Bradford assay [40].

### 3.5. Volatile Analysis

Volatile compounds extraction and analysis were proposed by Duan et al [20]. Five milliliters of icewine, 1.5 g sodium chloride (NaCl), and ten microliter (10 μL) of 2-octanol (internal standard, with the concertation of 88.2 mg/L) were loaded together in a fifteen milliliter vial, and a septum made by polytetrafluoroethylene–silicon was immediately capped. The vial was equilibrated at 40 °C under agitation condition for 30 min, and the volatile components in headspace were adsorbed by a DVB/carboxen/PDMS fiber (50/30 μm, fiber length 1 cm, Supelco, Bellefonte, PA, USA) coupled with a manual holder at above condition [20]. A gas chromatography and a mass spectrometer system (TRACE 1310- ISQ, Thermo Fisher Scientific, San Jose, CA, USA) coupled with a DB-WAX column (60 m × 0.25 mm × 0.25 μm, Agilent Technology, Santa Clara, CA, USA) were used for volatile analysis. The injection port of GC was 250 °C and the volatile compounds were released from the fiber in ten minutes with splitless mode. Helium (99.9% purity) was used as the carrier gas and the flow rate was 1.0 mL/min. Oven temperature was programed as: isothermal at 40 °C for 5 min, heating at 4 °C /min until 200 °C, and kept for 20 min. Mass spectra conditions: electron impact mode (EI) with a 70 eV ionization energy, the ion source and transfer line temperature were set at 250 °C and 200 °C, respectively; mass range was 50–350 and operated with full scan mode (3 scans/s). Volatile identification was confirmed by comparing GC retention time and mass spectra with pure standards. Volatiles without standards were identified by comparing the mass spectra with NIST 11 library ($p > 90\%$). The retention indices calculated with C7–C24 n-alkane series (Supelco, Bellefonte, PA, USA) were also compared to the NIST standard reference database. Volatile concentration with standards (reported in Section 3.2) which were calculated by standard curve. Other compounds concentrations were calculated as μg/L of internal standard 2-octanol.

### 3.6. Odour Activity Values and Aroma Series

The odor activity value (OAV) was used to indicate the contribution of different volatile compounds to the overall aroma in the wine. Generally, it is defined as the ratio of concentration over its perception threshold of a volatile compound [30]. According to Simonetta et al., the overall aroma in wine can be categorized into thirteen aroma series including chemical, pungent, oxidized, microbiologic, floral, spicy, fruity, vegetative, nutty, caramelized, woody, earthy and fatty [30]. Aroma series was calculated by summing the OAVs of volatile compounds (OAV > 0.1) in different aromatic categories.

### 3.7. Sensory Analysis

The icewines were evaluated by a panel of 8 judges, comprising members of the staff in viticulture and enology department (age range from 29 to 57 years old, 6 men and 2 women), certificated in wine tasting. The tasting was conducted from 10:00 a.m. to 12:00 p.m. in a standard sensory analysis laboratory (wine tasting room of the College of Food Science and Engineering, Gansu Agricultural University, Lanzhou City, Gansu Province, China) with an individual booth. The samples were served in a random order according to standardized procedures. Twenty milliliters (20 mL) of icewine samples were served into an odor-free tasting glass and the serving temperature was set at 8 ± 2 °C. Prior to the sensory analysis, ten attributes were selected: visual (limpidity and color intensity), aroma (aroma intensity, complicacy, fruity, and floral) and taste (sweetness, acidity, persistence, and balance). The attributes were quantified using a ten-point intensity scale. Low values were "attributed not perceptible" and on the contrast, high values were "attributed strongly perceptible". The final score was calculated for each wine as the sum of an average score of visual, aroma, and taste attributes [41].

*3.8. Statistical Analysis*

Data were analyzed by using SPSS 19.0 (IBM SPSS, Inc., Chicago, IL, USA). Significance was judged at $p < 0.05$ by analysis of variance (ANOVA) and followed by Duncan's range test. Furthermore, a principal component analysis (PCA) was carried out on the detected volatile compounds with OAV > 1 by MetaboAnalyst 4.0 after normalization (normalization by median) of data.

## 4. Conclusions

All treatments improved the limpidity of wines. Microfiltration and centrifugation achieved better results, while bentonite and soy protein fining showed higher color intensity. Bentonite fining showed the most significant decrease in total sugar and total acidity content while membrane filtration showed the biggest influence in total phenol content.

Volatile analysis indicated that clarification decreased the concentration of aroma compounds in wines. Among the various clarification treatments, MF treatment showed the most significant influence while SP fining showed much lower impact. Odor activity values evaluation demonstrated that β-damascenone, ethyl hexanoate, rose oxide, isoamyl acetate and linalool presented higher odor activity (>10) being active compounds in icewines. The floral and fruity aromas were found to be the dominant aromatic series of 'Italian Riesling' ice-wine, followed by fatty, earthy, spicy, vegetative and pungent flavors. These series decreased significantly in clarified wines where membrane filtration showed the most significant influence in floral, fruity, fatty and spicy aromas while earthy were mainly influenced by soybean protein fining.

Sensory evaluation demonstrated that clarification could improve limpidity and decrease the color intensity of wine samples. The aroma and taste properties of the wine samples were more influenced by bentonite fining, while membrane filtration mainly influenced color and aroma. SP and CF treatments achieved better sensory quality. These results suggested that clarification possessed significant influence on wine composition, thus soybean protein and centrifugation could be an alternative method in icewine clarification to minimize possible negative effects on wine quality. Moreover, clarification is a continuous and fast technique that do not produce adjuvant wastes as the other techniques.

**Supplementary Materials:** The following are available online, Figure S1: Principal component analysis score plot of volatile compounds (OAV > 1) in wine samples with different clarification treatments, Figure S2: Principal component analysis loading plot of volatile compounds (OAV > 1) in wine samples with different clarification treatments, Figure S3: The diagram of experiment.

**Author Contributions:** T.-Z.M., performed the experiments and wrote the study; P.-F.G. performed the experiments; R.-R.L. and B.Z. assisted in data analysis; A.M. and S.-Y.H. conceived and designed the experiments. All authors have read and agreed to the published version of the manuscript.

**Funding:** This research was funded by the Gansu Development and Reform Commission Program (Grant No. GSSKS-2015-153-3) and the College Research Plan of Gansu Agricultural University (Grant No. GSAU-ZL-2018-8 and GSPTJ-2017-02).

**Acknowledgments:** The author would like to thank Chen Yanxiong from Qilian winery (Gansu, China) for providing icewine samples and Sam Faisal Eudes for the language corrections.

**Conflicts of Interest:** The authors declare no conflict of interest.

## References

1. Lapuente, L.; Guadalupe, Z.; Ayestaran, B. Effect of egg albumin fining, progressive clarification and crossflow microfiltration on the polysaccharide and proanthocyanidin composition of red varietal wines. *Food Res. Int.* **2017**, *96*, 235–243. [CrossRef] [PubMed]

2. Mierczynska, V.A.; Smith, P.A. Current state of knowledge and challenges in wine clarification. *Aust. J. Grape Wine Res.* **2015**, *21*, 615–626. [CrossRef]

3.  Ribereau, G.; Glories, Y.; Maujean, A.; Dubourdieu, D. Clarifying Wine by Filtration and Centrifugation. In *Handbook of Enology: The Chemistry of Wine and Stabilization and Treatments*, 2nd ed.; John Wiley: Hoboken, NJ, USA, 2006; Volume 2, pp. 333–367. [CrossRef]

4.  Carrazana, A.; Saeznavarrete, C.; Bordeu, E. Membrane filtration effects on aromatic and phenolic quality of Cabernet Sauvignon wines. *J. Food Eng.* **2005**, *68*, 363–368. [CrossRef]

5.  Lambri, M.; Dordoni, R.; Silva, A.; De Faveri, D.M. Comparing the impact of bentonite addition for both must clarification and wine fining on the chemical profile of wine from Chambave Muscat grapes. *Int. J. Food Sci. Technol.* **2012**, *47*, 1–12. [CrossRef]

6.  Smith, P.A.; Mcrae, J.M.; Bindon, K.A. Impact of winemaking practices on the concentration and composition of tannins in red wine. *Aust. J. Grape Wine Res.* **2015**, *21*, 601–614. [CrossRef]

7.  Gil, M.; Louazil, P.; Iturmendi, N.; Moine, V.; Cheynier, V.; Saucier, C. Effect of polyvinylpolypyrrolidone treatment on rosés wines during fermentation: Impact on color, polyphenols and thiol aromas. *Food Chem.* **2019**, *295*, 493–498. [CrossRef]

8.  Granato, T.M.; Nasi, A.; Ferranti, P.; Iametti, S.; Bonomi, F. Fining white wine with plant proteins: Effects of fining on proanthocyanidins and aroma components. *Eur. Food Res. Technol.* **2014**, *238*, 265–274. [CrossRef]

9.  Horvat, I.; Radeka, S.; Plavša, T.; Lukić, I. Bentonite fining during fermentation reduces the dosage required and exhibits significant side-effects on phenols, free and bound aromas, and sensory quality of white wine. *Food Chem.* **2019**, *285*, 305–315. [CrossRef]

10. Maury, C.; Sarnimanchado, P. Influence of Fining with Plant Proteins on Proanthocyanidin Composition of Red Wines. *Am. J. Enol. Vitic.* **2003**, *54*, 105–111. [CrossRef]

11. González, N.G.; Favre, G.; Gil, G. Effect of fining on the color and pigment composition of young red wines. *Food Chem.* **2014**, *157*, 385–392. [CrossRef]

12. Granato, T.M.; Ferranti, P.; Iametti, S.; Bonomi, F. Affinity And Selectivity Of Plant Proteins For Red Wine Components Relevant To Color And Aroma Traits. *Food Chem.* **2018**, *256*, 235–243. [CrossRef] [PubMed]

13. Patrick, W.; Hans, S.; Angelika, P. Determination of the Bovine Food Allergen Casein in White Wines by Quantitative Indirect ELISA, SDS-PAGE, Western Blot and Immunostaining. *J. Agric. Food Chem.* **2009**, *57*, 8399–8405. [CrossRef] [PubMed]

14. Wang, J.; Li, M.; Li, J.; Ma, T.; Han, S.; Morata, A.; Lepe, J.A.S. Biotechnology of ice wine production. In *Advances in Biotechnology for Food Industry*; Elsevier: Amsterdam, The Netherlands, 2018; pp. 267–300. [CrossRef]

15. Bowen, A.J.; Reynolds, A.G. Aroma compounds in Ontario Vidal and Riesling icewines. I. Effects of harvest date. *Food Res. Int.* **2015**, *76*, 540–549. [CrossRef] [PubMed]

16. Simonato, B.; Mainente, F.; Selvatico, E.; Violoni, M.; Pasini, G. Assessment of the fining efficiency of zeins extracted from commercial corn gluten and sensory analysis of the treated wine. *LWT-Food Sci. Technol.* **2013**, *54*, 549–556. [CrossRef]

17. Cosme, F.; Capão, I.; Filipe-Ribeiro, L.; Bennett, R.N.; Mendes-Faia, A. Evaluating potential alternatives to potassium caseinate for white wine fining: Effects on physicochemical and sensory characteristics. *LWT-Food Sci. Technol.* **2012**, *46*, 382–387. [CrossRef]

18. Laborde, B.; Moine-Ledoux, V.; Richard, T.; Saucier, C.; Dubourdieu, D.; Monti, J.P. PVPP-polyphenol complexes: A molecular approach. *J. Agric. Food Chem.* **2006**, *54*, 4383–4389. [CrossRef]

19. Sánchez-Palomo, E.; Delgado, J.A.; Ferrer, M.A.; Viñas, M.G. The aroma of La Mancha Chelva wines: Chemical and sensory characterization. *Food Res. Int.* **2019**, *119*, 135–142. [CrossRef]

20. Duan, W.P.; Zhu, B.Q.; Song, R.R.; Zhang, B.; Lan, Y.B.; Zhu, X.; Duan, C.Q.; Han, S.Y. Volatile composition and aromatic attributes of wine made with, *Vitisvinifera* L. cv Cabernet Sauvignon grapes in the Xinjiang region of China: Effect of different commercial yeasts. *Int. J. Food Prop.* **2018**, *21*, 1423–1441. [CrossRef]

21. Moio, L.; Ugliano, M.; Gambuti, A.; Genovese, A.; Piombino, P. Influence of clarification treatment on concentrations of selected free varietal aroma compounds and glycoconjugates in Falanghina (*Vitis vinifera* L.) must and wine. *Am. J. Enol. Vitic.* **2004**, *55*, 7–12. [CrossRef]

22. Dufour, C.; Bayonove, C.L. Interactions between Wine Polyphenols and Aroma Substances. An Insight at the Molecular Level. *J. Agric. Food Chem.* **1999**, *47*, 678–684. [CrossRef]

23. Robinson, A.L.; Boss, P.K.; Solomon, P.S.; Trengove, R.D.; Heymann, H.; Ebeler, S.E. Origins of Grape and Wine Aroma. Part 1. Chemical Components and Viticultural Impacts. *Am. J. Enol. Vitic.* **2014**, *65*, 1–24. [CrossRef]

24. Vincenzi, S.; Panighel, A.; Gazzola, D.; Flamini, R.; Curioni, A. Study of Combined Effect of Proteins and Bentonite Fining on the Wine Aroma Loss. *J. Agric. Food Chem.* **2015**, *63*, 2314–2320. [CrossRef] [PubMed]

25. Swiegers, J.H.; Bartowsky, E.J.; Henschke, P.A.; Pretorius, I. Yeast and bacterial modulation of wine aroma and flavour. *Aust. J. Grape Wine Res.* **2005**, *11*, 139–173. [CrossRef]

26. Mateo, J.J.; Jiménez, M. Monoterpenes in grape juice and wines. *J. Chromatogr. A* **2000**, *881*, 557–567. [CrossRef]

27. Lan, Y.B.; Xiang, X.F.; Qian, X.; Wang, J.M.; Ling, M.Q.; Zhu, B.Q.; Liu, T.; Sun, L.B.; Shi, Y.; Reynolds, A.G.; et al. Characterization and differentiation of key odor-active compounds of 'Beibinghong' icewine and dry wine by gas chromatography-olfactometry and aroma reconstitution. *Food Chem.* **2019**, *287*, 186–196. [CrossRef]

28. Parker, M.; Capone, D.L.; Francis, I.L.; Herderich, M.J. Aroma Precursors in Grapes and Wine: Flavor Release during Wine Production and Consumption. *J. Agric. Food Chem.* **2018**, *66*, 2281–2286. [CrossRef]

29. Oberholster, A.; Carstens, L.; Toit, W. Investigation of the effect of gelatine, egg albumin and cross-flow microfiltration on the phenolic composition of Pinotage wine. *Food Chem.* **2013**, *138*, 1275–1281. [CrossRef]

30. Simonetta, C.; Maria, T.; Siciliano, P. Analytical characterisation of Negroamaro red wines by "Aroma Wheels". *Food Chem.* **2013**, *141*, 2906–2915. [CrossRef]

31. Cai, J.; Zhu, B.Q.; Wang, Y.H.; Lu, L.; Lan, Y.B.; Reeves, M.J.; Duan, C.Q. Influence of pre-fermentation cold maceration treatment on aroma compounds of Cabernet Sauvignon wines fermented in different industrial scale fermenters. *Food Chem.* **2014**, *154*, 217–229. [CrossRef]

32. Peng, C.T.; Wen, Y.; Tao, Y.S.; Lan, Y.Y. Modulating the Formation of Meili Wine Aroma by Prefermentative Freezing Process. *J. Agric. Food Chem.* **2013**, *61*, 1542–1553. [CrossRef]

33. Ferreira, V.; López, R.; Cacho, J.F. Quantitative determination of the odorants of young red wines from different grape varieties. *J. Sci. Food Agric.* **2000**, *80*, 1659–1667. [CrossRef]

34. Crandles, M.; Reynolds, A.G.; Khairallah, R.; Bowen, A. The effect of yeast strain on odor active compounds in Riesling and Vidal blanc icewines. *LWT-Food Sci. Technol.* **2015**, *64*, 243–258. [CrossRef]

35. Bowen, A.J.; Reynolds, A.G. Odor Potency of Aroma Compounds in Riesling and Vidal blanc Table Wines and Icewines by Gas Chromatography–Olfactometry–Mass Spectrometry. *J. Agric. Food Chem.* **2012**, *60*, 2874–2883. [CrossRef] [PubMed]

36. Styger, G.; Prior, B.; Bauer, F.F. Wine flavor and aroma. *J. Ind. Microbiol. Biotechnol.* **2011**, *38*, 1145–1159. [CrossRef] [PubMed]

37. Ghanem, C.; Taillandier, P.; Rizk, M.; Rizk, Z.; Nehme, N.; Souchard, J.P.; El Rayess, Y. Analysis of the impact of fining agents types, oenological tannins and mannoproteins and their concentrations on the phenolic composition of red wine. *LWT-Food Sci. Technol.* **2017**, *83*, 101–109. [CrossRef]

38. *Analytical Methods of Wine and Fruit Wine*; GB/T 15038-2006; Standards Press of China: Beijing, China, 2006.

39. International Organization of Vine and Wine (OIV). *Compendium of International Methods of Wine and Must Analysis*; Standards Press of China: Beijing, China, 2014; pp. 86–87, 98–99.

40. Bradford, M.M. A rapid and sensitive method for the quantitation of microgram quantities of protein utilizing the principle of protein-dye binding. *Anal. Biochem.* **1976**, *72*, 248–254. [CrossRef]

41. Ribeiro, T.; Fernandes, C.; Nunes, F.M.; Filipe-Ribeiro, L.; Cosme, F. Influence of the structural features of commercial mannoproteins in white wine protein stabilization and chemical and sensory properties. *Food Chem.* **2014**, *159*, 47–54. [CrossRef]

**Sample Availability:** Samples of the icewines are available from the authors.

*Article*

# Fresh and Aromatic Virgin Olive Oil Obtained from Arbequina, Koroneiki, and Arbosana Cultivars

**Alfonso M. Vidal, Sonia Alcalá, Antonia De Torres, Manuel Moya, Juan M. Espínola and Francisco Espínola ***

Department of Chemical, Environmental and Materials Engineering, Universidad de Jaén, Campus Las Lagunillas, 23071 Jaén, Spain; amvidal@ujaen.es (A.M.V.); salcala@ujaen.es (S.A.); antorres@ujaen.es (A.D.T.); mmoya@ujaen.es (M.M.); juan1411994@hotmail.com (J.M.E.)
* Correspondence: fespino@ujaen.es; Tel.: + 34-953-21-29-48

Academic Editor: Eugenio Aprea
Received: 14 September 2019; Accepted: 3 October 2019; Published: 5 October 2019

**Abstract:** Three factors for the extraction of extra virgin olive oil (EVOO) were evaluated: diameter of the grid holes of the hammer-crusher, malaxation temperature, and malaxation time. A Box–Behnken design was used to obtain a total of 289 olive oil samples. Twelve responses were analyzed and 204 mathematical models were obtained. Olives from super-intensive rainfed or irrigated crops of the Arbequina, Koroneiki, and Arbosana cultivars at different stages of ripening were used. Malaxation temperature was found to be the factor with the most influence on the total content of lipoxygenase pathway volatile compounds; as the temperature increased, the content of volatile compounds decreased. On the contrary, pigments increased when the malaxation temperature was increased. EVOO from irrigated crops and from the Arbequina cultivar had the highest content of volatile compounds. Olive samples with a lower ripening degree, from the Koroneiki cultivar and from rainfed crops, had the highest content of pigments.

**Keywords:** super-high-density orchard; aroma; color; flavor; response surface methodology

---

## 1. Introduction

Spain, and particularly, the Spanish province of Jaén, is one of the great world powers in the production of virgin olive oil (VOO). Spanish olive oil production represents approximately 45% of the world total. Furthermore, due to its great extension, olive culture represents one of the principal economic sectors in Jaén. VOO is a fat that is known worldwide for its benefits to human health. Olive oil consumption in the Mediterranean diet is associated with the prevention of several diseases and low mortality from cardiovascular disease [1].

Furthermore, consumers greatly appreciate extra virgin olive oil (EVOO) for its aromatic characteristics and intense color. The specific flavor of EVOO defines its quality and uniqueness [2]. Consumers demand EVOO with a balanced composition. To achieve a balanced final product, several studies to optimize the extraction conditions and agronomic factors have been conducted [3–5].

The first aspects of olive oil that the consumer detects are its color and smell, which are important attributes in terms of the perception of food products. Chlorophylls and carotenoids are the main pigments responsible for the color of green olives and these photosynthetic pigments are present in olive oil [6]. Therefore, olive oil produced from green olives with a low degree of maturation will take on a green color, while more mature purple olives impart a golden color. The volatile compounds present in VOO play an important role in the perception of the odor of VOO. The volatile C6 and C5 aldehydes, alcohols, and esters in VOO are mainly produced via the lipoxygenase (LOX) pathway; the formation of these compounds begins due to cell disruption during the crushing of the olives and continues throughout the extraction process [7,8]. More than 180 different aromatic compounds have been identified in olive oils [9] but only a small fraction of the large number of volatile compounds present in olive oil actually contribute to its overall aroma. In addition, these compounds are responsible for the positive attributes of olive oils and are indicators of a high-quality EVOO [10]. Phenolic compounds play an important role in the intensity of the release of certain aroma compounds during the consumption of EVOO. These compounds interact with certain volatile compounds; the resulting complexes probably reduce their volatility during the organoleptic perception of olive oil [11].

During olive oil production, many factors influence the extraction process and product quality. The cultivar of olives used is one of the most important factors and significantly influences the volatile compound composition. Growing the same olive cultivar in different locations results in oils with different volatile profiles [9]. In addition to the cultivar, the irrigation management, stage of maturity, and conditions used in the extraction process are other principal factors. The malaxation time and temperature are among the main technological variables during the extraction process. Modifying the EVOO extraction conditions can change the oil composition [12]. Agronomic practices are also a relevant factor. Traditionally, olive trees have been grown under rainfed conditions. However, new crops are adapted to irrigation techniques, increasing the total production of VOO [13].

The objective of this work was to determine the optimum extraction conditions in terms of maximizing the content of volatile compounds and pigments in the oil using a response surface methodological approach. The obtained mathematical models allow mathematical optimization, producing virgin olive oil that is desirable to consumers. Mathematical models were developed for agronomic factors, namely, the use of irrigation or rain-feeding and the ripening time of the olives, for the Arbequina, Koroneiki, and Arbosana cultivars.

## 2. Results and Discussion

Table 1 shows the results of the characterization of the samples. The samples had maturity index (MI) values ranging from 0 to 3. As expected, the oil content increased with an increasing MI of the olive samples. Samples from rainfed crops produced a higher percentage of olive oil than irrigated ones, probably due to the higher proportion of moisture present in the irrigated samples. On the other hand, irregular rainfall that occurred during the olive crop harvest due to climate change was probably the reason that the moisture and solids content did not follow a linear trend.

Considering the number of olive samples in Table 1 and the runs of each sample, a total of 289 oils were obtained and 12 responses, including volatile compounds and photosynthetic pigments, were analyzed for each oil. The tables and figures presented in this paper are only an example of the results obtained.

As an example, the experimental values corresponding to a sample of the Arbequina cultivar (irrigated and MI = 1.11) obtained from the twelve responses studied for each olive sample are shown in Table 2. Seventeen similar tables were produced in this work. These results were processed using the Design-Expert software and mathematical models were obtained. These mathematical models can help to predict the optimal values of the technological factors to maximize each response within the range of factors studied. Table 3 shows the proposed models in terms of the actual factors with the statistical parameters and optimal values of the technological factors to obtain maximum response values for a sample of the Arbequina cultivar (irrigated, MI = 1.11). Seventeen tables similar to Table 3 were produced in this work.

Table 4 shows the maximum values for each of the volatile compound and pigment responses for all samples. Finally, Table 5 shows the optimal values of the technological factors related to the extraction for two important responses: the total volatile compound and chlorophyll contents.

### 2.1. LOX Pathway Volatile Compounds

Among the agronomic factors studied, the stage of maturation had only a small effect on the total volatile compounds, probably because the maturation stages studied were very similar (Table 4). However, the differences between the oils obtained from irrigated and rainfed olives were significant, as were those among the oils derived from different cultivars. Similar differences were reported by Gómez-Rico et al. [14] in a study involving the Arbequina, Cornicabra, Morisca, Picolimon, Picudo, and Picual cultivars. Olive oils originating from irrigated crops had a higher content of volatile compounds than those from rainfed crops for a given cultivar. Comparing the three cultivars, the irrigated Arbequina (MI = 1.11) sample had the highest content of volatile compounds at 34.94 mg/kg; the sample with the lowest content was the rainfed Koroneiki sample (MI = 1.68) at 6.77 mg/kg, which was five times smaller. According to the results obtained by García et al. [15], the oils from irrigated orchards had higher contents of (*E*)-2-hexenal, the major volatile compound, than oils from rainfed orchards.

According to Fregapane et al. [16], the volatile compounds most affected by irrigation are (*E*)-2-hexenal, (*Z*)-3-hexenol, and hexanol. The increase in the amount of water supplied to the olive trees significantly increases the contents of these volatile compounds. It is noteworthy that the (*E*)-2-hexenal content of the Arbequina cultivar samples was very high (29.39 mg/kg for the irrigated sample with MI = 1.11), compared with that of the Koroneiki samples (4.37 mg/kg for the irrigated sample with MI = 0.67); these compounds are very important contributors to the delicate green perception of EVOO. Our results are similar to those obtained by Cherfaoui et al. [17], which showed that the contents of volatile compounds and (*E*)-2-hexenal increased with maturity, reaching a maximum concentration when the color of the olive fruit skin changed from purple to black. As the MI increased, the contents of linoleic and linolenic FA decreased, whereas the contents of volatile compounds from the LOX pathway increased. This could be attributed to the consumption of these two FA, which are used as substrates for the LOX enzyme present in the olive pulp [18].

**Table 1.** Compositional characteristics and maturity index of the processed olives *.

| Variety | Irrigation | Maturity Index (MI) | Moisture (%) | Oil (%) | Solids (%) |
|---|---|---|---|---|---|
| Arbequina | Rainfed | 0.20 | 53.72 ± 1.13 [a] | 15.89 ± 0.49 [a] | 30.39 ± 0.68 [a] |
| | | 1.31 | 55.88 ± 0.38 [b] | 17.78 ± 0.10 [b] | 26.34 ± 0.39 [b] |
| | | 2.56 | 54.96 ± 0.78 [b] | 20.05 ± 0.59 [c] | 24.99 ± 0.81 [c] |
| | Irrigated | 0.16 | 60.93 ± 0.70 [c] | 12.36 ± 0.25 [d] | 26.71 ± 0.63 [b,d] |
| | | 1.11 | 61.78 ± 0.43 [c,d] | 15.13 ± 0.19 [e] | 23.08 ± 0.47 [e] |
| | | 2.52 | 61.96 ± 0.28 [d,e] | 16.19 ± 0.28 [a] | 21.85 ± 0.23 [f] |
| Koroneiki | Rainfed | 0.16 | 53.11 ± 0.85 [a,f] | 16.06 ± 0.18 [a] | 30.82 ± 0.81 [a] |
| | | 1.68 | 53.45 ± 0.56 [a,f] | 21.73 ± 0.15 [f] | 24.82 ± 0.59 [c] |
| | | 2.05 | 52.60 ± 0.59 [f] | 20.41 ± 0.19 [c] | 26.99 ± 0.62 [b,d] |
| | Irrigated | 0.07 | 59.91 ± 0.78 [g] | 11.65 ± 0.29 [g] | 28.44 ± 0.83 [g] |
| | | 0.67 | 59.03 ± 0.40 [g] | 16.25 ± 0.17 [a] | 24.71 ± 0.53 [c] |
| | | 2.30 | 53.29 ± 0.56 [a,f] | 19.38 ± 0.34 [h] | 27.33 ± 0.87 [d] |
| Arbosana | Rainfed | 0.15 | 55.38 ± 0.87 [b] | 17.23 ± 0.38 [i] | 27.39 ± 0.65 [d] |
| | | 0.95 | 59.89 ± 0.64 [g] | 15.92 ± 0.38 [a] | 24.19 ± 0.54 [c,h] |
| | | 2.11 | 54.99 ± 0.41 [b] | 21.64 ± 0.17 [f] | 23.38 ± 0.54 [e,h] |
| | Irrigated | 0.07 | 63.19 ± 0.32 [h] | 12.67 ± 0.10 [d] | 24.14 ± 0.38 [c,h] |
| | | 0.58 | 62.8 ± 0.54 [e,h] | 14.53 ± 0.3 [j] | 22.67 ± 0.58 [e,f] |
| | Fisher's LSD | | 0.94 | 0.42 | 0.88 |

* Means of five replicates ± SD. The letters represent different groups. There were no statistically significant differences among those groups that share the same letter. The method used to discriminate between the means was Fisher's least significant difference (LSD).

**Table 2.** Experimental results for the irrigated Arbequina cultivar (MI = 1.11).

| Run | (E)-2-Hexenal | Hexanal | (Z)-3-Hexenol | Hexanol | (E)-2-Hexenol | (Z)-2-Pentenol | 1-Penten-3-ol | 1-Penten-3-ona | (Z)-3-Hexenyl acetate | TOTAL | Chlorophylls | Carotenoids |
|---|---|---|---|---|---|---|---|---|---|---|---|---|
| | Lipoxygenase (LOX) Pathway Volatile Compounds (mg/kg) | | | | | | | | | | Pigments (mg/kg) | |
| 1 | 23.27 | 1.24 | 0.73 | 0.34 | 0.34 | 0.52 | 0.35 | 0.55 | 0.52 | 27.87 | 5.07 | 4.38 |
| 2 | 28.48 | 2.10 | 0.77 | 0.49 | 0.32 | 0.56 | 0.43 | 0.62 | 0.43 | 34.19 | 4.60 | 4.61 |
| 3 | 23.89 | 1.82 | 0.58 | 0.41 | 0.38 | 0.64 | 0.45 | 0.62 | 0.60 | 29.39 | 5.88 | 4.80 |
| 4 | 19.82 | 1.75 | 1.10 | 0.78 | 0.63 | 0.58 | 0.37 | 0.52 | 0.47 | 26.03 | 5.02 | 4.09 |
| 5 | 15.81 | 0.97 | 1.09 | 0.51 | 0.67 | 0.53 | 0.33 | 0.47 | 0.33 | 20.70 | 7.22 | 4.58 |
| 6 | 26.49 | 1.84 | 0.74 | 0.61 | 0.21 | 0.53 | 0.38 | 0.69 | 0.44 | 31.93 | 2.83 | 3.36 |
| 7 | 29.16 | 2.05 | 1.19 | 0.30 | 0.40 | 0.54 | 0.38 | 0.64 | 0.29 | 34.96 | 3.36 | 3.55 |
| 8 | 14.58 | 0.94 | 0.77 | 0.62 | 0.53 | 0.53 | 0.33 | 0.45 | 0.47 | 19.22 | 4.02 | 5.01 |
| 9 | 19.36 | 0.80 | 0.55 | 0.33 | 0.37 | 0.52 | 0.35 | 0.50 | 0.32 | 23.10 | 8.28 | 5.53 |
| 10 | 21.14 | 0.83 | 0.78 | 0.46 | 0.36 | 0.55 | 0.39 | 0.57 | 0.48 | 25.55 | 5.39 | 4.45 |
| 11 | 25.23 | 0.96 | 1.09 | 0.49 | 0.41 | 0.51 | 0.33 | 0.58 | 0.31 | 29.92 | 3.76 | 3.44 |
| 12 | 24.77 | 0.96 | 0.52 | 0.46 | 0.21 | 0.56 | 0.40 | 0.66 | 0.46 | 29.01 | 5.50 | 4.86 |
| 13 | 24.88 | 1.28 | 0.71 | 0.43 | 0.34 | 0.53 | 0.36 | 0.55 | 0.50 | 29.58 | 4.20 | 4.00 |
| 14 | 26.23 | 1.93 | 0.65 | 0.56 | 0.21 | 0.61 | 0.49 | 0.73 | 0.57 | 31.98 | 4.18 | 4.26 |
| 15 | 23.86 | 0.91 | 0.72 | 0.35 | 0.35 | 0.52 | 0.35 | 0.54 | 0.50 | 28.11 | 4.94 | 4.08 |
| 16 | 24.13 | 1.27 | 0.71 | 0.32 | 0.37 | 0.55 | 0.37 | 0.58 | 0.45 | 28.76 | 5.90 | 4.59 |
| 17 | 23.35 | 1.14 | 0.66 | 0.43 | 0.36 | 0.54 | 0.35 | 0.56 | 0.51 | 27.90 | 6.36 | 4.86 |

**Table 3.** Proposed models in terms of actual factors with statistical parameters (analysis of variance (ANOVA) for the fit of experimental data was used) and optimal values of the technological factors for the response maximum values. Irrigated Arbequina cultivar sample (MI = 1.11).

| Response | Model * | $p$-Value | $R^2$ | Std. Dev. | Maximum Value (mg/kg) | Diameter (mm) | Temperature (°C) | Time (min) |
|---|---|---|---|---|---|---|---|---|
| LOX pathway volatile compounds | | | | | | | | |
| (E)-2-Hexenal (mg/kg) | $-8.407 + 6.601\,D + 0.8259\,T + 0.3760\,T - 0.1622\,D\,T - 0.03773\,D\,t - 7.124E{-}03\,T\,t$ | <0.0001 | 0.967 | 0.91 | 29.38 | 6.46 | 20.01 | 41.34 |
| Hexanal (mg/kg) | $+2.48071\, {-}0.054697\,T + 8.33412E{-}003\,t$ | <0.0001 | 0.841 | 0.20 | 2.11 | In range | 20.26 | 88.51 |
| (Z)-3-Hexenol (mg/kg) | $+3.080 - 0.9373\,D - 0.03426\,T + 0.1099\,D^2 + 5.392E{-}04\,T^2$ | <0.0001 | 0.982 | 0.03 | 1.16 | 6.50 | 20.00 | In range |
| Hexanol (mg/kg) | $+3.843 - 0.4551\,D - 0.07405\,T - 0.04065\,t + 0.01079\,D\,T + 2.845E{-}03\,D\,t + 2.366E{-}04\,T\,t + 1.589E{-}04\,T^2$ | 0.0101 | 0.813 | 0.07 | 0.81 | 6.41 | 37.65 | 89.94 |
| (E)-2-Hexenol (mg/kg) | $+1.5301 - 0.5926\,D - 8.738E{-}03\,T + 1.022E{-}03\,t + 2.729E{-}03\,D\,T + 5.891E{-}05\,T\,t + 0.0571\,D^2 + 0.5495$ | <0.0001 | 0.987 | 0.02 | 0.71 | 6.49 | 37.62 | 86.61 |
| (Z)-2-Pentenol (mg/kg) | | - | – | 0.03 | 0.55 | In range | In range | In range |
| 1-Penten-3-ol (mg/kg) | $+1.978 - 0.4505\,D - 0.01744\,T + 8.377E{-}04\,t + 2.275E{-}03\,D\,T + 0.03150\,D^2$ | <0.0001 | 0.988 | 0.01 | 0.52 | 4.50 | 20.00 | 90.00 |
| 1-Penten-3-one (mg/kg) | $+1.125 - 0.03819\,D - 8.740E{-}03\,T - 1.235E{-}03\,t$ | <0.0001 | 0.895 | 0.03 | 0.73 | 4.61 | 20.34 | 31.62 |
| (Z)-3-Hexenyl acetate (mg/kg) | $+0.45091$ | - | – | 0.09 | 0.45 | In range | In range | In range |
| TOTAL (mg/kg) | $+1.171 + 5.934\,D + 0.6129\,T + 0.3793\,t - 0.1344\,D\,T - 0.03562\,D\,t - 7.165E{-}03\,T\,t$ | <0.0001 | 0.978 | 0.81 | 34.94 | 6.50 | 20.00 | 90.00 |
| *Pigments* | | | | | | | | |
| Chlorophylls (mg/kg) | $+1.825 - 0.5600\,D + 0.1709\,T + 0.02394\,t$ | 0.0002 | 0.803 | 0.70 | 8.30 | 4.50 | 40.00 | 90.00 |
| Carotenoids (mg/kg) | $+4.960 - 0.4732\,D + 0.04738\,T + 0.01001\,t$ | 0.0001 | 0.792 | 0.29 | 5.54 | 4.50 | 39.91 | 81.92 |

* According to coefficients of model (Equation (1)), only significant values are included ($p$-value < 0.005); D is the grid hole diameter of the crusher (mm), T is the malaxation temperature (°C), and t is the malaxation time (min); $R^2$ is the determination coefficient, Std. Dev. is standard deviation.

Table 4. Maximum value for volatile compounds and pigments response *.

| Cultivar | Irrigation | MaturityIndex (MI) | (E)-2-Hexenal | Hexanal | (Z)-3-Hexenol | Hexanol | (E)-2-Hexenol | (Z)-2-Pentenol | 1-Penten-3-ol | 1-Penten-3-one | (Z)-3-Hexenyl acetate | TOTAL | Chlorophylls | Carotenoids |
|---|---|---|---|---|---|---|---|---|---|---|---|---|---|---|
| | | | | | | | | | | LOX Pathway Volatile Compounds (mg/kg) | | | Pigments (mg/kg) | |
| Arbequina | Rainfed | 0.20 | 9.66 ± 0.85 | 4.24 ± 0.58 | 0.87 ± 0.04 | 0.33 ± 0.04 | 1.18 ± 0.06 | 0.79 ± 0.05 | 0.76 ± 0.04 | 0.71 ± 0.04 | 0.73 ± 0.09 | 17.29 ± 1.66 | 32.24 ± 2.06 | 13.61 ± 0.66 |
| | | 1.31 | 13.93 ± 0.47 | 1.56 ± 0.10 | 0.87 ± 0.03 | 0.52 ± 0.16 | 0.63 ± 0.04 | 0.64 ± 0.01 | 0.54 ± 0.02 | 0.71 ± 0.03 | 0.51 ± 0.06 | 18.48 ± 0.58 | 11.23 ± 0.54 | 6.91 ± 0.29 |
| | | 2.56 | 13.23 ± 0.36 | 1.53 ± 0.08 | 0.65 ± 0.05 | 0.49 ± 0.14 | 0.64 ± 0.01 | 0.52 ± 0.03 | 0.47 ± 0.03 | 0.70 ± 0.03 | 0.48 ± 0.07 | 17.90 ± 0.68 | 13.65 ± 1.15 | 5.52 ± 0.30 |
| | Irrigated | 0.16 | 18.53 ± 1.09 | 6.87 ± 0.10 | 1.48 ± 0.08 | 0.31 ± 0.07 | 0.59 ± 0.04 | 0.62 ± 0.08 | 0.52 ± 0.11 | 0.53 ± 0.07 | 0.53 ± 0.08 | 27.83 ± 0.98 | 18.42 ± 0.63 | 9.14 ± 0.38 |
| | | 1.11 | 29.39 ± 0.91 | 2.10 ± 0.20 | 1.16 ± 0.03 | 0.80 ± 0.07 | 0.67 ± 0.02 | 0.55 ± 0.03 | 0.52 ± 0.01 | 0.74 ± 0.03 | 0.45 ± 0.09 | 34.94 ± 0.81 | 8.30 ± 0.70 | 5.56 ± 0.29 |
| | | 2.52 | 26.52 ± 0.76 | 2.05 ± 0.05 | 0.87 ± 0.02 | 0.36 ± 0.03 | 0.67 ± 0.03 | 0.51 ± 0.01 | 0.39 ± 0.01 | 0.58 ± 0.02 | 0.74 ± 0.09 | 31.80 ± 1.20 | 18.27 ± 0.63 | 9.06 ± 0.39 |
| Koroneiki | Rainfed | 0.16 | 2.85 ± 0.09 | 2.55 ± 0.40 | 0.89 ± 0.03 | 0.27 ± 0.06 | 0.63 ± 0.02 | 0.78 ± 0.04 | 0.64 ± 0.05 | 0.64 ± 0.04 | 1.15 ± 0.08 | 10.03 ± 0.52 | 60.13 ± 4.06 | 23.50 ± 1.34 |
| | | 1.68 | 1.84 ± 0.09 | 0.47 ± 0.04 | 0.57 ± 0.03 | 0.79 ± 0.05 | 0.57 ± 0.03 | 0.85 ± 0.02 | 0.59 ± 0.08 | 0.90 ± 0.05 | 0.59 ± 0.01 | 6.77 ± 0.16 | 50.06 ± 0.91 | 15.95 ± 0.76 |
| | | 2.05 | 2.17 ± 0.11 | 0.54 ± 0.01 | 0.67 ± 0.03 | 0.54 ± 0.18 | 0.48 ± 0.07 | 0.93 ± 0.02 | 0.90 ± 0.01 | 0.97 ± 0.03 | 0.68 ± 0.02 | 7.53 ± 0.32 | 49.20 ± 1.61 | 18.27 ± 1.62 |
| | Irrigated | 0.07 | 4.22 ± 0.46 | 3.96 ± 0.61 | 1.01 ± 0.07 | 0.47 ± 0.02 | 0.82 ± 0.03 | 0.53 ± 0.10 | 0.47 ± 0.17 | 0.40 ± 0.07 | 0.61 ± 0.13 | 10.66 ± 1.35 | 42.83 ± 3.59 | 18.03 ± 1.11 |
| | | 0.67 | 4.37 ± 0.24 | 0.62 ± 0.17 | 1.17 ± 0.03 | 0.43 ± 0.07 | 0.79 ± 0.03 | 0.62 ± 0.04 | 0.66 ± 0.03 | 0.70 ± 0.03 | 0.49 ± 0.04 | 9.71 ± 0.40 | 46.15 ± 1.58 | 14.38 ± 1.03 |
| | | 2.30 | 5.44 ± 0.24 | 0.79 ± 0.04 | 0.83 ± 0.04 | 0.46 ± 0.08 | 0.68 ± 0.01 | 0.74 ± 0.05 | 0.72 ± 0.04 | 0.90 ± 0.04 | 0.91 ± 0.11 | 10.24 ± 0.64 | 54.38 ± 1.94 | 18.15 ± 0.61 |
| Arbosana | Rainfed | 0.15 | 11.77 ± 0.29 | 0.99 ± 0.09 | 0.68 ± 0.02 | 0.47 ± 0.05 | 0.33 ± 0.11 | 0.63 ± 0.01 | 0.53 ± 0.02 | 0.72 ± 0.02 | 1.00 ± 0.07 | 16.40 ± 0.33 | 12.08 ± 0.82 | 6.96 ± 0.29 |
| | | 0.95 | 10.29 ± 0.34 | 0.56 ± 0.08 | 0.63 ± 0.02 | 0.47 ± 0.10 | 0.53 ± 0.02 | 0.55 ± 0.01 | 0.49 ± 0.03 | 0.76 ± 0.05 | 0.41 ± 0.07 | 14.49 ± 0.60 | 10.97 ± 0.83 | 6.65 ± 0.34 |
| | | 2.11 | 11.21 ± 0.29 | 0.92 ± 0.04 | 0.69 ± 0.08 | 0.44 ± 0.09 | 0.48 ± 0.12 | 0.56 ± 0.06 | 0.41 ± 0.03 | 0.87 ± 0.05 | 1.84 ± 0.24 | 17.11 ± 0.31 | 10.43 ± 0.55 | 6.60 ± 0.32 |
| | Irrigated | 0.07 | 17.45 ± 0.62 | 2.19 ± 0.06 | 1.14 ± 0.05 | 0.50 ± 0.14 | 0.62 ± 0.01 | 0.57 ± 0.03 | 0.48 ± 0.02 | 0.64 ± 0.02 | 0.48 ± 0.08 | 22.62 ± 0.72 | 12.45 ± 0.85 | 7.10 ± 0.35 |
| | | 0.58 | 16.86 ± 0.54 | 0.80 ± 0.25 | 1.17 ± 0.05 | 0.44 ± 0.12 | 0.62 ± 0.02 | 0.57 ± 0.01 | 0.43 ± 0.03 | 0.65 ± 0.02 | 0.49 ± 0.06 | 22.73 ± 0.43 | 11.04 ± 0.81 | 6.63 ± 0.28 |

* Value ± Standard deviation of the model.

Table 5. Optimal values for volatile compound and chlorophylls response.

| Cultivar | Irrigation | Maturity Index (MI) | Total LOX Pathway Volatile Compounds | | | | | Chlorophylls | | | |
|---|---|---|---|---|---|---|---|---|---|---|---|
| | | | Maximum Value * (mg/kg) | Diameter (mm) | Temperature (°C) | Time (min) | | Maximum Value * (mg/kg) | Diameter (mm) | Temperature (°C) | Time (min) |
| Arbequina | Rainfed | 0.20 | 17.29 ± 1.66 | In range | 20.00 | In range | | 32.24 ± 2.06 | 4.50 | 40.00 | 90.00 |
| | | 1.31 | 18.48 ± 0.58 | In range | 25.91 | 50.47 | | 11.23 ± 0.54 | 4.50 | 40.00 | 90.00 |
| | | 2.56 | 17.90 ± 0.68 | 4.50 | 25.24 | 90.00 | | 13.65 ± 1.15 | 4.50 | 40.00 | 90.00 |
| | Irrigated | 0.16 | 27.83 ± 0.98 | In range | 20.00 | 90.00 | | 18.42 ± 0.63 | 4.50 | 40.00 | 90.00 |
| | | 1.11 | 34.94 ± 0.81 | 6.50 | 20.00 | 90.00 | | 8.30 ± 0.70 | 4.50 | 40.00 | 90.00 |
| | | 2.52 | 31.80 ± 1.20 | In range | 20.00 | 90.00 | | 18.27 ± 0.63 | 4.50 | 40.00 | 90.00 |
| Koroneiki | Rainfed | 0.16 | 10.03 ± 0.52 | 6.50 | 20.00 | 90.00 | | 60.13 ± 4.06 | 4.50 | 40.00 | 90.00 |
| | | 1.68 | 6.77 ± 0.16 | 6.50 | 20.00 | 90.00 | | 50.06 ± 0.91 | 4.50 | 40.00 | 90.00 |
| | | 2.05 | 7.53 ± 0.32 | 6.50 | 20.00 | In range | | 49.20 ± 1.61 | 4.50 | 40.00 | 88.56 |
| | Irrigated | 0.07 | 10.66 ± 1.35 | In range | 20.00 | In range | | 42.83 ± 3.59 | 4.50 | 40.00 | 90.00 |
| | | 0.67 | 9.71 ± 0.40 | In range | 20.00 | 30.00 | | 46.15 ± 1.58 | In range | 40.00 | 90.00 |
| | | 2.30 | 10.24 ± 0.64 | In range | 21.53 | In range | | 54.38 ± 1.94 | 4.50 | 40.00 | 90.00 |
| Arbosana | Rainfed | 0.15 | 16.40 ± 0.33 | In range | 22.36 | 49.77 | | 12.08 ± 0.82 | 4.50 | 40.00 | 90.00 |
| | | 0.95 | 14.49 ± 0.60 | 4.50 | 20.00 | 55.97 | | 10.97 ± 0.83 | 4.50 | 40.00 | 89.99 |
| | Irrigated | 2.11 | 17.11 ± 0.31 | 4.50 | 26.41 | 90.00 | | 10.43 ± 0.55 | 4.50 | 40.00 | 90.00 |
| | | 0.07 | 22.62 ± 0.72 | 4.50 | 20.00 | 90.00 | | 12.45 ± 0.85 | 4.50 | 40.00 | 90.00 |
| | | 0.58 | 22.73 ± 0.43 | 4.50 | 20.00 | 90.00 | | 11.04 ± 0.81 | 4.50 | 40.00 | 90.00 |

* Value ± Standard Deviation of the model.

In terms of the technological factors, the malaxation temperature had the greatest influence on the total content of LOX pathway volatile compounds; as the temperature increased, the volatile compound contents decreased. Thus, low temperatures are required to obtain the highest content of volatile compounds (Figures 1–3). This enzymatic pool is sensitive to its environmental conditions; the temperature, in particular, can affect the level and the activity of the enzymes involved in the LOX pathway [19]. According to the mathematical models obtained, 20 °C (the lower limit of the range studied) was the optimum working temperature for most of the samples (Table 5). However, Ridolfi et al. [20] studied the kinetic constants of the olive LOX enzyme and reported that the maximum LOX activity was recorded at 30 °C. Also, in agreement with the results of Angerosa et al. [21] and Kalua et al. [9], we observed that increased malaxing temperature increased the content of hexanol and (*E*)-2-hexenol (Table 3), which is considered to impart a highly unpleasant odor.

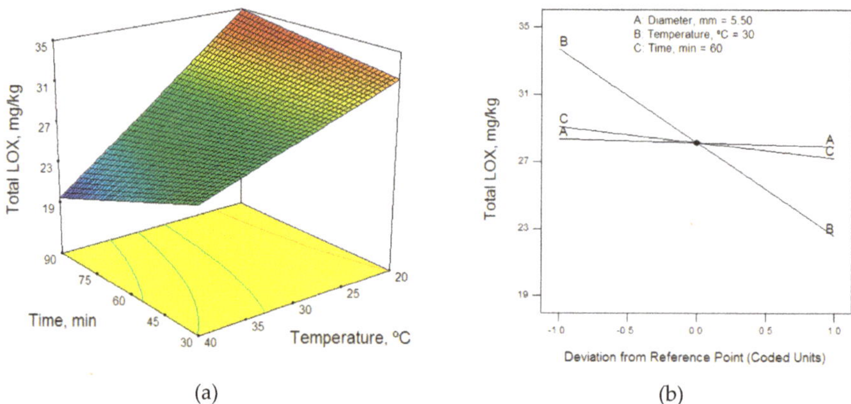

(a)         (b)

**Figure 1.** (**a**) Response surface and (**b**) perturbation graphic of the effects of temperature and time on the total LOX volatile content for the Arbequina cultivar (irrigated, MI = 1.11), for a diameter of 5.5 mm.

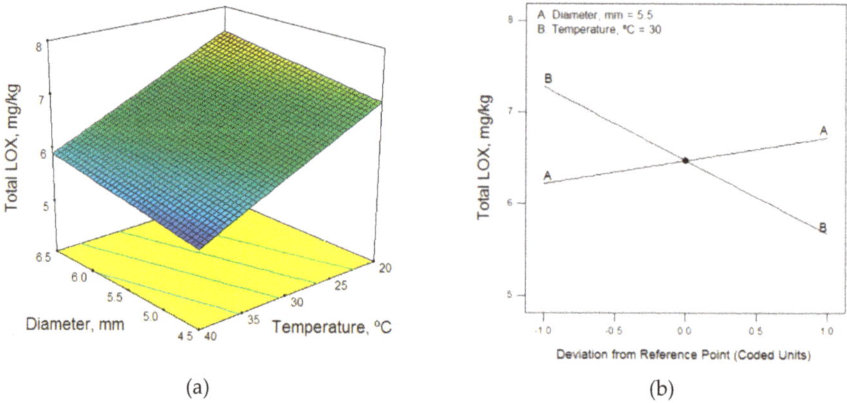

(a)         (b)

**Figure 2.** (**a**) Response surface and (**b**) perturbation graphic of the effects of temperature and time on the total LOX volatile content for the Koroneiki cultivar (rainfed, MI = 2.05); time does not have an influence.

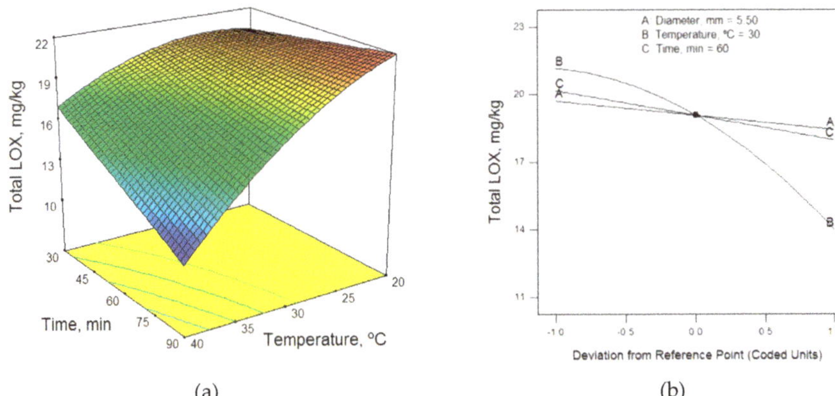

**Figure 3.** (a) Response surface and (b) perturbation graphic of the effects of temperature and time on the total LOX volatile content for the Arbosana cultivar (irrigated, MI = 0.07), for a diameter of 5.5 mm.

Different authors have reported widely different conclusions regarding the effect of malaxing time on the volatile compound content, with some saying that the malaxing time should be short, while others recommend increasing malaxing time. The malaxing time influences the activity of the enzymes and the vaporized compounds in the ambient atmosphere [22]. The recommendations for malaxing time differ because the interactions with malaxing temperature have not been sufficiently studied. At low temperatures, the content of volatile compounds increases with increasing time, while at high temperatures it decreases, as shown in Figures 1 and 3. Thus, to obtain the highest content of volatile compounds at low temperatures, a long malaxing time of 90 min was required (upper limit of the range studied), as shown for most of the samples in Table 5. However, the malaxing time was not a significant factor in four mathematical models.

Just as the malaxing time had little influence on the mathematical models, as seen in the perturbation graphs of Figures 1–3, the diameter of the crusher holes did not appear in eight of the models. In the models in which it did appear, the maximum volatile compound content was obtained using the upper limit size, 6.5 mm, in some cases and the smallest size, 4.5 mm, in others. This would seem to indicate that the diameter did not have a significant influence on the volatile content of the olive oils.

According to Cevik et al. [10] to obtain the maximum quantity of (*E*)-2-hexenal, a lower temperature and long malaxation time should be used. We do not agree with the results obtained by Lukić et al. [5], which indicated that (*E*)-2-hexenal and 1-penten-3-one content increased with higher malaxation temperatures.

### 2.2. Photosynthetic Pigments

Table 4 shows the maximum values of the chlorophyll and carotenoid content predicted by the models. In addition to being responsible for the color of olive oil, these natural pigments also play an important role in the oxidative stability of the oil [23]. The sample with the highest content of chlorophylls was the Koroneiki rainfed sample and that with the lowest content was the Arbequina irrigated sample. The sample with the highest content of carotenoids was the Koroneiki rainfed sample, while the Arbequina sample had the lowest carotenoid content. García et al. [15] reported that EVOO produced from rainfed crops exhibited a higher level of photosynthetic pigments than EVOO from irrigated crops. Samples with lower MIs had a higher content of these photosynthetic compounds. During the ripening process, the chlorophyll and carotenoid contents decreased, which was similar to the results observed by Benito et al. [24]. To obtain olive oils with a green color, i.e., with a high chlorophyll content, green or less mature olives should be used in the extraction. In addition, the rainfed samples contained more chlorophylls than the irrigated ones. These results also coincided with those obtained by Benito et al. [24,25].

The decrease in the chlorophyll and carotenoid content during the ripening process is due to the involvement of an enzymatic system in the degradation of chlorophylls during the maturation of olive fruits, as described by Vergara-Domínguez et al. [26]. The ripeness stage of the olives at harvesting was correlated to the amount of pigments in the resulting EVOOs. Interestingly, a decrease in the ratio of chlorophyll derivatives to carotenoids in the olive oils was observed with increasing maturity of the olives at harvesting [25].

In terms of the technological factors, in general, the adjustment of the mathematical models for both photosynthetic pigments was good and the three factors studied had a great influence on the contents of chlorophylls and carotenoids, as can be seen in the perturbation graph in Figure 4. Essentially, to obtain a high content of photosynthetic compounds, a malaxation temperature of 40 °C, a malaxation time of 90 min, and a hammer-crusher grid hole diameter of 4.5 mm should be used. The same conditions were obtained for the Arbequina, Koroneiki, and Arbosana cultivars. The temperature and malaxation time had a great influence; when both were increased, the chlorophyll content of the resulting oil increased. On the contrary, the diameter had less influence, with increasing diameter decreasing pigment content.

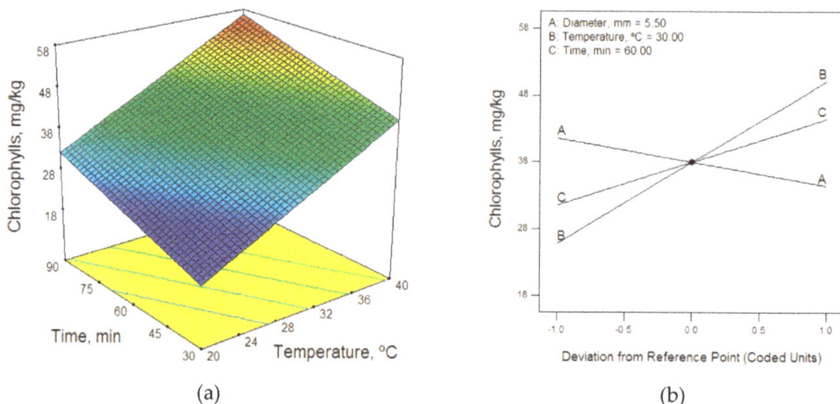


**Figure 4.** (**a**) Response surface and (**b**) perturbation graphic of the effects of temperature and time on the chlorophyll content for the Koroneiki cultivar (rainfed, MI = 0.16), for a diameter of 5.5 mm.

Our models are different from those obtained by Brahim et al. [27] using the same methodology (RSM); their model did not indicate an influence of the temperature on the chlorophyll content, probably due to the fact that they used a different cultivar, Chemlali, at a different maturity index, 4.7.

## 3. Materials and Methods

### 3.1. Raw Material

Olive fruits (*Olea europaea* L.) were picked by hand from super-intensive crops located in El Carpio (Córdoba, Spain) between October and December 2015. In total, 17 samples of olives were collected. Each sample was composed of approximately 10 kg of olives. The olives were harvested at different ripening degrees from olive trees of the Arbequina, Koroneiki, and Arbosana cultivars that had been either irrigated or rainfed. The maturity index (MI) or ripening degree was determined according to the method of Uceda and Frías, as described in Espínola et al. [28]. The olives were harvested with maturity indices between 0 and 3 to prevent the fruits from falling onto the ground. The oil content was analyzed by the Soxhlet method and the moisture content was determined by drying at 105 °C. All measurements were carried out quadruplicate; the results are presented in Table 1.

### 3.2. Olive Oil Extraction

Olive oils were obtained using an Abencor centrifugal system (Abencor analyser, MC2 Ingeniería y Sistemas S.L., Seville, Spain) under laboratory-scale conditions, as previously described by Espínola et al. [29]. The obtained olive oils were decanted in a graduated tube for at least three hours, filtered using paper, and stored in amber glass bottles at −18 °C under a nitrogen atmosphere until their analyses.

### 3.3. Analysis of Photosynthetic Pigments: Chlorophylls and Carotenoids

The composition of the pigments was determined following the procedure proposed by Minguez-Mosquera et al. [30]. The absorbance was measured using a Shimadzu UV-1800 spectrophotometer (Shimadzu Corp., Kyoto, Japan); a wavelength of 470 nm was used for the carotenoid pigments and 670 nm for the chlorophyllic pigments. Cyclohexane was used as the solvent. Equation (1) has been used to obtain the pigments concentration.

$$C_p = \frac{A_\lambda \cdot V_f}{\varepsilon \cdot m_a} \times 10000 \tag{1}$$

where $C_p$ is the concentration of the pigment (mg of pigment / kg of oil); $A_\lambda$, is the absorbance at 670 nm for chlorophyll and 470 nm for carotenoids; $\varepsilon$ is the specific absorbance, for chlorophyll is 613 and for carotenoids is 2000; $m_a$ is the weight of the sample (g) and $V_f$ is the volume of the solution. The pigment concentration of the samples is expressed as mg of pigment per kg of oil.

### 3.4. Analysis of Volatile Compounds

The procedure for the determination of the volatile compounds was similar to that used by Vidal et al. [31]. The volatile compounds were analyzed by headspace solid-phase microextraction (HS-SPME) and gas chromatography-flame ionization detection (GC-FID). Two grams of the sample was placed in a 20 mL amber glass vial tightly capped with a polytetrafluoroethylene (PTFE)/silicone septum and a magnetic cap. The vial was heated to 40 °C for 10 min to allow the volatile compounds to reach equilibrium in the headspace. Subsequently, the SPME needle was inserted through the septum and the fiber was exposed for 40 min. The SPME fiber (2 cm length and 50/30 µm film thickness) was purchased from Supelco (Bellefonte, PA, USA) and was composed of Carboxen/DVB/polydimethylsiloxane (PDMS). The fiber had been previously conditioned following the instructions of the manufacturer.

GC-FID analysis was performed using a gas chromatograph (model 7890B, Agilent Technologies, CA, USA) equipped with a split/splitless injector and a flame ionization detector. The volatile compounds adsorbed on the fiber were desorbed in the injector port for 1 min in splitless mode. A DB-WAXetr polyethylene glycol capillary column (30 m length, 0.25 mm internal diameter, 0.25 µm coating) (Agilent Technologies, USA) was used for the chromatographic separation. The carrier gas was helium at a flow rate of 1 mL/min. The injector temperature was 260 °C and that of the detector was 280 °C. The oven temperature was initially set to 40 °C and was held at this temperature for 10 min. Subsequently, the temperature was increased to 160 °C with a temperature ramp of 3 °C/min, followed by a ramp of 15 °C/min to 200 °C; the sample was then held at 200 °C for 5 min.

The chromatographic peaks were quantified using the internal standard method. 4-Methyl-2-pentanol was used as the internal standard, and all the compounds were used as external standards. The internal standard solution was prepared in previously deodorized oil. Then, the standard was added to each oil sample and stirred so that the mixture is homogeneous. The results are expressed as mg of standard compound per kg of oil.

*3.5. Experimental Design and Statistical Analysis*

The statistical design of experiments (SDE) and response surface methodology (RSM) statistical tools were used for planning the methodology and data analysis. The optimal experimental strategy was used to obtain the most information with the minimum cost. RSM facilitates the evaluation of results and generates reliable conclusions [32]. Different experimental designs can be used with RSM; in this case, a Box–Behnken design was used for the optimization of three factors: the diameter of the grid holes of the hammer-crusher, the malaxation temperature, and the malaxation time, with 17 runs including five central points. The extraction conditions were a crusher hole diameter of 4.5, 5.5, or 6.5 mm; a malaxation temperature of 20, 30, or 40 °C; and a malaxation time of 30, 60, or 90 min. Table 6 shows the values of the factors.

**Table 6.** Actual factors of the experimental design (coded factors).

| Run | Diameter (mm) | Temperature (°C) | Time (min) |
|---|---|---|---|
| 1 | 5.5 (0) | 30 (0) | 60 (0) |
| 2 | 5.5 (0) | 20 (−1) | 90 (+1) |
| 3 | 4.5 (-1) | 30 (0) | 90 (+1) |
| 4 | 6.5 (+1) | 30 (0) | 90 (+1) |
| 5 | 6.5 (+1) | 40 (+1) | 60 (0) |
| 6 | 5.5 (0) | 20 (−1) | 30 (−1) |
| 7 | 6.5 (+1) | 20 (−1) | 60 (0) |
| 8 | 5.5 (0) | 40 (+1) | 90 (+1) |
| 9 | 4.5 (−1) | 40 (+1) | 60 (0) |
| 10 | 5.5 (0) | 40 (+1) | 30 (−1) |
| 11 | 6.5 (+1) | 30 (0) | 30 (−1) |
| 12 | 4.5 (−1) | 30 (0) | 30 (−1) |
| 13 | 5.5 (0) | 30 (0) | 60 (0) |
| 14 | 4.5 (−1) | 20 (−1) | 60 (0) |
| 15 | 5.5 (0) | 30 (0) | 60 (0) |
| 16 | 5.5 (0) | 30 (0) | 60 (0) |
| 17 | 5.5 (0) | 30 (0) | 60 (0) |

The response variables studied were: 9 individual volatile compounds, the sum of the compounds (TOTAL), and 2 pigments (chlorophylls and carotenoids).

The experimental results were analyzed using the software Design-Expert v. 8.0.7.1 (Stat-Ease, Inc., Minneapolis, MN, USA). The coefficient of determination ($R^2$), the lack of fit, and the Fisher value (F-value) were obtained from the analysis of variance (ANOVA) and were used to determine the adequacy of the proposed model. Equation (2), a quadratic model, was used for each response studied.

$$Y = \beta_0 + \beta_1\,D + \beta_2\,T + \beta_3\,t + \beta_{12}\,D\,T + \beta_{13}\,D\,t + \beta_{23}\,T\,t + \beta_{11}\,D^2 + \beta_{22}\,T^2 + \beta_{33}\,t^2 \pm SD \quad (2)$$

where D is the grid hole diameter of the crusher (mm), T is the malaxation temperature (°C), and t is the malaxation time (min). Y is the predicted response and was correlated with the set of coefficients ($\beta$): the intercept ($\beta_0$), linear ($\beta_1$, $\beta_2$, $\beta_3$), interaction ($\beta_{12}$, $\beta_{13}$, $\beta_{23}$), and quadratic ($\beta_{11}$, $\beta_{22}$, $\beta_{33}$). The p-value was established as 5% (p-value $\leq 0.05$). SD is the standard deviation.

## 4. Conclusions

The results of this study suggest that the samples harvested from irrigated crops had the highest total content of LOX pathway volatile compounds. Comparing the three cultivars analyzed, the Arbequina cultivar had the highest content of these compounds. Samples harvested from rainfed crops had a higher content of photosynthetic pigments than irrigated samples. The Koroneiki cultivar had the highest content of chlorophylls and carotenoids. Therefore, it is not possible to optimize the irrigation management strategy to obtain an olive oil that is both rich in volatile compounds and high in photosynthetic pigments. Olive samples with a lower ripening degree had the highest content of these photosynthetic compounds, as well as of the total LOX pathway volatile compounds. The maturation stage had only a small influence, probably because the maturation stages studied were very similar.

In relation to the optimal conditions of each cultivar, it has been determined that they are practically the same for the three cultivars. So for maximum volatile compound content, we should work with irrigated olives at the lowest malaxation temperature, 20 °C, malaxation time, 90 min, and a small grid hole diameter, 4.5 mm, for a higher chlorophyll content, the same conditions except the temperature that must be the highest malaxation temperature, 40 °C.

The malaxation temperature was the factor that had the greatest influence on the total LOX pathway volatile compounds. Lower malaxation temperature resulted in a higher quantity of volatile compounds. On the other hand, the chlorophyll content increased with increasing temperature and malaxation time. Therefore, it is not possible to optimize the malaxing temperature to obtain an olive oil that is both rich in volatile compounds and high in photosynthetic pigments.

**Author Contributions:** Conceptualization, F.E. and M.M.; methodology, A.M.V., S.A., and A.D.T.; software, F.E and M.M.; formal analysis, F.E and J.M.E.; investigation, A.M.V., S.A., and A.D.T.; data curation, A.M.V. and J.M.E.; writing—original draft preparation, F.E.; writing—review and editing, A.M.V. and J.M.E.; supervision, F.E.; project administration, F.E.; funding acquisition, F.E.

**Funding:** This research was funded by Consejería de Economía, Innovación, Ciencia y Empleo, Junta de Andalucia, Spain (Research Project of Excellence PI11-AGR-7726).

**Acknowledgments:** The authors would also like to acknowledge Todolivo S.L. and all their staff for their kindness and attention.

**Conflicts of Interest:** The authors declare no conflict of interest.

## References

1. Covas, M.; Fitó, M.; De la Torre, R. Minor Bioactive Olive Oil Components and Health: Key Data for Their Role in Providing Health Benefits in Humans. In *Olive and Olive Oil Bioactive Constituents*; Boskou, D., Ed.; AOCS Press: Urbana, IL, USA, 2015; pp. 31–52.

2. Reboredo-Rodríguez, P.; González-Barreiro, C.; Cancho-Grande, B.; Simal-Gándara, J. Dynamic headspace/GC–MS to control the aroma fingerprint of extra-virgin olive oil from the same and different olive varieties. *Food Control* **2012**, *25*, 684–695. [CrossRef]

3. Vidal, A.; Alcalá, S.; Ocaña, M.; De Torres, A.; Espínola, F.; Moya, M. Modeling of volatile and phenolic compounds and optimization of the process conditions for obtaining balanced extra virgin olive oils. *Grasas Y Aceites* **2018**, *69*, 250. [CrossRef]

4. Vidal, A.M.; Alcalá, S.; De Torres, A.; Moya, M.; Espínola, F. Industrial production of a balanced virgin olive oil. *LWT Food Sci. Technol.* **2018**, *97*, 588–596. [CrossRef]

5. Lukić, I.; Krapac, M.; Horvat, I.; Godena, S.; Kosić, U.; Brkić Bubola, K. Three-factor approach for balancing the concentrations of phenols and volatiles in virgin olive oil from a late-ripening olive cultivar. *LWT Food Sci. Technol.* **2018**, *87*, 194–202. [CrossRef]

6. Rallo, L.; Díez, C.M.; Morales-Sillero, A.; Miho, H.; Priego-Capote, F.; Rallo, P. Quality of olives: A focus on agricultural preharvest factors. *Sci. Hortic.* **2018**, *233*, 491–509. [CrossRef]

7. Angerosa, F.; Servili, M.; Selvaggini, R.; Taticchi, A.; Esposto, S.; Montedoro, G. Volatile compounds in virgin olive oil: Occurrence and their relationship with the quality. *J. Chromatogr. A* **2004**, *1054*, 17–31. [CrossRef]

8. Sánchez-Ortiz, A.; Pérez, A.G.; Sanz, C. Synthesis of aroma compounds of virgin olive oil: Significance of the cleavage of polyunsaturated fatty acid hydroperoxides during the oil extraction process. *Food Res. Int.* **2013**, *54*, 1972–1978. [CrossRef]

9. Kalua, C.M.; Allen, M.S.; Bedgood Jr, D.R.; Bishop, A.G.; Prenzler, P.D.; Robards, K. Olive oil volatile compounds, flavour development and quality: A critical review. *Food Chem.* **2007**, *100*, 273–286. [CrossRef]

10. Cevik, S.; Ozkan, G.; Kiralan, M. Optimization of Malaxation Process using Major Aroma Compounds in Virgin Olive Oil. *Brazilian Archives of Biology and Technology* **2016**, *59*, e16160356. [CrossRef]
11. Genovese, A.; Yang, N.; Linforth, R.; Sacchi, R.; Fisk, I. The role of phenolic compounds on olive oil aroma release. *Food Res. Int.* **2018**, *112*, 319–327. [CrossRef]
12. Cevik, S.; Aydin, S.; Sermet, O.S.; Ozkan, G.; Karacabey, E. Optimization of Olive Oil Extraction Process by Response Surface Methodology. *Akad. Gida* **2017**, *15*, 337–343.
13. Fregapane, G.; Salvador, M.D. Production of superior quality extra virgin olive oil modulating the content and profile of its minor components. *Food Res. Int.* **2013**, *54*, 1907–1914. [CrossRef]
14. Gómez-Rico, A.; Fregapane, G.; Salvador, M.D. Effect of cultivar and ripening on minor components in Spanish olive fruits and their corresponding virgin olive oils. *Food Res. Int.* **2008**, *41*, 433–440.
15. García, J.; Morales-Sillero, A.; Pérez-Rubio, A.G.; Diaz-Espejo, A.; Montero, A.; Fernández, J.E. Virgin olive oil quality of hedgerow 'Arbequina' olive trees under deficit irrigation: Oil quality from deficit irrigated olive hedgerow. *J. Sci. Food Agric.* **2016**, *97*, 1018–1026. [CrossRef]
16. Fregapane, G.; Gómez-Rico, A.; Salvador, M.D. *Chapter 6: Influence of Irrigation Management and Ripening on Virgin Olive Oil Quality and Composition*; San Diego, Ed.; Academic Press: Cambridge, MA, USA, 2010; pp. 51–58.
17. Cherfaoui, M.; Cecchi, T.; Keciri, S.; Boudriche, L. Volatile compounds of Algerian extra-virgin olive oils: Effects of cultivar and ripening stage. *Int. J. Food Prop.* **2018**, *21*, 36–49. [CrossRef]
18. Salas, J.J.; Williams, M.; Harwood, J.L.; Sánchez, J. Lipoxygenase activity in olive (Olea europaea) fruit. *J. Am. Oil Chem. Soc.* **1999**, *76*, 1163–1168. [CrossRef]
19. Angerosa, F. Influence of volatile compounds on virgin olive oil quality evaluated by analytical approaches and sensor panels. *Eur. J. Lipid Sci. Technol.* **2002**, *104*, 639–660. [CrossRef]
20. Ridolfi, M.; Terenziani, S.F.; Patumi, M.F.; Fontanazza, G. Characterization of the lipoxygenases in some olive cultivars and determination of their role in volatile compounds formation. *J. Agric. Food Chem.* **2002**, *50*, 835–839. [CrossRef]
21. Angerosa, F.; Mostallino, R.; Basti, C.; Vito, R. Influence of malaxation temperature and time on the quality of virgin olive oils. *Food Chem.* **2001**, *72*, 19–28. [CrossRef]
22. Clodoveo, M.L.; Hbaieb, R.H.; Kotti, F.; Mugnozza, G.S.; Gargouri, M. Mechanical strategies to increase nutritional and sensory quality of virgin olive oil by modulating the endogenous enzyme activities. *Compr. Rev. Food Sci. Food Saf.* **2014**, *13*, 135–154. [CrossRef]
23. Criado, M.; Romero, M.; Casanovas, M.; Motilva, M. Pigment profile and colour of monovarietal virgin olive oils from Arbequina cultivar obtained during two consecutive crop seasons. *Food Chem.* **2008**, *110*, 873–880. [CrossRef] [PubMed]
24. Benito, M.; Lasa, J.M.; Gracia, P.; Oria, R.; Abenoza, M.; Varona, L.; Sánchez-Gimeno, A.C. Olive oil quality and ripening in super-high-density Arbequina orchard. *J. Sci. Food Agric.* **2013**, *93*, 2207–2220. [CrossRef] [PubMed]
25. Lazzerini, C.; Cifelli, M.; Domenici, V. Pigments in extra virgin olive oils produced in different mediterranean countries in 2014: Near UV-vis spectroscopy versus HPLC-DAD. *LWT Food Sci. Technol.* **2017**, *84*, 586–594. [CrossRef]
26. Vergara-Domínguez, H.; Ríos, J.J.; Gandul-Rojas, B.; Roca, M. Chlorophyll catabolism in olive fruits (var. Arbequina and Hojiblanca) during maturation. *Food Chem.* **2016**, *212*, 604–611. [CrossRef]
27. Ben Brahim, S.; Marrakchi, F.; Gargouri, B.; Bouaziz, M. Optimization of malaxing conditions using CaCO3 as a coadjuvant: A method to increase yield and quality of extra virgin olive oil cv. Chemlali. *LWT Food Sci. Technol.* **2015**, *63*, 243–252. [CrossRef]
28. Espínola, F.; Moya, M.; Fernández, D.G.; Castro, E. Improved extraction of virgin olive oil using calcium carbonate as coadjuvant extractant. *J. Food Eng.* **2009**, *92*, 112–118. [CrossRef]
29. Espínola, F.; Moya, M.; Fernández, D.G.; Castro, E. Modelling of virgin olive oil extraction using response surface methodology. *Int. J. Food Sci. Technol.* **2011**, *46*, 2576–2583. [CrossRef]
30. Minguez-Mosquera, I.M.; Rejano-Navarro, L.; Gandul-Rojas, B.; Sanchez-Gómez, A.H.; Garrido-Fernandez, J. Color-pigment correlation in virgin olive oil. *J. Am. Oil Chem. Soc.* **1991**, *68*, 332–336. [CrossRef]

31. Vidal, A.M.; Alcalá, S.; de Torres, A.; Moya, M.; Espínola, F. Use of talc in oil mills: Influence on the quality and content of minor compounds in olive oils. *LWT Food Sci. Technol.* **2018**, *98*, 31–38. [CrossRef]
32. Box, G.E.; Hunter, J.S.; Hunter, W.G. *Estadística Para Investigadores: Diseño, Innovación y Descubrimiento*, 2nd ed.; Ed. Reverté: Barcelona, Spain, 2008; pp. 475–487.

**Sample Availability:** Samples of the compounds are available from the authors.

 *molecules* 

*Article*

# Aroma Investigation of New and Standard Apple Varieties Grown at Two Altitudes Using Gas Chromatography-Mass Spectrometry Combined with Sensory Analysis

**Giulia Chitarrini, Nikola Dordevic, Walter Guerra, Peter Robatscher and Lidia Lozano ***

Laimburg Research Centre, Ora (BZ), 39040 Auer, Italy; giulia.chitarrini@laimburg.it (G.C.);
nikola.dordevic@laimburg.it (N.D.); walter.guerra@laimburg.it (W.G.); peter.robatscher@laimburg.it (P.R.)
* Correspondence: lidia.lozano@laimburg.it; Tel.: +39-0471-969-682

Academic Editor: Eugenio Aprea
Received: 29 May 2020; Accepted: 29 June 2020; Published: 30 June 2020

**Abstract:** The aromatic profile of apples constitutes important information for the characterization and description of local products. Apple flavor is determined by perception in mouth and aroma; while the first is mainly defined by sugars and organic acids, aroma is a complex mixture of many volatile organic compounds (VOCs) whose composition is often specific to the variety. Headspace-solid phase microextraction gas chromatography coupled with mass spectrometry (HS-SPME-GC-MS) allows for the detection of detailed information of volatile constituents. In this study, eleven apple varieties (Braeburn, Fuji, Gala, Golden Delicious, Coop 39-Crimson Crisp®, Dalinette-Choupette®, Fujion, CIV323-Isaaq®, Coop43-Juliet®, SQ159-Natyra®, UEB32642-Opal®) grown in two pedoclimatic locations at different altitudes in South Tyrol (Italy) (ca. 225 m and ca. 650 m a.s.l.) were investigated. Thirty-eight VOCs were identified and combined with sensory analysis results (from 11 trained panelist) to characterize the aroma of new and standard apple varieties with a special focus on pedoclimatic location differences. The study shows strong diversification of the varieties based on their VOC profiles and sensory attributes, as expected. Moreover, investigating how the pedoclimatic location at different altitudes can influence the apple aroma profile, we identified twelve VOCs involved in these differences and provided a deeper investigation on how different altitudes can influence the apple aroma composition and perceptions combining the analytical and sensory parameters.

**Keywords:** *Malus domestica*; flavor; sensory analysis; HS-SPME-GC-MS

## 1. Introduction

South Tyrol is Italy's northernmost province, located in the heart of the Alps in a very central position in Europe, and bordering Austria, Switzerland, and the Italian provinces of Sondrio, Trento, and Belluno. The territory comprises a total area of 7400 km$^2$. Nevertheless, the region is extremely mountainous, and only a small portion supports human settlement and farming.

The area used for agricultural purposes in South Tyrol amounts to 2670 km$^2$, corresponding to 36% of South Tyrol's total area. The greater part of income in the agricultural sector is earned by fruit growers [1]. South Tyrol is the largest single apple growing region in Europe; protected in the north by the Alps and open to the south, the region has the ideal climate for excellent harvests and highest product quality. The 300 days of sun per year guarantee ripe and succulent fruits of excellent flavor.

Flavor plays a very important role in consumer choice and perception of apple freshness, and a good relationship between consumer preference and sensory characteristics has been reported [2–4]. Consequently, to improve quality evaluation, sensory attributes assessed by trained panels should also be considered [5]. Therefore, the apples' aromatic profile constitutes an important information for the characterization and description of local products. Trained sensory judges are able to describe the perception of product attributes through descriptive sensory analysis; moreover, one notable strength of descriptive analysis is its ability to allow relationships between descriptive sensory and instrumental analyses [6]. Flavor is determined by perception in mouth (sweetness, acidity, or bitterness), mainly defined by sugars and organic acids, and aroma, a complex mixture of many volatile organic compounds (VOCs) whose composition is specific to the variety [7,8]. VOCs are low-molecular weight compounds derived from different biosynthetic pathways with fatty acids and amino acids as precursors [7]. In addition, the contribution of each volatile compound is dependent on its particular odor threshold and the presence of other compounds [9].

It is well known that the apple aroma profile changes during fruit maturation with a general aldehyde dominance during the first stage, moving on to an alcohol compound increase during maturation, and finally showing dominance of ester compounds at maturity [10]. Aroma is influenced not only by genetic differences and maturity but also by environmental factors [11–13]. Changing altitude of grow locations has an impact on temperature, humidity, pressure, ultraviolet radiation, sun exposure, wind, and geology [14]. Due to the complexity of the interactions between fruit development and environment, little information is available regarding the effects of these factors on fruit quality. Few works report a deep investigation on fruit quality and fruit aroma at different altitudes; however, this factor can be crucial for selecting orchard location towards improving fruit quality. Brat et al. showed that bananas at a comparable maturity stage cultivated at a high altitude may have had different sensory characteristics from those cultivated on a plain, particularly for the *Robusta cv.* (independent of the maturity stage) grown at a high altitude, which possessed the highest concentrations of volatile compounds [15]. In grape, Tomasi et al. reported the effect of vineyard altitude on monoterpenes and norisoprenoids [16]; consequently, differences have been reported in wine with a "bell pepper" aroma exalted in wines from higher altitudes, while wines from lower altitudes were correlated with "red fruit" aromas [17,18]. In apple, the effect of altitude has been focused on in studies of physical and chemical parameters [19–21], and only very few works concentrated on sensory evaluation by using consumers or trained panels, such as Charles et al., who studied sensory attributes at different locations on one single apple variety [22]. Altitudes had a reported influence on aroma compounds [13,14,23], but there was a lack of information on its influence on apple VOCs. Anyway, it seems that compared with environment, year of production, and storage time, the variety is a more important source of compositional variations [24]. Several works have been published investigating the apple cultivar aroma profile [25,26]. Compounds believed to be characteristic of specific varieties have been reported; ethyl 2-methyl butanoate, 2-methyl butyl acetate, and hexyl acetate have been attributed to Fuji apples aroma, and ethyl butanoate and ethyl 2-methyl butanoate are described as active compounds in the Elstar variety [8]; however, most of them involve one or a limited number of varieties. Interesting works have been done on apple juices, and ciders considering that the apple variety is an important factor for the aroma profile of the final product [27–29].

Analytical chemistry, and particularly gas-chromatography with head space solid-phase microextraction (HS-SPME-GC), is a suitable approach for the analysis of VOCs in food matrices, mainly due to its simplicity, reproducibility, selectivity, and sensitivity [30–33].

In the present work we investigated the chemical composition and sensory profile of 11 apple varieties grown in two different altitudes in South Tyrol. The dataset includes standard and new varieties still poorly investigated. The aim was to highlight the aroma profile results identifying differences between the varieties and possible differences correlated with the pedoclimatic locations.

## 2. Results and Discussion

### 2.1. Volatile Organic Compound Identification in Samples

The volatile organic compounds (VOCs) of 11 apple varieties grown in two pedoclimatic location at different altitudes (A = 225 m; B = 650 m) were investigated.

Using a headspace solid phase microextraction gas chromatography mass-spectrometry (HP-SPME-GC-MS) analysis with identification and integration of the peaks, we identified 38 compounds expressed as the peak height percentage compared with the total peak heights of identified compounds in the samples under investigation (Table S1, Supplementary Materials).

The identification level for each compound is reported in Table S1 in the Supplementary Materials; level A refers to the NIST 2017 database match, level B refers to the linear retention index (LRI) match, and level C refers to a match with the commercial standard.

The main representative class of compounds in our results were esters, with a variation between 72% and 92% of the total peak height percentage. Alcohols varied between 2% and 16% and aldehydes between 2% and 9% of the total height percentage.

Among the 20 esters reported in our study, the most abundant was hexyl acetate in all samples; it varied between 24.98% in the "Juliet®-Coop43" site A and 66.12% in the "Gala" site B compared with the total peak height of identified VOCs. Hexyl acetate has been mentioned to confer a green, green apple, fruity, pear, and banana odor characteristic of unripe Antei apples [34,35].

The other two esters mainly represented were butyl acetate and 2-methylbutyl acetate. The first one varied from 3.58% in "Choupette®-Dalinette" site A up to 19.98% in the "CIV323-Isaaq®" site A. It is described in the literature as contributing to the ethereal fruity, fruity, and ripe banana notes [35–37]; 2-methylbutyl acetate has been described as one of the most important volatile esters contributing to the characteristic apple, sweet, and banana aroma [36], and it was also reported as the most effective compound of the sensory attributes for the Royal Gala apples with a red apple perception [25]. It varied from 2.19% in "Gala" site B up to 33.67% in "Natyra®-SQ 159" site A. Butyl acetate and 2-methylbutyl acetate appeared to be inversely present in our varieties, suggesting that they can be inversely related and influencing a different group of apple varieties. Despite both being reported as giving sweet and banana perception [34], our results indicated that the two compounds could overall contribute to the typical apple aroma as reported in literature, adding some specific notes that diversified the varieties based on their greater or lesser presence in our samples. These results are understandable considering that we analyzed complex matrices and knowing that not all detected chemical compounds can influence flavor [38]. Moreover, the compound contribution depends on specific enzyme activity and substrate availability, odor threshold, and the presence of other compounds [8].

Looking at the other compounds belonging to the group of acids, alcohols, aldehydes, and others, they represented no more than around 27% of the apple VOC profile. Among them we found hexan-1-ol as the most abundant with a variation between 1.81% and 15.38% compared with the total height of identified VOCs, which was characteristic of the "Juliet®-Coop43" variety site A with 15.38% of content compared with site B with 7.57% of the same variety.

A very small relative amount of 1-methoxy-4-[(*E*)-prop-1-enyl]benzene (anethole) was found in our samples, which can give an aniseed-like typical flavor in food; the same flavor perception was given by the 1-methoxy-4-prop-2-enylbenzene (estragole) compound present in a higher amount in our whole dataset (Table S1, Supplementary Materials). It is well known that esters are the main VOCs in apple; in previous studies, a range of volatiles terpenes was also detected with the predominance of (3*E*,6*E*)-3,7,11-trimethyldodeca-1,3,6,10-tetraene (alpha-farnesene) [39]. In our results, we found alpha-farnesene influencing "Fuji", "Fujion" and "SQ 159" varieties; the biological function of alpha-farnesene is unclear and probably is not a key odor compound in apple as previously reported [39,40], even though it has been proposed together with ester compounds as responsible for cultivar classification [36].

### 2.2. Can the Altitude Influence the VOC Profile?

Excluding the differences due to the variety, we investigated the chemical differences between the pedoclimatic locations at different altitudes using a Wilcoxon signed rank test.

This test was used to order parameters according to their discriminative importance between A = low, 225 m a.s.l. and B = high, 650 m a.s.l. (Table S2, Supplementary Materials). According to statistical testing we selected the top 12 most discriminative VOCs reporting a *p* value < 0.05 (Figure 1). These compound results related to the different pedoclimatic locations independent of variety differences. Samples of the same cultivar have the same dot color connected with lines in Figure 1 to emphasize the general trend.

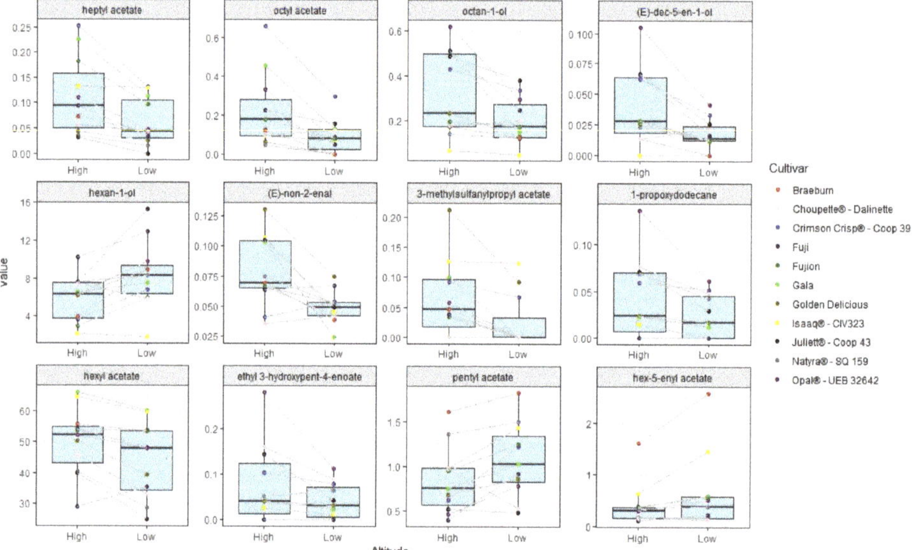

**Figure 1.** Boxplots according to statistical Wilcoxon signed rank test. Values represent the height percentage and altitude represents the different pedoclimatic locations (High = 650 m, Low = 225 m).

Heptyl acetate was the top 12 most significant compound reported, taking into account the *p* value (*p* value = 0.001) (Table S2, Supplementary Materials); it was higher in B site samples (650 m) compared with A site samples (250 m) in all varieties except for "CIV323-Isaaq®", in which it seemed stable. It was reported as giving a green, citrus, waxy, and fatty characteristic [34].

Another compound affected by the geographical location with a *p* value = 0.0186 was hexyl acetate, already described as the most abundant esters we found. It was higher in B site (650 m) samples compared with A site (250 m) samples in almost all varieties except for "Coop 39-Crimson Crisp®" (Figure 1); in the literature, this compound is associated with green, fruity, and sweet descriptors [34].

Other compounds that were higher in B site (650 m) samples compared with A site (250 m) samples included octyl acetate (*p* value = 0.0029), reportedly giving floral, herbaceous, and fruity notes [34], and octan-1-ol (*p* value = 0.0068), described in the literature as having citrus, floral, and fatty, waxy notes [28].

By contrast, 1-hexanol appeared to be decreasing in high altitude sites compared with low altitude site in almost all varieties studied in this work except for "CIV323-Isaaq®" and "Choupette®-Dalinette" (Figure 1). Another compound significantly more present in A site samples (225 m) compared to B site samples (650 m) was acetic acid pentyl ester, which was reported as a compound giving a sensation of sweet and fruity [11,35].

### 2.3. Can the Altitude Influence Sensory Attributes and What are the Relationship between These Attributes and VOCs?

Results of two-way ANOVA on sensory data indicated significant differences on factor variety on Overall Odor, Odor-Banana, Odor-Green grass, Odor-Honey, Flavor-Pear, Flavor-Banana, and Flavor-Lemon (Table 1). Non-significant effects of the altitude factor were found on sensory attributes. However, significant interactions between factor variety and factor altitude were detected for Odor-Kiwi, Odor-Pear, Odor-Banana, Odor-Vanilla, Flavor-Lemon, Flavor-Kiwi, and Flavor-Green grass, illustrating significant effects for some combinations of altitude and variety. Similarly, Charles et al. reported honey odor perception as the only sensory attribute to have significant differences at different altitudes on the "Golden Delicious" variety [22].

**Table 1.** *p* values from two-way ANOVA on sensory data, considering variety and altitude as factors.

| Attribute | Variety | | Altitude | | Variety * Altitude | |
|---|---|---|---|---|---|---|
| | F Value | *p* Value | F Value | *p* Value | F Value | *p* Value |
| Overall Odor | 8.1408 | $1.65 \times 10^{-9}$ | 0.0019 | 0.9657 | 1.5815 | 0.1214 |
| Odor-Apple | 1.1746 | 0.3169 | 0.0007 | 0.98 | 0.9453 | 0.3169 |
| Odor-Banana | 1.9456 | 0.0475 | 0.0142 | 0.9074 | 2.9763 | 0.0025 |
| Odor-Green grass | 2.3315 | 0.0163 | 4.2561 | 0.066 | 1.2612 | 0.2627 |
| Odor-Honey | 2.1175 | 0.0297 | 0.3214 | 0.5833 | 1.1793 | 0.3137 |
| Odor-Kiwi | 1.8719 | 0.0579 | 0.5908 | 0.4599 | 1.9911 | 0.042 |
| Odor-Lemon | 1.0516 | 0.4068 | 0.3059 | 0.5813 | 1.1356 | 0.3427 |
| Odor-Pear | 1.5726 | 0.1257 | 0.3695 | 0.5568 | 2.4756 | 0.0108 |
| Odor-Pineapple | 1.6284 | 0.1093 | 0.4685 | 0.5092 | 1.0795 | 0.3851 |
| Odor-Vanilla | 1.1786 | 0.3142 | 2.2739 | 0.1625 | 3.2509 | 0.0011 |
| Flavor-Apple | 0.4536 | 0.9156 | 0.0665 | 0.8018 | 1.388 | 0.1967 |
| Flavor-Banana | 3.2078 | 0.0013 | 1.4151 | 0.2617 | 1.3436 | 0.2181 |
| Flavor-Green grass | 3.7531 | 0.0003 | 0.1727 | 0.6865 | 3.1478 | 0.0015 |
| Flavor-Honey | 3.7253 | 0.0003 | 0.6803 | 0.4287 | 0.6625 | 0.7564 |
| Flavor-Kiwi | 1.5298 | 0.1398 | 0.702 | 0.4217 | 2.5304 | 0.0093 |
| Flavor-Lemon | 11.259 | $1.12 \times 10^{-12}$ | 0.3541 | $5.65 \times 10^{-1}$ | 4.9818 | $7.584 \times 10^{-6}$ |
| Flavor-Pear | 2.1496 | 0.0271 | 0.0523 | 0.8237 | 1.9076 | 0.0526 |
| Flavor-Pineapple | 1.4049 | 0.189 | 0.0818 | 0.7807 | 0.0818 | 0.1755 |
| Flavor-Vanilla | 2.4997 | 0.0101 | 1.4139 | 0.2619 | 0.5751 | 0.8308 |

Indeed, based on results of the non-parametric paired t-test, differences in the sensory attributes profile of the same variety were observed between low and high pedoclimatic locations (A-low 225 m and B-high 650 m).

Overall odor attribute was significantly different ($p = 0.001$) in the "Fuji" variety, showing more odor in the B site that in the A site.

For specifics odors attributes, Odor-Kiwi and Odor-Vanilla were significant higher ($p = 0.034$; $p = 0.0215$) in A site than in B site in the "UEB32642-Opal®" variety, while Odor-Banana and Odor-Honey showed significantly higher intensities ($p = 0.0091$; $p = 0.0298$) in the B site in the "Fuji" variety. In three varieties, Odor-Pear showed significant differences depending on the altitude, being higher in A site for "Gala" ($p = 0.0243$) and "SQ159-Natyra®" ($p = 0.042$) varieties and with higher intensity in high altitude site B for "Fujion" ($p = 0.0108$).

The relationships between sensory parameters and VOCs were visualized using PCA (Figure 2 and Table S3 and S4, Supplementary Materials); every arrow corresponded to one parameter in the data set. the longest arrows in the direction of a particular principal component had the largest impact and importance for that PC. The same direction of groups of arrows and their lengths indicated correlation of corresponding parameters in the data set, and the opposite directed arrows indicated a negative correlation between corresponding parameters.

In addition, Spearman correlations between sensory and VOC parameters were computed. And all significant results were evaluated (Figure 3 and Table S5, Supplementary Materials).

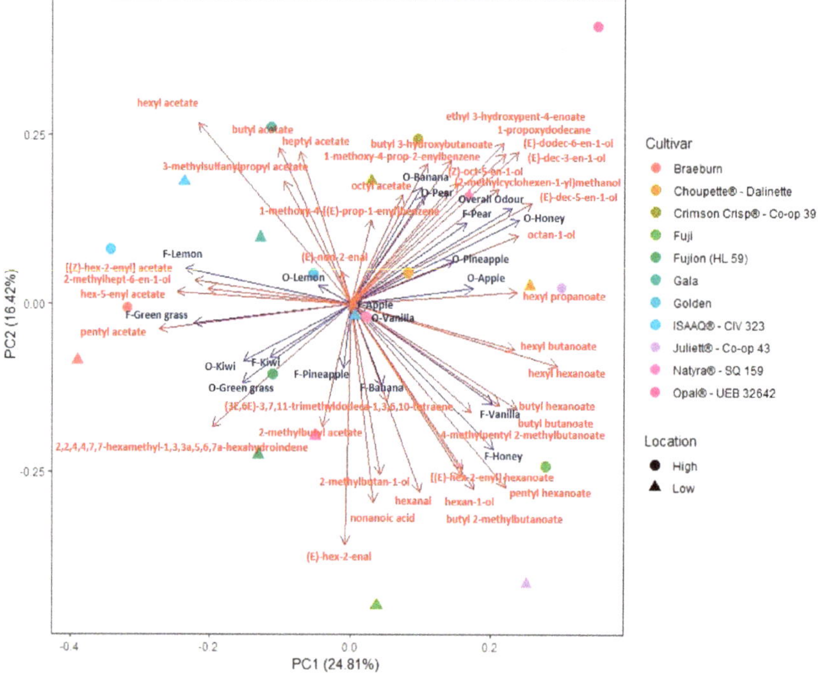

**Figure 2.** Principal component analysis reporting VOCs (red) and sensory parameters (blue) together. Varieties are reported with different colors and different shapes for low (225 m) and high (650 m) pedoclimatic locations.

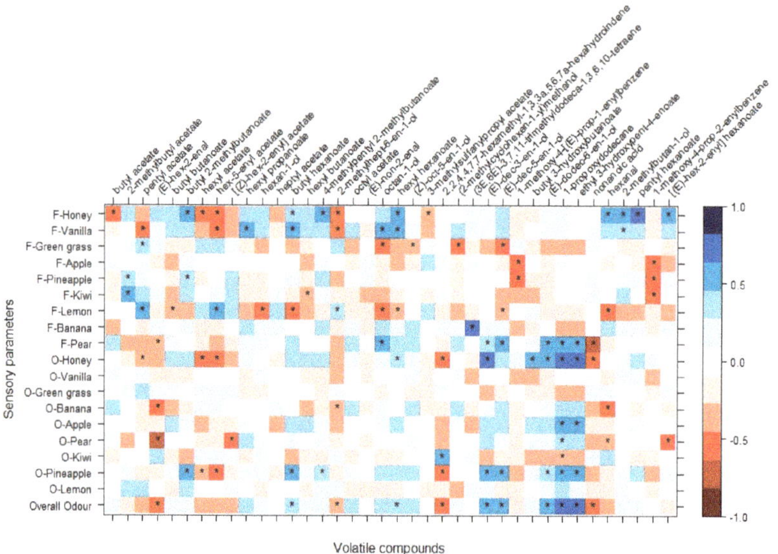

**Figure 3.** Correlation matrix. Sensory parameters and volatile organic compounds are reported as results of cross correlation analysis. Red colored squares indicate negative correlations, while blue colored squares indicate positive correlation. The significative correlations with a *p* value < 0.05 are reported with asterisks.

The correlation results are visualized in the correlation matrix figure (Figure 3) in which VOCs and sensory parameters are reported. The red color indicates a negative correlation between the two corresponding parameters of the matrix, while the blue color indicates a positive correlation.

The Overall odor parameter appeared to be positively correlated with butyl hexanoate, (*E*)-dec-3-en-1-ol, (*E*)-dec-5-en-1-ol, (*E*)-dodec-6-en-1-ol, 1-propoxydodecane, and ethyl 3-hydroxypent-4-enoate (Figure 3). It was projected in the PCA with a positive influence on the first PC together with Odor-Honey, Odor-Apple, and Odor-Pineapple (Table S4, Supplementary Materials). Odor-Vanilla did not show correlation with any identified VOCs; looking at the PCA it was reported with a very short arrow in the same direction of Flavor-Vanilla, which indicates a low influence of this parameter for the PCs in our dataset. Odor-Kiwi influenced negatively the first PC (Table S4, Supplementary Materials) with a projection in PC grouped with Odor-Green grass. Odor-Banana showed only negative correlations with (*E*)-hex-2-enal, 2-methylhept-6-en-1-ol, and hexanal (Figure 3); anyhow, it was projected in the upper right quadrant of the PCA, indicating a strong relation with the other parameters influencing the PC1, such as Overall odor, Odor-Pear, and Odor-Honey (Figure 3). Odor-Honey was positively correlated with hexyl hexanoate, (*E*)-dec-3-en-1-ol, butyl 3-hydroxybutanoate, (*E*)-dodec-6-en-1-ol, 1-propoxydodecane, and ethyl 3-hydroxypent-4-enoate; and negatively correlated with pentyl acetate (fruity), hexyl acetate (described as fruity, green apple, and banana [26] and significant in our Wilcoxon signed rank test (Figure 1)), hex-5-enyl acetate, 2,2,4,4,7,7-hexamethyl-1,3,3a,5,6,7a-hexahydroindene, and nonanoic acid, which gives a fruity, green apple, banana, and waxy odor [26] (Figure 3). In the same upper right quadrant of the PCA was also located Odor-Pear, which was positively correlated with 1-propoxydodecane (Figure 3), as was Odor-Banana, and very close to Odor-Banana in the PCA (Figure 2).

Regarding specific flavor attributes, results showed significantly greater value of Flavor-Pear in higher altitude (B) in "Fuji" ($p = 0.0243$) and "Coop43-Juliet®" ($p = 0.0141$); Flavor-Lemon was significantly superior in B site (high altitude) in "Coop39-Crimson Crisp®" ($p = 0.0165$) as well as in "Fujion" ($p = 0.008$) and contrary to "SQ159-Natyra®" ($p = 0.0089$), which showed higher values at the low altitude location (A) (Table 1).

Flavor-Kiwi and Flavor-Green grass presented significantly higher values of intensity at high altitude (B) in the "Choupette®-Dalinette" ($p = 0.0215$) and "Fujion" ($p = 0.0418$) varieties (Table 1).

Both varieties also showed significantly larger values of Flavor-Pineapple in low altitudes ($p = 0.0215$, $p = 0.0421$, respectively) (Table 1). For Flavor-Honey, results revealed a bigger significant value in higher altitude locations in the "Golden Delicious" variety. Non-distinguishing significant odor and flavors were found in two varieties, "Braeburn" and "CIV323-Isaaq®", in relation to the pedoclimatic location (Table S6, Supplementary Materials).

Flavor-Lemon and Flavor-Green grass were projected together influencing the PC 1 in a negative way, and they were grouped with compounds among which were pentyl acetate, hex-5-enyl acetate, [(*Z*)-hex-2-enyl] acetate, hexyl acetate, and others (Figure 2 and Table S4, Supplementary Materials). Flavor-Lemon and Flavor-Green grass results positively correlated with pentyl acetate, which was reported as a compound giving a sensation of sweet and fruity [11,27] (Figure 3). According to the PCA, we found associations between Flavor-Lemon and hex-5-enyl acetate, [(*Z*)-hex-2-enyl] acetate and, 2-methylhept-6-en-1-ol, with a significant positive correlation with the first and last one (Figures 2 and 3).

Flavor-Honey was positively correlated with butyl 2-methylbutanoate, butyl hexanoate, 4-methylpentyl 2-methylbutanoate, hexyl hexanoate, hexanal, 2-methylbutan-1-ol, pentyl hexanoate, and [(*E*)-hex-2-enyl] hexanoate (Figure 3 and Table S5, Supplementary Materials). Looking at the PCA, we found a correlated group of parameters that included butyl butanoate and butyl hexanoate; and hexan-1-ol, butyl 2-methylbutanoate, [(*E*)-hex-2-enyl] hexanoate, pentyl hexanoate, and hexanal (Figure 2). Flavor-Honey and Flavor-Vanilla were correlated and in the same direction in the PCA, grouped with 2-methylbutan-1-ol, butyl butanoate, butyl 2-methylbutanoate, hexyl propanoate, hexan-1-ol, 4-methylpentyl 2-methylbutanoate, 2-methylhept-6-en-1-ol, hexyl hexanoate, butyl hexanoate, hexyl butanoate, [(E)-hex-2-enyl] hexanoate, and pentyl hexanoate regarding loadings (Figure 2 and Table S3, Supplementary Materials). Flavor-Honey was instead negatively correlated with butyl acetate, one of the main esters found in our results (Figure 3). Butyl acetate appeared correlated with heptyl acetate, hexyl acetate (also not significantly correlated with Flavor-Honey), and 3-methylsulfanylpropyl acetate; in fact, their arrows were in the same direction of the PCA influencing the dimension 2 of our PC (Figure 1 and Table S4, Supplementary Materials).

Flavor-Kiwi and Flavor-Pineapple had a low influence in the PC (short arrows were reported); both were significantly correlated with 2-methylbutyl acetate, already described in our experiment as one of the major esters found.

## 3. Materials and Methods

### 3.1. Apple Samples

Eleven varieties from commercial apple orchards (Malus × domestica Borkh., ca. 3000–4000 trees ha$^{-1}$, rootstock M9, planted between 2006 and 2013) located in South Tyrol (Italy), managed according to the regional guidelines of integrated fruit production, and representing either valley (ca. 225 m a.s.l.) or hilly areas (ca. 650 m a.s.l.) were sampled during the 2014 growing season. At the valley site at the Laimburg Research Center, the soil was sandy loam with 1.7% of organic matter. In 2014, the average air temperature at 2 m was 12.8 °C with a minimum of −7.1 °C and a maximum of 34.5 °C, and 1151.8 mm yearly rainfall and 1721 yearly sunshine hours were registered [41]. At the hilly site in Vinschgau Valley, the soil was loamy sand with 5.2% organic matter. In 2014, the average air temperature at 2 m was 9.6 °C with a minimum of −10.1 °C and a maximum of 30.5 °C, and 593.3 mm yearly rainfall and 1940 yearly sunshine hours were registered. The new monogenic scab resistant varieties "Coop 39-Crimson Crisp®", "Dalinette-Choupette®", "Fujion", "CIV323-Isaaq®", "Coop43-Juliet®", "SQ159-Natyra®", and "UEB32642-Opal®", which are all promising and being introduced into commercial plantings [42], and the standard varieties "Braeburn", "Fuji", "Gala", "Golden Delicious" were harvested at optimal harvest stage (at a starch level between 5 and 7). Fruits were stored under normal atmosphere at 2 °C and 90% relative humidity (RH) for 60 days. VOC and sensory analysis were performed using 18 apple fruits from each of the eleven different varieties and for each of the two location (A-225 m and B-650 m) (18 fruits × 11 varieties × 2 altitudes). The eighteen fruits were selected and kept at 20 °C for 1 day prior to the analysis. Following Corollaro et al.'s method with some modification, briefly, each fruit was cut into three equal equatorial discs using an apple cutter; then, 5 to 7 cylinders (1.8 cm diameter, 1.3 cm height) per slice, depending on fruit size, were cut using a commercial apple corer (Tescoma, Brescia, Italy) [43]. Each cylinder was immediately treated with an antioxidant solution (0.2% citric acid, 0.2% ascorbic acid, 0.5% calcium chloride). Each sample was composed of 10 apple cylinders in a clear plastic cup encoded with a random three-digit code. Cylinders from the same apple fruits were used for sensory and VOC analyses.

### 3.2. Volatile Organic Compound Analysis

Volatile compounds were extracted with head space solid phase micro-extraction (HS-SPME) from the same apples used for sensory analysis. Apple cylinders (see Section 3.1) were ground with an IKA A11 analytical mill with liquid nitrogen; 2 g of fresh pulp powder were placed in 20 mL glass vials with 2 mL of milliq water, 0.75 g of NaCl, and 15 µL of 1-heptanol (15 mg/L, hexane solution) as internal standard. Following the method of Aprea et al. [44], samples were kept in agitation (250 rpm) at 40 °C for 10 min, and compounds in the headspace were captured for 30 min at 40 °C on a 2-cm DVB/CAR/PDMS 50/30 µm fiber from Supelco (Bellefonte, PA, USA). A GCMS-QP2010 SE gas chromatograph mass spectrometer (Shimadzu, Kyoto, Japan) was used to separate the compounds with a capillary ZB-WAX column (30 m × 0.25 mm i.d. × 0.25 µm) (Phenomenex, Torrance, CA, USA). The compounds were desorbed in the GC inlet at 250 °C for 5 min. The GC oven parameters were set as follow: 40 °C for 3 min, then up to 220 °C at 4 °C/min held for 1 min; then up to 250 °C at 10 °C/min held for 1 min. Helium was used as the carrier gas with a flow rate of 1.2 mL/min. The MS detector was operated in full scan mode (mass range 40–400 *m/z*) and the transfer line to the MS system was maintained at 250 °C. Data processing was performed using GC-MS Solution SOFTWARE (Shimadzu, Kyoto, Japan). Identification of volatile compounds was carried out by comparing mass spectra and retention indexes using the NIST 2017 database and our internal database consisting of MS spectra of commercial standards. The experimental linear temperature retention index of each compound was calculated using a series of n-alkanes (C8–C20) in the same experimental conditions as the samples. Results were expressed as peak height percentage of the individual compound compared with the total identified compounds' height. Analysis was carried out on three technical replicates.

### 3.3. Sensory Analysis

Sensory profiling based on the quantitative descriptive analysis (QDA) method [45] was performed by a trained panel composed of eleven judges/assessors (six females and five males). All panelists were trained over 10 h (five sessions of two hours), on nineteen selected odor and flavor descriptors (Table 2). Training consisted of several types of tests of all sensory categories in order to be sure that the panelists were able to rate attributes using the linear scales anchored at the extremities with the product references (for more details on panel training and standard references refer to Corollaro et al. 2013 [43]). During sample evaluation, the panel rated odors (orthonasal perceptions by smelling) and flavors-by-mouth (retronasal perceptions by tasting) on an intensity scale going from 0 (absence) to 100 (100 maximum intensity) with graduation halfway (50). The protocols for sample preparation and testing procedure were adapted from Corollaro et al. [43]. Samples were evaluated in one session per week in four consecutive weeks, with six samples per session presented to each panelist monadically following a William's Latin square design. Sensory evaluations were performed within 2 h of sample preparation Tests were carried out within 2 h of sample preparation in the sensory laboratory of the Laimburg Research Center equipped with individual booths and under artificial lighting.

**Table 2.** Sensory vocabulary used by the sensory panel.

| Attributes | Definition |
| --- | --- |
| Overall odor | Overall odor sensation perceived via the orthonasal route |
| Apple | Specific odor (O) or retronasal flavor (F) apple sensation [1] |
| Banana | Specific odor (O) or retronasal flavor (F) banana sensation |
| Green grass | Specific odor (O) or retronasal flavor (F) green grass sensation |
| Honey | Specific odor (O) or retronasal flavor (F) honey sensation |
| Kiwi | Specific odor (O) or retronasal flavor (F) kiwi sensation |
| Lemon | Specific odor (O) or retronasal flavor (F) lemon sensation |
| Pear | Specific odor (O) or retronasal flavor (F) pear sensation |
| Pineapple | Specific odor (O) or retronasal flavor (F) pineapple sensation |
| Vanilla | Specific odor (O) or retronasal flavor (F) vanilla sensation |

[1] "O" and "F" refer to codings in Figure 2.

### 3.4. Data Analysis

In this analysis, the same apples samples from 11 different varieties obtained from two different altitudes were measured in triplicate for VOC analysis, using 11 assessors for sensory analysis; missing values on one replicate of three were replaced with median values considering the other replicates for VOC analysis. To evaluate differences between the aroma profile from two different altitudes, the Wilcoxon signed-rank test was used, which considers the same apple varieties measured in two different locations. To investigate the variabilities in sensory parameters by apple variety, location, and their interaction simultaneously, a two-way repeated measures ANOVA was applied. This method takes into consideration the panelist IDs involved in the sensory panel. To assess the relationships between VOCs and sensory parameters, and climatic and geographical parameters, exploratory analysis was performed using principal component analysis (PCA) [46]. Prior to performing PCA, all parameters were normalized to 0 mean and standard deviation of 1. To evaluate the relationship between sensory and VOC parameters, Spearman correlation analysis was performed. All statistical analyses were performed using *R* software [47] with packages lattice [48] and ggplot2 [49] for the visualization and lmerTest for performing a two-way repeated measures ANOVA [50].

## 4. Conclusions

Aroma composition combined with sensory parameters of eleven apple varieties grown in two different pedoclimatic location at different altitudes in South Tyrol were investigated. The goal of this work was to analyze and characterize the aroma of new and standard apple varieties with a special focus on the differences due to the pedoclimatic locations at two different altitudes (site A = 225 m and site B = 650 m). As expected, our results show a strong differentiation on aroma profile due to the variety.

Anyhow, we were able to distinguish twelve volatile organic compounds that change in relationship with pedoclimatic locations, independent of the variety. Moreover, significant interactions were found between variety and altitude on sensory parameters. In particular, nine of the eleven varieties presented significant differences on their sensory profile, including six odors and six flavors, between low and high pedoclimatic locations. In general, we provided a deeper investigation on how different altitudes can influence the apple aroma composition and perceptions, filling the gap of available information. To the best of the authors' knowledge, no studies have evaluated the sensory profile at low and high pedoclimatic locations by using a training panel on several apple varieties, or the relationship between these different sensory attributes and volatile organic compounds with a high number of new and commercial varieties.

From a practical point of view, this information is useful to help the growers in the site specific choice of apple varieties. On one hand, the investigated new varieties are all monogenic scab resistant, which makes them interesting for more sustainable apple growing. On the other hand, the consumer's expectations are strongly related to the inner quality of fruits, including aroma. Anyway, further experiments would be necessary to better highlight the VOCs' influence on aroma differentiation and consumer perception.

**Supplementary Materials:** The following are available online. Table S1. VOC semi quantification in 11 apple varieties grown in two different pedoclimatic locations in South Tyrol. Results are expressed as the peak height percentage compared with the total identified peaks height. The identification level for each compound is reported: level A refers to the NIST 2017 database match, level B refers to the linear retention index (LRI) match, and level C refers to a match with an in-house pure standard. Table S2. Wilcoxon signed rank test results on VOCs. Table S3. PCA loading results. Table S4. PCA results: variable correlations with Dim1 and Dim2. Table S5. Spearman correlation results. Table S6. Wilcoxon signed rank test results on sensory parameters.

**Author Contributions:** Conceptualization, L.L.; methodology, G.C. and L.L.; software, G.C., L.L., and N.D.; formal analysis, G.C. and L.L.; investigation, L.L., G.C., and N.D.; resources, L.L. and P.R.; data curation, N.D., G.C., and L.L.; writing—original draft preparation, G.C. and L.L.; writing—review and editing, G.C., L.L., N.D., P.R., and W.G.; visualization, N.D., G.C., and L.L.; supervision, L.L. and P.R.; project administration, L.L.; funding acquisition, L.L. and P.R. All authors have read and agreed to the published version of the manuscript.

**Funding:** This research was funded by the Autonomous Province of Bolzano, Capacity building 1 (Decision no. 1472, 07.10.2013), project Step-Up (Decision no. 864, 04.09.2018), and the Incoming Researcher Project (decree 334, 16.01.2019)

**Acknowledgments:** The authors are grateful for the cooperation of the panel judges from the Laimburg Research Centre. Elisa Piaia, Elisa Vanzo, and Philipp Coassin are gratefully acknowledged for administrative support.

**Conflicts of Interest:** All authors have read and agreed to the published version of the manuscript.

## References


1. Lechner, O.; Moroder, B. *Economic Portrait of South Tyrol*; Chamber of Commerce, Industry, Crafts and Agriculture of Bolzano: Bolzano, Italy, 2012.
2. Durán, L.; Costell, E. Revision: Percepción del gusto. Aspectos fisicoquímicos y psicofísicos/Review: Perception of taste. Physiochemical and psychophysical aspects. *Food Sci. Technol. Int.* **1999**, *5*, 299–309. [CrossRef]
3. Echeverria, G.; Fuentes, M.; Graell, J.; López, M.; López, L. Relationships between volatile production, fruit quality and sensory evaluation of Fuji apples stored in different atmospheres by means of multivariate analysis. *J. Sci. Food Agric.* **2003**, *84*, 5–20. [CrossRef]
4. Echeverría, G.; Fuentes, T.; Graell, J.; Lara, I.; López, L. Aroma volatile compounds of 'Fuji' apples in relation to harvest date and cold storage technology. *Postharvest Boil. Technol.* **2004**, *32*, 29–44. [CrossRef]
5. Echeverria, G.; Graell, J.; Lara, I.; López, M.; Puy, J.; López, L. Panel consonance in the sensory evaluation of apple attributes: Influence of mealiness on sweetness perception. *J. Sens. Stud.* **2008**, *23*, 656–670. [CrossRef]
6. Murray, J.; Delahunty, C.; Baxter, I. Descriptive sensory analysis: Past, present and future. *Food Res. Int.* **2001**, *34*, 461–471. [CrossRef]
7. Espino-Díaz, M.; Sepúlveda, D.R.; González-Aguilar, G.; Orozco, G.I.O. Biochemistry of Apple Aroma: A Review. *Food Technol. Biotechnol.* **2016**, *54*, 375–397. [CrossRef] [PubMed]
8. El Hadi, M.A.M.; Zhang, F.-J.; Wu, F.-F.; Zhou, C.-H.; Tao, J. Advances in Fruit Aroma Volatile Research. *Molecules* **2013**, *18*, 8200–8229. [CrossRef]
9. Buttery, R.G. Quantitative and sensory aspects of flavor of tomato and other vegetable and fruits. Flavor Science: Sensible Principles and Techniques. *Amer. Chem. Soc.* **1993**, 259–286.
10. Fellman, J.K.; Miller, T.; Mattinson, D.; Mattheis, J. Factors That Influence Biosynthesis of Volatile Flavor Compounds in Apple Fruits. *HortScience* **2000**, *35*, 1026–1033. [CrossRef]
11. Dixon, J.; Hewett, E.W. Factors affecting apple aroma/flavor volatile concentration: A Review. *New Zealand J. Crop. Hortic. Sci.* **2000**, *28*, 155–173. [CrossRef]
12. Li, G.; Jia, H.; Wu, R.; Teng, Y. Changes in volatile organic compound composition during the ripening of 'Nanguoli' pears (Pyrus ussuriensis Maxim) harvested at different growing locations. *J. Hortic. Sci. Biotechnol.* **2013**, *88*, 563–570. [CrossRef]
13. Mphahlele, R.; Caleb, O.J.; Fawole, O.A.; Opara, U.L. Effects of different maturity stages and growing locations on changes in chemical, biochemical and aroma volatile composition of 'Wonderful' pomegranate juice. *J. Sci. Food Agric.* **2015**, *96*, 1002–1009. [CrossRef] [PubMed]
14. Naryal, A.; Dolkar, D.; Bhardwaj, A.K.; Kant, A.; Chaurasia, O.P.; Stobdan, T. Effect of Altitude on the Phenology and Fruit Quality Attributes of Apricot (*Prunus armeniaca* L.) Fruits. *Def. Life Sci. J.* **2020**, *5*, 18–24. [CrossRef]
15. Brat, P.; Yahia, A.; Chillet, M.; Bugaud, C.; Bakry, F.; Reynes, M.; Brillouet, J.-M. Influence of cultivar, growth altitude and maturity stage on banana volatile compound composition. *Fruits* **2004**, *59*, 75–82. [CrossRef]
16. Tomasi, D.; Calo, A.; Costacurta, A.; Aldighieri, R.; Pigella, E.; Di Stefano, R. Effects of the microclimate on vegetative and aromatic response of the vine variety Sauvignon blanc, clone R3. *RiV. Vitic. Enol.* **2000**, *53*, 27–44.
17. Falcão, L.D.; De Revel, G.; Perello, M.C.; Moutsiou, A.; Zanus, M.C.; Bordignon-Luiz, M. A Survey of Seasonal Temperatures and Vineyard Altitude Influences on 2-Methoxy-3-isobutylpyrazine, C13-Norisoprenoids, and the Sensory Profile of Brazilian Cabernet Sauvignon Wines. *J. Agric. Food Chem.* **2007**, *55*, 3605–3612. [CrossRef]
18. Yue, T.-X.; Chi, M.; Song, C.-Z.; Liu, M.-Y.; Meng, J.-F.; Zhang, Z.-W.; Li, M.-H. Aroma Characterization of Cabernet Sauvignon Wine from the Plateau of Yunnan (China) with Different Altitudes Using SPME-GC/MS. *Int. J. Food Prop.* **2014**, *18*, 1584–1596. [CrossRef]
19. Jing, C.; Feng, D.; Zhao, Z.; Wu, X.; Chen, X. Effect of environmental factors on skin pigmentation and taste in three apple cultivars. *Acta Physiol. Plant.* **2020**, *42*, 1–12. [CrossRef]
20. Singh, S.R.; Sharma, A.K.; Sharma, M.K. Effect of different NPK combinations on fruit yield, quality and leaf nutrient composition of apple (*Malus domestica* Borkh) cv. Red Delicious at different altitudes. *Environ. Ecol.* **2006**, *24*, 71–75.

21. Comai, M.; Dorigoni, A.; Fadanelli, L.; Piffer, I.; Micheli, F.; Dallabetta, N. Influenza della carica e dei siti di produzione sulle caratteristiche fisico-chimiche di Golden Delicious in Val di Non. *Riv. Fruttic. Ortofloric.* **2005**, *67*, 52–57. (In Italian)

22. Charles, M.; Corollaro, M.L.; Manfrini, L.; Endrizzi, I.; Aprea, E.; Zanella, A.; Grappadelli, L.C.; Gasperi, F. Application of a sensory-instrumental tool to study apple texture characteristics shaped by altitude and time of harvest. *J. Sci. Food Agric.* **2017**, *98*, 1095–1104. [CrossRef] [PubMed]

23. Kritioti, A.; Paikousis, L.; Drouza, C. Characterization of the volatile profile of virgin olive oils of Koroneiki and Cypriot cultivars, and classification according to the variety, geographical region and altitude. *LWT* **2020**, *129*, 109543. [CrossRef]

24. Belton, P.S.; Colquhoun, I.J.; Kemsley, E.K.; Delgadillo, I.; Roma, P.; Dennis, M.; Sharman, M.; Holmes, E.; Nicholson, J.K.; Spraul, M. Application of chemometrics to the 1H NMR spectra of apple juices: Discrimination between apple varieties. *Food Chem.* **1998**, *61*, 207–213. [CrossRef]

25. Young, H.; Gilbert, J.M.; Murray, S.H.; Ball, R. Causal Effects of Aroma Compounds on Royal Gala Apple Flavours. *J. Sci. Food Agric.* **1996**, *71*, 329–336. [CrossRef]

26. Zhu, D.; Ren, X.; Wei, L.; Cao, X.; Ge, Y.; Liu, H.; Li, J. Collaborative analysis on difference of apple fruits flavor using electronic nose and electronic tongue. *Sci. Hortic.* **2020**, *260*, 108879. [CrossRef]

27. Nicolini, G.; Roman, T.; Carlin, S.; Malacarne, M.; Nardin, T.; Bertoldi, D.; Larcher, R. Characterisation of single-variety still ciders produced with dessert apples in the Italian Alps. *J. Inst. Brew.* **2018**, *124*, 457–466. [CrossRef]

28. Rosend, J.; Kuldjärv, R.; Rosenvald, S.; Paalme, T. The effects of apple variety, ripening stage, and yeast strain on the volatile composition of apple cider. *Heliyon* **2019**, *5*, e01953. [CrossRef]

29. Maragò, E.; Michelozzi, M.; Calamai, L.; Camangi, F.; Sebastiani, L. Antioxidant properties, sensory characteristics and volatile compounds profile of apple juices from ancient Tuscany (Italy) apple varieties. *Eur. J. Hortic. Sci.* **2016**, *81*, 255–263. [CrossRef]

30. Merkle, S.; Kleeberg, K.K.; Fritsche, J. Recent Developments and Applications of Solid Phase Microextraction (SPME) in Food and Environmental Analysis—A Review. *Chromatograph* **2015**, *2*, 293–381. [CrossRef]

31. Zhang, H.; Chen, H.; Wang, W.; Jiao, W.; Chen, W.; Zhong, Q.; Yun, Y.-H.; Chen, W. Characterization of Volatile Profiles and Marker Substances by HS-SPME/GC-MS during the Concentration of Coconut Jam. *Foods* **2020**, *9*, 347. [CrossRef]

32. Kebede, B.; Ting, V.; Eyres, G.; Oey, I. Volatile Changes during Storage of Shelf Stable Apple Juice: Integrating GC-MS Fingerprinting and Chemometrics. *Foods* **2020**, *9*, 165. [CrossRef] [PubMed]

33. Guo, J.; Yue, T.; Yuan, Y.; Sun, N.; Liu, P. Characterization of volatile and sensory profiles of apple juices to trace fruit origins and investigation of the relationship between the aroma properties and volatile constituents. *LWT* **2020**, *124*, 109203. [CrossRef]

34. The Good Scent Company. The Good Scent Company. Available online: http://thegoodscentscompany.com (accessed on 29 May 2020).

35. Mehinagic, E.; Royer, G.; Symoneaux, R.; Jourjon, F.; Prost, C. Characterization of Odor-Active Volatiles in Apples: Influence of Cultivars and Maturity Stage. *J. Agric. Food Chem.* **2006**, *54*, 2678–2687. [CrossRef] [PubMed]

36. Holland, D.; Larkov, O.; Bar-Ya'Akov, I.; Bar, E.; Zax, A.; Brandeis, E.; Ravid, U.; Lewinsohn, E. Developmental and Varietal Differences in Volatile Ester Formation and Acetyl-CoA: Alcohol Acetyl Transferase Activities in Apple (*Malus domestica* Borkh.) Fruit. *J. Agric. Food Chem.* **2005**, *53*, 7198–7203. [CrossRef]

37. Xiao, Z.; Wang, R.; Xiao, Z.; Zhu, J.; Sun, X.; Wang, P. Characterization of ester odorants of apple juice by gas chromatography-olfactometry, quantitative measurements, odor threshold, aroma intensity and electronic nose. *Food Res. Int.* **2019**, *120*, 92–101. [CrossRef]

38. Echeverria, G.; Correa, E.-C.; Altisent, M.R.; Graell, J.; Puy, J.; López, L. Characterization of Fuji Apples from Different Harvest Dates and Storage Conditions from Measurements of Volatiles by Gas Chromatography and Electronic Nose. *J. Agric. Food Chem.* **2004**, *52*, 3069–3076. [CrossRef]

39. Souleyre, E.; Bowen, J.K.; Matich, A.J.; Tomes, S.; Chen, X.; Hunt, M.B.; Wang, M.Y.; Ileperuma, N.R.; Richards, K.; Rowan, D.D.; et al. Genetic control of α-farnesene production in apple fruit and its role in fungal pathogenesis. *Plant. J.* **2019**, *100*, 1148–1162. [CrossRef]

40. Plotto, A.; McDaniel, M.R. Tools of sensory analysis applied to apples. In Proceedings of the Washington Tree Fruit Postharvest Conference 2001, Wenatchee, WA, USA, 13–14 March 2001.

41. Thalheimer, M.; Paoli, N. Die Witterung im Jahr 2014. *Obstbau Weinbau* **2015**, *52*, 19–24.

42. Guerra, W. Apfelsorten mit Resistenzeigenschaften-Derzeitiger Anbau und Perspektiven in der Züchtung. *Obstbau Weinbau* **2015**, *52*, 153–156.

43. Corollaro, M.L.; Endrizzi, I.; Bertolini, A.; Aprea, E.; Demattè, M.L.; Costa, F.; Biasioli, F.; Gasperi, F. Sensory profiling of apple: Methodological aspects, cultivar characterization and postharvest changes. *Postharvest Boil. Technol.* **2013**, *77*, 111–120. [CrossRef]

44. Aprea, E.; Gika, H.; Carlin, S.; Theodoridis, G.; Vrhovsek, U.; Mattivi, F. Metabolite profiling on apple volatile content based on solid phase microextraction and gas-chromatography time of flight mass spectrometry. *J. Chromatogr. A* **2011**, *1218*, 4517–4524. [CrossRef] [PubMed]

45. Stone, H.; Sidel, J.L. *Sensory evaluation practices*; Academic Press: London, UK, 2006.

46. Richards, L.E.; Jolliffe, I.T. Principal Component Analysis. *J. Mark. Res.* **1988**, *25*, 410. [CrossRef]

47. R Core Team. R: A Language and Environment for Statistical Computing. Foundation for Statistical Computing. 2019. Available online: https://www.r-project.org/index.html (accessed on 29 May 2020).

48. Deepayan, S. *Lattice: Multivariate Data Visualization with R*; Springer: New York, NY, USA, 2008.

49. Gómez-Rubio, V. ggplot2-Elegant Graphics for Data Analysis (2nd Edition). *J. Stat. Softw.* **2017**, *77*. [CrossRef]

50. Kuznetsova, A.; Brockhoff, P.B.; Christensen, R.H.B. lmerTest Package: Tests in Linear Mixed Effects Models. *J. Stat. Softw.* **2017**, *82*, 1–26. [CrossRef]

**Sample Availability:** Samples of the compounds are not available from the authors.

*Article*

# Volatile Profile of Mead Fermenting Blossom Honey and Honeydew Honey with or without *Ribes nigrum*

Giulia Chitarrini [1], Luca Debiasi [1], Mary Stuffer [1,2], Eva Ueberegger [1], Egon Zehetner [2], Henry Jaeger [2], Peter Robatscher [1] and Lorenza Conterno [1,*]

[1]   Laimburg Research Centre, Ora (BZ), 39040 Auer, Italy; giulia.chitarrini@laimburg.it (G.C.); luca.debiasi@laimburg.it (L.D.); mary.stuffer@students.boku.ac.at (M.S.); eva.ueberegger@laimburg.it (E.U.); peter.robatscher@laimburg.it (P.R.)

[2]   Institute of Food Technology, University of Natural Resources and Life Sciences (BOKU), 1190 Vienna, Austria; egon.zehetner@boku.ac.at (E.Z.); henry.jaeger@boku.ac.at (H.J.)

*   Correspondence: lorenza.conterno@laimburg.it; Tel.: +39-0471969591

Academic Editor: Eugenio Aprea
Received: 27 March 2020; Accepted: 11 April 2020; Published: 15 April 2020

**Abstract:** Mead is a not very diffused alcoholic beverage and is obtained by fermentation of honey and water. Despite its very long tradition, little information is available on the relation between the ingredient used during fermentation and the aromatic characteristics of the fermented beverage outcome. In order to provide further information, multi-floral blossom honey and a forest honeydew honey with and without the addition of black currant during fermentation were used to prepare four different honey wines to be compared for their volatile organic compound content. Fermentation was monitored, and the total phenolic content (Folin–Ciocalteu), volatile organic compounds (HS-SPME-GC-MS), together with a sensory evaluation on the overall quality (44 nontrained panelists) were measured for all products at the end of fermentation. A higher total phenolic content resulted in honeydew honey meads, as well as the correspondent honey wine prepared with black currant. A total of 46 volatile organic compounds for pre-fermentation samples and 62 for post-fermentation samples were identified belonging to higher alcohols, organic acids, esters, and terpenes. The sensory analysis showed that the difference in meads made from blossom honey and honeydew honey was perceptible by the panelists with a general greater appreciation for the traditional blossom honey mead. These results demonstrated the influences of different components in meads, in particular, the influence of honey quality. However, further studies are needed to establish the relationship between the chemical profile and mead flavor perception.

**Keywords:** gas chromatography-mass spectrometry; fermentation; honey; black currant

## 1. Introduction

Mead, also called honey wine, is traditionally an alcoholic beverage obtained through yeast fermentation of diluted honey. Mead is found in the history of many countries all around the world, and it is one of the oldest alcoholic beverages with variable alcohol content (8–18% alcohol *v/v*), mostly depending on the honey to water dilution ratio. Besides the traditional mead (the fermented diluted honey), many variations can be found, containing also herbs and spices (metheglin) or fruit (melomel) [1].

Fructose and glucose are generally the most abundant simple sugar found in honey, and fructose is the dominant one (on average from 32% to 42% depending on the honey origin [2]). The mead fermentation process is usually longer than most alcoholic fermentation, where other sugars are present and in higher concentrations. In fact, this fermentation often takes several months to complete, depending on the type of honey, yeast strain, and honey-must composition. Mead contains ethanol and many other compounds, such as sugars, acids, vitamins, phenolic compounds, and minerals, also in dependence on the added ingredient beside honey (reviewed [3]).

The three main factors mead flavor depends are the honey, the yeast strain carrying out the alcoholic fermentation, and the fermentation conditions. Flavor perception may also be influenced by the final alcohol content, the residual sugars, and the acidic value.

Besides the botanical and the geographical origin, honey can be divided into two main groups: blossom or floral honey and honeydew honey. The former is produced starting from nectar from the flower of blossoming plants, while the latter from the exudates from certain plants (such as *Pinus, Abies, Castanea,* and *Quercus,* among others), usually with the concourse of insects, mainly from the family Aphididae [4]. Honeydew has a stronger taste than blossom honey and is perceived as less sweet. It has higher antioxidant activity and a higher concentration of oligosaccharides. Many researchers have found out that honeydew honey with a darker color has a higher concentration of total phenolic compounds and a higher antioxidative capacity [5,6]. For honey wine production, wine yeast strains are usually used because the sugar, pH, and nitrogen characteristics in mead are similar to the ones of grape must [1]. Yeast produces during fermentation many metabolites, which have a large impact on the beverage flavor. Most of the unique flavors of mead depend on the type of honey. For every additional ingredient, additional flavor molecules may be developed [7]. The formation of metabolites during fermentation not only depends on the raw materials used but also on the yeast strain. Most by-products are synthesized when there is a high rate of sugar and another nutrient uptake by yeasts, which all join the catabolizing pathway at the level of pyruvate. Many chemically different metabolites in diverse concentrations contribute to the flavor of alcoholic beverages, such as organic acids, fusel alcohols, aromatic alcohols, esters, carbonyls, and various sulfur-containing compounds [8,9]. Yeast activity behavior during fermentation caused by stressful conditions can lead to the production of unwanted flavors, for example, high volatile acidity or undesired esters [10].

The acids in the mead are coming from the honey, the added fruits, and the acids used for acidification of the must [11]. The acids in honey are usually citric, malic, succinic, formic, proglutamic, acetic, gluconic, and lactic [12]. Organic acids have a very important function in alcoholic beverages influencing organoleptic characteristics and product stability. The addition of organic acids is regulated by the European legislation for wine and recommended because low pH helps to minimize the risk of microbial spoilage by preventing bacterial growth. On the other hand, pH values below 3 could make a stressful environment for the yeast, which leads to the production of undesirable by-products [13].

Volatile organic compounds (VOCs) in mead are present due to raw materials or produced by yeast during fermentation. VOCs belong to various chemical classes, such as esters, higher alcohols, acids, aldehydes, ketones, etc. They have an impact on the aroma and odor of mead and especially contribute to the fruity and floral nuances [7]. Alcohols, such as n-propanol, iso-butanol, 2-phenylethanol, amyl alcohol, and others, influence the flavor of alcoholic beverages. In high concentrations, their flavor is usually described as a solvent, and they contribute to the intensification of an alcoholic taste, which creates a warm mouth feeling [8,14]. In lower concentrations, they may have a positive effect, increasing the complexity of fermented beverages [15]. Esters contribute to a floral and fruity flavor of mead. Ethyl acetate, for example, is considered as fruity or solvent (depending on concentration and combination with other volatiles), and isoamyl acetate has a banana or apple flavor. Ethyl esters are the most present in mead because ethanol is the most available substrate [8]. Terpenic compounds are mostly produced by plants and some insects but also from yeast [16] and have been found in remarkable concentrations in mead [7,13,17]. Organic acids have a sour flavor, and additionally, they can have individual characteristic flavors. Short-chain fatty acids have a mostly negative influence on the flavor, depending on the concentration and combination of the product. Furthermore, they affect foam performance [8]. In the mead fermentation process, acetic acid and succinate acid are formed in considerable amounts. They reduce the pH, increase the total acidity, and reduce the dissociation of fatty acids. High amounts of acetic acid, succinic acid, and a high concentration of fatty acids may cause a slowdown of the fermentation process [18].

Despite the long tradition of mead making, a limited scientific background is available in this field, which may be due to the medium and small-scale production. Therefore, systematic information and knowledge are needed in order to be able to define those parameters necessary to understand how to master mead quality, develop adequate formulations, and optimize the fermentation conditions as reviewed by Iglesias et al. [3].

The present work aimed to contribute to filling some of the gaps on the influence of different ingredients used for mead making on the final flavor profile. To reach this aim, multi-floral blossom honey and a forest honeydew honey, with or without black currant (*Ribes nigrum*), added before fermentation, were used to prepare four different honey wines to be compared for their volatile organic compounds content. Due to the high complexity of the *Ribes nigrum*, its addition was decided to originate a mead variant rich in polyphenols, minerals, and vitamins, which might have a positive effect on the fermentation process or the end product quality [19]. Polyphenol content and consumer acceptance were explored, as well.

## 2. Results and Discussion

Four different meads of honey wine were prepared either with multi-floral blossom honey and forest honeydew honey alone (B and H, respectively) or with black currant added (BC and HC, respectively).

### 2.1. Fermentation Kinetics

Fermentation kinetics were monitored by measuring the weight loss due to $CO_2$ effluence. HC was the first product reaching the stationary phase in fermentation after approx. 14 days, whereas H and BC took approx. 20 days. B had the longest fermentation time with approx. 30 days (Figure S1). According to these results, the products prepared with blossom honey had a longer fermentation time, and the addition of black currants accelerated the fermentation. Differences in honey composition [20] and black currant [19] might provide factors influencing the yeast rate of sugar depletion.

Total $CO_2$ production in g/L was equal to 86.1 ($\pm$0.3), 90.6 ($\pm$0.7), 75.5 ($\pm$0.2), and 79.2 ($\pm$0.5) for B, BC, H, and HC, respectively, in theory corresponding to 176.3 ($\pm$0.7), 185.5 ($\pm$1.4), 154.5 ($\pm$0.5), and 162.1 ($\pm$0.9) of fermented glucose in g/L. Measurement of sugar content in the original honey showed 690 $\pm$10 g/L for the blossom honey and 650 $\pm$10 g/L for the honeydew honey, expressed as total glucose (sum of glucose, fructose, and sucrose contribution). Glucose and fructose measured in the prepared product before fermentation showed a lower amount of glucose and fructose in H and HC (177.0 and 177.1 g/L, respectively), compared to B and BC (202.5 and 201.5 g/L, respectively). Further measurement of honey sugars besides the one measured might provide a more detailed explanation.

Being the residual sugar content in all honey wines below 5 g/L (Table 1), even if differences in fermentation length were observed, all the fermentation could be considered successfully completed.

**Table 1.** Physiochemical parameters characterizing the four products prepared with blossom honey (B), blossom honey and blackcurrant (BC), honeydew honey (H), and honeydew honey and black currant (HC). Parameters were measured before fermentation (t0) and/or at the end of the fermentation process (END). END data are reported as the average of the three biological replicates ± standard deviation (in brackets). Apex letter in the same row shows the results of the statistic evaluation, and different letters correspond to significant different parameters ($p \leq 0.05$).

| | Analysis Time | B | BC | H | HC |
|---|---|---|---|---|---|
| pH | t0 | 3.17 | 3.15 | 3.16 | 3.22 |
| pH | END | 3.13 (±0.03) a | 3.33 (±0.01) b | 3.27 (±0.01) bc | 3.29 (±0.01) c |
| Brix (%) | t0 | 21.6 | 22.9 | 21.7 | 21.5 |
| Glu+Fru (g/L) | t0 | 202.5 | 201.5 | 177.0 | 177.1 |
| Residual sugar (g/L) | END | 4.3 (±0.6) a | 0.4 (±0.1) bc | 1.0 (±0.1) b | 0.1 (±0.1) c |
| Ethanol (% vol/vol) | END | 11.32 (±0.44) a | 10.63 (±0.67) a | 8.60 (±0.13) b | 8.66 (±0.18) b |
| Acetic acid (g/L) | END | 0.40 (±0.02) a | 0.26 (±0.02) a | 0.32 (±0.01) a | 0..20 (±0.01) a |
| Acetaldehyde (mg/L) | END | 6.2 (±1.4) a | 11.3 (±2.1) ab | 10.4 (±3.4) ab | 16.8 (±2.7) b |
| L-lactic acid (g/L) | END | 0.14 (±0) a | 0.23 (±0.02) a | 0.30 (±0) a | 0.36 (±0.01) a |

## 2.2. Physiochemical Parameters

The mean values for all parameters measured in the fermented product are listed in Table 1. B, which had a residual sugar value of 4.3 ± 0.6 g/L, significantly differed from the others, although usually meads can be considered dry [11]. Ethanol concentration in the four products was 11.32% *v/v*, 10.63% *v/v*, 8.60% *v/v*, and 8.66% *v/v*, respectively, measured in B, BC, H, and HC. Ethanol content was significantly higher in varieties with blossom honey. Even if prepared at a similar honey dilution rate, blossom honey used in B and BC products had a higher fermentable sugar content, explaining both the higher ethanol content and the higher carbon dioxide loss observed in the B and BC products.

The pH did not change to a great extent during the fermentation, and the lower one was recorded for the B mead, and L-lactic acid content was not significantly different in the four products. Acetic acid content at the end of fermentation ranged between 0.20 g/L and 0.40 ±0.02 g/L. Besides being yeast strain-dependent and related to the amount of sugar fermented, acetic acid might be also providing information on the yeast stress status or be the symptoms of microbial spoilage. This latter event seemed not to have occurred in the investigated products, and the acetic acid was detectable in amounts comparable to previous studies [3,13,18]. Acetaldehyde concentration ranged between 6.2 ± 1.2 and 16.8 ± 2.2 mg/L. Acetaldehyde concentration in meads usually ranges between 18.2 and 125.5 mg/L, as reported in the literature [21]. In general, 0.5–286 mg/L is the concentration of acetaldehyde produced by *Saccharomyces cerevisiae* in white wine [22]. A maximum of 0.5 g/L has been indicated in beer [8]. High concentrations of acetaldehyde would lead to a pungent, green, and grassy flavor, and it is associated with microbial spoilage of the fermented beverages. The acetaldehyde concentration found in this study testified the absence of microbial spoilage. Lactic acid was measured in very few quantities of 0.14–0.36 g/L. This was an indication that no lactic fermentation from bacteria took place.

## 2.3. Total Polyphenolic Content in Honey Wine

Total polyphenolic content measured using Folin–Ciocalteu reagent was 54.91 mg/L (±2.16) in B honey wine, 289.09 mg/L (±14.97) in BC, 101.95 mg/L (±5.78) in H, and 304.44 mg/L (±14.13) in HC, all expressed as mg/L of gallic acid. Results for traditional meads (B and H) were similar to the results reported in the literature [23,24]. Blossom honey varieties had less polyphenolic content than honeydew honey varieties, which corresponded to the result found in the literature [5,6]. Products prepared with currants (BC and HC) had much higher concentrations than the traditional meads as expected. The black currants added about 200 mg/L of polyphenols to the mead. Together with the red color, black currant imparted to the honey wine a higher capacity to counteract oxidation, due to the higher amount of compound recognized to protect from oxidation [25] and as found in polish mead by Socha [26].

### 2.4. Volatile Organic Compounds

Meads, fermented from honeydew honey and blossom honey with and without the addition of *Ribes nigrum*, were analyzed to evaluate the influence of the starting honey and blackcurrant addition on the volatile organic compounds profile using HS-SPME-GC-MS method. The flavor and aroma of the final product depend on the type of honey and the floral source, the fermenting yeasts and conditions, and the presence of additives and fruits [27–29].

In this work, a semi quantification of 62 compounds was reported as an average of three biological replicates (Table S1); the table includes the compound name and class, Chemical Abstracts Service (CAS) number, retention index, retention time, and level of identification. Peaks in the chromatograms, acquired in full scan mode, had been integrated and reported as the area ratio of the peak with the internal standard (2-octanol). An ANOVA test was performed with a Turkey posthoc method to identify significant differences between the samples. In total, 4 acids, 13 alcohols, 4 aldehydes, 14 esters, 1 ketone, 16 terpenes and derivatives, 7 others, and 3 unknown compounds were identified. A general view of the results showed a higher compound formation or increase in samples analyzed after the fermentation.

To explore the dataset, a principal component analysis (PCA) was performed using the three replicates for each mead (Figure 1). The first and second components explained 80% of the total variance; the eigenvalues and the correlation results between variables and PCs are reported in Supplementary File 1. The first component allowed the separation between pre-fermentation and post-fermentation samples; the second component allowed to separate meads based on the honey (honeydew honey or blossom honey) and the blackcurrant addition. To better understand which metabolite influenced the diversification of the products, a heatmap was used (Figure 2) with a log10 transformed data (average of three replicates).

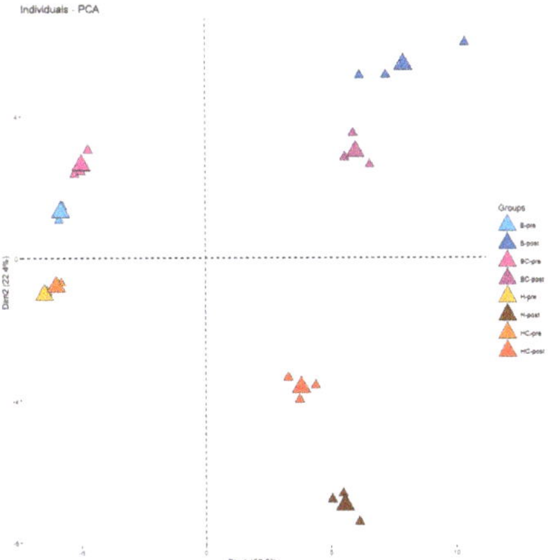

**Figure 1.** Principal component analysis (PCA) of the volatile compounds found in four honey wines made with blossom honey (B), blossom honey with blackcurrant (BC), honeydew honey (H), and honeydew honey with blackcurrant (HC), analyzed before (pre) and at the end of the fermentation process (post). The first component explained 57.5%, and the second component explained 22.4% of the total variance. Samples replicates are shown by the smaller triangle-shaped dots, while the average is represented by the bigger shaped dots.

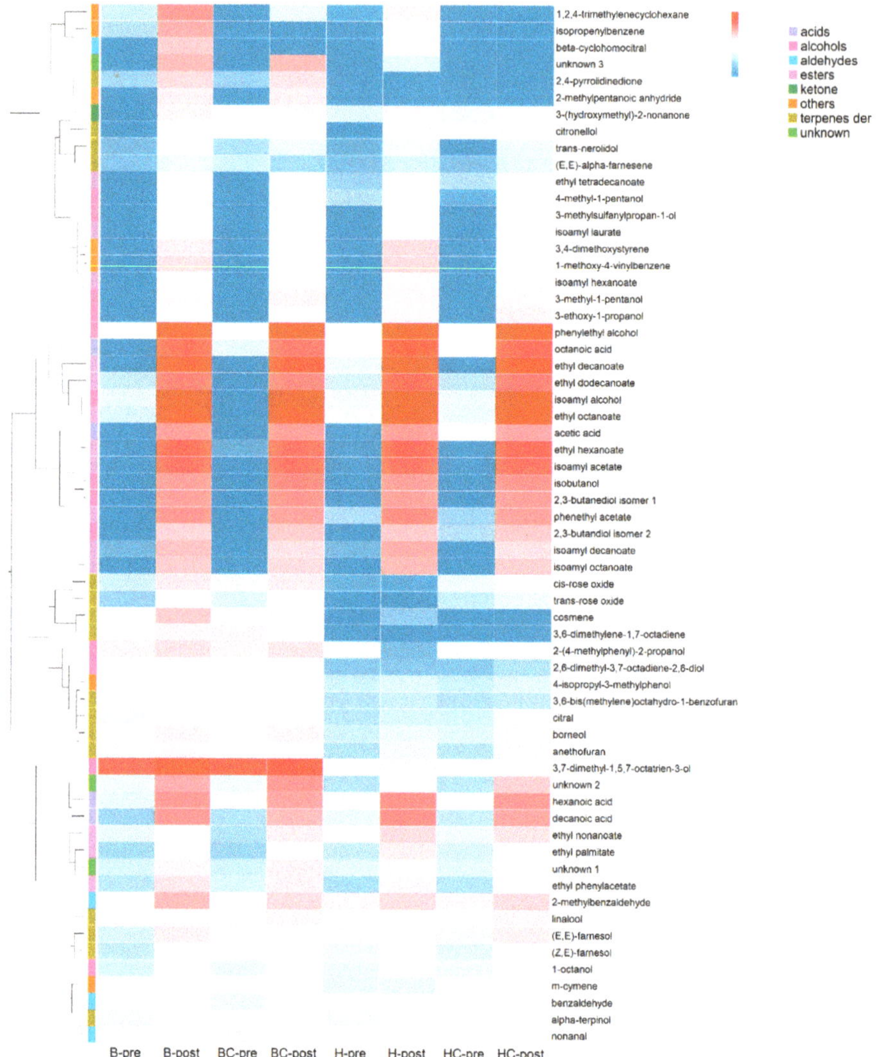

**Figure 2.** Heatmap and one-dimensional hierarchical dendrogram of the volatile compounds found in the investigated samples. Heatmap represents a log10 transformed data (average of three replicates). The heatmap color represents the magnitude of each compound. Dark red color indicates the higher magnitude, and then the magnitude gradually decreases to light red, white, light blue, up to dark blue (one order each), with the latter, indicating the lower magnitude. The magnitude represents transformed data (log10 of the ratio) to fit them in the same range. On the left: the colored sidebar indicates the class of metabolites. Sample legend: B = blossom honey, BC = blossom honey with blackcurrant, H = honeydew honey, HC = honeydew honey with blackcurrant, pre = pre-fermentation, and post = post-fermentation.

### 2.4.1. Volatile Profile in the Products before and after Fermentation with *S. cerevisiae*

Looking at the PCA, a clear separation between the pre- and post-fermentation samples was noticeable due to the volatile organic compounds produced during the fermentation. The number of identified VOCs increased in post-fermentation samples since some VOCs are produced by the yeasts during the alcoholic fermentation [30]. At least 18 compounds produced by the yeasts during the fermentation process were found, being mainly alcohols and esters (Figure 2), already described as the products of *S. cerevisiae* EC1118 fermentation [31,32]. As reported in the literature, it is well known that yeasts are VOCs producers; in wine, the main groups of compounds that form the fermentation bouquet are the acids, alcohols, and esters and, to a lesser extent, aldehydes and ketones [33].

Furthermore, it is known that the compounds influencing the aroma of alcoholic beverages are mainly higher alcohols, esters, volatile acids, and aldehydes [34], making yeasts strain the main actor in establishing the sensory characteristics. Looking at our results, B and BC post-fermentation had a higher number of volatile compounds than H and HC post-fermentation. Being that the yeast strains are the same in all fermentation processes, the different numbers are due to the different starter matrix. In the product B-post, 62 VOCs were detected, while BC-post was characterized by 60 different compounds. In H-post and in HC-post, 56 and 54 VOCs were, respectively, identified. The product prepared with blackcurrants, in pre- and post-fermentation, had a lower VOCs content than those produced with the same honey but without fruit addition. The compounds that contributed to this difference were 1,2,4-trimethylenecyclohexane found in B-post and H-post and a very low amount in BC-post, isopropenylbenzene found in B-post and H-post, and beta-cyclohomocitral only found in B-post (Figure 3). These three compounds were not confirmed by the standard injection; for this reason, their identification could be only considered as putative. It is reasonable to assume that *Ribes nigrum* added the nutrients, shaping the yeast metabolism; however, there is no evidence in the literature that which metabolic pathway leads to the synthesis of these compounds that might be regulated during fermentation by yeast.

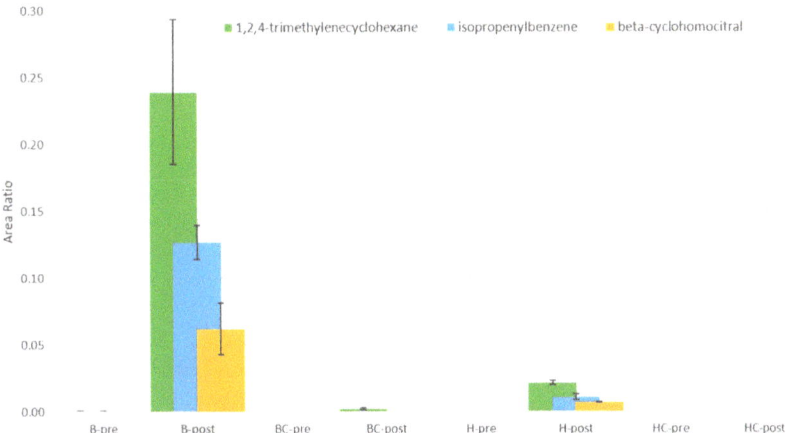


**Figure 3.** 1,2,4-trimethylencyclohexane, isopropenylbenzene, and beta-ciclohomocitral in honey wine before and after fermentation by *S. cerevisiae* EC1118. Samples were named as follows: B = blossom honey, BC = blossom honey with blackcurrant, H = honeydew honey, HC = honeydew honey with blackcurrant, pre = pre-fermentation, and post = post-fermentation.

The post-fermentation volatile organic compounds had been highlighted in the heatmap (Figure 2). Among them, isoamyl alcohol was found featuring in many fermented alcoholic beverages and already reported in mead [10,17,35], giving a solvent, sweet, and nail polish aroma and also part of fusel oil [36]. Another detected fusel oil representative was isobutanol, only found in post-fermentation samples (Figure 2), which lead to green notes in the flavor of beverages [37]. Esters are contributing to the fruity and floral nuances of meads [38]. In the samples, 14 esters were found; these compounds appearing in different concentrations in the samples, as shown in Figure 4, with a visible magnitude increased at the end of the fermentation process. H-pre had a higher content of esters and a slightly different profile of this class of compounds compared with the other pre-fermentation samples (Figure 4a). H-pre was, in fact, characterized by the presence of ethyl octanoate end ethyl nonanoate in higher amount and by the presence of ethyl decanoate not revealed in the other worts. In the same product, at the end of the fermentation process, the esters profile seemed to be more similar, although H-post exhibited a higher content in ethyl octanoate and ethyl decanoate compared to the other honey wines (Figure 4b). All found esters are common components in alcoholic beverages and are often found in fruits. In detail, the most abundant compounds determined in our fermented samples were isoamyl acetate, a characteristic compound of banana flavor, ethyl hexanoate and ethyl octanoate that contribute to a sweet and strawberry-like aroma, and ethyl decanoate with a sweet, fruity apple flavor [39]. Ethyl octanoate and ethyl decanoate aroma have also been described as waxy and soapy [40].

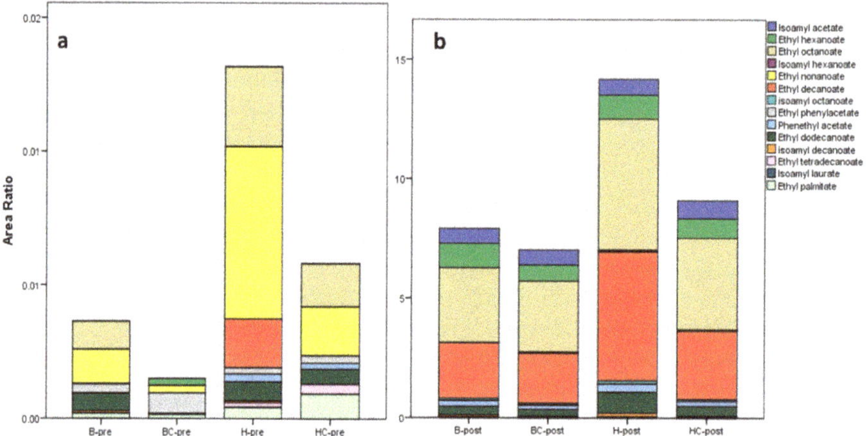


**Figure 4.** Esters composition in pre (**a**) and post (**b**) fermentation samples. Values represent the area ratio with the internal standard (n = 3). B-pre: blossom honey pre-fermentation; B-post: blossom honey post-fermentation; BC-pre: blossom honey with blackcurrant pre-fermentation; BC-post: blossom honey with blackcurrant post-fermentation; H-pre: honeydew honey pre-fermentation; H-post: honeydew honey post-fermentation; HC-pre: honeydew honey with blackcurrant pre-fermentation; HC-post: honeydew honey with blackcurrant post-fermentation.

### 2.4.2. Honey Influence on the Aromatic Characteristics of the Product

The honey used for the fermented product represented another important factor influencing the final aroma of the beverages. In our study, two types of honey were used: blossom honey (B) and honeydew honey (H). The PCA showed a separation on the second component, based on the honey used for the fermented beverage (Figure 1); looking at the heatmap, we identified some compounds potentially related to the types of used honey (Figure 2).

Few compounds seemed to be higher in B-pre product compared to the others and were accumulated in B-post products in comparison to the H samples; they were 2-(4-methylphenyl)-2-propanol, 2,6-dimethyl-3,7-octadiene-2,6-diol, 3,6-bis(methylene)octahydro-1-benzofuran, 3,6-dimethylene-1,7-octadiene, 3,7-dimethyl-1,5,7-octatrien-3-ol, 4-isopropyl-3-methylphenol, citral, and m-cymene.

This cluster showed to be more abundant in all B meads (with and without blackcurrant addition) compared with H samples. Other characteristic compounds of B meads were *trans*-nerolidol, *cis* and *trans*-rose oxide, anethofuran, borneol, and ethyl phenylacetate. These compounds produced during the fermentation process (Figure 2) could be responsible for a floral, rose, and balsamic camphor perception in the final product. Ethyl phenylacetate can not only contribute to a positive note, but it is also considered as off-flavor formed in beer during aging from precursors, which are produced during the fermentation [41].

Regarding the honeydew honey samples, no typical compounds were found in pre-fermentation samples. However, the final beverage seemed to be richer in acids, such as hexanoic acid, octanoic acid, decanoic acid, and esters, such as ethyl octanoate, isoamyl hexanoate, ethyl nonanoate, ethyl decanoate, ethyl dodecanoate. Short-chain fatty acids, such as octanoic acid (caprylic acid) and hexanoic acid (caproic acid), were also reported [17]. These compounds were present in higher concentrations and are associated with negative characteristics of "rancid," "cheese," and "fatty" aroma [7].

Terpene compounds were found in both pre- and post-fermentation samples, and, as expected, these compounds mostly originate from the raw materials. Many of the detected terpenes are known for their positive influences on mead aroma. Citral, for example, has a lemon-like pleasant odor [42], linalool a floral and spicy odor [43]. The stereoisomers *cis*-rose oxide and *trans*-rose oxide are found in flowers, and fruit and essential oils can contribute to fruit and floral notes in fruits and grapes. Mentofuran is a constituent of peppermint oil [44].

### 2.4.3. Influence of *Ribes nigrum* Addition

Blackcurrant addition, with its peculiar composition, had the capability to modify the fermentation environment for the yeast, leading to different volatiles, influencing, therefore, the aroma profile of the final product. Blackcurrant aroma is characterized by various volatile components, including esters and terpenoids. As reported, cultivars and growing and storage conditions can affect the flavor component [45,46]. Among the most reported compounds for the characteristic of blackcurrant fruit, 2-methylbutyl acetate, methyl butanoate, ethyl butanoate, and ethyl hexanoate, belonging to the esters class, that confer fruity and sweet notes are mentioned. Besides, nonanal, beta-damascenone, and monoterpenes, ketones, and sulfur compounds, such as 4 methoxy-2-methyl-butanethiol (catty note flavor), are reported. In our results, we found few compounds with significantly higher amounts in BC and HC post-fermentation: citronellol, a-terpinol, and nonanal, as shown in Figure 2. It was noted that besides the esters nonanal, an aldehyde C-9 could give the classical aldehyde note of waxy, citrus, floral, and green.

### 2.5. Sensory Test

The sensory test was carried out in order to find a possible link between the volatile compound profile of the product and their sensory properties (Table S2). Among the 44 nontrained participants to the sensory panel, 77% correctly paired the honey wine prepared with the same honey; the sensory differences linked to the honey used for mead production were, therefore, significantly perceivable.

A nine-point hedonic evaluation scale was structured, point 1 stated for "dislike extremely", while point 9 stated for "like extremely". Using this scale, panelists described the samples as likable as on average; most judged the product as like slightly, like moderately, or like very much. There was no significant difference between the samples in relation to the overall impression and no significant difference between the samples in relation to the olfactory impression. The acceptance level for the honey wine for the sensorial analysis was expressed as a mean value. None of the means presented significant differences. Overall impression and flavor attributes for all beverages variated from 64 to 68% and between 65 and 72%, respectively. According to Dutcosky [47], the acceptance factor (AF) ≥ 70% represented good acceptability for the attribute analyzed in a sensorial analysis. Honey wine obtained with blossom honey showed an AF of 72%.

Honey wine B ranked first, followed by BC, HC, and H, for both the overall impression and the odor. According to the Friedman test, only the ranking for the odor impression showed to be significant. This seemed to be in line with the AF. This might be related to the higher amount of hexanoic acid, octanoic acid, decanoic acid, or esters, such as ethyl octanoate, isoamyl hexanoate, ethyl nonanoate, ethyl decanoate, ethyl dodecanoate found in H and HC product compared to B and BC. Short-chain fatty acids, such as octanoic acid (caprylic acid) and hexanoic acid (caproic acid), have been associated with negative characteristics, such as "rancid," "cheese," and "fatty" aroma [7]. The odor of esters like ethyl octanoate and ethyl decanoate odors have also been described as waxy and soapy [40].

In general, it seems that the VOCs profile imparted by the honey, described above, also has some impact on the sensory perception. However, for a more descriptive sensory evaluation, a trained panel will be able to provide a clearer link between the sensory perception and the specific volatiles or group of volatile as distinguished in the heatmap.

## 3. Materials and Methods

### 3.1. Mead Ingredient

Honeydew honey (Bosco, Mieli Thun, Vigo di Ton, Italy) and blossom honey (mixed honey with a prevalence of *Ailanthus altissima*) (Mieli Thun, Vigo di Ton, Italy) were used. Honeydew honey was collected in the northern Italian wood; dark amber color was described by the supplier as spicy (black pepper, juniper berries, and cloves) with a note of fresh vegetables, carob, rhubarb, and liquorice stick. Blossom honey had a golden color and a creamy consistency and was characterized, according to the supplier, by the smell of muscat grapes and peach syrup; lychee was conferred by the prevalence of *Ailanthus altissima*, also known as the tree of heaven. One part of honey (*w/w*) was used in all the preparation. Warm tap water was used to dissolve the honey. Black currant (*Ribes nigrum*), common berry fruit in South Tyrol, Italy, had been purchased from a local producer and stored at −80 °C until used. Half part of berry fruit (w/w) was used for the *Ribes nigrum* added recipes. The berries were crushed before the addition of other ingredients. The final mixture had a temperature of about 30 °C before acidification. Acidification was carried out using a citrate buffer in order to reach a pH value between 3–3.2. A preliminary test was carried out to establish the amount of citrate buffer to be added to each mixture.

*Saccharomyces cerevisiae* yeast strain EC1118 (Lallemand Inc., Montreal, QC, Canada) was used in the dry active form at the ratio of 25 g/hl after rehydration, according to the manufacturer's instructions. As a yeast protectant in the rehydration step (GoFerm Lallemand Inc., Montreal, QC, Canada), it was used at a 1:1 ratio with the weighted yeast. This product contained all essential vitamins, minerals, and amino acids required to create a non-stressful environment for yeast rehydrating in water.

To ensure the necessary amount of nitrogen and avoid a stuck fermentation, "FermaidE" (Lallemand Inc., Montreal, QC, Canada) was added to the must as a vitamin, organic, and inorganic nitrogen source at the ratio of 30 g/hl.

### 3.2. Honey Wine Wort Preparation and Fermentation Follow Up

Four different recipes for honey wine preparation were used. Blossom honey with and without black currant recipes was compared with honeydew honey with and without black currant (B, BC, H, and HC, respectively). Each recipe was tested in triplicate, carrying out the fermentation in a 5 L glass flask filled up to 3.5 L. All flasks were closed with an air-lock valve. All recipes are described in Table 2. Nitrogen exogen source was added, according to the manufacturer's instructions: two-third at the yeast inoculum and the remaining one-third after one-third of the fermentation was completed. Fermentation was monitored, measuring the weight loss once or twice per day. All the flasks were incubated at 18 °C. Fermentation end was detected when two subsequent weight measurements did not differ for more than approx. one gram: at this stage, flasks were left overnight at 4 °C. After overnight cold storage, berry solids were separated with the aid of a strainer, honey wine was separated by the yeast sediment, and samples were collected for further analysis. The remain was bottled and left at 4 °C until the sensory test. Samples were immediately analyzed for pH and total acidity. For the other analyses, samples were stored at −80 °C until used.

**Table 2.** Ingredients used for the four recipes tested for honey wine production. Weight is referring to the total amount of 11 kg prepared for each recipe before aliquoting the 3.5 L of each replicate (n = 3). B: blossom honey; BC: blossom honey with black currant; H: honeydew honey; HC: honeydew honey with black currant.

| Recipe | Component | Weight (kg) | Ratio |
|---|---|---|---|
| B | Honey | 2.750 | 1.00 |
| | Water | 8.075 | 2.94 |
| | citrate buffer | 0.175 | 0.06 |
| BC | Honey | 2.750 | 1.00 |
| | Water | 6.871 | 2.50 |
| | Buffer | 0 | 0 |
| | black currants | 1.379 | 0.50 |
| H | Honey | 2.750 | 1.00 |
| | Water | 7.750 | 2.82 |
| | citrate buffer | 0.500 | 0.18 |
| HC | Honey | 2.763 | 1.00 |
| | Water | 6.682 | 2.43 |
| | Buffer | 0.180 | 0.07 |
| | Black currants | 1.375 | 0.50 |

*3.3. Honey Wine Analysis*

3.3.1. Physicochemical Parameters

The total soluble solids determined as Brix were measured using a digital refractometer (PAL-BX/RI, Atago, WA, USA). For the pH, a pH electrode was used (pH70+ DNH pH meter with Digital pH electrode mod 201 T, XS instruments, Carpi, Italy).

The content was the sum of glucose and fructose after inversion was measured, according to the OIV-MA-AS311-02 R2009 + OIV-MA-AS2-03B R2012. The alcohol content in volume percent (% vol) was measured following the international methods of wine and must analysis (OIV-MA-AS312-01A R2016 par 4.B). Fructose, glucose, and sucrose in honey were expressed as g/L of glucose, acetic, and lactic acid, and residual sugars were measured enzymatically with the CDR BeerLab® Touch (CDR s.r.l., Ginestra Fiorentina, Italy), according to the manufacturer's instructions.

Acetaldehyde content was measured with the aid of a spectrophotometric-enzymatic-based kit for acetaldehyde, according to the manufacturer's instructions (K-ACHYD 06/18, Megaenzyme International Ireland INC. Bray, Co. Wicklow, Ireland), measuring the absorbance variation at 340 nm with the aid of a microtiter plate reader spectrophotometer (Multiscan Sky Spectrophotometer, Thermo Fisher Scientific Life Technologies Italia, Monza, Italy) where a 96-well microplate was used.

3.3.2. Total Polyphenolic Content

The polyphenolic content of the meads was measured using Folin–Ciocalteu reagent (Merck KGaA, Darmstadt, Germany) in a colorimetric assay. The method was adapted from a published method [48]: Folin-Ciocalteu reagent was added to the sample, and after a reaction time of 3 min at room temperature, 2 M sodium carbonate solution was added. After two hours at 21 °C, the absorbance at 765 nm was read. For the quantification, a standard curve ranging from 50–500 mg/L of gallic acid was prepared. The polyphenol content was determined by linear regression from the standard curve, and the results were expressed as mg/L of gallic acid.

*3.4. Volatile Organic Compounds*

For the volatile organic compounds (VOCs) analysis, a headspace solid-phase microextraction coupled with gas chromatography-mass spectrometry (HS-SPME-GC-MS) was used (QP2010 SE

Shimadzu, Kyoto, Japan). The fermented mead, as well as the unfermented musts, were examined. The method was adapted from Ravasio et al. [49]. Briefly, 2.5 mL of sample was placed in 20 mL glass vials with 0.75 g of NaCl and 10 µL of internal standard (2-octanol, 50 µg/mL). Samples were incubated for 10 min at 40 °C and 250 rpm. Compounds in the headspace were adsorbed for 40 min at 40 °C using 2 cm DVB/CAR/PDMS 50/30 µm fiber from Supelco (Bellefonte, PA, USA). Compounds were desorbed in the GC inlet at 250 °C for 4 min. Chromatographic separation was carried out using a ZB-WAX (30 m × 0.25 mm × 0.25 µm, 40–260 °C) Capillary GC-Column Zebron (Phenomenex, Torrance, CA, USA), using helium as carrier gas at 1.2 mL/min.

The temperature program for the oven was set as follows: 40 °C for 4 min, then up to 250 °C, at 6 °C/min held for 5 min. The total run time was 44 min. The mass spectrometer (quadrupole) was operating in full scan mode, detecting fragments in a mass range of 35 to 350 m/z. Data processing was performed using GC-MS solution Software from Shimadzu. The most intensive ion was used as a quantifier and the ratio of the second and third as a qualifier. The identification of VOCs was carried out by comparing mass spectra using the NIST 2017 database, retention indexes, and with the standard injection when available. The experimental linear temperature retention index of each compound was calculated using a series of n-alkanes (C8-C20) in the same experimental conditions as the samples. Results were expressed as area ratio between compound and internal standard.

### 3.5. Sensory Analysis

A sensory consumer test was carried out with the aid of 44 nontrained consumer panelists; each participant was asked to evaluate all four types of honey wine. Forty milliliters of samples of each product were served for each panelist. Samples were coded with the number served in different sequences and arrangements randomly. The panelists were asked to rate each sample, with a nine-point hedonistic scale, for the (a) the overall taste impression and (b) for the odor overall impression. The nine-point scale was structured as follows: 9: like extremely, 8: like very much, 7: like moderately, 6: like slightly, 5: like neither nor, 4: dislike slightly, 3: dislike moderately, 2: dislike very much, 1: dislike extremely. The acceptability of the tested beverages was assessed by calculating the acceptability factor (AF) using standardized criteria: AF = A*100/B, where A is the mean value obtained for each attribute, and B is the maximum value used to judge each attribute [47,50]. They were asked also to pair the product made with the same honey. Participants were asked to rank the four samples for both the overall impression and the odor overall, assigning the number one to the best one and number four the worst.

### 3.6. Statistical Methods

For statistical analyses, SPSS Statistics software Version 26 (IBM Inc., Armonk, NY, USA) and Microsoft Excel 2019 software were used. Means for every data are expressed as arithmetic mean ± standard deviation of the three replicates for every product. To determine if there was a significant difference between results, a one-way analysis of variance (ANOVA) and a Tukey posthoc test were performed. For all analyses, $p < 0.05$ was considered statistically significant. Used test and corresponding $p$-value were indicated together with the result in each specific session. In addition, Microsoft Excel 2019 was used to verify the significance of the pairing test [51]. Friedmann-Test for statistical analysis with n = 44 test subjects and k = 4 samples was carried out using Microsoft Excel 2019 [48]. The R FactoMineR package was used to perform the PCA [52], and the factoextra package for extracting and visualizing the results. The data had been scaled to unit variance before the analysis to avoid some variables to become dominant just because of their large measurement units. The NMF package was used for the heatmap and hierarchical cluster analysis with a Euclidean distance [53].

## 4. Conclusions

Two different kinds of honey in the presence or absence of black currant were tested for honey wine production. Using these ingredients, no stressful condition seemed to be occurring for the yeast, leading to fermentation delay or arrest, producing a medium or low alcoholic drink. The fermented products were described by a large number of volatile organic compounds capable of allowing the distinctive metabolic contribution of the yeast, also as a response to the honey and the fruit added in fermentation. In particular, the honey contributed to shaping a specific volatile profile, somehow perceivable during sensory analysis. To a lesser extent, also using berry fruit, such as black currant, provided a way to shape flavor and polyphenols content of the final drink. Further investigation would be necessary to evaluate the specific sensory contribution of every single volatile organic compound alone or in association with others found in this study. More information on volatile metabolites associated with mead had been provided that might help to develop alternative medium to low alcoholic drinks at a reasonable cost, adding value to beehive products.

**Supplementary Materials:** The following are available online. Figure S1. Fermentation kinetic; Figure S2. Volatile organic compound measured in the honey wines before and after fermentation; Table S1. Volatile organic compound measurement and descriptions; File S1. Eigenvalues and correlation of the PC analysis; Table S2. Sensory evaluation of the honey wine.

**Author Contributions:** Conceptualization, L.C.; methodology, G.C. and L.C.; validation, G.C., L.C., and P.R.; formal analysis L.D., M.S., and E.U.; investigation, G.C., L.D., and M.S.; resources, L.C. and P.R.; data curation, G.C. and L.C.; writing—original draft preparation, G.C., L.C., and M.S.; writing—review and editing, G.C., L.C., L.D., H.J., P.R., E.U., and E.Z.; visualization, G.C. and L.C.; supervision, L.C., H.J., and P.R.; project administration, L.C.; funding acquisition, L.C. and P.R. All authors have read and agreed to the published version of the manuscript.

**Funding:** This research was funded by the Autonomous Province of Bolzano, Capacity building 1 (Decision n. 1472, 07.10.2013), project Step-Up (Decision n. 864, 04.09.2018), and Aktionplan 2016–2022 (Decision n. 1016, 01.09.2015).

**Acknowledgments:** We sincerely thank Norbert Kofler, Francesca Martinelli, Christof Sanoll, and Andreas Sölva for their technical support. Elisa Piaia, Kathrin Plunger, and Elisa Vanzo are gratefully acknowledged for administrative support. The authors thank the Department of Innovation, Research and University of the Autonomous Province of Bozen/Bolzano for covering the Open Access publication costs.

**Conflicts of Interest:** The authors declare no conflict of interest.

## References

1. Schramm, K. *The Complete Meadmaker. Home Production of Honey Wine from Your First Batch to Award-Winning Fruit and Herb Variations*; Brewers Publications: Boulder, CO, USA, 2003; pp. 8–60.
2. Escuredo, O.; Dobre, I.; Fernández-González, M.; Seijo, M.C. Contribution of botanical origin and sugar composition of honeys on the crystallization phenomenon. *Food Chem.* **2014**, *149*, 84–90. [CrossRef] [PubMed]
3. Iglesias, A.; Pascoal, A.; Choupina, A.B.; Carvalho, C.A.; Feás, X.; Estevinho, L.M. Developments in the fermentation process and quality improvement strategies for mead production. *Molecules* **2014**, *19*, 12577–12590. [CrossRef] [PubMed]
4. Iglesias, M.; De Lorenzo, C.; Polo, M.; Martin-Alverez, P.; Pueyo, E. Usefulness of amino acid composition to discriminate between honeydew and floral honeys. Application to honeys from a small geographic area. *J. Agric. Food Chem.* **2004**, *52*, 84–89. [CrossRef] [PubMed]
5. Vela, L.; Lorenzo, C.; Pérez, R.A. Antioxidant capacity of Spanish honeys and its correlation with polyphenol content and other physicochemical properties. *J. Sci. Food Agric.* **2007**, *87*, 1069–1075. [CrossRef]
6. Pita-Calvo, C.; Vázquez, M. Differences between honeydew and blossom honeys. A review. *Trends Food Sci. Tech.* **2017**, *59*, S79–S87. [CrossRef]
7. Li, R.; Sun, Y. Effects of honey variety and non- *Saccharomyces cerevisiae* on the flavor volatiles of mead. *J. Am. Soc. Brew. Chem.* **2019**, *77*, 40–53. [CrossRef]
8. Boulton, C.; Quain, D. *Brewing Yeast and Fermentation*; Blackwell Science Ltd.: Malden, MA, USA, 2001; pp. 69–142.
9. Stewart, G.G.; Russell, I.; Anstruther, A. *Handbook of Brewing*, 3rd ed.; CRC Press: Boca Raton, FL, USA, 2017; pp. 10–155.
10. Pereira, A.P.; Mendes-Ferreira, A.; Oliveira, J.M.; Estevinho, L.M.; Mendes-Faia, A. High-cell-density fermentation of Saccharomyces cerevisiae for the optimisation of mead production. *Food Microbiol.* **2013**, *33*, 114–123. [CrossRef]

11. Stückler, K. *Met. Honigwein-Bereitung Leicht Gemacht!* 2nd ed.; Stocker: Graz, Austria, 2013; pp. 20–55.

12. Akalin, H.; Bayram, M.; Anli, R.E. Determination of some individual phenolic compounds and antioxidant capacity of mead produced from different types of honey. *J. Inst. Brew.* **2017**, *123*, 167–174. [CrossRef]

13. Mendes-Ferreira, A.; Cosme, F.; Barbosa, C.; Falco, V.; Inês, A.; Mendes-Faia, A. Optimization of honey-must preparation and alcoholic fermentation by Saccharomyces cerevisiae for mead production. *Int. J. Food Microb.* **2010**, *144*, 193–198. [CrossRef]

14. Antón, M.J.; Suárez Valles, B.; García Hevia, A.; Picinelli Lobo, A. Aromatic profile of ciders by chemical quantitative, gas chromatography-olfactometry, and sensory analysis. *J. Food Sci.* **2014**, *79*, S92–S99. [CrossRef]

15. Meilgaard, M. Flavour chemistry of beer. Part 2 Flavour and treshold of beer volatiles. *Tech. Q. Master Brew. Assoc. Am.* **1975**, *12*, 161–168.

16. Alves, Z.; Melo, A.; Raquel Figueiredo, A.; Coimbra, M.A.; Gomes, A.C.; Rocha, S.M. Exploring the *Saccharomyces cerevisiae* volatile metabolome: Indigenous versus commercial strains. *PLoS ONE* **2015**, *10*, e0143641. [CrossRef] [PubMed]

17. Pascoal, A.; Oliveira, J.M.; Pereira, A.P.; Féas, X.; Anjos, O.; Estevinho, L.M. Influence of fining agents on the sensorial characteristics and volatile composition of mead. *J. Inst. Brew.* **2017**, *123*, 562–571. [CrossRef]

18. Sroka, P.; Tuszyński, T. Changes in organic acid contents during mead wort fermentation. *Food Chem.* **2007**, *104*, 1250–1257. [CrossRef]

19. Staszowska-Karkut, M.; Materska, M. Phenolic composition, mineral content, and beneficial bioactivities of leaf extracts from black currant (*Ribes nigrum* L.), raspberry (*Rubus idaeus*), and aronia (*Aronia melanocarpa*). *Nutrients* **2020**, *12*, 463. [CrossRef]

20. Da Silva, P.M.; Gauche, C.; Gonzaga, L.V.; Costa, A.C.; Fett, R. Honey: Chemical composition, stability and authenticity. *Food Chem.* **2016**, *196*, 309–323. [CrossRef]

21. Steinkraus, K.H.; Morse, R.A. Chemical analysis of honey wines. *J. Apic. Res.* **1973**, *12*, 191–195. [CrossRef]

22. Liu, S.-Q.; Pilone, G.J. An overview of formation and roles of acetaldehyde in winemaking with emphasis on microbiological implications. *Int. J. Food Sci. Tech.* **2000**, *35*, 49–61. [CrossRef]

23. Kahoun, D.; Řezková, S.; Královský, J. Effect of heat treatment and storage conditions on mead composition. *Food Chem.* **2017**, *219*, 357–363. [CrossRef]

24. Švecová, B.; Bordovská, M.; Kalvachová, D.; Hájek, T. Analysis of Czech meads. Sugar content, organic acids content and selected phenolic compounds content. *J. Food Comp. Anal.* **2015**, *38*, 80–88. [CrossRef]

25. Cheynier, V. Phenolic compounds: From plants to foods. *Phytochem. Rev.* **2012**, *11*, 153–177. [CrossRef]

26. Socha, R.; Pająk, P.; Fortuna, T.; Buksa, K. Phenolic profile and antioxidant activity of Polish meads. *Int. J. Food Prop.* **2015**, *18*, 2713–2725. [CrossRef]

27. Gaglio, R.; Alfonzo, A.; Francesca, N.; Corona, O.; Di Gerlando, R.; Columba, P.; Moschetti, G. Production of the Sicilian distillate "Spiritu re fascitrari" from honey by-products: An interesting source of yeast diversity. *Int. J. Food Microbiol.* **2017**, *261*, 62–72. [CrossRef] [PubMed]

28. Chen, C.H.; Wu, Y.L.; Lo, D.; Wu, M.C. Physicochemical property changes during the fermentation of longan (*Dimocarpus longan*) mead and its aroma composition using multiple yeast inoculations. *J. Inst. Brew.* **2013**, *4*, 303–308. [CrossRef]

29. Gupta, J.K.; Sharma, R.K. Production thecnology and quality characteristics of mead and fruit-honey wines: A review. *Nat. Prod. Radiance.* **2009**, *8*, 345–355.

30. Alba-Lois, L.; Segal-Kischinevzky, C. Beer and wine makers. *Nat. Educ.* **2010**, *3*, 17.

31. Del Barrio-Galán, R.; Úbeda, C.; Gil, M.; Medel-Marabolí, M.; Sieczkowski, N.; Peña-Neira, A. Evaluation of yeast derivative products developed as an alternative to lees: The effect on the polysaccharide, phenolic and volatile content, and colour and astringency of red wines. *Molecules* **2019**, *24*, 1478. [CrossRef]

32. Romani, C.; Lencioni, L.; Gobbi, M.; Mannazzu, I.; Ciani, M.; Domizio, P. *Schizosaccharomyces japonicus*: A polysaccharide-overproducing yeast to be used in winemaking. *Fermentation* **2018**, *4*, 14. [CrossRef]

33. Lambrechts, M.G.; Pretorius, I.S. Yeast and its Importance to Wine Aroma—A Review. *S. Afr. J. Vitic.* **2000**, *21*, 97–129. [CrossRef]

34. Procopio, S.; Krause, D.; Hofmann, T.; Becker, T. Significant amino acids in aroma compound profiling during yeast fermentation analyzed by PLS regression. *LWT Food Sci. Technol.* **2013**, *51*, 423–432. [CrossRef]

35. Pascoal, A.; Anjos, O.; Feás, X.; Oliveira, J.M.; Estevinho, L.M. Impact of fining agents on the volatile composition of sparkling mead. *J. Inst. Brew.* **2019**, *125*, 125–133. [CrossRef]

36. Noguerol-Pato, R.; González-Barreiro, C.; Cancho-Grande, B.; Simal-Gándara, J. Quantitative determination and characterisation of the main odourants of Mencía monovarietal red wines. *I Food Chem.* **2009**, *117*, 473–484. [CrossRef]

37. Reynolds, A.G. *(Hg.) Managing Wine Quality*; Woodhead Publishing: Cambridge, PA, USA; New Delhi, India, 2014.

38. Saerens, S.M.G.; Delvaux, F.; Verstrepen, K.J.; Van Dijck, P.; Thevelein, J.M.; Delvaux, F.R. Parameters affecting ethyl ester production by Saccharomyces *cerevisiae* during fermentation. *Appl. Environ. Microbiol.* **2018**, *74*, 454–461. [CrossRef] [PubMed]

39. Torrens, J.; Riu-Aumatell, M.; Vichi, S.; López-Tamames, E.; Buxaderas, S. Assessment of volatile and sensory profiles between base and sparkling wines. *J. Agric. Food Chem.* **2010**, *58*, 2455–2461. [CrossRef] [PubMed]

40. The Good Scent Company. Available online: http://www.thegoodscentscompany.com/odor/waxy.html (accessed on 27 March 2020).

41. Li, H.; Liu, F.; He, X.; Cui, Y.; Hao, J. A study on kinetics of beer ageing and development of methods for predicting the time to detection of flavour changes in beer. *J. Inst. Brew.* **2015**, *121*, 38–43. [CrossRef]

42. Berk, Z. Morphology and Chemical Composition. In *Citrus Fruit Processing*; Elsevier: Amsterdam, The Netherlands, 2016; pp. 9–54.

43. Burdock, G.A.; Fenaroli, G. *Fenaroli's Handbook of Flavor Ingredients*, 5th ed.; CRC Press: Boca Raton, FL, USA, 2005; p. 243.

44. Markus Lange, B.; Seyed Mahmoud, S.; Wildung, M.R.; Turner, G.W.; Davis, E.M.; Lange, I.; Baker, R.C.; Boydston, R.A.; Croteau, R.B. Improving peppermint essential oil yield and composition by metabolic engineering. *Proc. Natl. Acad. Sci. USA* **2011**, *108*, 16944–16949. [CrossRef]

45. Kampuss, K.; Christensen, L.P.; Lindhard Pedersen, H. Volatile Composition of Black Currant Cultivars. In *ISHS Acta Horticulturae 777: IX International Rubus and Ribes Symposium*; Dale, A., Bañados, P., Eds.; ISHS: Pucon, Chile, 2008; pp. 525–530.

46. Marsol-Vall, A.; Kortesniemi, M.; Karhu, S.T.; Kallio, H.; Yang, B. Profiles of volatile compounds in blackcurrant (*Ribes nigrum*) cultivars with a special focus on the influence of growth latitude and weather conditions. *J. Agric. Food Chem.* **2018**, *66*, 7485–7495. [CrossRef]

47. Dutcosky, S.D. *Análise Sensorial de Alimentos*, 4th ed.; Champagnat: Curitiba, Brazil, 2013; p. 531.

48. Thompson-Witrick, K.A.; Goodrich, K.M.; Neilson, A.P.; Kenneth Hurley, E.; Peck, G.M.; Stewart, A.C. Characterization of the polyphenol composition of 20 cultivars of cider, processing, and dessert apples (Malus × domestica Borkh.) grown in Virginia. *J. Agric. Food Chem.* **2014**, *62*, 10181–10191. [CrossRef]

49. Ravasio, D.; Carlin, S.; Boekhout, T.; Groenewald, M.; Vrhovsek, U.; Walther, A.; Wendland, J. Adding flavor to beverages with non-conventional yeasts. *Fermentation* **2018**, *4*, 15. [CrossRef]

50. Prado, F.C.; Lindner, J.D.D.; Inaba, J.; Thomaz-Soccol, V.; Kau-Brar, S.; Soccol, C.R. Development and evaluation of a fermented coconut water beverage with potential health benefits. *J. Funct. Foods* **2015**, *12*, 489–497. [CrossRef]

51. Lawless, H.; Heymann, H. *Sensory Evaluation of Food*; Food Science Text Series; Springer: New York, NY, USA, 2010; pp. 303–347.

52. Le, S.; Josse, J.; Husson, F. FactoMineR: An R Package for Multivariate Analysis. *J. Stat. Softw.* **2008**, *25*, 1–18. [CrossRef]

53. Gaujoux, R.; Seoighe, C. A flexible R package for non negative matrix factorization. *BMC Bioinform.* **2010**, *11*, 367. [CrossRef] [PubMed]


**Sample Availability:** Samples of the compounds are not available from the authors.

 *molecules*

*Article*

# Identification of Aroma Differences in Refined and Whole Grain Extruded Maize Puffs

Kenneth Smith [1] and Devin G. Peterson [2],*

[1] Department of Food Science and Nutrition, 1334 Eckles Avenue, 145 FScN Building, University of Minnesota, St. Paul, MN 55108, USA; smit4423@umn.edu
[2] Department of Food Science and Technology, 2015 Fyffe Rd., 317 Parker Food Science & Technology Building, The Ohio State University, Columbus, OH 43210, USA
* Correspondence: peterson.892@osu.edu; Tel.: +01-614-688-2723

Academic Editors: Eugenio Aprea and Chiara Emilia Cordero
Received: 30 March 2020; Accepted: 5 May 2020; Published: 11 May 2020

**Abstract:** Differences in the aroma profiles of extruded maize puffs made from refined grain and whole grain flour were investigated. Gas chromatography/mass spectrometry/olfactometry (GC/MS/O) analysis reported 13 aroma compounds with a flavor dilution (FD) value ≥16. Quantitative analysis identified eight compounds as statistically different, of which seven compounds were higher in concentration in the whole grain sample. Sensory recombination and descriptive analysis further supported the analytical data, with higher mean aroma intensities for cooked, corn chip, roasted, and toasted attributes for the whole grain sample. Generally, the compounds responsible for perceived differences in whole grain maize extruded puffs were associated with increased levels of Maillard reaction products, such as 2-ethyl-3,5-dimethylpyrazine and 2-acetyl-2-thiazoline.

**Keywords:** whole grain; refined grain; GC/O; Maillard reaction; maize; aroma; flavor

---

## 1. Introduction

The consumption of whole grain has been associated with a range of health benefits such as body-weight regulation, reduced risk of chronic pathological conditions, and reduced blood glucose levels [1–3]. However, most Americans fail to consume the recommended whole grain intake (48 g/day), which has a direct effect on health and was recently identified as a main contributor to suboptimal diets that are responsible for 1 out of 5 deaths globally [4,5]. In cereal-based foods, refined grains are often preferred in comparison to their whole grain counterparts. The negative flavor attributes associated with whole grain products including bitter taste and vegetative aromas have been reported as one of the most influential factors limiting consumption [6,7].

Breakfast cereals are cooked products introduced in the human diet at a young age and constitute an excellent opportunity for early exposure to whole grain flavor [8]. Maize is a common grain used for cereal production and the impact of extrusion (cooking) parameters on physico-chemical and sensory properties of extruded cereals has been largely studied [9–12]. Flavor generation during the extrusion of cereals involves thermally induced reactions, such as the Maillard reaction and lipid degradation. Extrusion conditions such as heat, water content, and residence time have been shown to exert significant effects on the flavor profiles of extruded products [9,12] with cooking temperature identified as a main influential factor for the formation of flavor compounds. Flavor development in extruded products has been investigated with a focus on processing conditions and has overlooked the impact of whole grain versus refine grain flour formulation. In wheat bread, the utilization of whole versus refined grain flour had a significant impact on flavor generation [13]. Several key compounds that give refined wheat bread its typical aroma attributes were less abundant in whole wheat bread due to the suppression of key Maillard-type flavor formation pathways caused by the elevated levels of phenolic compounds. In addition to aroma generation, taste compounds are thermally generated by Maillard-type pathways during bread making [14]. Whole grain flour, as compared to refined grain flour, has elevated levels of phenolic compounds, lipids, vitamins and is composed of a unique protein composition, all of which can significantly alter the thermal generation of flavor compounds [13,15–17]. During extrusion processing, high temperature and short time conditions favor Maillard and lipid oxidation flavor-formation pathways [9]; however, the characterization of the flavor differences between whole and refined grain maize products has not been reported.

The objective of this work was to investigate the influence of whole versus refined maize flour on the aroma of extruded puffs and the sensory impact of the identified differences. Aroma-active compounds were identified using gas chromatography/mass spectrometry (GC/MS) and sensory differences were characterized using sensory descriptive analysis.

## 2. Results and Discussion

To characterize the main differences in the aroma profiles of extruded maize puffs made from whole grain versus refined grain flour, odorants were identified using gas chromatography/mass spectrometry/olfactometry (GC/MS/O) and selected based on cut-off flavor dilution (FD) values ≥16 (see Table 1). Thirteen compounds were selected based on this criterion and all the odor compounds identified have been previously reported in extruded maize products [9,12]. However, the influence of flour type on their generation and their impact on the aroma profile of extruded maize whole grain puffs (WGP) and refined grain puffs (RGP) has not been previously reported. Further quantitative analysis of the 13 compounds was conducted and is reported in Table 2.

**Table 1.** Identified aroma compounds in extruded maize refined grain puffs (RGP) and whole grain puffs (WGP).

| Compound [a] | Odor Descriptor [b] | LRI [c] | | Flavor Dilution Value ≥ 16 [d] | |
|---|---|---|---|---|---|
| | | DB-Wax | DB-5 | WGP | RGP |
| hexanal | Green/Oxidized | 1084 | 801 | 64 | 32 |
| 2-methylpyrazine | Roasted | 1176 | 827 | 128 | 64 |
| 2,3-dimethylpyrazine | Roasted | 1240 | 911 | 128 | 32 |
| 2,5-dimethylpyrazine | Roasted | 1253 | 912 | 16 | 32 |
| 2-methyl-2-thiazoline | Roasted/Toasted | 1436 | 933 | 128 | 64 |
| 2-pentylfuran | Earthy/Oxidized | 1240 | 993 | 128 | 32 |
| 2-ethyl-3,5-dimethylpyrazine | Roasted | 1457 | 1081 | 64 | 32 |
| 3-hydroxy-2-methyl-4*H*-pyran-4-one | Caramel/Toasted | 1955 | 1087 | 128 | 64 |
| 2-methoxyphenol | Smokey | 1872 | 1088 | 128 | 64 |
| 2-acetyl-2-thiazoline | Popcorn/Corn Chip | 1772 | 1103 | 64 | 32 |
| (*E*,*E*)-2,4-decadienal | Oxidized | 1815 | 1312 | 64 | 32 |
| 2-methoxy-4-vinylphenol | Clove | 2189 | 1322 | 128 | 64 |
| 4-hydroxy-3-methoxybenzaldehyde | Vanilla-like | 2589 | 1410 | 64 | 16 |

[a] Compounds positively identified by linear retention index (LRI), mass spectrometry (MS), and authentic compound; [b] Odor described at sniffing port during gas chromatography/olfactometry (GC/O), [c] LRIs calculated using GC-MS on DB-WAX and DB-5 columns, values relatively to the n-alkane ladder, [d] Flavor dilution based on the average of two panelists.

Eight compounds including six Maillard reaction products and two phenolic compounds were found to be statistically different between WGP and RGP. All the compounds were found in greater amounts in the WGP samples except for 2,5-dimethylpyrazine, which was found in higher concentration in the RGP with 140 µg/kg compared to 100 µg/kg in the WGP. A higher formation of Maillard reaction aroma compounds, in general, in the WGP can be explained by compositional differences between whole and refined grain flours. Milling cereals alters the concentration and composition of proteins and lipids in the flour. The milling process used to produce refined maize flour removes the protein-rich pericarp/germ leaving primarily the starchy endosperm flour [16]. Protein and amino acid content in whole grain maize flour is altered compared to refined maize flour [16,17]. Amino acids are very influential for the progression of Maillard reaction pathways [21,22] and key precursors for the formation of key odorants such as 2-acetyl-2-thiazoline (cysteine [23]) and 2-ethyl-3,5-dimethylpyrazine (alanine [24]). These two aroma compounds showed concentrations 2.2 and 2.1-fold higher in WGP when compared to RGP, respectively. A lower concentration of amino acids (i.e. cysteine and alanine) in refined maize flour could have resulted in the observed changes in the aroma generation noted or perhaps are due to differences in sugar fragmentation. The ratio of precursors, i.e., reducing sugar to *N*-containing compounds, has been demonstrated to selectively favor formation pathways through the modification of the intermediate reactive chemistry [25,26]. In glucose model mixtures in particular, changes in the glucose to amino acid ratio ultimately modulate the generation of pyrazines; a greater ratio of sugars:amino acids in the RGP could have favored the generation of reactive intermediate species involved in the formation of 2,5-dimethylpyrazine, while suppressing other products, such as 2-acetyl-2-thiazoline or 2-ethyl-3,5-dimethylpyrazine.

**Table 2.** Concentration of aroma compounds in extruded maize refined grain puffs (RGP) and whole grain puffs (WGP).

| Compound | Mean ± CV [2] | | Concentration Ratio (WGP/RGP) | Odor Threshold in Water (µg/L) [18–20] |
|---|---|---|---|---|
| | WGP (µg/kg) | RGP (µg/kg) | | |
| hexanal [1] | 470± 8 | 436 ± 24 | 1.1 | 4.5–5.0 |
| 2-methylpyrazine | 363± 7 [b] | 292± 3 [a] | 1.2 | 60,000–105,000 |
| 2,3-dimethylpyrazine | 653± 7 [b] | 280± 2 [a] | 2.3 | 2,500–35,000 |
| 2,5-dimethylpyrazine | 100± 8 [b] | 141± 5 [a] | 0.7 | 800–1800 |
| 2-methyl-2-thiazoline [1] | 147± 10 | 137± 13 | 1.1 | 2 |
| 2-pentylfuran [1] | 183± 12 | 153± 9 | 1.2 | 6 |
| 2-ethyl-3,5-dimethylpyrazine [1] | 260± 16 [b] | 124± 18 [a] | 2.1 | 1 |
| 3-hydroxy-2-methyl-4*H*-pyran-4-one | 370± 9 [b] | 321± 10 [a] | 1.2 | 35,000 |
| 2-methoxyphenol [1] | 317± 5 | 297± 15 | 1.1 | 3–21 |
| 2-acetyl-2-thiazoline [1] | 377± 5 [b] | 168± 10 [a] | 2.2 | 1 |
| (*E,E*)-2,4-decadienal [1] | 293± 14[b] | 243± 3 [a] | 1.2 | 0.07 |
| 2-methoxy-4-vinylphenol [1] | 3600± 4 [b] | 843± 3 [a] | 4.3 | 3 |
| 4-hydroxy-3-methoxybenzaldehyde [1] | 3517± 7 [b] | 1218± 5 [a] | 2.9 | 20–200 |

[1] Concentration above aqueous odor threshold values; [2] Different letters (a, b) indicate a significant difference between samples using a t-test, $p < 0.05$, $n = 5$.

Two ferulic acid degradation products 2-methoxy-4-vinylphenol and 4-hydroxy-3-methoxybenzaldehyde were quantified in higher amounts in WGP (Table 2) and described in GC/O with clove and vanilla aroma descriptors (Table 1), respectively. In grains, the phenolic material is mainly distributed in the bran layer. For example, in sweet maize the pericarp and germ contain approximately 325 and 702 ug/g of ferulic acid, respectively while the endosperm, the primary component of refined maize flour, contains approximately 13 mg/g [27]. In general, the total phenolic content of the pericarp is approximately 30-fold higher than the endosperm [28]. In bread, the liberation of phenolic compounds (i.e., ferulic acid) from whole wheat flour during baking was reported to suppress the generation of Maillard aroma compounds through carbonyl trapping mechanisms [13]. The noted increase in Maillard aroma compound generation in the WGP versus RGP suggested the phenolic-carbonyl reaction mechanisms that suppressed aroma formation in bread were not relevant in extruded maize products, albeit the heating profile of extrusion cooking is drastically different than baking bread. Others have shown that the addition of the phenolic compound, rutin, during the preparation of baked rye-buckwheat biscuits resulted in higher levels of Maillard-type aroma compounds, such as alkyl-pyrazines [29].

Two lipid oxidation compounds, hexanal and 2-pentylfuran were not found to be statistically different in concentration between RGP and WGP, however (*E,E*)-2,4-decadienal was significantly higher (approximately 20%) in the WGP sample. Thus, the higher content of lipid in the whole grain maize did not have a major impact on the formation of these lipid oxidation aroma compounds, perhaps because of the elevated levels of antioxidative components of the pericarp.

To draw further insight regarding the impact of quantitative differences of aroma compounds of the maize extruded puffs (Table 2) on the aroma profile, sensory descriptive analysis (DA) was conducted on both the WGP and RGP samples (Figure 1), as well as aroma recombination models (Figure 2).

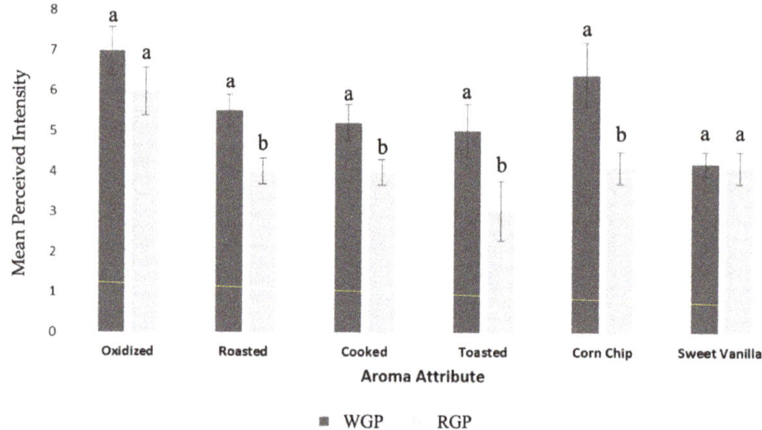



**Figure 1.** Mean aroma attribute intensity scores of extruded maize refined grain puffs (RGP) and whole grain puffs (WGP); Different letters (a, b) indicate a significant difference between samples for each attribute according to Tukey's HSD, $p < 0.05$

The perceived intensities of the aroma descriptors were significantly higher in whole grain versus refined maize puffs for four out of six attributes (Figure 1). In the whole grain sample, the highest reported mean intensities were for the attributes cooked, corn chip, roasted, and toasted and likely were associated with the increased concentration of the Maillard-derived compounds (Table 2). All the Maillard compounds, with the exception of 2,5-dimethylpyrazine, were statistically higher in concentration in the WGP compared to the RGP. The oxidized attribute was not rated as significantly different in intensity between WGP and RGP. The lipid oxidation compounds typically associated with oxidized sensory properties including 2-pentylfuran and hexanal were indeed found at similar levels in both samples, whereas 2,4-decadienal was approximately 20% higher in the WGP (Table 2). These results indicated that lipid oxidation did not play a major role in aroma differences between WGP and RGP. Finally, the intensity of the vanilla attribute was rated similarly in both samples, and thus was not established as a discriminant sensory trait common to both WGP and RGP despite higher levels of these phenolic degradation compounds in the WGP sample.

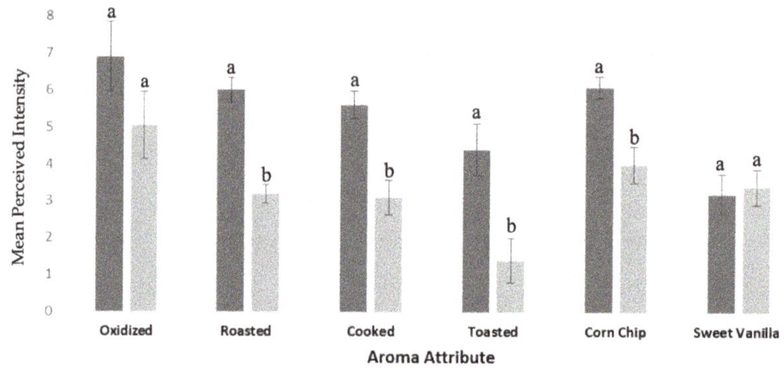



**Figure 2.** Mean aroma attribute intensity scores of aroma recombination models for extruded maize refined grain puffs (RGP) and whole grain puffs (WGP); Different letters (a, b) indicate a significant difference between samples for each attribute according to Tukey's HSD, $p < 0.05$.

Aroma recombination models were developed to determine the contribution of the aroma composition (Table 2) on the sensory attributes of the WGP and RGP samples (Figure 1). In general, the recombination models of the maize puff samples (Figure 2) agreed with the authentic samples (Figure 1) and showed that toasted, roasted, corn chip, and cooked attributes had significantly higher mean intensities in WGP in comparison to RGP. The odor threshold values are also shown in Table 2, with 10 of the 13 compounds being reported at concentrations above the threshold. Odor thresholds provide a basis to understand sensory relevance; however, some caution would be warranted when extrapolating these threshold values in water to the aroma attributes perceived by the orthonasal evaluation of a low-moisture high-starch puffed cereal product. Nonetheless, when focused on compounds above their aqueous odor threshold values that were also significantly different in concentration (Table 2) and considering the odor properties (Table 1), two Maillard reaction products, 2-ethyl-3,5-dimethylpyrazine and 2-acetyl-2-thiazoline were indicated as the main contributors to the noted sensory differences in the aroma profile of the WGP and RGP samples (Figure 1).

In summary, the Maillard reaction products were established as the main aroma differences between WGP and RGP. This study showed a predominance of Maillard aroma compounds and sensory traits in the WGP likely induced by the heat conditions of the extrusion process. Historically, the aroma attributes of Maillard compounds associated with roasted, toasted, corn chip, and cooked are viewed as positive traits in heat-processed foods. Thus, the aroma attributes of extruded whole grain maize did not appear to negatively alter the flavor profile (compared to refined grain product). However, further work is needed to understand if these changes could contribute to an unbalanced aroma profile when present at higher intensities. Moreover, the impact of whole grain maize on the taste profile (i.e., bitterness) could also play a role in product acceptance.

## 3. Materials and Methods

### 3.1. Materials

Hexanal, 2-methylpyrazine, 2,3-dimethylpyrazine, 2,5-dimethylpyrazine, 2-methyl-2-thiazoline, 2-pentylfuran, 2-ethyl-3,5-dimethylpyrazine, 3-hydroxy-2-methyl-4*H*-pyran-4-one, 2-methylphenol, 2-acetyl-2-thiazoline, (*E,E*)-2,4-decadienal, 2-methoxy-4-vinylphenol, 4-hydroxy-3-methoxybenzaldehyde, 4-heptanone, 2-methyl-3-heptanone, trisodium phosphate, calcium carbonate, anhydrous sodium sulfate, corn starch and sodium chloride were purchased from Sigma-Aldrich (St. Louis, MO, USA). Methanol and methylene chloride, in GC-Resolv® grade, were purchased from Fischer Scientific (Pittsburgh, PA, USA). Maizewise™ whole grain and Innovasure™ refined maize flours were purchased from Cargill (Minneapolis, MN, USA). All the sensory reference materials, dry uncooked Bob's Red Mill Steel Cut Oats, Bergen Unsalted Lightly Roasted Almonds, Organic Valley Whole UHT Milk, Toasted Wonder Bread™, Nilla® Wafers, and Old Dutch Restaurante® Style Yellow Corn Tortillas Chips, were purchased from a local grocery store.

### 3.2. Twin-Screw Extrusion

Extrusion conditions were designed to yield uniform cell structure throughout each puff [10]. Briefly, extrusion processing was carried out using The Joseph J. Wartheson pilot plant (Department of Food Science and Nutrition, University of Minnesota, St. Paul, MN) with a Buhler DNDL-44 twin-screw extruder (Uzwil, Switerland). Two formulations were produced, a refined maize flour formulation and a whole grain maize flour formulation. The refined maize flour dry formulation consisted of 970 g (97%) refined maize flour with 10 g (1%) trisodium phosphate, 10 g (1%) calcium carbonate, and 10 g (1%) sodium chloride. The whole grain maize formulation consisted of 465 g (48%) refined maize flour and 505 g (52%) whole grain maize flour with 10 g (1%) trisodium phosphate (TSP), 10 g (1%) calcium carbonate, and 10 g (1%) sodium chloride. The ingredients were added to a mixer and mixed for 10 min. The mixture was added with 14% (*w/w*) water into the extruder with a low work screw configuration via a feeder and processed per the following extrusion parameters:

computer-controlled shaft speed of 350 rpm, measured die pressure of 10.1 ± 0.5 bar, die temperature of 160 ± 1 °C, material throughput of 50.8 ± 0.1 kg/h with 7 kg/h water, and a cutter speed of 1200 rpm. Due to differences in the physical and chemical characteristics of the refined and the whole grain flour mixes, the refined maize flour formulation showed an increased shaft torque of 224 N·m over the whole grain maize flour formulation, which had a shaft torque of 215 N·m. The specific mechanical energy for refined maize flour formulation was 164 kW/h, while the whole grain maize flour was 159 kW/h. Other parameters were constant across both formulations. The puffed product was collected, dried on a liquid air bed, and stored in high-density polyethylene bags at −40 °C for later analysis.

### 3.3. Solvent Extraction

Briefly, 300 g of maize puffs were ground and placed in a 1 L Erlenmeyer flask. Next, 600 g of methanol spiked with 0.1 mg/L 4-heptanone were added to the flask, which was then shaken for 24 h on an orbital shake table set at 200 rpm. Methanol was collected and the ground maize puffs were re-extracted for 2 h using 400 g of methanol at 200 rpm. Organic layers were pooled, and 600 g of the methanol collected was subsequently combined with 600 mL of reverse osmosis purified water. The water-methanol mixture was then poured into three 1 L separatory funnels and extracted using 500 g of methylene chloride (DCM) spiked with 0.1 mg/L of 2-methyl-3-heptanone. DCM was added in 100 mL aliquots to each funnel for a total of 5 extractions. The DCM extract was then placed in a −20 °C freezer overnight to separate and remove any residual water-methanol. The DCM extract was collected and then dried using sodium sulfate and subsequently concentrated via distillation to 1.0 g. The concentrated extract was stored at −80 °C until analysis. Additionally, internal standards used were analyzed for reproducibility during extraction. Methanol was spiked with 4-heptanone and DCM was spiked with 2-methyl-3-heptanone to achieve 100 and 150 mg/L, respectively, in the concentrated solvent. This protocol, when compared to DCM extraction, resulted in 30% more aroma actives detected during GC/O (data not shown).

### 3.4. Gas Chromatography/Olfactometry/Mass Spectrometry (GC/O/MS): Aroma Extraction Dilution Analysis (AEDA)

GC/O analyses were performed on an HP6890 GC (Agilent Technologies, Santa Clara, CA, USA) equipped with a DB-5 column (30 m × 0.25 mm i.d. × 0.25 µm film thickness (Agilent Technologies)) coupled with a 5973 MS (Agilent Technologies) operated in electron impact mode as similarly described by Moskowitz et al. [13]. The system was also equipped with an olfactometry port (Gerstel, Mülheim an der Ruhr, Germany). The effluent was divided 1:1 between the MS and the olfactometry port. The GC conditions were as follows: 0.5 µL sample was injected via air sandwich technique into the inlet which was held at 250 °C set to splitless mode, helium carrier gas was at a constant pressure of 180 kPa. The GC oven temperature program was as follows: initial conditions 40 °C held for 2 min, followed by a 7 °C/min ramp until 250 °C, which was held for 10 min. Each sample was diluted by half-volume in dichloromethane until the dilution had been carried out to a concentration of 128th of the original extraction had been achieved. The largest dilution at which each compound was detected was defined as the FD value. Each dilution was analyzed in triplicate by two panelists. Compound identification was performed using mass spectral data, odor descriptors, and the linear retention index (LRI) of the authentic compound. LRI values were calculated using an *n*-alkane ladder.

### 3.5. GC/MS Identification and Quantitation

The GC/MS analysis was performed using a 7890 GC (Agilent Technologies) coupled to a time of flight (TOF) MS (LECO Pegasus 4D, St. Joseph, MI, USA). The isolate was analyzed on two alternate column chemistries, namely DB-5 and DB-Wax. For the DB-5 analysis analogous column and oven conditions were as previously described. For the DB-Wax (60 m × 0.25 mm i.d. × 0.25 µm film thickness, Agilent Technologies) the GC conditions were as follows: 0.5 µL was injected into an inlet heated to 250 °C. The GC oven temperature program was as follows: initial conditions 40 °C followed by a 5 °C/min ramp to 250 °C and then held for 10 min, flow at 1 mL/min.

Quantification was carried out using five-point calibration curves for each of the 18 compounds in the following concentration ranges (µg/kg) listed, hexanal (50–800), 2-methylpyrazine (55–80), 2,3-dimethylpyrazine (51–815), 2,5-dimethylpyrazine (54–860), 2-methyl-2thiazoline (61–975), 2-pentylfuran (43.5–775), 2-ethyl-3,5-dimethylpyrazine (52.5–840), 3-hydroxy-2-methyl-4*H*-pyran-4-one (44–700), 2-methylphenol (60–965), 2-acetyl-2-thiazoline (61–975), (*E*,*E*)-2,4-decadienal (92.5–1480), 2-methoxy-4-vinylphenol (438–7000), 4-hydroxy-3-methoxybenzaldehyde (612–10000), all curves had high linearity ($R^2 > 0.98$ for all compounds) as similarly described by Trikusuma et al. [30].

*3.6. Sensory Evaluation*

The aroma of the maize puffs was evaluated by 12 trained panelists (4 male and 8 female, ages 22–32) from the University of Minnesota Department of Food Science and Nutrition (St. Paul, MN, USA). Training consisted of 10 sessions of 1 h. The first training session was dedicated to lexicon development and selection of references. Panelists generated the six following descriptive terms: oxidized, roasted, cooked, toasted, vanilla, and corn chip. Representative food samples were selected as references for sensory attributes: dry uncooked Bob's Red Mill Steel Cut Oats represented the oxidized aroma, Bergen Unsalted Lightly Roasted Almonds represented the roasted aroma, Organic Valley Whole UHT (ultra-heat treated) Milk represented the cooked aroma, toasted Wonder Bread™ represented the toasted aroma, Nilla® Wafers represented the vanilla aroma, and Old Dutch Restaurante® Style Yellow Corn Tortillas Chip represented the corn chip aroma. The recombination samples were prepared by adding 10 µL of the aroma compound mixture (in ethanol) to 15 g corn starch at the levels quantified in refined and whole grain maize puff samples (Table 2) in sealed 50 mL amber glass containers with Teflon® lined lids. The recombination samples were allowed to equilibrate for 12 h and mixed in a drum tumbler prior to evaluation. Panelists were asked to assess the intensity of the six aroma descriptors orthonasally on a 0–15 pt scale with 0 being not noticeable and 15 being intense. All samples and recombination samples were evaluated in duplicate. For each replicate, a new sample bottle was analyzed. Data were analyzed using analysis of variance and Tukey's HSD test with a probability of $p \leq 0.05$. The effect of replicate and the panelist–sample interaction was not significant, indicating that data collected were reproducible and that panel was aligned toward the sensory attributes. Data were processed using SPSS Statistics (IBM, Armonk, NY, USA).

**Author Contributions:** Conceptualization, D.G.P.; Methodology, D.G.P.; Validation, D.G.P. and K.S.; Formal Analysis, D.G.P. and K.S.; Investigation, K.S.; Resources, D.G.P.; Data Curation, D.G.P. and K.S.; Writing, Original Draft Preparation, K.S.; Writing, Review and Editing, D.G.P. and K.S.; Visualization, D.G.P.; Supervision, D.G.P.; Project Administration, D.G.P.; Funding Acquisition, D.G.P. All authors have read and agreed to the published version of the manuscript.

**Funding:** This research received no external funding.

**Acknowledgments:** The authors would like to acknowledge the financial support provided by the Flavor Research and Education Center at The Ohio State University and its supporting members.

**Conflicts of Interest:** The authors declare no conflicts of interest.

## References

1. Kasum, C.M.; Jacobs, D.R.J., Jr.; Nicodemus, K.; Folsom, A.R. Dietary risk factors for upper aerodigestive tract cancers. *Int. J. Cancer* **2002**, *99*, 267–272. [CrossRef]
2. Slavin, J.L. Whole grains and human health. *Nutr. Res. Rev.* **2004**, *17*, 99–110. [CrossRef] [PubMed]
3. McKeown, N.M.; Meigs, J.B.; Liu, S.; Wilson, P.W.F.; Jacques, P.F. Whole-grain intake is favorably associated with metabolic risk factors for type 2 diabetes and cardiovascular disease in the Framingham Offspring Study. *Am. J. Clin. Nutr.* **2002**, *76*, 390–398. [CrossRef] [PubMed]
4. GBD 2017 Diet Collaborators. Health effects of dietary risks in 195 countries, 1990-2017: A systematic analysis for the Global Burden of Disease Study. *Lancet* **2019**, *393*, 1958–1972. [CrossRef]

5. Liu, S.; Stampfer, M.; Hu, F.B.; Giovannucci, E.; Rimm, E.; Manson, J.E.; Hennekens, C.H.; Willett, W.C. Whole-grain consumption and risk of coronary heart disease: Results from the Nurses' Health Study. *Am. J. Clin. Nutr.* **1999**, *70*, 412–419. [CrossRef]

6. Bakke, A.; Vickers, Z. Consumer Liking of Refined and Whole Wheat Breads. *J. Food Sci.* **2007**, *72*, S473–S480. [CrossRef]

7. Heiniö, R.L.; Noort, M.; Katina, K.; Alam, S.; Sozer, N.; De Kock, R.; Hersleth, M.; Poutanen, K. Sensory characteristics of wholegrain and bran-rich cereal foods – A review. *Trends Food Sci. Technol.* **2016**, *47*, 25–38. [CrossRef]

8. Haro-Vicente, J.F.; Bernal-Cava, M.-J.; Lopez-Fernandez, A.; Ros-Berruezo, G.; Bodenstab, S.; Sanchez-Siles, L.M. Sensory Acceptability of Infant Cereals with Whole Grain in Infants and Young Children. *Nutrients* **2017**, *9*, 65. [CrossRef]

9. Bredie, W.L.; Mottram, N.S.; Guy, R.C.E. Aroma Volatiles Generated during Extrusion Cooking of Maize Flour. *J. Agric. Food Chem.* **1998**, *46*, 1479–1487. [CrossRef]

10. Chinnaswamy, R.; Hanna, M. Optimum Extrusion-Cooking Conditions for Maximum Expansion of Corn Starch. *J. Food Sci.* **1988**, *53*, 834–836. [CrossRef]

11. Gómez, M.H.; Aguilera, J.M. A Physicochemical Model for Extrusion of Corn Starch. *J. Food Sci.* **1984**, *49*, 40–43. [CrossRef]

12. Nair, M.; Shi, Z.; Karwe, M.V.; Ho, C.-T.; Daun, H. Collection and Characterization of Volatile Compounds Released at the Die during Twin Screw Extrusion of Corn Flour. In *Thermally Generated Flavors*; American Chemical Society (ACS): Washington, DC, USA, 1993; Volume 543, pp. 334–347.

13. Moskowitz, M.R.; Bin, Q.; Elias, R.J.; Peterson, D.G. Influence of Endogenous Ferulic Acid in Whole Wheat Flour on Bread Crust Aroma. *J. Agric. Food Chem.* **2012**, *60*, 11245–11252. [CrossRef] [PubMed]

14. Jiang, D.; Peterson, D.G. Identification of bitter compounds in whole wheat bread. *Food Chem.* **2013**, *141*, 1345–1353. [CrossRef] [PubMed]

15. Delgado, R.M.; Hidalgo, F.J.; Zamora, R. Antagonism between lipid-derived reactive carbonyls and phenolic compounds in the Strecker degradation of amino acids. *Food Chem.* **2016**, *194*, 1143–1148. [CrossRef] [PubMed]

16. Naves, M.M.V.; De Castro, M.V.L.; De Mendonça, A.L.; Santos, G.G.; Silva, M.S. Corn germ with pericarp in relation to whole corn: nutrient contents, food and protein efficiency, and protein digestibility-corrected amino acid score. *Food Sci. Technol.* **2011**, *31*, 264–269. [CrossRef]

17. Gwirtz, J.A.; Garcia-Casal, M.N. Processing maize flour and corn meal food products. *Ann. New York Acad. Sci.* **2013**, *1312*, 66–75. [CrossRef]

18. Leffingwell & Associates. Odor & Flavor Detection Thresholds in Water. Available online: https://www.leffingwell.com/odorthre.htm (accessed on 3 March 2020).

19. Schieberle, P.; Güntert, M.; Sommer, H.; Werkhoff, P. Structure determination of 4-methyl-3-thiazoline in roasted sesame flavour. *Food Chem.* **1996**, *56*, 369–372. [CrossRef]

20. Kerler, J.; Van Der Ven, J.G.; Weenen, H. $\alpha$-Acetyl-N-heterocycles in the Maillard reaction. *Food Rev. Int.* **1997**, *13*, 553–575. [CrossRef]

21. Mottram, N.S. Flavor Compounds Formed during the Maillard Reaction. In *Thermally Generated Flavors*; American Chemical Society (ACS): Washington, DC, USA, 1993; Volume 543, pp. 104–126. [CrossRef]

22. Izzo, H.; Ho, C.-T. Peptide-specific Maillard reaction products: a new pathway for flavor chemistry. *Trends Food Sci. Technol.* **1993**, *3*, 253–257. [CrossRef]

23. Hofmann, T.; Schieberle, P. Studies on the formation and stability of the roast-flavor compound 2-acetyl-2-thiazoline. *J. Agric. Food Chem.* **1995**, *43*, 2946–2950. [CrossRef]

24. Amrani-Hemaimi, M.; Cerny, C.; Fay, L. Mechanisms of formation of alkylpyrazines in the Maillard reaction. *J. Agric. Food Chem.* **1995**, *43*, 2818–2822. [CrossRef]

25. Scalone, G.L.L.; Cucu, T.; De Kimpe, N.; De Meulenaer, B. Influence of free amino acids, oligopeptides, and polypeptides on the formation of pyrazines in maillard Model Systems. *J. Agric. Food Chem.* **2015**, *63*, 5364–5372. [CrossRef] [PubMed]

26. Paravisini, L.; Peterson, D.G. Reactive carbonyl species as key control point for optimization of reaction flavors. *Food Chem.* **2019**, *274*, 71–78. [CrossRef] [PubMed]

27. Das, A.K.; Singh, V. Antioxidative free and bound phenolic constituents in botanical fractions of Indian specialty maize (*Zea mays* L.) genotypes. *Food Chem.* **2016**, *201*, 298–306. [CrossRef]

28. Das, A.K.; Singh, V. Antioxidative free and bound phenolic constituents in pericarp, germ and endosperm of Indian dent (Zea mays var. indentata) and flint (*Zea mays* var. indurata) maize. *J. Funct. Foods* **2015**, *13*, 363–374. [CrossRef]
29. Starowicz, M.; Koutsidis, G.; Zieliński, H. Determination of Antioxidant Capacity, Phenolics and Volatile Maillard Reaction Products in Rye-Buckwheat Biscuits Supplemented with 3β-d-Rutinoside. *Molecules* **2019**, *24*, 982. [CrossRef]
30. Trikusuma, M.; Paravisini, L.; Peterson, D.G. Identification of aroma compounds in pea protein UHT beverages. *Food Chem.* **2020**, *312*, 126082. [CrossRef]


**Sample Availability:** Samples of the compounds are not available from the authors.

 *molecules*

*Article*

# Proteolytic Volatile Profile and Electrophoretic Analysis of Casein Composition in Milk and Cheese Derived from Mironutrient-Fed Cows

Andrea Ianni [1,†], Francesca Bennato [1,†], Camillo Martino [2], Lisa Grotta [1], Nicola Franceschini [3] and Giuseppe Martino [1,*]

[1] Faculty of BioScience and Technology for Food, Agriculture and Environment, University of Teramo, Via Renato Balzarini 1, 64100 Teramo, Italy; aianni@unite.it (A.I.); fbennato@unite.it (F.B.); lgrotta@unite.it (L.G.)

[2] Istituto Zooprofilattico Sperimentale dell'Abruzzo e del Molise "G. Caporale", Via Campo Boario 37, 64100 Teramo, Italy; c.martino@izs.it

[3] Department of Biotechnological and Applied Clinical Sciences, University of L'Aquila, Via Vetoio 1, 67100 L'Aquila, Italy; nicola.franceschini@univaq.it

[*] Correspondence: gmartino@unite.it; Tel.: +39-0861-266950

[†] These authors contributed equally to this work.

Academic Editor: Eugenio Aprea
Received: 31 March 2020; Accepted: 9 May 2020; Published: 10 May 2020

**Abstract:** The aim of the study was to evaluate the proteolytic process in Caciocavallo cheese obtained from Friesian cows fed zinc, selenium, and iodine supplementation. Thirty-six Friesian cows, balanced for parity, milk production, and days in milk, were randomly assigned to four groups. The control group (CG) was fed with a conventional feeding strategy, while the three remaining groups received a diet enriched with three different trace elements, respectively zinc (ZG), selenium (SG), and iodine (IG). At the end of the experimental period, samples of milk were collected and used to produce Caciocavallo cheese from each experimental group. Cheese samples were then analyzed after 7 and 120 days from the cheese making in order to obtain information on chemical composition and extent of the proteolytic process, evaluated through the electrophoretic analysis of caseins and the determination of volatiles profile. Both milk and cheese samples were richer in the amount of the microelement respectively used for the integration of the cattle's diet. The zymographic approach was helpful in evaluating, in milk, the proteolytic function performed by endogenous metalloenzymes specifically able to degrade gelatin and casein; this evaluation did not highlight significant differences among the analyzed samples. In cheese, the electrophoretic analysis in reducing and denaturing condition showed the marked ability of β-casein to resist the proteolytic action during ripening, whereas the dietary selenium supplementation was shown to perform a protective action against the degradation of S1 and S2 isoforms of α-casein. The analysis of the volatile profile evidenced the presence of compounds associated with proteolysis of phenylalanine and leucine. This approach showed that selenium was able to negatively influence the biochemical processes that lead to the formation of 3-methyl butanol, although the identification of the specific mechanism needs further investigation.

**Keywords:** proteolysis; microelement; dairy cow; caciocavallo cheese; casein; volatile compound

## 1. Introduction

High-yielding animals require feeding strategies that guarantee the right contribution of all the necessary microelements, such as zinc, manganese, copper, cobalt, iodine, and selenium. Dietary microelements deficiency in livestock commonly leads to a wide range of disorders especially associated to growth depression, inefficient feed utilization, lower production performance, and depressed immunocompetence that may increase animals' susceptibility to infectious diseases [1].

Zinc is a ubiquitous element in cells and represents an essential component of several metalloenzymes [2] and transcription factors, with relevant roles in the metabolism of essential nutrients in animals. Zinc is not stored in the animal body; therefore, a constant dietary supply is necessary to avoid the onset of a wide range of pathological conditions, such as skin parakeratosis, reduction or cessation of growth, general debility, lethargy, and increased susceptibility to infection [3]. Selenium is involved in numerous biological mechanisms, including cellular response to oxidative stress, redox signaling, cellular differentiation, immune response, and protein folding [4]. Selenium was also reported to improve rumen fermentation, milk yields, and feed digestion in Holstein dairy cows [5]. Iodine is the main component of the thyroid hormones and when its requirement is not satisfied, a reduced functionality of the thyroid gland could occur (hypothyroidism) with consequences for proper mental development, body growth, and decreased fertility. In animal husbandry, iodine supplementation is needed because the native iodine content of plant straight feed-stuffs is low; moreover, the increasing use of rapeseed meal in livestock diets is associated with the intake of glucosinolates, which are known to be iodine antagonists, inhibiting the activity of sodium iodide symporter [6].

Different studies have been carried out in order to evaluate the effect of dietary microelements intake on ruminants' metabolism [7–9] and chemical-nutritional quality of dairy products [10–12], but the topic concerning the microelements' influence on ripened cheese flavor has received less attention. Conversion of lactose and citrate, lipolysis, and proteolysis represent the main chemical processes involved in the development of aroma in dairy products. Among these processes, proteolysis of caseins is an important biochemical pathway responsible for the formation of flavor and texture in hard- and semi hard-type cheeses [13]. Proteolysis in cheese can be divided into the primary and the secondary phase. Primary proteolysis is performed by rennet, native milk enzymes and induces degradation of caseins into large, well-defined polypeptides. Further proteolytic processes operated by starter and nonstarter bacteria during ripening contribute to secondary proteolysis, which cause formation of small polypeptides and free amino acids responsible for cheese aroma and taste [14]. Branched chain amino acids (leucine, isoleucine, and valine), aromatic amino acids (phenylalanine, tyrosine, and tryptophan), and methionine are thought to be the precursors of important volatile compounds in dairy products [15]. Therefore, the aim of the present study was to investigate the effect of dietary microelements' intake on the development of proteolysis in fresh and ripened dairy products obtained from lactating dairy cows. Specifically, the study was conducted on Caciocavallo cheese, a dairy product of bovine origin, which is generally subjected to seasoning for fairly long intervals of time compared to other products, and which is therefore more exposed to both lipolytic and proteolytic processes. The study in any case did not concern only the cheese but was also extended to the milk used for cheesemaking, in order to verify the actual enrichment with the microelements respectively used for dietary supplementation and also to evaluate the presence and the activity of native milk metalloenzymes, which exploit these microelements, especially zinc, as cofactors.

## 2. Results

### 2.1. Microelements Quantification in Milk and Cheese

At the end of the 56 days of the trial, milk samples obtained from each experimental group (zinc (ZG), selenium (SG), and iodine (IG)) in the feeding strategy were found to be effective in inducing an enrichment of the microelement respectively used for the dietary supplementation (Table 1).

**Table 1.** Microelements quantification in milk samples obtained from lactating dairy cows fed control diet (CG) and control diet supplemented with zinc (ZG), selenium (SG), and iodine (IG).

| Microelement | Milk Samples | | | |
|---|---|---|---|---|
| | CG | ZG | SG | IG |
| Zinc [1] | 4.18 [a] ± 0.37 | 5.76 [b] ± 0.41 | 3.98 [a] ± 0.33 | 4.04 [a] ± 0.40 |
| Selenium [1] | 0.036 [a] ± 0.004 | 0.041 [a] ± 0.005 | 0.049 [b] ± 0.005 | 0.039 [a] ± 0.004 |
| Iodine [1] | 0.12 [a] ± 0.03 | 0.11 [a] ± 0.02 | 0.10 [a] ± 0.02 | 0.17 [b] ± 0.02 |

[1] Data are reported on a dry matter basis, as mean (mg·kg$^{-1}$) ± standard deviation (S.D.). [a,b] Different letters in the same row indicate significant differences ($p < 0.05$).

The finding concerning the enrichment with the microelements used for the dietary supplementation was also found in samples of Caciocavallo cheese, both fresh ($T_7$) and after 120 days of repining ($T_{120}$). The results concerning the quantification performed on individual samples for the two ripening times is shown in Table 2.

**Table 2.** Microelements content in cheese samples obtained from lactating dairy cows fed control diet (CG) and control diet supplemented with different trace elements: zinc (ZG), selenium (SG), and iodine (IG).

| Trace Element | Ripening Time [1] | | | | | | | |
|---|---|---|---|---|---|---|---|---|
| | T$_7$ | | | | T$_{120}$ | | | |
| | CG | ZG | SG | IG | CG | ZG | SG | IG |
| Zinc [2] | 41.34 [a] ± 2.03 | 52.61 [b] ± 2.37 | 42.77 [a] ± 2.19 | 40.77 [a] ± 1.98 | 43.21 [a] ± 2.41 | 54.74 [b] ± 2.39 | 41.82 [a] ± 3.09 | 42.91 [a] ± 2.93 |
| Selenium [2] | 0.21 [a] ± 0.03 | 0.19 [a] ± 0.03 | 0.32 [b] ± 0.04 | 0.22 [a] ± 0.03 | 0.22 [a] ± 0.02 | 0.18 [a] ± 0.03 | 0.31 [b] ± 0.04 | 0.19 [a] ± 0.03 |
| Iodine [2] | 0.21 [a] ± 0.03 | 0.24 [a] ± 0.03 | 0.19 [a] ± 0.03 | 0.31 [b] ± 0.04 | 0.20 [a] ± 0.03 | 0.22 [a] ± 0.03 | 0.18 [a] ± 0.02 | 0.29 [b] ± 0.04 |

[1] 7 and 120 days of ripening ($T_7$ and $T_{120}$ respectively); [2] Data are reported in mg·kg$^{-1}$ on a dry matter basis. [a,b] Different letters in the same row indicate significant differences ($p < 0.05$).

## 2.2. Zymographic Evaluation of Gelatinolytic and Caseinolytic Activity in Milk

Enzymatic activities able to induce the degradation of gelatin and casein in milk samples have been evaluated using a zymographic approach.

The gelatin-zymography (Figure 1) was helpful in highlighting the enzymatic activity closely associated with the two major gelatinases: matrix metalloproteinase 2 (MMP-2) and matrix metalloproteinase 9 (MMP-9).

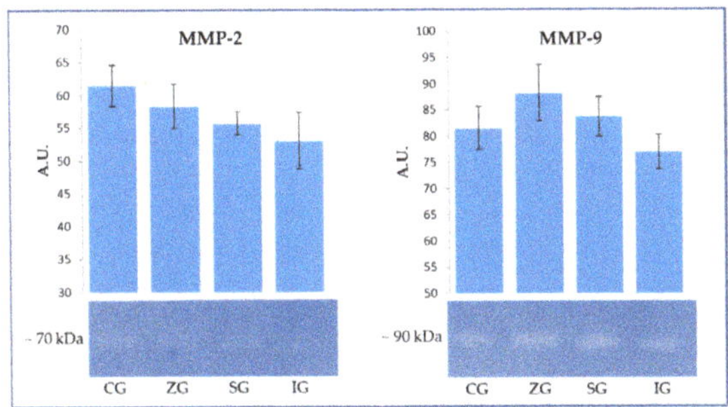


**Figure 1.** Gelatin-zymography on milk samples obtained from lactating dairy cows fed control diet (CG) and control diet supplemented with zinc (ZG), selenium (SG), and iodine (IG). Analysis was performed in order to obtain information on the enzymatic activities associated to matrix metalloproteinase 2 (MMP-2) and matrix metalloproteinase 9 (MMP-9). The ImageJ software was used to perform the quantitative analysis of visualized spots. Data are reported as mean values expressed in arbitrary unit (A.U.) ± standard deviation.

With specific regard to MMP-2, no significant variations were detected ($p > 0.05$), although the quantitative analysis of the spots highlighted a tendency in the samples from the experimental groups (ZG, SG, and IG) to degrade gelatin with less efficacy. In the case of MMP-9, all samples showed greater activity than that found for MMP-2. However, similarly to what was observed for MMP-2, no significant differences between the various samples were identified ($p > 0.05$). The only noteworthy phenomenon concerns the slight tendency of the ZG sample to degrade the substrate more effectively.

Total caseinolytic activity was assessed through casein-zymography. The data shown in Figure 2 showed a picture quite similar to that observed for MMP-9 with the ZG milk sample, which seemed to have a greater ability to degrade the substrate, although this difference compared to the control (CG) and to the other experimental samples (SG and IG) was not significant ($p > 0.05$).

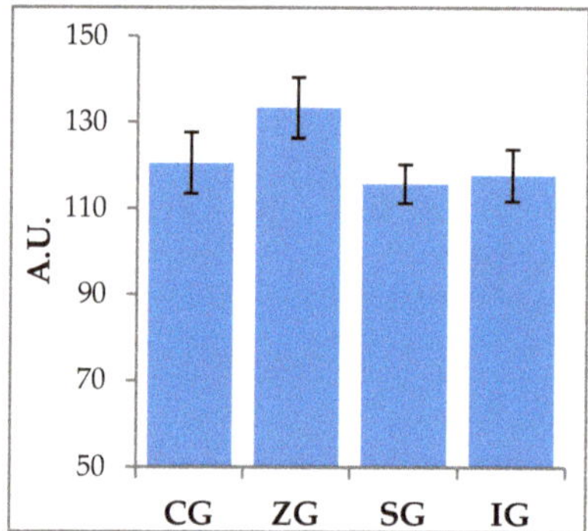

**Figure 2.** Casein-zymography on milk samples obtained from lactating dairy cows fed control diet (CG) and control diet supplemented with zinc (ZG), selenium (SG), and iodine (IG). The ImageJ software was used to perform the quantitative analysis. Data are reported as mean values expressed in arbitrary unit (A.U.) ± standard deviation.

### 2.3. Caseins Separation by Sodium Dodecyl Sulfate Polyacrylamide Gel Electrophoresis (SDS-PAGE)

The sodium dodecyl sulfate polyacrylamide gel electrophoresis (SDS-PAGE) was performed with the aim of monitoring the degradation of caseins by the rennet and indigenous milk enzymes. As showed in Figure 3, cheese proteins have been separated into three major casein components ($\alpha$S1-CN, $\alpha$S2-CN, and $\beta$-CN).

Under the applied experimental conditions, the protein profile of both fresh ($T_7$) and ripened ($T_{120}$) cheese showed a major $\beta$-CN band and less intensive bands corresponding to $\alpha$S1-CN and $\alpha$S2-CN. In all samples, five low molecular weight peptides (from 25 kDa to 10 kDa) were also identified as proteolysis products (Table 3). Dietary supplementation with zinc and iodine did not generate significant changes compared to the CG samples both at $T_7$ and at $T_{120}$, while selenium influenced the proteolytic process, partly protecting $\alpha$S2-CN, as evidenced by the lack of significant differences in the proteolysis products corresponding to bands 2, 3, and 4 in SG samples ($p > 0.05$).

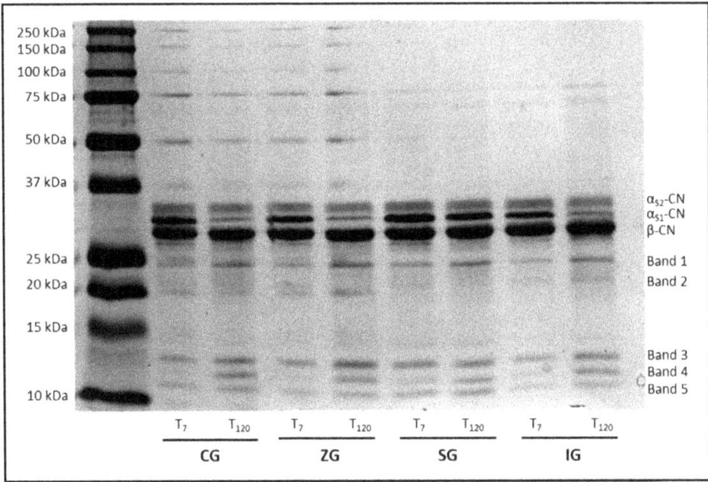



**Figure 3.** Sodium dodecyl sulfate polyacrylamide gel electrophoresis (SDS-PAGE) pattern of caseins and peptides resulting from protein degradation in fresh ($T_7$) and 120-days ripened ($T_{120}$) cheese samples obtained from lactating dairy cows fed control diet (CG) and control diet supplemented with zinc (ZG), selenium (SG), and iodine (IG) deviation.

## 2.4. Identification of Volatile Compounds in Fresh and Ripened Cheese

Volatile compounds resulting from secondary proteolysis during ripening were identified in all the analyzed samples (Table 4). Two volatile compounds, phenylacetaldehyde and 2-phenylethyl alcohol, derived from phenylalanine catabolism were identified, while only the 3-methyl-1-butanol was identified as a derivative of leucine degradation.

In both $T_7$ and $T_{120}$ samples no significant variations were found in the phenylacetaldehyde content ($p > 0.05$). In the case of 2-phenylethyl alcohol, dietary supplementation with zinc, selenium, and iodine, seems to have led to a significant reduction of this compound in $T_7$ samples (6499 AU in CG vs. 3125, 1364, and 3037 AU in ZG, SG, and IG, respectively; $p < 0.01$). In ripened cheese, the phenomenon was confirmed only in ZG samples, while an increase of 2-phenylethyl alcohol was detected in IG. The only identified compound deriving from the leucine degradation was 3-methyl butanol, which tends to be synthesized in the various cheese samples with a comparable pattern both after 7 and 120 days of ripening. Specifically, no significant differences were observed between CG, ZG, and IG samples, while both at $T_7$ and $T_{120}$, the SG samples were characterized by a lower concentration of this compound ($p < 0.01$ at $T_7$ and $p < 0.05$ at $T_{120}$).

**Table 3.** Densitometric analysis of SDS-PAGE protein bands (Figure 3) in fresh ($T_7$) and 120-days ripened ($T_{120}$) cheese samples obtained from lactating dairy cows fed control diet (CG) and control diet supplemented with zinc (ZG), selenium (SG), and iodine (IG).

| Protein | CG | | | ZG | | | SG | | | IG | | |
|---|---|---|---|---|---|---|---|---|---|---|---|---|
| | $T_7$ | $T_{120}$ | $p$ | $T_7$ | $T_{120}$ | $p$ | $T_7$ | $T_{120}$ | $p$ | $T_7$ | $T_{120}$ | $p$ |
| $\alpha_{S2}$-CN | 22.93 ± 2.07 | 26.59 ± 2.78 | ns | 20.98 ± 1.87 | 19.60 ± 1.79 | ns | 22.30 ± 2.13 | 21.02 ± 2.04 | ns | 23.80 ± 2.21 | 25.10 ± 2.33 | ns |
| $\alpha_{S1}$-CN | 33.16 ± 2.88 | 10.82 ± 0.85 | ** | 29.66 ± 2.83 | 8.30 ± 0.78 | ** | 31.82 ± 2.97 | 21.14 ± 1.99 | ** | 29.98 ± 2.83 | 14.66 ± 1.42 | ** |
| β-CN | 26.92 ± 2.54 | 27.94 ± 2.41 | ns | 30.57 ± 2.92 | 30.93 ± 2.82 | ns | 23.39 ± 2.18 | 27.21 ± 2.51 | ns | 31.33 ± 2.86 | 33.60 ± 3.11 | ns |
| Band 1 | 5.04 ± 0.56 | 8.41 ± 0.86 | * | 5.19 ± 0.53 | 10.70 ± 0.96 | ** | 5.48 ± 0.55 | 8.09 ± 0.78 | * | 4.10 ± 0.42 | 6.36 ± 0.61 | * |
| Band 2 | 3.86 ± 0.44 | 6.36 ± 0.67 | ** | 5.18 ± 0.52 | 8.76 ± 0.83 | ** | 4.68 ± 0.47 | 5.96 ± 0.61 | ns | 3.09 ± 0.32 | 6.27 ± 0.62 | ** |
| Band 3 | 3.51 ± 0.39 | 6.77 ± 0.69 | ** | 3.64 ± 0.38 | 8.20 ± 0.77 | ** | 4.39 ± 0.45 | 4.27 ± 0.43 | ns | 2.76 ± 0.29 | 4.30 ± 0.42 | * |
| Band 4 | 1.72 ± 0.23 | 6.70 ± 0.68 | ** | 0.96 ± 0.11 | 6.77 ± 0.66 | ** | 3.11 ± 0.33 | 3.99 ± 0.40 | ns | 2.10 ± 0.22 | 4.50 ± 0.44 | ** |
| Band 5 | 1.87 ± 0.21 | 6.41 ± 0.65 | ** | 3.82 ± 0.39 | 6.74 ± 0.65 | ** | 4.84 ± 0.50 | 8.32 ± 0.81 | ** | 2.84 ± 0.28 | 5.21 ± 0.51 | ** |

Data are reported as mean (%) ± S.D. of the total proteins found in the electrophoretic profile of each sample. $\alpha_S$-CN = $\alpha_S$-casein and β-CN = β-casein. Bands 1, 2, 3, 4, and 5 = fragments of protein degradation. * $p < 0.05$; ** $p < 0.01$; and ns = not significant.

**Table 4.** Proteolytic volatile compounds in fresh ($T_7$) and 120-days ripened ($T_{120}$) cheese obtained from lactating cows fed control diet (CG) and control diet supplemented with zinc (ZG), selenium (SG), and iodine (IG).

| Volatile Compounds | T.I. [2] | Ripening Time [1] | | | | | | | | | |
|---|---|---|---|---|---|---|---|---|---|---|---|
| | | $T_7$ | | | | | $T_{120}$ | | | | |
| | | CG | ZG | SG | IG | SEM | CG | ZG | SG | IG | SEM |
| Phenylalanine | | | | | | | | | | | |
| Phenylacetaldehyde | 91 | 750 | 862 | 1083 | 480 | 227 | 3015 | 3658 | 5748 | 4334 | 1222 |
| 2-phenylethyl alcohol | 91 | 6499 [a] | 3125 [b] | 1364 [c] | 3037 [b] | 399 | 9596 [a] | 6833 [b] | 7564 [a] | 14815 [c] | 2643 |
| Leucine | | | | | | | | | | | |
| 3-methyl-1-butanol | 55 | 7481 [a] | 6526 [a] | 2697 [b] | 6091 [a] | 987 | 7648 [a] | 6327 [a] | 5921 [b] | 7597 [a] | 1179 |

[a,b,c] Different letters in the same row indicate significant differences between groups ($p < 0.05$). [1] 7 and 120 days of ripening ($T_7$ and $T_{120}$ respectively); [2] Target ion. Data are reported as arbitrary units.

## 3. Discussion

Preventive analysis performed to determine the chemical composition of cheese samples obtained from the various experimental groups did not show significant changes both in relation to the feeding strategy (CG, ZG, SG, and IG) and in relation to the ripening time ($T_7$ and $T_{120}$). In particular, there were no differences in the protein content (Supplementary Table S1), testifying that the proteolytic process took place in the presence of equal substrate concentrations among the analyzed cheese samples. This finding is in complete agreement with what has been observed in other studies, which tested the dietary supplementation with essential trace elements in dairy cows [10,16]. With regards to the dosage of zinc, selenium, and iodine in milk and cheese samples, a significant increase of the micronutrient respectively used for the integration of the animal diet was highlighted. Limited to selenium, this result is in agreement with several studies [17], while in the case of zinc and iodine, there are discrepancies with what was previously reported. Pechová et al. [18] evidenced the inability of dietary zinc supplementation to influence its amount in bovine milk and cheese; these authors discussed such phenomenon advancing the hypothesis of an impaired incidence of rumen acidosis in the herd before the start of the experiment. With regard to iodine, Moschini et al. [19] indicated this microelement suitable for milk fortification through feed supplementation but its poor ability to interact with protein structures seems to compromise its transfer to cheese. Given the relevance of zinc, selenium, and iodine in human biochemical mechanisms, their general enrichment in milk and dairy products at the concentrations detected in this study should represent a positive aspect. However, it must be taken into account that an excess of these trace elements can constitute a technological risk factor for dairy products, both for the commercial image of the products and, above all, for consumers' health. This consideration assumes particular value especially in the case of nonessential or toxic metals, such as lead and cadmium, that even in low concentrations are responsible for metabolic disorders with extremely serious consequences [20].

The milk enrichment with the microelements respectively used for dietary supplementation made it necessary to verify the possibility that this event could influence the function of endogenous enzymes. In particular, attention was focused on the activity of matrix metalloproteinases (MMPs), which, following cheesemaking, can be incorporated into the dairy matrix, actively participating in the proteolytic events that characterize the ripening process, especially in the initial stages. MMPs represent a family of calcium-dependent endopeptidases, with a catalytic domain containing a zinc ion coordinated by three histidines. These enzymes are involved in the physiological degradation of the extracellular matrix (ECM) in mammals, a fundamental process for tissue development, morphogenesis, remodeling, and repair. Based on their specificity for the substrate, MMPs are divided into four main subgroups: gelatinases, collagenases, stromelysins, and membrane type MMPs (MT-MMPs) [21]. In this study, the activity of gelatinases MMP-2 (Gelatinase A) and MMP-9 (Gelatinase B) have been specifically evaluated; furthermore, the caseinolytic potential of milk samples was also assayed, since casein can be hydrolyzed by different metalloenzymes, such as MMP-1 (Collagenase-1) and MMP-3 (Stromelysin-1) [22]. Although MMPs are constitutively present in all tissues, their release in milk can be partly influenced by the health status of the mammary gland. As previously reported, the inflammatory process during mastitis results in the release of a wide range of proteolytic enzymes, which are mainly secreted by polymorphonuclear cells recruited into the mammary gland from blood circulation [23].

The dairy cows involved in this study maintained good health conditions for the entire duration of the trial, and the preliminary evaluations carried out on milk used for cheesemaking showed particularly low values associated with somatic cells count (SCC) and without significant changes in the data associated with the total bacterial count (data not shown). The zymographic approach did not evidence significant variations both for gelatinolytic (MMP-2 and MMP-9) and caseinolytic activities. Because these enzymes depend on the presence of zinc, an increase in the substrate hydrolysis capacity in ZG samples would have been expected. The justification for this evidence could be sought in the fact that dietary supplementation was performed by using zinc oxide (ZnO), which showed, in alternative research fields, the ability to even block the activity of MMP-9 [24]. It is therefore presumable that this organic form of zinc is difficult to use by this family of enzymes, which therefore do not undergo an improvement in the hydrolysis kinetics of the respective substrates.

Regarding the analysis of primary proteolysis in cheese, the SDS-PAGE has proved to be particularly useful in monitoring the degradation of caseins by the rennet and indigenous milk enzymes. Cheese proteins have been separated into three major casein components ($\alpha$S1-CN, $\alpha$S2-CN, and $\beta$-CN) in all the analyzed samples, with molecular weights and electrophoretic mobility that were consistent with those reported in the literature [25]. Under the applied experimental conditions, the protein profile of both fresh ($T_7$) and ripened ($T_{120}$) cheese showed a major $\beta$-CN band and less intensive bands corresponding to $\alpha$S1-CN and $\alpha$S2-CN. In all samples, five low molecular weight peptides (from 25 kDa to 10 kDa) were all identified as proteolysis products. Dietary supplementation with zinc and iodine did not generate significant changes compared to the CG samples both at $T_7$ and at $T_{120}$, while selenium influenced the proteolytic process, partly protecting $\alpha$S2-CN, as evidenced by the lack of significant differences in the proteolysis products corresponding to bands 2, 3, and 4 in SG samples. According to other studies, most of these peptides with molecular weight in the range 10–20 kDa were generated by rennet and plasmin following the degradation of $\alpha$S-CN and $\beta$-CN [26,27]. Selenium has been reported to inhibit the expression of urokinase-type plasminogen activator (uPA), a serine protease, which can convert plasminogen to plasmin, which is capable of degrading extracellular matrix proteins and activating latent forms of MMPs [28]. The $\alpha$S1 and $\beta$ caseins did not show significant changes both in relation to the feeding strategy and in relation to the maturing time; this finding is consistent with what was previously reported in other studies and represents an added value if we consider the growing interest on $\beta$-casein micelles as a nano vehicle for solubility enhancement of natural compounds in the food industry [29].

Several studies have shown how changes in the diet of ruminants can be effective in inducing significant changes in the volatile profile of milk and dairy products [30]. Volatile compounds (VOCs) resulting from secondary proteolysis during ripening were identified in all the analyzed samples. Two volatile compounds, phenylacetaldehyde and 2-phenylethyl alcohol, derived from phenylalanine catabolism, while only the 3-methyl-1-butanol was identified as a derivative of leucine degradation. In both $T_7$ and $T_{120}$ samples, no significant variations were found in the phenylacetaldehyde content. As reported by McSweeney and Sousa [14], the phenylacetaldehyde metabolism in dairy products may occur by the nonenzymatic Strecker synthesis, through the degradation of phenylalanine, or by enzymatic transamination of phenylalanine as imide that is subsequently degraded to aldehyde. The presence of 2-phenylethyl alcohol is instead due to the phenylacetaldehyde reduction. In this case, the dietary supplementation with zinc, selenium, and iodine seems to have led to a significant reduction of this compound in fresh cheese samples, while in ripened samples, this finding was confirmed only for ZG samples. Aromatic compounds from phenylalanine have a relevant impact on the aroma of cheese. In particular, phenylethyl alcohol is reported to be one of the most odorous aromatic compounds, associated with a rose flower note [31]. With regard to leucine, this amino acid can undergo several biochemical processes in cheese during ripening. After an extracellular enzymatic degradation of casein by proteases, released from starter bacteria, leucine is converted to the corresponding $\alpha$-keto acid ($\alpha$-ketoisocaproic acid) by an intracellular transamination; then, such compound can be converted to $\alpha$-hydroxy acids, aldehydes, or CoA-esters [32]. These leucine metabolites are not considered to be important for the development of cheese flavor, while the 3-methyl-1-butanol, derived from the hydrogenation of the corresponding aldehyde (3-methylbutanal), was reported to be responsible for alcoholic and fruity odors in Swiss-type cheese [33]. In this study, the only identified compound deriving from the leucine degradation was 3-methyl butanol, which tends to be synthesized in the various cheese samples with a comparable pattern both after 7 and 120 days of ripening. Specifically, no significant differences were observed between CG, ZG, and IG samples, while both at $T_7$ and $T_{120}$ of the SG, samples were characterized by a lower concentration of this compound. A similar behavior was also observed in a 30-days ripened caciotta cheese obtained from dairy cows fed organic selenium [34], as well as in a 90-days ripened Pecorino cheese deriving from ewes that received a dietary supplementation with organic zinc [35]. In addition to this, the dietary intake of organic zinc by dairy cows was even effective in inducing the reduction of 3-methyl butanol in a typical Italian soft cheese, the Giuncata cheese, after only 7 days storage at 4 °C [36]. Probably, the microelements administered in organic forms negatively influence the biochemical processes that lead to the formation of 3-methyl butanol, although the identification of the specific biochemical mechanisms needs further and more targeted investigations.

## 4. Materials and Methods

### 4.1. Experimental Design, Feeding Strategies, Cheesemaking, and Sampling

The experimental plan was performed according to Directive 2010/63/EU of the European Parliament (European Union, 2010) and Directive 86/609/EEC (European Economic Community, 1986), which deal with the protection of animals used for scientific purposes [37,38].

Thirty six healthy Friesian cows, homogeneous for age (41 ± 1.5 months) and lactation days (74 ± 12 days) have been used in this study. Animals, belonging to the same commercial farm, have been randomly divided into four groups of nine cows each. The control group (CG) was fed with a standard diet formulated taking into account the nutritional needs of cows in midlactation and guaranteeing each animal the daily requirement of each microelement, while the three experimental groups received the dietary supplementation with zinc (ZG), selenium (SG), and iodine (IG), respectively. The ZG received an additional total intake of about 100 mg of Zn. For the preparation of the rations, ZnO as a powder was used, and the dose management was performed according to Regulation (EC) No. 1831/2003 of the European Parliament and of the Council of 22 September 2003 (European Commission, 2003) on additives for use in animal nutrition [39]. With regards to SG, animals received a total daily supplementation of 0.47 mg of Se; current EU regulations have limited the use of Se supplementation to the overall content not exceeding 0.5 mg in complete feed daily administered. For the preparation of the ration organic Se as a crystalline powder was used that was incorporated into feed in the form of a premixture according to the recommendation reported in the Regulation No. 121/2014 of the European Commission concerning the authorization of ʟ-selenomethionine as a feed additive for all animal species [40]. Finally, the IG animals were fed with a daily total iodine content of 4.5 mg/cow. For the dietary supplementation, potassium iodide (KI) as a powder was used and the total iodine was set not to exceed the maximum daily amount of 5 mg allowed by law (Reg. 1459/2005) [41].

The study had an overall duration of 70 days, characterized by 14 initial days of adaptation in which all the involved animals received the standard diet, followed by 56 days of dietary supplementation, in which animals of each group were housed in separate areas of free housing with an access to an identical feeding area in which each animal had an individual feeding bin with water freely available all throughout the study. All animals received about 23 kg/head/day of dry matter of total mixed rations (TMR) whose composition (Supplementary Table S2) was defined taking into account the parameters reported on the seventh edition of Nutrient Requirements of Dairy Cattle (2001) [42]. Samples of TMR were analyzed, according to AOAC methods (1990), for crude protein (CD; method 930.15), ether extract (EE; method 920.39), crude fiber (CF; method 962.09), and ash (method 942.05) [43]; detergent procedures reported by Van Soest et al. [44] were used for the determination of neutral detergent fiber (NDF) and acid detergent fiber (ADF). On the 70th day of the trial, individual raw milk samples were collected separately from each group and immediately analyzed for chemical composition. The remaining milk from each group was pooled and manipulated in the same way during the cheese-making process to obtain the Caciocavallo cheese, according to the protocol previously described [45]. In order to evaluate changes in the chemical composition and quality attributes due to ripening, sampling and analyses on Caciocavallo cheese were carried out after 7 days ($T_7$) and 120 days ($T_{120}$) from the cheese-making batches. Caciocavallo cheese is a dairy product that lends itself to being aged for quite long periods and on average it is consumed even after 6 months from cheesemaking.

The analysis of the samples at the indicated times ($T_7$ and $T_{120}$) therefore allows us to compare the fresh product with a ripened product ready to be consumed. Samples, collected in triplicate from three different cheese-makings, were partly immediately analyzed and partly packed under vacuum and frozen at −20 °C until analysis.

### 4.2. Microelements Determination in Milk and Cheese

For the determination of the total amount of zinc in milk and cheese, samples were first mineralized by dry incineration, then subjected to atomic absorption spectrophotometry using an air/acetylene flame [46]. Selenium and iodine content was determined instead by inductively coupled plasma mass spectrometry (ICP-MS) by using an Agilent 7500ce (Agilent Technologies, Santa Clara, CA, USA) and following the procedures respectively reported by Gerber et al. and Fecher et al. [47,48].

### 4.3. Gelatin and Casein Zymography of Milk Samples

The evaluation of the activity of the zinc-dependent proteases in raw milk was performed through a zymographic approach. A total of 10 mL of each sample were placed in 15 mL tubes and centrifuged at 10,000 rpm for 20 min at 4 °C. This allowed us to obtain a separation in three distinct phases: an upper layer consisting of fat, a central serum fraction containing proteins, and a lower layer characterized by the cellular component and any interfering residues. The central phase was then carefully recovered and filtered through with 0.22 μm syringe filters before the total protein dosage that was calculated by the Bradford protein concentration assay [49].

Volumes of each sample corresponding to 10 μg of total proteins were diluted in a nonreducing sample buffer without heating, and resolved by 8% sodium dodecylsulphate polyacrylamide gel electrophoresis (SDS-PAGE) containing 0.2 mg/mL type B gelatin (Sigma Aldrich, Milan, Italy) [50]. The gels were then incubated for 45 min in a renaturation buffer (50 mM Tris-HCl pH 8.0, containing 2.5% Triton X-100) to remove SDS. Subsequently, incubation of 24 h in a developer buffer (50 mM Tris-HCl pH 8.0, containing 5 mM $CaCl_2$, 200 mM NaCl, and 0.02% Brij 35) was performed to allow enzymes renaturation and activity. The gels were then stained in a 0.1% solution of Coomassie Blue R250 in 40% (*v/v*) methanol and 10% (*v/v*) acetic acid to allow the spot visualization.

The evaluation of caseinolytic activity was instead carried out by 10% SDS-PAGE with the addition of 0.3% of bovine casein (Sigma Aldrich, Milan, Italy). Additionally, in this case, volumes of each sample corresponding to 10 μg of total proteins were diluted in a nonreducing sample buffer without heating, before the electrophoretic resolving. After electrophoresis, gels were subjected to the same protocol reported for gelatin zymography, with the only variations regarding the final incubation period that was prolonged to 48 h.

The ImageJ software (version 1.44, National Institutes of Health, Bethesda, MD, USA) [51] was used to perform the quantitative analysis of visualized spots for both gelatin and casein zymography.

### 4.4. Caseins Extraction and Separation by Sodium Dodecyl Sulfate Polyacrylamide Gel Electrophoresis (SDS-PAGE)

Casein degradation in $T_7$ and $T_{120}$ cheeses was evaluated by sodium dodecyl sulfate polyacrylamide gel electrophoresis (SDS-PAGE). Each cheese sample (1 g) was dissolved in 20 mL 0.01 M Tris-Glicina pH 8.3 and 6 M Urea and homogenized for 2 min. The cheese extract was incubated for 2 h at 37 °C to induce the solubilization of casein fraction. The solution was then centrifugated for 15 min at 10,000 g (4 °C), and the supernatant was recovered and filtered through Whatman filter paper to remove fat and other insoluble solids.

Volumes of each sample, containing 10 μg of total extracted proteins, were diluted in a 2X sample buffer (62.5 mM Tris-HCl pH 6.8, 2% SDS, 10% glycerol, 5% β-mercaptoethanol, and traces of bromophenol blue), boiled for 3 min, and loaded onto a 15% polyacrylamide gel. The electrophoresis was performed in a mini-protean III dual slab cell (Bio-Rad Laboratories, Wartford, UK) at 30 mA constant amperage. Immediately after ending of the run, gels were placed at room temperature for

1 hr in a staining solution containing 40% methanol, 10% acetic acid, and 0.1% Coomassie Brilliant Blue G-250. Finally, the gels were destained by several washings in distilled water containing 40% methanol and 7% acetic acid. Densytometric analysis of displayed bands was performed by using ImageJ software [51].

### 4.5. Volatile Compounds Extraction and GC-MS Analysis

The extraction of volatile compounds from milk and cheese was performed by a solid-phase microextraction (SPME) and then analyzed using a gas-chromatograph coupled with a mass spectrophotometry (GC-MS) following the procedure previously described by Bennato et al. [52] with slight modifications. Grated cheese (4.5 g) was mixed with a saturated NaCl solution containing the internal standard, the 4-methyl-2-heptanone. The SPME was performed by exposing a 50/30 μm of divinylbenzene/carboxen/polydimethylsiloxane fiber (DVB/CAR/PDMS Supelco, Bellefonte, PA, USA) into the headspace of capped vials with a PTFE septum for 40 min at 50 °C in stirring conditions. Desorption of volatile compounds was obtained into the splitless injector of the GC system set at 250 °C for 1 min. The gas-chromatograph (Clarus 580; Perkin Elmer, Waltham, MA, USA) was coupled with a mass spectrometer (SQ8S; Perkin Elmer, Waltham, MA, USA) and equipped with a PE-ELITE-5MS 30 × 0.25 mm, 0.25 μm column (Perkin Elmer, Waltham, MA, USA). The oven temperature was set at 50 °C for 1 min, then was increased to 200 °C at 3 °C/min for 1 min, and to 250 °C at 15 °C/min, held for 15 min. The carrier gas was helium at 1 mL/min. Source and interface temperature were held at 250 °C. The mass detector operated in electronic impact mode (70 eV) and data were acquired in full scan mode (range 35–350 *m/z*, dwell time 0.2 s/scan). The volatile compounds were identified by comparing their mass spectra with those of the National Institute of Standards and Technology library (NIST, Gaithersburg, MD, USA) and comparing the eluting order with Kovats retention indexes. The isoamyl butyrate and isoamyl isobutyrate were identified comparing the mass spectra and the retention time of authentic standard compounds (Sigma Aldrich, St. Louis, MO, USA). Samples were analyzed in triplicate and quantification was carried out considering the relative peak area expressed as arbitrary unit (AU, target ion area $\times 10^{-3}$).

### 4.6. Statistical Analysis

The statistical analysis of data was carried out by using SAS software, version 9.2 (SAS Institute Inc., Cary, NY, USA). Data on volatile compounds have been processed with two-way ANOVA, considering diet, ripening time, and their interaction as fixed effects, according to the following statistical model: $y_{ijk} = \mu + D_i + T_j + D_i \times T_j + e_{ijk}$, in which $y_{ijk}$ = volatile compound, $\mu$ = population average, $D_i$ = effect of dietary supplementation (CG, ZG, SG and IG), $T_j$ = effect of ripening time ($T_7$ vs. $T_{120}$), $D_i \cdot T_j$ = interaction between dietary supplementation and ripening time, and $e_{ijk}$ = error. Means separation was assessed by Tukey's test and differences were declared significant at $p < 0.05$.

## 5. Conclusions

The results highlighted the possibility of fortification with zinc, selenium, and iodine in cheese through animal feeding. The increase in concentration of essential trace elements in dairy products, in addition to representing an advantage for the consumers health, undoubtedly influences the biochemical mechanisms that characterize cheese during aging, also contributing to the development of flavor. No variations were evidenced in the caseinolytic activity in raw milk, whereas in cheese, the electrophoretic analysis in denaturing and reducing conditions of cheese showed the ability of selenium to preserve αS1-CN from primary proteolysis during ripening. With regard to the evaluation of the proteolytic volatile profile, only three compounds, resulting from the degradation of phenylalanine and leucine, have been identified. In this case, further studies are needed to clearly understand the relationship between the dietary microelements supplementation and the characterization of the free amino acids pattern in fresh and ripened cheese, as well as the mechanisms involved in the production of proteolytic compounds.

**Supplementary Materials:** The following are available online at http://www.mdpi.com/1420-3049/25/9/2249/s1, Table S1: Chemical composition of fresh ($T_7$) and 120-days ripened ($T_{120}$) Caciocavallo cheese obtained from lactating dairy cows fed control diet (CG) and control diet supplemented with zinc (ZG), selenium (SG), and iodine (IG). Table S2: Ingredients and chemical composition of complete diets administered to lactating dairy cows belonging to control group (CG), zinc group (ZG), selenium group (SG), and iodine group (IG).

**Author Contributions:** Conceptualization, G.M.; methodology, A.I., F.B., C.M. and L.G.; investigation, A.I. and F.B.; resources, G.M. and N.F.; data curation, A.I., F.B. and C.M.; writing—original draft preparation, A.I.; writing—review and editing, F.B. and N.F.; supervision, G.M.; project administration, G.M.; and funding acquisition, G.M. All authors have read and agreed to the published version of the manuscript.

**Funding:** This research is part of the project "Innovazione della filiera bovina da latte in Abruzzo per produzioni lattiero-casearie ad elevato contenuto salutistico ed ecosostenibile" supported by a grant from Rural Development Plan 2007–2013, MISURA 1.2.4, Regione Abruzzo (Italy), project manager: Giuseppe Martino.

**Acknowledgments:** Authors are grateful to the company "ANSAPE—Terrantica" (Raiano (AQ), Italy) for their kind cooperation.

**Conflicts of Interest:** The authors declare no conflict of interest. The funders had no role in the design of the study; in the collection, analyses, or interpretation of data; in the writing of the manuscript, or in the decision to publish the results.

## References


1. Reffett, J.K.; Spears, J.W.; Brown Jr, T.T. Effect of dietary selenium on the primary and secondary immune response in calves challenged with infectious bovine rhinotracheitis virus. *J. Nutr.* **1988**, *118*, 229–235. [CrossRef] [PubMed]
2. Coleman, J.E. Zinc enzymes. *Curr. Opin. Chem. Biol.* **1998**, *2*, 222–234. [CrossRef]
3. Miller, W.J. Zinc nutrition of cattle: A review. *J. Dairy Sci.* **1970**, *53*, 1123–1135. [CrossRef]
4. Steinbrenner, H.; Speckmann, B.; Klotz, L.O. Selenoproteins: Antioxidant selenoenzymes and beyond. *Arch Biochem. Biophys.* **2016**, *595*, 113–119. [CrossRef]
5. Wang, C.; Liu, Q.; Yang, W.Z.; Dong, Q.; Yang, X.M.; He, D.C.; Zhang, P.; Dong, K.H.; Huang, Y.X. Effects of selenium yeast on rumen fermentation, lactation performance and feed digestibilities in lactating dairy cows. *Livestock Sci.* **2009**, *126*, 239–244. [CrossRef]
6. Flachowsky, G.; Franke, K.; Meyer, U.; Leiterer, M.; Scho, F. Influencing factors on iodine content of cow milk. *Eur. J. Nutr.* **2014**, *53*, 351–365. [CrossRef]
7. Iannaccone, M.; Elgendy, R.; Ianni, A.; Martino, C.; Palazzo, F.; Giantin, M.; Grotta, L.; Dacasto, M.; Martino, G. Whole-transcriptome profiling of sheep fed with a high iodine-supplemented diet. *Animal* **2019**, *14*, 745–752. [CrossRef]
8. Elgendy, R.; Palazzo, F.; Castellani, F.; Giantin, M.; Grotta, L.; Cerretani, L.; Dacasto, M.; Martino, G. Transcriptome profiling and functional analysis of sheep fed with high zinc-supplemented diet: A nutrigenomic approach. *Anim. Feed Sci. Technol.* **2017**, *234*, 195–204. [CrossRef]
9. Biscarini, F.; Palazzo, F.; Castellani, F.; Masetti, G.; Grotta, L.; Cichelli, A.; Martino, G. Rumen microbiome in dairy calves fed copper and grape-pomace dietary supplementations: Composition and predicted functional profile. *PLoS ONE* **2018**, *13*. [CrossRef]
10. Calamari, L.; Petrera, F.; Bertin, G. Effects of either sodium selenite or Se yeast (Sc CNCM I-3060) supplementation on selenium status and milk characteristics in dairy cows. *Livestock Sci.* **2010**, *128*, 154–165. [CrossRef]
11. Iannaccone, M.; Ianni, A.; Elgendy, R.; Martino, C.; Giantin, M.; Cerretani, L.; Dacasto, M.; Martino, G. Iodine Supplemented Diet Positively Affect Immune Response and Dairy Product Quality in Fresian Cow. *Animals* **2019**, *9*, 866. [CrossRef] [PubMed]
12. Schirone, M.; Tofalo, R.; Perpetuini, G.; Manetta, A.C.; Di Gianvito, P.; Tittarelli, F.; Battistelli, N.; Corsetti, A.; Suzzi, G.; Martino, G. Influence of iodine feeding on microbiological and physico-chemical characteristics and biogenic amines content in a raw ewes' milk cheese. *Foods* **2018**, *7*, 108. [CrossRef] [PubMed]
13. Van Kranenburg, R.; Kleerebezem, M.; Van Hylckama Vlieg, J.; Ursing, B.M.; Boekhorst, J.; Smit, B.A.; Ayad, E.H.E.; Smit, G.; Siezen, R.J. Flavour formation from amino acids by lactic acid bacteria: Predictions from genome sequence analysis. *Int. Dairy J.* **2002**, *12*, 111–121. [CrossRef]
14. McSweeney, P.L.H.; Sousa, M.J. Biochemical pathways for the production of flavour compounds in cheeses during ripening: A review. *Lait* **2000**, *80*, 293–324. [CrossRef]

15. Yvon, M.; Rijnen, L. Cheese flavor formation by amino acid catabolism. *Int. Dairy J.* **2001**, *11*, 185–201. [CrossRef]

16. Ianni, A.; Innosa, D.; Martino, C.; Grotta, L.; Bennato, F.; Martino, G. Zinc supplementation of Friesian cows: Effect on chemical-nutritional composition and aromatic profile of dairy products. *J. Dairy Sci.* **2019**, *102*, 2918–2927. [CrossRef]

17. Ling, K.; Henno, M.; Jõudu, I.; Püssa, T.; Jaakson, H.; Kass, M.; Anton, D.; Ots, M. Selenium supplementation of diets of dairy cows to produce Se-enriched cheese. *Int. Dairy. J.* **2017**, *71*, 76–81. [CrossRef]

18. Pechová, A.; Pavlata, L.; Lokajova, E. Zinc supplementation and somatic cell count in milk of dairy cows. *Acta Vet. Brno* **2006**, *75*, 355–361. [CrossRef]

19. Moschini, M.; Battaglia, M.; Beone, G.M.; Piva, G.; Masoero, F. Iodine and selenium carry over in milk and cheese in dairy cows: Effect of diet supplementation and milk yield. *Animal* **2010**, *4*, 147–155. [CrossRef]

20. Meshref, A.M.; Moselhy, W.A.; Hassan, N.E.H.Y. Heavy metals and trace elements levels in milk and milk products. *J. Food Meas. Charact.* **2014**, *8*, 381–388. [CrossRef]

21. Nagase, H.; Visse, R.; Murphy, G. Structure and function of matrix metalloproteinases and TIMPs. *Cardiovasc. Res.* **2006**, *69*, 562–573. [CrossRef] [PubMed]

22. Manicourt, D.H.; Lefebvre, V. An assay for matrix metalloproteinases and other proteases acting on proteoglycans, casein, or gelatin. *Anal. Biochem.* **1993**, *215*, 171–179. [CrossRef] [PubMed]

23. Caggiano, N.; Smirnoff, A.L.; Bottini, J.M.; De Simone, E.A. Protease activity and protein profile in milk from healthy dairy cows and cows with different types of mastitis. *Int. Dairy J.* **2019**, *89*, 1–5. [CrossRef]

24. Santos, M.C.L.G.; De Souza, A.P.; Gerlach, R.F.; Trevilatto, P.C.; Scarel-Caminaga, R.M.; Line, S.R.P. Inhibition of human pulpal gelatinases (MMP-2 and MMP-9) by zinc oxide cements. *J. Oral Rehabil.* **2004**, *31*, 660–664. [CrossRef]

25. Costa, F.F.; Brito, M.A.V.P.; Furtado, M.A.M.; Martins, M.F.; De Oliveira, M.A.L.; De Castro Barra, P.M.; Lourdes, A.G.; Dos Santos, A.S.D.O. Microfluidic chip electrophoresis investigation of major milk proteins: Study of buffer effects and quantitative approaching. *Anal. Met.* **2014**, *6*, 1666–1673. [CrossRef]

26. Trujillo, A.J.; Guamis, B.; Carretero, C. Hydrolysis of Bovine and Caprine Caseins by Rennet and Plasmin in Model Systems. *J. Agric. Food Chem.* **1998**, *46*, 3066–3072. [CrossRef]

27. Van Hekken, D.L.; Tunick, M.H.; Soryal, K.A.; Zeng, S.S. Proteolytic and Rheological Properties of Aging Cheddar-Like Caprine Milk Cheeses Manufactured at Different Times during Lactation. *J. Food Sci.* **2007**, *72*, E115–E120. [CrossRef]

28. Zeng, H.; Combs, G.F., Jr. Selenium as an anticancer nutrient: Roles in cell proliferation and tumor cell invasion. *J. Nutr. Biochem.* **2008**, *19*, 1–7. [CrossRef]

29. Esmaili, M.; Ghaffari, S.M.; Moosavi-Movahedi, Z.; Atri, M.S.; Sharifizadeh, A.; Farhadi, M.; Yousefi, R.; Chobert, J.M.; Haertlé, T.; Moosavi-Movahedi, A.A. Beta casein-micelle as a nano vehicle for solubility enhancement of curcumin; food industry application. *LWT—Food Sci. Technol.* **2011**, *44*, 2166–2172. [CrossRef]

30. Ianni, A.; Bennato, F.; Martino, C.; Grotta, L.; Martino, G. Volatile Flavor Compounds in Cheese as Affected by Ruminant Diet. *Molecules* **2020**, *25*, 461. [CrossRef]

31. Suriyaphan, O.; Drake, M.; Chen, X.Q.; Cadwallader, K.R. Characteristic aroma components of British farmhouse Cheddar cheese. *J. Agric. Food Chem.* **2001**, *49*, 1382–1387. [CrossRef] [PubMed]

32. Smit, B.A.; Engels, W.J.M.; Wouters, J.T.M.; Smit, G. Diversity of L-leucine catabolism in various microorganisms involved in dairy fermentations, and identification of the rate-controlling step in the formation of the potent flavour component 3-methylbutanal. *Appl. Microbiol. Biotechnol.* **2004**, *64*, 396–402. [CrossRef] [PubMed]

33. Thierry, A.; Maillard, M.B. Production of cheese flavour compounds derived from amino acid catabolism by *Propionibacterium freudenreichii. Lait* **2002**, *82*, 17–32. [CrossRef]

34. Ianni, A.; Martino, C.; Pomilio, F.; Di Luca, A.; Martino, G. Dietary selenium intake in lactating dairy cows modifies fatty acid composition and volatile profile of milk and 30-day-ripened caciotta cheese. *Eur. Food Res. Technol.* **2019**, *245*, 2113–2121. [CrossRef]

35. Martino, C.; Ianni, A.; Grotta, L.; Pomilio, F.; Martino, G. Influence of Zinc Feeding on Nutritional Quality, Oxidative Stability and Volatile Profile of Fresh and Ripened Ewes' Milk Cheese. *Foods* **2019**, *8*, 656. [CrossRef]

36. Ianni, A.; Iannaccone, M.; Martino, C.; Innosa, D.; Grotta, L.; Bennato, F.; Martino, G. Zinc supplementation of dairy cows: Effects on chemical composition, nutritional quality and volatile profile of Giuncata cheese. *Int. Dairy J.* **2019**, *94*, 65–71. [CrossRef]

37. European Union (2010) Directive 2010/63/EU of the European parliament and of the council of 22 September 2010 on the protection of animals used for scientific purposes. Available online: https://eur-lex.europa.eu/legal-content/EN/TXT/PDF/?uri=CELEX:32010L0063&from=EN (accessed on 25 March 2020).

38. European Economic Community (1986) Council Directive 86/609/EEC of 24 November 1986 on the approximation of laws, regulations and administrative provisions of the Member States regarding the protection of animals used for experimental and other scientific purposes. Available online: https://eur-lex.europa.eu/legal-content/EN/TXT/PDF/?uri=CELEX:31986L0609&from=EN (accessed on 25 March 2020).

39. European Commission (2003) Regulation (EC) No. 1831/2003 of the European Parliament and of the Council of 22 September 2003 on Additives for Use in Animal Nutrition. Available online: https://eur-lex.europa.eu/legal-content/EN/TXT/PDF/?uri=CELEX:32003R1831&from=EN (accessed on 25 March 2020).

40. European Commission (2014) Regulation No. 121/2014 of the concerning the authorization of L-selenomethionine as a feed additive for all animal species. Available online: https://eur-lex.europa.eu/legal-content/EN/TXT/PDF/?uri=CELEX:32014R0121&from=EN (accessed on 25 March 2020).

41. European Commission (2005) Regulation No. 1459/2005 amending the conditions for authorisation of a number of feed additives belonging to the group of trace elements. Available online: https://eur-lex.europa.eu/LexUriServ/LexUriServ.do?uri=OJ:L:2005:233:0008:0010:EN:PDF (accessed on 25 March 2020).

42. NRC. *Nutrient Requirements of Dairy Cattle*; Natl Acad Press: Washington, DC, USA, 2001.

43. AOAC. *Official Methods of Analysis*, 15th ed.; Association of Official Analytical Chemists: Arlington, VA, USA, 1990; Volume 1.

44. Van Soest, P.J.; Robertson, J.B.; Lewis, B.A. Methods for dietary fiber, neutral detergent fiber, and nonstarch polysaccharides in relation to animal nutrition. *J. Dairy Sci.* **1991**, *74*, 3583–3597. [CrossRef]

45. Ianni, A.; Bennato, F.; Martino, C.; Innosa, D.; Grotta, L.; Martino, G. Effects of selenium supplementation on chemical composition and aromatic profiles of cow milk and its derived cheese. *J. Dairy Sci.* **2019**, *102*, 6853–6862. [CrossRef]

46. Nascentes, C.C.; Arruda, M.A.Z.; Nogueira, A.R.A.; Nóbrega, J.A. Direct determination of Cu and Zn in fruit juices and bovine milk by thermospray flame furnace atomic absorption spectrometry. *Talanta* **2004**, *64*, 912–917. [CrossRef]

47. Gerber, N.; Brogioli, R.; Hattendorf, B.; Scheeder, M.R.L.; Wenk, C.; Günther, D. Variability of selected trace elements of different meat cuts determined by ICP-MS and DRC-ICPMS. *Animal* **2009**, *3*, 166–172. [CrossRef]

48. Fecher, P.; Goldmann, I.; Nagengast, A. Determination of iodine in food samples by inductively coupled plasma mass spectrometry after alkaline extraction. *J. Anal. At. Spectrom.* **1998**, *13*, 977–982. [CrossRef]

49. Kruger, N.J. The Bradford method for protein quantitation. In *The Protein Protocols Handbook*; Humana Press: Totowa, NJ, USA, 2009; pp. 17–24.

50. Ianni, A.; Celenza, G.; Franceschini, N. Oxaprozin: A new hope in the modulation of matrix metalloproteinase 9 activity. *Chem. Biol. Drug Des.* **2019**, *93*, 811–817. [CrossRef] [PubMed]

51. Rasband, W.S. *ImageJ Software*; National Institutes of Health: Bethesda, MD, USA, 1997.

52. Bennato, F.; Ianni, A.; Innosa, D.; Grotta, L.; D'Onofrio, A.; Martino, G. Chemical-nutritional characteristics and aromatic profile of milk and related dairy products obtained from goats fed with extruded linseed. *Asian-Australas. J. Anim. Sci.* **2020**, *33*, 148. [CrossRef] [PubMed]

**Sample Availability:** Samples analysed during the current study are available from the corresponding author on reasonable request.

*Article*

# Volatile Profile in Yogurt Obtained from Saanen Goats Fed with Olive Leaves

Francesca Bennato [1,†], Denise Innosa [1,†], Andrea Ianni [1], Camillo Martino [2], Lisa Grotta [1] and Giuseppe Martino [1,*]

[1] Faculty of BioScience and Technology for Food, Agriculture and Environment, University of Teramo, Via Renato Balzarini 1, 64100 Teramo, Italy; fbennato@unite.it (F.B.); dinnosa@unite.it (D.I.); aianni@unite.it (A.I.); lgrotta@unite.it (L.G.)

[2] Istituto Zooprofilattico Sperimentale dell'Abruzzo e del Molise "G. Caporale", Via Campo Boario 37, 64100 Teramo, Italy; c.martino@izs.it

* Correspondence: gmartino@unite.it

† These authors contributed equally to this work.

Academic Editor: Eugenio Aprea

Received: 24 April 2020; Accepted: 13 May 2020; Published: 14 May 2020

**Abstract:** The aim of this study was to evaluate the development of volatile compounds in yogurt samples obtained from goats fed a dietary supplementation with olive leaves (OL). For this purpose, thirty Saanen goats were divided into two homogeneous groups of 15 goats each: a control group that received a standard diet (CG) and an experimental group whose diet was supplemented with olive leaves (OLG). The trial lasted 28 days, at the end of which the milk of each group was collected and used for yogurt production. Immediately after production, and after 7 days of storage at 4 °C in the absence of light, the yogurt samples were characterized in terms of fatty acid profile, oxidative stability and volatile compounds by the solid-phase microextraction (SPME)–GC/MS technique. Dietary OL supplementation positively affected the fatty acid composition, inducing a significant increase in the relative proportion of unsaturated fatty acids, mainly oleic acid (C18:1 *cis*9) and linolenic acid (C18:3). With regard to the volatile profile, both in fresh and yogurt samples stored for 7 days, the OL supplementation induced an increase in free fatty acids, probably due to an increase in lipolysis carried out by microbial and endogenous milk enzymes. Specifically, the largest variations were found for C6, C7, C8 and C10 free fatty acids. In the same samples, a significant decrease in aldehydes, mainly heptanal and nonanal, was also detected, supporting—at least in part—an improvement in the oxidative stability. Moreover, alcohols, esters and ketones appeared lower in OLG samples, while no significant variations were observed for lactones. These findings suggest the positive role of dietary OL supplementation in the production of goats' milk yogurt, with characteristics potentially indicative of an improvement in nutritional properties and flavor.

**Keywords:** goats' milk yogurt; Saanen goat; olive leaves; unsaturated fatty acid; volatile compound; free fatty acid; aldehyde

---

## 1. Introduction

The flavor development in dairy products is an important factor determining its acceptability and preference and is strongly affected by the combination of a wide range of compounds, mostly produced by a series of biochemical events involving the metabolism of residual lactose, lactate, and citrate, lipolysis and proteolysis. Among the mentioned catabolic processes, lipolysis is undoubtedly the mechanism that, more than others, contributes to the release of flavor-affecting compounds, mainly free fatty acids (FFAs), aldehydes, alcohols, esters, methyl ketones, γ- and δ-lactones. Milk and cheese are, in fact, characterized by relevant levels of short- and medium-chain fatty acids, whose specific composition is strongly influenced by the animal species and the administered dietary treatment [1].

In recent years, an increase in consumer demand for goats' milk and its derivatives, due to the high nutritional features and health-promoting benefits of these products, has been observed. Goat dairy products were, in fact, reported to be characterized by proteins with lower allergenicity in comparison to bovine products, and higher concentrations of bioactive compounds, attributed to their greater digestibility [2,3]. Specifically, a growing interest has been developed towards yogurt, commonly considered a safe and nutritious food and not only being rich in vitamins and minerals, but also in calcium and proteins [4]. Furthermore, it was demonstrated that yogurt consumption is helpful for consumers affected by specific gastrointestinal conditions, mainly those associated with lactose intolerance and constipation [5].

In the past, several studies were conducted on the inclusion of different natural sources of bioactive compounds in goats' milk yogurt in order both to improve nutritional value and reduce the unpleasant "goaty" aroma and aftertaste, commonly associated with decreased acceptance by consumers [6,7]. In time, studies performed in the sector of ruminant husbandry led to the development of feeding strategies able to influence the chemical composition of animal products. In this regard, several experiments have been performed by supplementing animals' diets with plant matrices, especially agro-industrial byproducts rich in bioactive compounds with interesting functions from a biochemical point of view. These strategies have shown positive effects on several fronts, preserving, in many cases, animal welfare [8], improving the nutritional quality of animal products [9,10], and inducing, especially in dairy products, the development of volatile compounds capable of influencing the aroma and taste [11].

Byproducts derived from olive oil production are accumulated yearly in high amounts in the Mediterranean area, and their disposal represents an issue of great importance both from an environmental and economic point of view. Several studies focused their attention on the valorization of these byproducts as feeding supplements for farm animals, both ruminant and monogastric [12,13]. Specifically, olive leaves (OL) have been reported to be a cheap raw material that can be used as a source of phenolic bioactive compounds with antioxidant, antihypertensive and anti-inflammatory functions, such as oleuropein, hydroxytyrosol, verbascoside, apigenin-7-glucoside and luteolin-7-glucoside [14].

Considering the biological relevance of this byproduct, the purpose of this work is to evaluate the effects of dietary OL supplementation of lactating dairy goats on the development of volatile compounds in fresh and stored yogurt samples, with specific attention paid to compounds derived from lipolytic events.

## 2. Results

### 2.1. Chemical Properties of Yogurt Samples

The evaluation of chemical properties in yogurt samples did not evidence significant variations between the control group that received a standard diet (CG) and the experimental group whose diet was supplemented with olive leaves (OLG) in relation to moisture ($85.74 \pm 0.26\%$ vs. $85.17 \pm 0.49\%$ in CG and OLG respectively, $p > 0.05$) and the amount of total lipids ($19.81 \pm 1.22\%$ vs. $19.68 \pm 1.35\%$ in CG and OLG respectively, $p > 0.05$). The only significant difference was observed for pH values that decreased in yogurt obtained from goats fed the dietary OL supplementation ($4.69 \pm 0.02$ vs. $4.43 \pm 0.04$ in CG and OLG samples respectively, $p < 0.05$).

### 2.2. Fatty Acid Composition

The dietary OL supplementation in the goats' diet induced several modifications in the fatty acid composition of yogurt samples. As reported in Table 1, in OLG samples, we observed a significant increase in the relative proportion of total monounsaturated fatty acids (MUFA; $22.25 \pm 0.92\%$ vs. $24.80 \pm 0.77\%$ for CG and OLG respectively, $p < 0.05$) and total polyunsaturated fatty acids (PUFA; $4.62 \pm$

0.14% and 4.84 ± 0.07% for CG and OLG respectively, $p < 0.05$). With specific regard to the sum of the saturated fatty acids (SFA), a slight decrease was instead observed in OLG samples, although this variation was not significant ($p > 0.05$).

**Table 1.** Fatty acid composition of yogurt samples obtained from goats fed a standard diet (CG) and goats fed a dietary supplementation of olive leaves (OLG).

| Fatty Acids [1] | CG | OLG | *p*-Value |
|---|---|---|---|
| C4:0 | 0.54 ± 0.22 | 0.83 ± 0.39 | ns |
| C6:0 | 1.03 ± 0.26 | 1.31 ± 0.39 | ns |
| C8:0 | 1.73 ± 0.29 | 1.92 ± 0.34 | ns |
| C10:0 | 7.65 ± 0.69 | 7.33 ± 0.58 | ns |
| C12:0 | 4.31 ± 0.21 | 3.48 ± 0.12 | * |
| C14:0 | 11.72 ± 0.44 | 10.75 ± 0.33 | * |
| C15:0 | 0.93 ± 0.07 | 0.92 ± 0.03 | ns |
| C16:0 | 29.58 ± 0.80 | 26.67 ± 0.66 | * |
| C17:0 | 0.67 ± 0.02 | 0.66 ± 0.01 | ns |
| C18:0 | 11.04 ± 0.42 | 13.10 ± 0.74 | ** |
| C20:0 | 0.27 ± 0.03 | 0.29 ± 0.01 | ns |
| C22:0 | 0.09 ± 0.03 | 0.09 ± 0.01 | ns |
| total SFA | 69.54 ± 5.80 | 67.35 ± 5.61 | ns |
| C14:1 | 0.42 ± 0.02 | 0.43 ± 0.01 | ns |
| C16:1 | 0.33 ± 0.01 | 0.34 ± 0.01 | ns |
| C18:1 *trans*11 | 0.43 ± 0.01 | 0.52 ± 0.15 | * |
| C18:1 *cis*9 | 20.89 ± 0.59 | 23.22 ± 0.61 | ** |
| C18:1 *cis*11 | 0.38 ± 0.02 | 0.30 ± 0.01 | ns |
| total MUFA | 22.25 ± 0.92 | 24.80 ± 0.77 | * |
| C18:2 | 2.92 ± 0.18 | 2.60 ± 0.14 | * |
| CLA | 0.89 ± 0.07 | 1.12 ± 0.09 | * |
| C18:3 | 0.78 ± 0.02 | 1.13 ± 0.05 | ** |
| total PUFA | 4.62 ± 0.14 | 4.84 ± 0.07 | * |
| other FAs | 3.20 ± 0.31 | 3.01 ± 0.30 | ns |
| MUFA/SFA | 0.32 ± 0.02 | 0.37 ± 0.03 | * |
| PUFA/SFA | 0.07 ± 0.01 | 0.07 ± 0.02 | ns |
| UFA/SFA | 0.39 ± 0.02 | 0.44 ± 0.04 | * |
| DI C14:1 *cis*-9/(C14:0+C14:1*cis*-9) | 0.04 ± 0.01 | 0.04 ± 0.01 | ns |
| DI C16:1*cis*-9/(C16:0+C16:1*cis*-9) | 0.01 ± 0.01 | 0.01 ± 0.01 | ns |
| DI C18:1*cis*-9/(C18:0+C18:1*cis*-9) | 0.65 ± 0.01 | 0.64 ± 0.01 | ns |
| DI CLA/(C18:1*trans*-11+CLA) | 0.62 ± 0.09 | 0.69 ± 0.02 | ns |

[1] Data are reported as mean percentage of total fatty acid methyl esters ± S.D. Saturated fatty acid (SFA); monounsaturated fatty acid (MUFA); polyunsaturated fatty acid (PUFA); conjugated linoleic acids (CLA); desaturation index (DI); not significant (ns); * $p < 0.05$; ** $p < 0.01$.

With reference to individual fatty acids, the OLG yogurt samples were characterized by a decrease in the relative proportion of lauric acid (C12:0, $p < 0.05$), myristic acid (C14:0, $p < 0.05$), palmitic acid (C16:0, $p < 0.05$), and linoleic acid (C18:2, $p < 0.05$), while a significant increase in the relative proportion was observed for stearic acid (C18:0, $p < 0.01$), oleic acid (C18:1 *cis*-9, $p < 0.01$), linolenic acid (C18:3, $p < 0.01$), and conjugated linoleic acids (CLA, $p < 0.05$). In addition to this, in OLG samples, the ratio MUFA/SFA and UFA/SFA was significantly higher ($p < 0.05$), whereas no variations were evidenced in the individual desaturation indices calculated for C14:1, C16:1, C18:1 and CLA.

*2.3. Evaluation of the Oxidative Stability in Fresh and Stored Yogurt*

The evaluation of the overall antioxidant potential in fresh yogurt highlighted a higher antioxidant capacity in OLG samples (79.69 ± 4.76 μmol TEAC/g and 89.77 ± 5.03 μmol TEAC/g for CG and OLG respectively, $p < 0.05$).

The oxidative process on the lipid component was instead determined by a Thiobarbituric Acid Reactive Species (TBARS) test, both after 1 day (T1) and 7 days (T7) of storage at +4 °C and the results are shown in Figure 1. In T1 samples, the amount of malondialdehyde (MDA) was lower in OLG samples compared with CG samples ($p < 0.01$). In T7 samples, the same trend was observed, although a marked increase in the difference between CG and OLG samples was found ($p < 0.01$). By comparing the obtained results on the basis of the storage time (T1 vs. T7), a significant increase in oxidation was evidenced in CG yogurt ($p < 0.01$), while no variations were induced by the OL dietary enrichment.

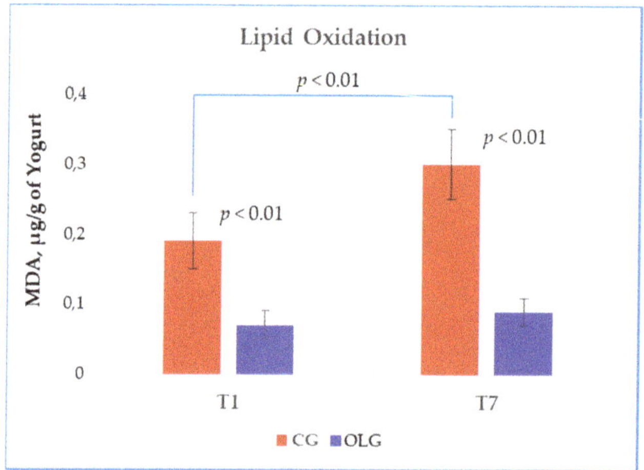


**Figure 1.** Lipid oxidation evaluated in yogurt samples obtained from goats fed the standard diet (CG) and goats that received the dietary supplementation with olive leaves (OLG). Samples were analyzed after 1 day (T1) and 7 days of storage (T7) at 4 °C. Malondialdehyde (MDA).

*2.4. Identification of Volatile Compounds*

In all samples, we found 26 volatile compounds (VOCs): seven free fatty acids (FFAs), five alcohols, five aldehydes, one ester, six ketones and two lactones. As schematized in Figure 2, both after 1 and 7 days of storage, the dietary OL supplementation was effective in inducing an increase in the relative proportion of FFAs ($p < 0.05$ and $p < 0.01$, for T1 and T7 respectively) and a significant decrease in the relative proportion of alcohols ($p < 0.05$ and $p < 0.01$, for T1 and T7 respectively), aldehydes ($p < 0.01$), esters ($p < 0.05$), and ketones ($p < 0.05$).

Table 2 reports the specific compositions of the listed VOC families. After 1 day after yogurt production, all the alcohols found were lower in OLG samples in comparison with CG samples: 1-hexanol ($p < 0.01$), 2-ethyl-hexan-1-ol, ($p < 0.05$), 1-heptanol ($p < 0.001$), 1-octanol ($p < 0.01$), and 1-nonanol ($p < 0.001$). After 7 days of storage, such a condition persisted, with the only exception being 1-hexanol, 2-ethyl, for which no significant variations were evidenced. With regard to the aldehydes, only nonanal was significantly lower in T1 samples derived from the dietary OL supplementation ($p < 0.01$), while, at T7, this behavior was also observed for heptanal ($p < 0.05$). Among the FFAs, the OLG samples stored for 1 day after the yogurt production were characterized by higher relative proportions of octanoic acid ($p < 0.05$) and decanoic acid ($p < 0.05$). The finding concerning the octanoic acid was also confirmed after 7 days of yogurt storage ($p < 0.01$) and, in addition to this, the same samples also showed a decrease in the relative proportions of hexanoic acid ($p < 0.05$) and heptanoic acid ($p < 0.01$). Concerning ketones, both after 1 and 7 days of storage, we observed lower relative proportions for 2-heptanone ($p < 0.05$) and 2-nonanone ($p < 0.01$). Butyl heptanoate was the only ester to be detected and was less represented in OLG yogurt compared to CG, both at T1 and T7 ($p < 0.01$). Finally, no significant variations were evidenced for the two identified lactones: δ-decalactone and δ-dodecalactone ($p > 0.05$).

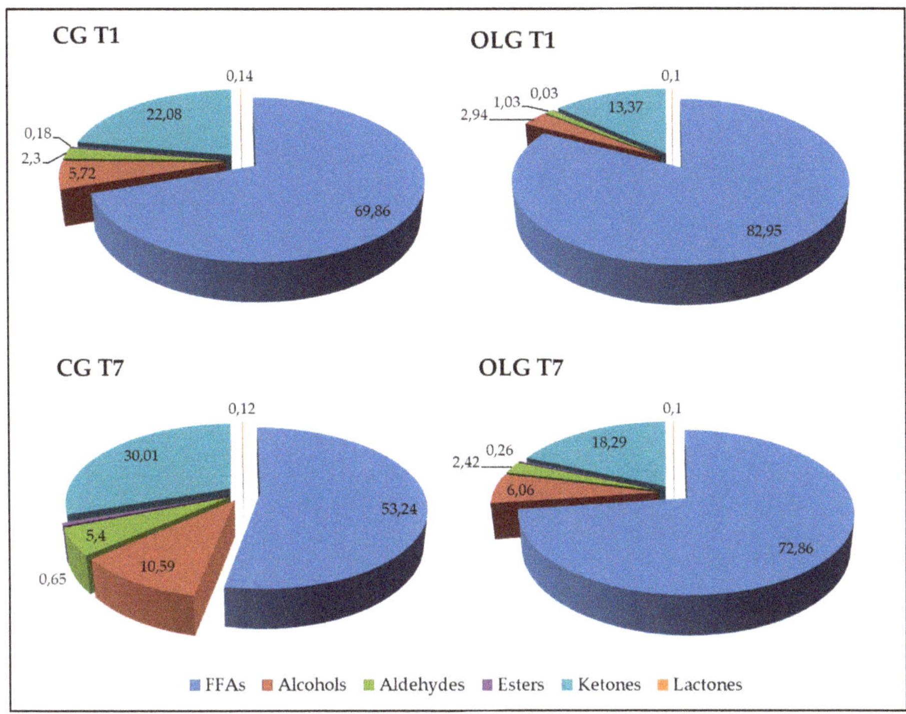


**Figure 2.** Schematic representation of relative proportions (%) of volatile compound families detected in yogurt samples obtained from goats fed the standard diet (CG) and goat that received the dietary supplementation with olive leaves (OLG). Samples were analyzed after 1 day (T1) and 7 days of storage (T7) at 4 °C. Free fatty acids (FFAs).

**Table 2.** Volatile compounds (VOC) identified in yogurt samples obtained from goats fed a standard diet (CG) and goats fed a dietary supplementation of olive leaves (OLG). Samples were analyzed after 1 day (T1) and 7 days of storage (T7) at 4 °C.

| VOC [1] | T1 | | | T7 | | |
|---|---|---|---|---|---|---|
| | CG | OLG | $p$ | CG | OLG | $p$ |
| **FFAs** | | | | | | |
| Acetic acid | 0.26 ± 0.14 | 0.40 ± 0.15 | ns | nd | 0.48 ± 0.20 | ns |
| Butanoic acid | 0.27 ± 0.23 | 0.16 ± 0.06 | ns | 0.10 ± 0.03 | 0.11 ± 0.03 | ns |
| Hexanoic acid | 18.03 ± 1.69 | 20.80 ± 2.30 | ns | 15.39 ± 0.99 | 13.94 ± 0.74 | * |
| Heptanoic acid | 0.56 ± 0.05 | 0.60 ± 0.06 | ns | 0.62 ± 0.02 | 0.43 ± 0.01 | ** |
| Octanoic acid | 38.91 ± 3.31 | 45.70 ± 4.28 | * | 22.99 ± 2.61 | 43.79 ± 4.46 | ** |
| Nonanoic acid | 0.29 ± 0.14 | 0.18 ± 0.03 | ns | 0.11 ± 0.03 | 0.14 ± 0.04 | ns |
| Decanoic acid | 11.54 ± 0.64 | 15.11 ± 1.50 | * | 14.03 ± 2.77 | 13.97 ± 2.26 | ns |

**Table 2.** *Cont.*

| VOC [1] | T1 | | | T7 | | |
|---|---|---|---|---|---|---|
| | CG | OLG | *p* | CG | OLG | *p* |
| **Alcohols** | | | | | | |
| 1-Hexanol | 1.80 ± 0.13 | 0.88 ± 0.06 | ** | 3.88 ± 0.71 | 1.72 ± 0.01 | * |
| 2-Ethyl-hexan-1-ol | 1.76 ± 0.14 | 1.13 ± 0.11 | * | 3.41 ± 0.25 | 2.45 ± 0.19 | ns |
| 1-Heptanol | 0.95 ± 0.13 | 0.24 ± 0.03 | ** | 1.06 ± 0.18 | 0.67 ± 0.09 | * |
| 1-Octanol | 0.39 ± 0.04 | 0.17 ± 0.02 | ** | 0.92 ± 0.03 | 0.39 ± 0.01 | ** |
| 1-Nonanol | 0.81 ± 0.05 | 0.52 ± 0.02 | ** | 1.32 ± 0.11 | 0.83 ± 0.10 | ** |
| **Aldehydes** | | | | | | |
| Heptanal | 0.05 ± 0.03 | 0.02 ± 0.01 | ns | 0.20 ± 0.10 | 0.04 ± 0.02 | * |
| Nonanal | 1.84 ± 0.17 | 0.64 ± 0.05 | ** | 4.53 ± 0.25 | 2.06 ± 0.10 | * |
| 2-Heptenal | 0.09 ± 0.01 | 0.08 ± 0.04 | ns | 0.09 ± 0.01 | 0.02 ± 0.01 | ns |
| 2-Octenal | 0.12 ± 0.02 | 0.07 ± 0.02 | ns | 0.28 ± 0.01 | 0.11 ± 0.03 | ns |
| 2-Decenal | 0.20 ± 0.01 | 0.22 ± 0.02 | ns | 0.30 ± 0.04 | 0.19 ± 0.03 | ns |
| **Esters** | | | | | | |
| Butyl heptanoate | 0.18 ± 0.01 | 0.03 ± 0.01 | ** | 0.65 ± 0.05 | 0.26 ± 0.03 | ** |
| **Ketones** | | | | | | |
| Acetoin | 1.99 ± 0.08 | 3.20 ± 0.31 | ** | 3.22 ± 1.59 | 2.10 ± 0.39 | ns |
| 2-Pentanone | 0.10 ± 0.01 | 0.04 ± 0.01 | ns | 0.13 ± 0.02 | 0.59 ± 0.02 | ns |
| 2,3-Pentanedione | 0.06 ± 0.02 | 0.04 ± 0.01 | ns | 0.67 ± 0.01 | 0.03 ± 0.05 | ns |
| 2-Heptanone | 9.30 ± 3.11 | 4.43 ± 1.02 | * | 11.32 ± 2.40 | 6.81 ± 1.11 | * |
| 2-Nonanone | 9.25 ± 0.74 | 4.68 ± 0.41 | ** | 12.73 ± 0.99 | 7.38 ± 0.53 | ** |
| 2-Undecanone | 1.38 ± 0.25 | 0.98 ± 0.17 | ns | 1.94 ± 0.46 | 1.38 ± 0.20 | ns |
| **Lactones** | | | | | | |
| δ-Decalactone | 0.09 ± 0.01 | 0.03 ± 0.02 | ns | 0.05 ± 0.03 | 0.07 ±0.05 | ns |
| δ-Dodecalactone | 0.05 ± 0.01 | 0.07 ± 0.01 | ns | 0.07 ± 0.01 | 0.03 ± 0.01 | ns |

[1] Data are reported as mean percentage of total volatile compounds (VOCs) ± S.D; Free fatty acids (FFAs); not detectable (nd); not significant (ns); * $p < 0.05$; ** $p < 0.01$.

## 3. Discussion

Flavor is an important factor determining the acceptability of food products. With specific regard to dairy products, the sensory properties are largely dependent on the development of volatile flavor compounds mainly derived from biochemical processes that degrade fat and proteins present in milk. Considerable knowledge has been accumulated on the wide range of aromatic compounds contributing to the development of flavor in yogurt. Most of these compounds include VOCs already present in milk and compounds derived from fermentation processes [15]. Dietary OL supplementation was able to influence VOC development in goats' milk yogurt, with variations particularly evident in the production of lipolytic catabolites.

The experimental feeding strategy was, first of all, able to induce significant changes in the fatty acid profile of yogurt samples. The relationship established between fat and flavor development in dairy products is complex; however, it should be considered that fat is a rich reservoir of flavors, as a consequence of the tendency of many aromatic compounds to be soluble in fat rather than in water [16]. Furthermore, the proportion and structure of fat may affect the rheology of the product, and thus its dispersion in the mouth. It is also well known that lipids present in food represent precursors of several flavor compounds, since may undergo spontaneous oxidation or enzymatic hydrolysis, which are essential for flavor development [17]. It is therefore plausible that variations at the substrate level could have influenced the extent of these biochemical mechanisms, thus influencing VOC production. The most relevant variation was associated to the increase in the relative proportion of MUFAs and PUFAs in OLG samples. Since the calculated desaturation indices (for C14:1, C16:1, C18:1 *cis*-9 and CLA) did not show significant differences, it is presumable that the changes observed in the fatty acid composition are totally ascribable to the different diets administered to goats. Variations found in the calculation of these desaturation indices are, in fact, generally attributed to a greater expression or activity of stearoyl-CoA desaturase (SCD), also known as $\Delta^9$-desaturase [18]. This enzyme plays a leading role in the lipid metabolism of the mammary gland, due to its ability to catalyze the addition of a double bond in the cis-$\Delta^9$-position in a large spectrum of medium- and long-chain fatty acids [19]. The substrates with which this enzyme interacts with greater affinity are precisely the acyl-CoA of C14, C16, C18 and trans-11 C18:1, which are respectively converted into C14:1, C16:1, C18:1 cis-9, and C18:2 *cis*-9 *trans*-11 [20].

The presence in the OLG samples of lower relative proportions of SFA lends itself to be discussed from different points of view. First of all, the dietary intake of SFA for humans is notoriously associated with an increased risk of developing cardiovascular diseases [21], and the importance of limiting the concentration of these compounds in foods acquires particular relevance in ruminant products because of the biohydrogenation mechanisms that are responsible for dietary PUFA conversion into SFA or MUFA [22]. Additionally, it must be specified that the improvement in the health indices of a food product is more specifically associated with the increase in omega-3 fatty acids and, over time, different feeding strategies have been tested in the zootechnical field, with the aim of obtaining the enrichment of such compounds in goats' milk and its derived dairy products [23]. Therefore, in this study, the increase in the relative proportion of linolenic acid (C18:3) observed in yogurt samples obtained from goat fed the OL supplementation acquires particular value. This finding is consistent with what was recently evidenced in a similar study focused on the evaluation of nutritional quality of Ricotta cheese made from goats' milk [24], and can be fully justified by the fact that C18:3 has been reported to be the major fatty acid in OL [25].

In addition to what has been reported, it must be considered that the increase in the relative proportion of unsaturated forms exposes these food products to greater susceptibility towards oxidative processes, which take place in most cases as an effect of the action of reactive species that are able to interact with C=C double bonds, producing peroxides [26]. The PUFA tendency to undergo oxidation is an aspect of remarkable importance for the food industry, since high concentrations of these compounds can induce detrimental effects on food nutritional quality and may represent a cause of concern for food safety [27]. Despite what has been reported, it should be said that, in the present study, the antioxidant potential in OLG yogurt samples was higher than in the CG samples—a finding also consistent with the results obtained by the TBARS test, which showed a greater resistance to lipid peroxidation in yogurt obtained from goats fed the dietary OL supplementation. The enrichment of ruminants' diets with plant matrices rich in compounds of high value from a biological point of view is generally associated with an improvement in the oxidative stability of animal productions [28,29]. Benavente-Garcia et al. [30] performed an accurate qualitative and quantitative characterization of the phenolic compounds present in OL extracts, assigning each compound its own antioxidant potential. Oleuropein was indicated as the most represented phenolic compound in the extracts. However, the greatest antioxidant function was attributed to hydroxytyrosol which is derived from the oleuropein hydrolysis. There is not much information on the bioavailability of these compounds in ruminants and, above all, there is a lack of information on their possible transfer to milk in the form of secondary metabolites able to act as antioxidants. However, even other studies conducted in the past on dairy cows and goats have highlighted the better oxidative stability of dairy products obtained by feeding animals with olive oil byproducts [13,24].

With regard to the volatile profile, numerous studies have been conducted in recent years, which highlight the diet's role in influencing the production of volatile flavor compounds in ruminant products [11]. Most of the compounds identified in all samples belong to the FFA group, which is testament to the prevalence of lipolytic processes compared to other catabolic mechanisms. Both in the freshly produced yogurt (T1) and in the one kept at 4 °C for 7 days (T7), carboxylic acids are present in higher relative proportions than in the OLG samples. In T1, this finding is mainly due to the greater presence of octanoic and decanoic acids. Both of these compounds have been reported to contribute to the characteristic animal-like, rancid and "soapy" flavor notes [31]. At T7, the data regarding octanoic acid is confirmed, but it is nevertheless interesting to note the slight but significant reduction in the relative proportion of hexanoic and heptanoic acids. These compounds are associated with pungent, rancid and flowery notes [32]; however, their relative concentrations place them in second place with respect to octanoic acid as contributors to the overall flavor. Apart from the question concerning the flavor development, the greater presence of octanoic acid in OLG samples also lends to be discussed from the perspective of food safety. In fact, the study conducted by Kinderlerer and Lund [33] reported the ability of this compound to inhibit the growth of 10 strains of *Listeria monocytogenes* and two strains of *L. innocua*, with a minimum inhibitory concentration (MIC), which was comparable with that determined in several cheeses. In particular, in T7 samples, the dietary OL supplementation induced significant differences between volatile short-chain fatty acids, while no significant changes were found for total long-chain fatty acids (nonanoic and decanoic). Since no variations were detected in the total lipid content between CG and OLG, it is plausible that this data is the result of the differential expression or activity of the bacterial lipases, whose reaction kinetics could have been influenced by secondary OL metabolites, which reached the milk through the mammary gland. However, in this study, we did not characterize of the OL metabolites in milk; therefore, further and more specific assessments are needed to clarify this aspect.

FFAs also contribute to the formation of cheese flavor indirectly, since they represent the precursors of other chemical families: methyl ketones, secondary alcohols, esters, aldehydes and lactones [34]. In this study, the second group of VOCs in order of importance was that of methyl ketones, which are formed following the oxidation of FFAs to β-ketoacids and subsequent decarboxylation to corresponding methyl ketones [16]. These compounds are reported to be the major determinants for the characteristic flavor of blue-veined and surface-mold-ripened cheeses, as a consequence of their typical odors and low perception thresholds [35]. However, in yogurt, many of these compounds have also been associated with well-defined aromatic notes. In this study, a specific pattern has been observed with regard to 2-heptanone and 2-nonanone, to which the ability to confer floral and green-fruity notes is attributed [15]. These compounds were found to be predominant in several dairy products, and their concentrations are reported to increase during the early stage of ripening, after which levels undergo reduction [36]. This behavior is generally observed over a period of a few weeks, and since the yogurt samples considered in the present study have been stored for few days, presumably only the initial part of the phenomenon, characterized by the increase in the concentrations of these compounds, has been highlighted. The peculiarity concerns the fact that both in T1 and T7 samples, the dietary OL supplementation seems to negatively influence the production of methyl ketones. The most plausible explanation could be precisely the greater antioxidant potential that should characterize the OLG samples and which would have protected the FFAs from oxidation to β-ketoacids.

In the presence of free radicals, the unsaturated fatty acids characterizing the matrix of dairy products can undergo non-enzymatic intrachain oxidation, giving origin to hydroperoxides, which rapidly decompose to form compounds such as propanal, hexanal, heptanal, octanal, nonanal or unsaturated aldehydes, whose presence is commonly associated with a "green grass-like" aroma [16,37]. The accumulation of aldehydes as a consequence of lipid oxidation is responsible for the off-flavor development during yogurt storage. For this reason, Carrillo-Carrion et al. [38] proposed to use the total concentration of volatile aldehydes (mainly from C5 to C9) as a marker of yogurt deterioration during storage. The dietary OLG supplementation was effective in curbing such mechanisms, especially after 7 days of storage, and this evidence also correlates with a presumed improvement in the yogurt's oxidative stability. This finding is consistent with what has been already observed in other dairy products obtained from ruminants fed diets supplemented with vegetable matrices rich in bioactive compounds, mainly polyphenols [11]. In addition to this, it should be mentioned that the oxidation of unsaturated fatty acids can even be mediated by microbial lypoxygenases. These enzymes are described as non-heme iron enzymes able to catalyze the dioxygenation of PUFAs to hydroperoxy fatty acids. Although limited information is available on the bacterial isoforms of these enzymes, it is, however, known that the preferred substrate is represented by linoleic acid [39]. Therefore, the higher concentration of linoleic acid found in the CG yogurt samples could, in part, have favored a greater accumulation of aldehydes; however, it should also be considered that the extent of this enzymatic phenomenon should have been limited in any case, given the relative percentage of linoleic acid compared to other fatty acids.

Aldehydes that accumulate in dairy products during storage or ripening may be reduced into the corresponding primary alcohols, which are considered transitory compounds [16]. Both in T1 and T7 samples, the dietary OL intake led to a limitation of this biochemical process, resulting in lower relative proportions of all the identified alcohols, precisely 1-hexanol, 1-heptanol, 1-octanol, 1-nonanol and 2-ethyl-hexan-1-ol. Contrary to what was expected, in all samples, the presence of ethanol, which is considered one of the most characteristically volatile compounds participating in the formation of yogurt flavor, was not evidenced. Ethanol can be alternatively released during yogurt manufacture by the starter lactic acid bacteria (LAB), through a mechanism that involves the conversion of lactose into lactate with consequent pH lowering, and lactate metabolization into formate, acetaldehyde and ethanol [15,34]. Therefore, in this case, any secondary metabolites derived from bioactive compounds consumed by the goats through the OL intake could have already influenced the metabolism of the microbial forms involved in the yogurt production in the early stages of the procedure.

Alcohols that accumulate in yogurt can also react with free acids to produce esters such as ethyl-acetate and butyl-acetate. In all the analyzed samples, the only identified ester was butyl heptanoate, which showed significantly lower concentrations in the OLG yogurt. This is a compound not commonly found in yogurt, whose synthesis mechanisms and specific contributions to the development of flavor require further and more specific evaluations.

## 4. Materials and Methods

### 4.1. Experimental Design, Yogurt Manufacturing Protocol and Sampling

Thirty Saanen goats, homogeneous in age, lactation period, and number of births were used for the experiment. The trial lasted 28 days, in which the nutritional needs of goats in lactation were satisfied. Animals were randomly divided into two groups, a control group (CG) and an experimental group (EG) that received the dietary enrichment of 350 g/die/goat of olive leaves.

Regarding the yogurt production, whole raw goats' milk was pasteurized at 92 °C. Then, the milk was cooled at 40 °C, inoculated with a lactic starter mixture (*Streptococcus thermophilus* and *Lactobacillus delbrueckii* subsp. *Bulgaricus*; Santamaria Srl, Burago di Molgora (MB), Italy), portioned into 300-mL glass containers and incubated at 45 °C for 12 h. For each group of goats, 12 yogurts were produced; six yogurts for each treatment were sampled one day after production (T1) and the remaining six for each group were sampled after seven days (T7) of storage at 4 °C. All the samples were collected and maintained at −20 °C until the analysis.

### 4.2. Chemical Analysis

The evaluation of the chemical properties of yogurt samples were performed according to AOAC methods [40]. Briefly, pH was measured at 25 °C by using a pH-meter HD 8705 (Delta OHM, Caselle di Selvazzano (PD), Italy) and moisture content (method 933.05) was determined on 5 g of sample left in the stove for 6 h at 105 °C.

With regard to the evaluation of total lipids, 3 g of yogurt was treated with 5 mL of ethanol (Sigma Aldrich, Milan, Italy) and 750 μL of ammonium hydroxide 25% in water to obtain protein precipitation. Then, three consecutive extractions with diethyl ether and petroleum ether (Sigma Aldrich, Milan, Italy) were performed and, every time, the supernatant was recovered in a previously calibrated flask. At the end of the extraction, a rotary evaporator was used to remove the solvent, and the flask containing the lipids was then moved into the stove at 40 °C for 20 min to eliminate humidity.

After cooling at room temperature in a dryer, all flasks were weighed to calculate the total fat percentage for each sample. Results were expressed on a dry matter (DM) basis.

### 4.3. Fatty Acid Composition

The evaluation of the fatty acids (FA) profile was performed as previously described with slight modifications [41]. Briefly, 60 mg of total fat was solubilized in 1 mL of hexane and methylated with 500 µL of sodium methoxide 2 N in water to obtain the fatty acid methyl esters (FAMEs). Then, 1 µL of methylated extract was injected in the gas chromatograph (GC; ThermoScientific, Waltham, MA, USA) coupled with a flame ionization detector (FID) and equipped with a column Restek Rt-2560 (100 m, 0.25 mm ID, 0.20 µm df). The injector and detector temperature were both set at 280 °C and hydrogen was used as carrier gas at a flow rate of 1 mL/min. The chromatographic run lasted 56 min and the oven temperature was first held at 80 °C for 10 min, then increased from 80 °C to 172 °C at 4 °C/min and held for 30 min, and finally increased from 172 °C to 190 °C at 4 °C/min and held for 10 min. The identification of each FAME was performed by comparing the peak retention time with those obtained from a mix of analytical standards (F.A.M.E. Mix C8-C24, Supelco). The peak area was quantified using ChromeCard Software and the results were expressed as mean percentages of total FA.

### 4.4. Total Antioxidant Capacity and Lipid Peroxidation

Antioxidant compounds were extracted by mixing 5 g of yogurt with 15 mL of methanol. The solution was gently shaken for 40 min in the dark and then centrifuged for 15 min at 4000 rpm. Then, the supernatants were filtered and used for analysis. The antioxidant capacity was evaluated through the 2,2-azinobis-3-ethylbenzothiazoline-6-sulfonic acid (ABTS) method according to the protocol described by Chen et al. [42]. Initially, 100 µL of extract was mixed with 1 mL of opportunely diluted ABTS solution, and the colorimetric evaluations were performed at 734 nm after 4 min of incubation at room temperature. An external seven-point calibration curve for Trolox (ranging from 1 to 50 µmol·g$^{-1}$; R$^2$ = 0.9961) was prepared for the quantification and the results were reported on a dry matter basis as µmol·g$^{-1}$ Trolox equivalent antioxidant capacity (TEAC).

The TBARs test was used to evaluate the lipid peroxidation through the identification of acid-reactive substances. The analysis was performed in accordance with the procedure previously adopted by Ianni et al. [43], with slight modifications. Five grams of yogurt were mixed with 500 µL of butylated hydroxytoluene (BHT) 0.1% in methanol to stop the oxidation process. At this point, the sample was distilled, and 2 mL of distillate were mixed with an equal volume of thiobarbituric acid (TBA) solution (0.02 M). Finally, the sample was heated for 1 h at 80 °C in a thermostatic bath; after cooling at room temperature, the absorbance at 534 nm was determined for each sample. Data were expressed in µg of malondialdehyde (MDA) equivalent per gram of yogurt.

### 4.5. Determination of Volatile Compounds

The volatile compounds (VOCs) were extracted through a solid-phase microextraction (SPME), and then separated and identified as previously described [44] with the use of a gas chromatograph (Clarus 580; Perkin Elmer, Walthman, MA, USA) coupled with a mass spectrometry (SQ8S; Perkin Elmer) and equipped with an Elite-5MS column (length: 30 m; internal diameter: 0.25 mm; film thickness: 0.25 µm; Perkin Elmer, Waltham, MA, USA). Ten grams of yogurt were transferred in vials and mixed with 5 mL of NaCl water solution (360 g/L), and 10 µL of 4-methyl-2-heptanone, which was used as internal standard with the aim of evaluating the extraction efficiency downstream of the analysis. After sealing the vial, the sample was stirred at 60 °C in a thermostatic bath, and the adsorption of VOCs was performed with a divinybenzene–carboxen–polydimethylsiloxane SPME fiber exposed for 45 min in the headspace. At this point, the extracted VOCs were thermally desorbed into the GC injector for 1 min in a splitless mode at 250 °C. Helium was used as carrier gas with a flow rate of 1 mL/min and the oven temperature program was started at 50 °C, held for 1 min, then increased up to 200 °C with a ratio of 3 °C/min, then held for 1 min and finally increased up to 250 °C with a ratio of 15 °C/min, then held for 15 min. The mass spectrometer operates in electronic impact ionization mode at 70 eV. Volatile compounds were identified by a comparison with the mass spectra included in the library database (NIST Mass Spectral library (2014), Search Program version 2.0, National Institute of Standards and Technology, US Department of Commerce, Gaithersburg, MD, USA) and by comparing the eluting order with modified Kovats indices, according to the method of Van Den Dool and Kratz [45]. The data were expressed as a percentage of the relative abundance of each compound in relation to the sum of total VOCs.

*4.6. Statystical Analysis*

All the listed evaluations were performed on 12 samples per group (six samples for T1 and six samples for T7), and the analyses on individual samples were performed in triplicate. Results were reported as mean values with corresponding standard deviations (SD). The analysis of statistically significant differences between the two groups of data were performed by using the SigmaPlot 12.0 Software (Systat software, Inc., San Jose, CA, US) for the Windows operating system (ANOVA, Student's t-test); $p$ values lower than 0.05 were considered statistically significant.

## 5. Conclusions

The results shown in this study suggest the positive role of dietary OL supplementation on the nutritional characteristics and volatile profile of goats' milk yogurt. Dietary OL intake was effective in inducing an increase in the concentration of unsaturated fatty acids, in addition to the general improvement in oxidative stability. This finding could justify an improvement in the shelf-life of the product and was also confirmed by a reduction in the concentration of volatile aldehydes. The characterization of the volatile profile was also useful in highlighting the accumulation of compounds that could justify an improvement in the yogurt flavor, although further sensorial analysis is necessary to evaluate any variations in consumer acceptability.

**Author Contributions:** Conceptualization, G.M. and A.I.; methodology, D.I., C.M.; F.B. and L.G.; validation, D.I., C.M. and F.B.; formal analysis, A.I. and F.B.; investigation, A.I., D.I. and C.M.; resources, G.M.; data curation, F.B. and D.I.; writing—original draft preparation, A.I. and F.B.; writing—review and editing, L.G. and G.M.; supervision, G.M.; project administration, G.M.; funding acquisition, G.M. All authors have read and agreed to the published version of the manuscript.

**Funding:** This research is part of the project "Applicazione di nuove strategie gestionali per il miglioramento quanti-qualitativo della produzione di carne e sviluppo di nuovi prodotti per il potenziamento del settore ovicaprino", supported by a grant from Rural Development Plan 2014–2020, MISURA 16.2, Regione Abruzzo (Italy). Project manager: Giuseppe Martino.

**Acknowledgments:** The authors are grateful to "La Fattoria del Nonno" of Giampietro De Vitis for the kind cooperation.

**Conflicts of Interest:** The authors declare no conflict of interest. The funders had no role in the design of the study; in the collection, analyses, or interpretation of data; in the writing of the manuscript, or in the decision to publish the results.

## References

1. Ceballos, L.S.; Morales, E.R.; Adarve, G.D.L.T.; Diaz-Castro, J.; Martínez, L.P.; Sampelayo, M.R.S. Composition of goat and cow milk produced under similar conditions and analyzed by identical methodology. *J. Food Compos. Anal.* **2009**, *22*, 322–329. [CrossRef]

2. Miller, B.A.; Lu, C.D. Current status of global dairy goat production: An overview. *Asian-Australasian J. Anim. Sci.* **2019**, *32*, 1219–1232. [CrossRef] [PubMed]

3. Chávez-Servín, J.L.; Andrade-Montemayor, H.M.; Vázquez, C.V.; Barreyro, A.A.; Garcia-Gasca, T.; Ferríz-Martínez, R.A.; Ramírez, A.M.O.; De La Torre-Carbot, K. Effects of feeding system, heat treatment and season on phenolic compounds and antioxidant capacity in goat milk, whey and cheese. *Small Rumin. Res.* **2018**, *160*, 54–58. [CrossRef]

4.	Robinson, R.; Tamime, A.Y. *Tamime and Robinson's Yoghurt Science and Technology, Third Edition*; Informa UK Limited: Cambridge, UK, 2007.

5.	Adolfsson, O.; Meydani, S.N.; Russell, R.M. Yogurt and gut function. *Am. J. Clin. Nutr.* **2004**, *80*, 245–256. [CrossRef]

6.	Ranadheera, C.S.; Evans, C.A.; Baines, S.K.; Balthazar, C.F.; Cruz, A.G.; Esmerino, E.A.; Freitas, M.Q.; Pimentel, T.C.; Wittwer, A.E.; Naumovski, N.; et al. Probiotics in Goat Milk Products: Delivery Capacity and Ability to Improve Sensory Attributes. *Compr. Rev. Food Sci. Food Saf.* **2019**, *18*, 867–882. [CrossRef]

7.	Silva, F.A.; De Oliveira, M.E.G.; De Figueirêdo, R.M.F.; Sampaio, K.; De Souza, E.L.; De Oliveira, C.E.V.; Pintado, M.; Queiroga, R.C. The effect of Isabel grape addition on the physicochemical, microbiological and sensory characteristics of probiotic goat milk yogurt. *Food Funct.* **2017**, *8*, 2121–2132. [CrossRef]

8.	Iannaccone, M.; Elgendy, R.; Giantin, M.; Martino, C.; Giansante, D.; Ianni, A.; Dacasto, M.; Martino, G. RNA Sequencing-Based Whole-Transcriptome Analysis of Friesian Cattle Fed with Grape Pomace-Supplemented Diet. *Animals* **2018**, *8*, 188. [CrossRef]

9.	Chedea, V.; Pelmus, R.S.; Lazar, C.; Marin, D.E.; Calin, L.G.; Toma, S.M.; Dragomir, C.; Taranu, I. Effects of a diet containing dried grape pomace on blood metabolites and milk composition of dairy cows. *J. Sci. Food Agric.* **2016**, *97*, 2516–2523. [CrossRef]

10.	Ianni, A.; Martino, G. Dietary Grape Pomace Supplementation in Dairy Cows: Effect on Nutritional Quality of Milk and Its Derived Dairy Products. *Foods* **2020**, *9*, 168. [CrossRef]

11.	Ianni, A.; Bennato, F.; Martino, C.; Grotta, L.; Martino, G. Martino Volatile Flavor Compounds in Cheese as Affected by Ruminant Diet. *Molecules* **2020**, *25*, 461. [CrossRef]

12.	Iannaccone, M.; Ianni, A.; Ramazzotti, S.; Grotta, L.; Marone, E.; Cichelli, A.; Martino, G. Whole Blood Transcriptome Analysis Reveals Positive Effects of Dried Olive Pomace-Supplemented Diet on Inflammation and Cholesterol in Laying Hens. *Animals* **2019**, *9*, 427. [CrossRef] [PubMed]

13.	Castellani, F.; Vitali, A.; Bernardi, N.; Marone, E.; Palazzo, F.; Grotta, L.; Martino, G. Dietary supplementation with dried olive pomace in dairy cows modifies the composition of fatty acids and the aromatic profile in milk and related cheese. *J. Dairy Sci.* **2017**, *100*, 8658–8669. [CrossRef] [PubMed]

14.	Özcan, M.M.; Matthäus, B. A review: Benefit and bioactive properties of olive (*Olea europaea* L.) leaves. *Eur. Food Res. Technol.* **2016**, *243*, 89–99. [CrossRef]

15.	Cheng, H. Volatile Flavor Compounds in Yogurt: A Review. *Crit. Rev. Food Sci. Nutr.* **2010**, *50*, 938–950. [CrossRef]

16.	Thierry, A.; Collins, Y.F.; Mukdsi, M.A.; McSweeney, P.L.H.; Wilkinson, M.G.; Spinnler, H.E. Lipolysis and Metabolism of Fatty Acids in Cheese. In *Cheese*; Elsevier BV: Amsterdam, The Netherlands, 2017; pp. 423–444.

17.	McSweeney, P.L. Biochemical pathways for the production of flavour compounds in cheeses during ripening: A review. *Le Lait* **2000**, *80*, 293–324. [CrossRef]

18.	Mele, M.; Conte, G.; Castiglioni, B.; Chessa, S.; Macciotta, N.P.; Serra, A.; Buccioni, A.; Pagnacco, G.; Secchiari, P. Stearoyl-Coenzyme A Desaturase Gene Polymorphism and Milk Fatty Acid Composition in Italian Holsteins. *J. Dairy Sci.* **2007**, *90*, 4458–4465. [CrossRef]

19.	Ntambi, J.M. Regulation of stearoyl-CoA desaturases and role in metabolism. *Prog. Lipid Res.* **2004**, *43*, 91–104. [CrossRef]

20.	Corl, B.A.; Baumgard, L.H.; Dwyer, D.A.; Griinari, J.; Phillips, B.S.; Bauman, D.E. The role of Δ9-desaturase in the production of cis-9, trans-11 CLA. *J. Nutr. Biochem.* **2001**, *12*, 622–630. [CrossRef]

21.	Livingstone, K.M.; Lovegrove, J.A.; Givens, D.I. The impact of substituting SFA in dairy products with MUFA or PUFA on CVD risk: Evidence from human intervention studies. *Nutr. Res. Rev.* **2012**, *25*, 193–206. [CrossRef]

22.	Jenkins, T.C.; Wallace, R.J.; Moate, P.; Mosley, E.E. BOARD-INVITED REVIEW: Recent advances in biohydrogenation of unsaturated fatty acids within the rumen microbial ecosystem. *J. Anim. Sci.* **2008**, *86*, 397–412. [CrossRef]

23.	Bennato, F.; Ianni, A.; Innosa, D.; Grotta, L.; D'Onofrio, A.; Martino, G. Chemical-nutritional characteristics and aromatic profile of milk and related dairy products obtained from goats fed with extruded linseed. *Asian-Australas. J. Anim. Sci.* **2019**, *33*, 148–156. [CrossRef] [PubMed]

24. Innosa, D.; Bennato, F.; Ianni, A.; Martino, C.; Grotta, L.; Pomilio, F.; Martino, G. Influence of olive leaves feeding on chemical-nutritional quality of goat ricotta cheese. *Eur. Food Res. Technol.* **2020**, *246*, 923–930. [CrossRef]

25. Bahloul, N.; Kechaou, N.; Mihoubi, N.B. Comparative investigation of minerals, chlorophylls contents, fatty acid composition and thermal profiles of olive leaves (Olea europeae L.) as by-product. *Grasas Aceites* **2014**, *65*, e035. [CrossRef]

26. Kolakowska, A. Lipid oxidation in food systems. *Chem. Funct. Prop. Food Lipids* **2003**, *8*, 133–165.

27. Innosa, D.; Ianni, A.; Palazzo, F.; Martino, F.; Bennato, F.; Grotta, L.; Martino, G. High temperature and heating effect on the oxidative stability of dietary cholesterol in different real food systems arising from eggs. *Eur. Food Res. Technol.* **2019**, *245*, 1533–1538. [CrossRef]

28. Ianni, A.; Di Maio, G.; Pittia, P.; Grotta, L.; Perpetuini, G.; Tofalo, R.; Cichelli, A.; Martino, G. Chemical–nutritional quality and oxidative stability of milk and dairy products obtained from Friesian cows fed with a dietary supplementation of dried grape pomace. *J. Sci. Food Agric.* **2019**, *99*, 3635–3643. [CrossRef]

29. Bennato, F.; Ianni, A.; Martino, C.; Di Luca, A.; Innosa, D.; Fusco, A.M.; Pomilio, F.; Martino, G. Dietary supplementation of Saanen goats with dried licorice root modifies chemical and textural properties of dairy products. *J. Dairy Sci.* **2019**, *103*, 52–62. [CrossRef]

30. Benavente-García, O.; Castillo, J.; Lorente, J.; Ortuño, A.; Del Rio, J. Antioxidant activity of phenolics extracted from Olea europaea L. leaves. *Food Chem.* **2000**, *68*, 457–462. [CrossRef]

31. Güler, Z. Changes in salted yoghurt during storage. *Int. J. Food Sci. Technol.* **2007**, *42*, 235–245. [CrossRef]

32. Ott, A.; Fay, L.B.; Chaintreau, A. Determination and Origin of the Aroma Impact Compounds of Yogurt Flavor. *J. Agric. Food Chem.* **1997**, *45*, 850–858. [CrossRef]

33. Kinderlerer, J.L.; Lund, B.M. Inhibition ofListeria monocytogenesandListeria innocuaby hexanoic and octanoic acids. *Lett. Appl. Microbiol.* **1992**, *14*, 271–274. [CrossRef]

34. Bertuzzi, A.S.; McSweeney, P.L.H.; Rea, M.; Kilcawley, K.N. Detection of Volatile Compounds of Cheese and Their Contribution to the Flavor Profile of Surface-Ripened Cheese. *Compr. Rev. Food Sci. Food Saf.* **2018**, *17*, 371–390. [CrossRef]

35. Collins, Y.F.; McSweeney, P.L.; Wilkinson, M.G. Lipolysis and free fatty acid catabolism in cheese: A review of current knowledge. *Int. Dairy J.* **2003**, *13*, 841–866. [CrossRef]

36. De Llano, D.G.; Ramos, M.; Polo, C.; Sanz, J.; Martínez-Castro, I. Evolution of the Volatile Components of an Artisanal Blue Cheese During Ripening. *J. Dairy Sci.* **1990**, *73*, 1676–1683. [CrossRef]

37. McGorrin, R.J.; Spanier, A.M.; Shahidi, F.; Parliment, T.H.; Mussinan, C.; Ho, C.-T.; Contis, E.T. Advances in dairy flavor chemistry. In *Special Publications*; Royal Society of Chemistry (RSC): Cambridge, UK, 2007; pp. 67–84.

38. Carrillo-Carrion, C.; Cardenas, S.; Valcarcel, M. Vanguard/rearguard strategy for the evaluation of the degradation of yoghurt samples based on the direct analysis of the volatiles profile through headspace-gas chromatography–mass spectrometry. *J. Chromatogr. A* **2007**, *1141*, 98–105. [CrossRef] [PubMed]

39. Hansen, J.; Garreta, A.; Benincasa, M.; Fusté, M.C.; Busquets, M.; Manresa, M. Ángeles Erratum to: Bacterial lipoxygenases, a new subfamily of enzymes? A phylogenetic approach. *Appl. Microbiol. Biotechnol.* **2013**, *97*, 7527. [CrossRef]

40. AOAC International. *Official Methods of Analysis*, 17th ed.; AOAC International: Gaithersburg, MD, USA, 2000; Volume II.

41. Ianni, A.; Bennato, F.; Martino, C.; Innosa, D.; Grotta, L.; Martino, G. Effects of selenium supplementation on chemical composition and aromatic profiles of cow milk and its derived cheese. *J. Dairy Sci.* **2019**, *102*, 6853–6862. [CrossRef]

42. Chen, J.; Lindmark-Månsson, H.; Gorton, L.; Åkesson, B. Antioxidant capacity of bovine milk as assayed by spectrophotometric and amperometric methods. *Int. Dairy J.* **2003**, *13*, 927–935. [CrossRef]

43. Ianni, A.; Martino, C.; Pomilio, F.; Di Luca, A.; Martino, G. Dietary selenium intake in lactating dairy cows modifies fatty acid composition and volatile profile of milk and 30-day-ripened caciotta cheese. *Eur. Food Res. Technol.* **2019**, *245*, 2113–2121. [CrossRef]

44. Ianni, A.; Iannaccone, M.; Martino, C.; Innosa, D.; Grotta, L.; Bennato, F.; Martino, G. Zinc supplementation of dairy cows: Effects on chemical composition, nutritional quality and volatile profile of Giuncata cheese. *Int. Dairy J.* **2019**, *94*, 65–71. [CrossRef]
45. Dool, H.V.D.; Kratz, P.D. A generalization of the retention index system including linear temperature programmed gas—liquid partition chromatography. *J. Chromatogr. A* **1963**, *11*, 463–471. [CrossRef]

**Sample Availability:** Samples analysed during the current study are available from the corresponding author on reasonable request.

*Article*

# Influence of Different Modalities of Grape Withering on Volatile Compounds of Young and Aged Corvina Wines

Davide Slaghenaufi [1], Anita Boscaini [2], Alessandro Prandi [1], Andrea Dal Cin [2], Vittorio Zandonà [2], Giovanni Luzzini [1] and Maurizio Ugliano [1,*]

[1]  Department of Biotechnology, University of Verona, Villa Lebrecht, via della Pieve 70, 37029 San Pietro, Cariano, Italy; davide.slaghenaufi@univr.it (D.S.); aleprandi96@libero.it (A.P.); giovanni.luzzini@univr.it (G.L.)

[2]  Masi Agricola, Via Monteleone, 26, Sant'Ambrogio di Valpolicella, 37015 Verona VR, Italy; anita.boscaini@masi.it (A.B.); andrea.dalcin@masi.it (A.D.C.); vittorio.zandona@masi.it (V.Z.)

*  Correspondence: maurizio.ugliano@univr.it

Academic Editor: Eugenio Aprea
Received: 31 March 2020; Accepted: 1 May 2020; Published: 3 May 2020

**Abstract:** Withering is a practice traditionally used in various regions to produce sweet or dry wines. During withering there is an increase in sugar content but also a modification in volatile compound profiles. Controlling metabolic changes through the dehydration process to obtain wines with desired characteristics is therefore a challenging opportunity. The effects of two different withering technologies, post-harvest or on-vine with blocked sap vessel flow, on the volatile profile of young and aged Corvina red wines was investigated. The results showed that modulation of wine aroma due to the withering process is associated with fermentative metabolites, such as esters, higher alcohols, and acids, as well as grape-related compounds such as C6 alcohols, terpenes and norisoprenoids. Significant differences were also found by comparing the two withering techniques. Post-harvest in a traditional "fruttaio" warehouse wines showed higher content of ethyl acetate, ethyl butanoate, β-citronellol and 3-oxo-α-ionol, whereas post-harvest withering on-vine increased β-damascenone in wines. The type of withering technique has an influence on the evolution of some aroma compounds during the aging of wine, among them linalool, (*E*)-1-(2,3,6-trimethylphenyl)buta-1,3-diene (TPB), n-hexyl acetate, ethyl acetate, ethyl 3-methylbutanoate, 3-oxo-α-ionol and β-damascenone.

**Keywords:** post-harvest; withering; on-vine; fruttaio; wine aroma; Corvina

## 1. Introduction

Valpolicella is a wine producing region characterized by the traditional practice of post-harvest withering for the production of dry and sweet red wines, among which Amarone is the most famous [1]. Valpolicella is located in the north-east of Italy close to Verona city, with Corvina, Corvinone and Rondinella grapes being the traditional varieties employed for local wines [1]. At ripening, grapes are harvested and stored in a specific warehouse traditionally called a "fruttaio", where they undergo slow dehydration [2]. The duration of withering varies depending on the wine type being produced, and it is generally monitored by assessing grape weight loss. In the case of Amarone or Recioto, withering generally lasts 2–3 months, with a weight loss of approximately 30% of the initial weight [3]. In the case of other wines such as Valpolicella Classico superiore as well as different IGT wines, a milder withering is usually carried out, lasting 4–8 weeks with weight loss of 10–15%.

Grape withering has a deep impact on the formation of the characteristic aroma of Amarone wine [4,5]. During the withering process, an increase in sugar content due to water loss is not the only transformation taking place. Phenolic and aromatic composition of grapes and wines is also affected [6–12], and skin wall composition as well as grape mechanical properties are also modified [13,14]. Interestingly, some of these processes are not due to dehydration but are the result of ongoing metabolic activities in the berry, resulting in peculiar gene expression patterns contributing to changes secondary metabolism [15–17].

Different grape withering techniques have been developed in the past, depending to local environment, grape variety and technical issues, entering in the local tradition and history as community heritage [18]. The different techniques can be divided into natural, forced and on-vine withering [18]. An example of a natural withering technique is the exposure of grapes to the sun, while in the forced method, grape dehydration is obtained using ventilated rooms like in the case of a modern fruttaio where withering conditions like temperature, humidity and air flow are controlled. On-vine withering can be obtained by practicing late harvest, cane cutting, or peduncle twist [18]. Though withering in Valpolicella is traditionally made in a fruttaio, there is an increasing interest to explore other postharvest methods to support traditional practices, in particular for mild withering processes requiring a short duration.

The aim of this research was to evaluate the effect of the two withering systems, in "fruttaio" and on-vine with peduncle twist, on the volatile profile of wines. The results were compared with those of wines obtained from not-withered grapes. Wines were also assessed after a period of aging, to evaluate the influence of the different withering practices on aging patterns of the resulting wines.

## 2. Results

### 2.1. Volatile Compounds in Young Wines

A total of 53 volatile compounds have been identified and quantified in wine samples (Table 1), including five alcohols, 3 $C_6$ alcohols, 10 esters, three acids, 18 terpenes, seven norisoprenoids, seven benzenoids. The analysis of the variance (ANOVA) made between the three modalities, showed statistically significant ($p < 0.05$) differences for 35 compounds. Wines from withered grapes in fruttaio and on-vine were characterized by higher content in terpinen-4-ol, β-citronellol, 1,4-cineole, 3-oxo-α-ionol, vinylguaiacol, ethyl acetate, benzyl alcohol and ethyl vanillate. At the same time, samples fruttaio and on-vine compared to control showed lower amounts of 1-pentanol, 1-hexanol, *cis*-3-hexenol, *trans*-3-hexenol, isoamyl acetate, *n*-hexyl acetate, phenylethyl acetate ethyl 3-methylbutanoate, ethyl hexanoate, ethyl octanoate, hexanoic acid, octanoic acid, *cis*-linalool oxide and β-damascenone. Compared to each other, statistical differences of the two withering techniques were observed for 19 volatile compounds, among them 1-butanol, 1-hexanol, *trans*-3-hexenol, ethyl butanoate, β-citronellol, limonene, α-phellandrene, terpinolene, benzyl alcohol and methyl vanillate were found in higher concentration in fruttaio samples, while on-vine samples showed higher content of 2-butanol, ethyl acetate, octanoic acid, β-damascenone, TDN and 4-vinyl guaiacol.

**Table 1.** Concentration (µg/L) of free compounds in young wine samples. Mean, standard deviation (SD) and ANOVA.

| Compounds | Odor Threshold [1] | Control Mean | SD | Fruttaio Mean | SD | On-Vine Mean | SD | *p*-Value |
|---|---|---|---|---|---|---|---|---|
| **Alcohols** | | | | | | | | |
| 2-Butanol | | 3283.8 [b] | ±207.9 | 2203.1 [a] | ±80.3 | 3467.7 [b] | ±93.1 | <0.0001 |
| 1-Butanol | | 269.7 [a] | ±6.7 | 506.3 [b] | ±48.8 | 228.0 [a] | ±48.3 | 0.000 |
| 1-Pentanol | | 58.3 [b] | ±5.3 | 43.0 [a] | ±1.5 | 42.8[a] | ±1.4 | 0.002 |
| Isoamyl alcool | 30,000 | 35,372.3 [a] | ±1853.6 | 33,820.8 [a] | ±922.9 | 34,951.5 [a] | ±2051.8 | 0.541 |
| Phenylethyl Alcohol | 14,000 | 16,240.6 [a] | ±721.1 | 20,779.8 [a] | ±466.8 | 16,568.9 [a] | ±6073.6 | 0.290 |
| **C6 Alcohols** | | | | | | | | |
| 1-Hexanol | 8000 | 3292.5 [c] | ±171.9 | 2292.4 [b] | ±84.2 | 1951.1 [a] | ±70.7 | <0.0001 |
| *trans*-3-Hexen-1-ol | | 39.4 [b] | ±2.4 | 32.0 [b] | ±6.59 | 21.4 [a] | ±2.0 | 0.006 |
| *cis*-3-Hexen-1-ol | 400 | 520.3 [b] | ±26.9 | 92.7 [a] | ±13.2 | 65.5 [a] | ±7.3 | <0.0001 |

Table 1. Cont.

| Compounds | Odor Threshold [1] | Control Mean | SD | Fruttaio Mean | SD | On-Vine Mean | SD | *p*-Value |
|---|---|---|---|---|---|---|---|---|
| **Acetate Esters** | | | | | | | | |
| Ethyl acetate | 12,000 | 19,537.4 [a] | ±1896.3 | 49,654.9 [c] | ±4413.3 | 33,649.2 [b] | ±7566.5 | 0.001 |
| Isoamyl acetate | 30 | 10,732.9 [b] | ±1503.1 | 3216.1 [a] | ±631.4 | 1877.3 [a] | ±293.0 | <0.0001 |
| n-Hexyl acetate | 1800 | 279.3 [b] | ±41.8 | 22.3 [a] | ±1.5 | 6.47 [a] | ±1.09 | <0.0001 |
| Phenylethyl acetate | 2400 | 106.2 [b] | ±11.0 | 35.3 [a] | ±4.5 | 30.7 [a] | ±1.9 | <0.0001 |
| **Ethyl Esters** | | | | | | | | |
| Ethyl butanoate | 20 | 254.7 [b] | ±14.7 | 265.6 [b] | ±28.1 | 210.1 [a] | ±16.4 | 0.036 |
| Ethyl 3-methylbutanoate | 3 | 215.4 [b] | ±32.5 | 64.6 [a] | ±14.1 | 36.7 [a] | ±6.3 | <0.0001 |
| Ethyl hexanoate | 14 | 548.5 [b] | ±33.4 | 331.0 [a] | ±9.5 | 302.1 [a] | ±24.9 | <0.0001 |
| Ethyl octanoate | 5 | 452.1 [b] | ±50.2 | 254.3 [a] | ±10.7 | 249.6 [a] | ±27.9 | 0.000 |
| Ethyl decanoate | 200 | 45.6 [a] | ±6.2 | 50.6 [a] | ±5.5 | 38.6 [a] | ±6.5 | 0.132 |
| Ethyl lactate | 100,000 | 1087.9 [b] | ±49.6 | 690.6 [a] | ±34.7 | 647.8 [a] | ±64.2 | <0.0001 |
| **Fatty Acids** | | | | | | | | |
| 3-Methylbutanoic acid | 250 | 511.9 [b] | ±38.5 | 402.3 [a] | ±6.6 | 449.1 [a] | ±19.0 | 0.005 |
| Hexanoic acid | 2080 | 2623.6 [c] | ±130.17 | 1286.5 [a] | ±103.3 | 1683.5 [b] | ±36.2 | <0.0001 |
| Octanoic acid | 2560 | 2044.8 [c] | ±184.0 | 820.0 [a] | ±36.8 | 1066.7 [b] | ±39.8 | <0.0001 |
| **Terpenes** | | | | | | | | |
| *cis*-Linalool oxide | 3000 | 6.57 [b] | ±3.12 | 1.29 [a] | ±0.78 | 1.33 [a] | ±0.19 | 0.038 |
| *trans*-Linalool oxide | 6000 | 0.415 [a] | ±0.211 | 0.330 [a] | ±0.130 | 0.50 [a] | ±0.282 | 0.797 |
| Linalool | 25 | 27.3 [b] | ±3.3 | 14.8 [a] | ±0.5 | 16.5 [a] | ±0.2 | 0.000 |
| Terpinen-1-ol | | 0.351 [a] | ±0.065 | 0.429 [a] | ±0.162 | 0.519 [a] | ±0.066 | 0.242 |
| Terpinen-4-ol | | 0.092 [a] | ±0.021 | 1.163 [b] | ±0.200 | 0.795 [b] | ±0.173 | 0.026 |
| Ho-trienol | 110 | 0.060 [a] | ±0.010 | 0.047 [a] | ±0.010 | 0.075 [a] | ±0.043 | 0.478 |
| α-Terpineol | 250 | 13.9 [b] | ±1.3 | 6.54 [a] | ±0.09 | 7.09 [a] | ±0.30 | <0.0001 |
| Nerol | 400 | 3.95 [a] | ±0.53 | 4.89 [a] | ±0.43 | 3.36 [a] | ±0.96 | 0.828 |
| Geraniol | 30 | 6.69 [a] | ±0.69 | 6.50 [a] | ±0.79 | 6.61 [a] | ±0.20 | 0.931 |
| β-Citronellol | 100 | 4.14 [a] | ±0.18 | 12.58 [c] | ±0.44 | 10.18 [b] | ±1.87 | 0.000 |
| p-Menthane-1,8-diol | | <LOQ | | <LOQ | | <LOQ | | |
| α-Phellandrene | | 0.035 [b] | ±0.013 | 0.040 [b] | ±0.006 | 0.016 [a] | ±0.002 | 0.029 |
| 1,4-Cineole | 0.54 | 0.110 [a] | ±0.013 | 0.205 [b] | ±0.023 | 0.182 [b] | ±0.024 | 0.003 |
| 1,8-Cineole | 1.1 | 0.215 [a] | ±0.022 | 0.153 [a] | ±0.040 | 0.176 [a] | ±0.068 | 0.336 |
| Limonene | | 0.228 [a] | ±0.028 | 0.388 [b] | ±0.060 | 0.225 [a] | ±0.040 | 0.006 |
| γ-Terpinen | | 1.10 [a] | ±0.51 | 1.15 [a] | ±0.09 | 1.03 [a] | ±0.19 | 0.904 |
| p-Cymene | | 0.083 [a] | ±0.008 | 0.145 [a] | ±0.077 | 0.093 [a] | ±0.021 | 0.289 |
| Terpinolene | | 0.137 [a] | ±0.024 | 0.257 [b] | ±0.045 | 0.125 [a] | ±0.041 | 0.010 |
| **Norisoprenoids** | | | | | | | | |
| β-Damascenone | 0.05 | 3.47 [c] | ±0.27 | 1.69 [a] | ±0.13 | 2.80 [b] | ±0.10 | <0.0001 |
| α-Ionone | | 2.27 [b] | ±0.70 | 0.85 [a] | ±0.34 | 1.03 [a] | ±0.67 | 0.050 |
| α-Ionol | | 0.233 [a] | ±0.019 | 0.283 [a] | ±0.026 | 0.25 [a] | ±0.010 | 0.955 |
| Vitispirane | | <LOQ | | <LOQ | | <LOQ | | |
| TPB | | 0.050 [a] | ±0.009 | 0.035 [a] | ±0.006 | 0.051 [a] | ±0.013 | 0.140 |
| TDN | 2 | 2.64 [b] | ±0.42 | 1.57 [a] | ±0.18 | 2.79 [b] | ±0.69 | 0.040 |
| 3-oxo-α-Ionol | | 1.58 [a] | ±0.24 | 2.65 [c] | ±0.20 | 2.10 [b] | ±0.23 | 0.003 |
| **Benzenoids** | | | | | | | | |
| 4-Ethyl guaiacol | 33 | <LOQ | | <LOQ | | <LOQ | | |
| 4-Vinyl guaiacol | 1100 | 7.36 [a] | ±0.44 | 9.40 [b] | ±0.53 | 11.63 [c] | ±1.63 | 0.006 |
| 2,6-Dimethoxyphenol | | 0.108 [a] | ±0.018 | 0.015 [a] | ±0.007 | 0.022 [a] | ±0.001 | 0.428 |
| Benzyl Alcohol | | 128.3 [a] | ±4.9 | 229.1 [c] | ±15.1 | 173.1 [b] | ±24.7 | 0.001 |
| Vanillin | 200 | 2.13 [a] | ±0.84 | 2.04 [a] | ±0.95 | 0.888 [a] | ±0.672 | 0.204 |
| Methyl vanillate | | 4.37 [a] | ±0.09 | 4.86 [b] | ±0.11 | 4.49 [a] | ±0.05 | 0.001 |
| Ethyl vanillate | | 16.4 [a] | ±1.1 | 21.6 [b] | ±2.0 | 31.4 [c] | ±5.2 | 0.004 |

Values in the same row with different letters indicate statistically significant differences, *p* < 0.05; [1] Data from: Ferreira et al. (2000) [19], Francis et al. (2005) [20], Sacks et al. (2012) [21] and Antalick et al. (2015) [22]. <LOQ: Values below the limit of quantification.

Eighteen glycosidically bound compounds have been quantified (Table 2). The ANOVA showed significant differences for eight of these compounds. Compared to control, fruttaio samples and on-vine showed significantly lower concentrations in *cis*-3-hexenol, benzyl alcohol precursor and higher content of methyl vanillate and ethyl vanillate precursors. Comparing fruttaio and on-vine samples, statistically significant differences have been observed for six glycosidically bound compounds. Fruttaio samples were richer in 1-hexanol, *trans*-3-hexenol, geraniol and vanillin precursors, while on-vine samples showed higher concentration only for the phenylethyl alcohol glycosidically bound precursor.

Principal component analysis (PCA) showed that 65.9% of the total variance was explained by the first and second component (Figure 1). In fact, first principal component (PC-1) explained 45.8% of the total variance, while PC-2 explained 20.1%. Samples were separated into three clusters according to their withering technique. PC-1 mostly discriminated withered samples from not-withered. Control samples were characterized principally by esters, fatty acids, $C_6$ compounds 1-hexanol, *trans*-3-hexenol, and *cis*-3-hexenol precursor. Instead withered compounds were characterized by terpenes like β-citronellol, linalool, terpinen-4-ol, 1,4-cineole; by the norisoprenoid 3-oxo-α-ionol; and by several benzenoids like phenylethyl alcohol, benzyl alcohol, methyl vanillate and its glycosidic precursor. PC-2 permitted to discriminate on-vine from fruttaio samples, the major drivers of this diversity were the ethyl butanoate, ethyl decanoate, ethyl vanillate, and the bound precursor of geraniol.

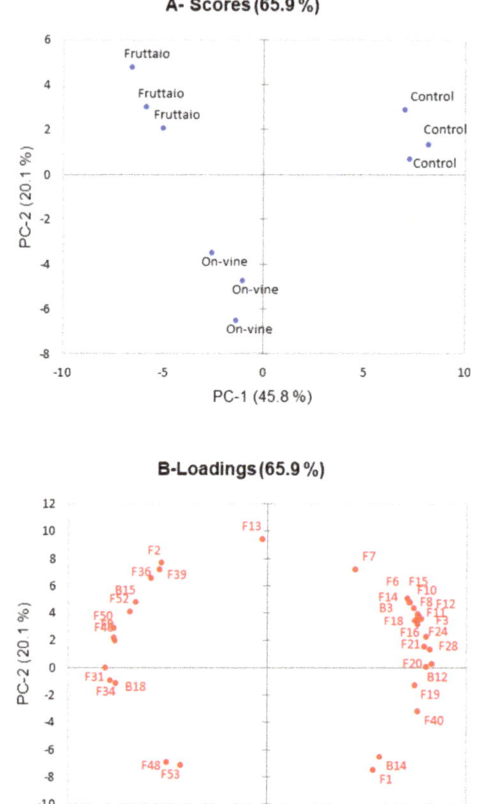


**Figure 1.** Principal component analysis showing aged wine samples scores (**A**) and loadings (**B**), only loadings with score value > 0.75 were shown. Loadings plot number correspond to: (F1) 2-Butanol, (F2) 1-Butanol, (F3) 1-Pentanol, (F6) 1-Hexanol, (F7) *trans*-3-Hexenol, (F8) *cis*-3-Hexenol, (F9) Ethyl acetate, (F10) Isoamyl acetate, (F11) n-Hexyl acetate, (F12) Phenylethyl acetate, (F13) Ethyl butanoate, (F14) Ethyl 3-methylbutanoate, (F15) Ethyl hexanoate, (F16) Ethyl octanoate, (F18) Ethyl lactate, (F19) 3-Methylbutanoic acid, (F20) Hexanoic acid, (F21) Octanoic acid, (F24) Linalool, (F28) α-Terpineol, (F31) β-Citronellol, (F34) 1,4-cineole, (F36) Limonene, (F39) Terpinolene, (F40) β-Damascenone, (F46) 3-oxo-α-Ionol, (F48) 4-Vinyl guaiacol, (F50) Benzyl alcohol, (F52) Methyl vanillate, (F53) Ethyl vanillate, (B3) bound cis-3-Hexenol, (B12) bound Benzyl alcohol, (B14) bound Phenylethyl alcohol, (B15) bound Vanillin, (B18) bound Ethyl vanillate.

**Table 2.** Concentration (µg/L) of glycosidically-bound compounds in young wine samples. Mean, standard deviation (SD) and ANOVA.

|  | Control |  | Fruttaio |  | On-Vine |  |  |
| --- | --- | --- | --- | --- | --- | --- | --- |
| **Compounds** | **Mean** | **SD** | **Mean** | **SD** | **Mean** | **SD** | **_p_-Value** |
| **Alcohols** |  |  |  |  |  |  |  |
| Phenylethyl alcohol | 284.9 [b] | ±22.5 | 191.7 [a] | ±11.0 | 282.0 [b] | ±47.8 | 0.017 |
| **C6 alcohols** |  |  |  |  |  |  |  |
| 1-Hexanol | 203.8 [b] | ±15.8 | 216.5 [b] | ±29.8 | 159.6 [a] | ±17.5 | 0.043 |
| *trans*-3-Hexen-1-ol | 1.80 [a] | ±0.28 | 3.27 [b] | ±0.41 | 2.03 [a] | ±0.29 | 0.003 |
| *cis*-3-Hexen-1-ol | 25.4 [b] | ±2.4 | 12.2 [a] | ±1.6 | 11.0 [a] | ±1.6 | 0.000 |
| **Terpenes** |  |  |  |  |  |  |  |
| *cis*-Linalooloxide | 1.93 [a] | ±0.71 | 2.15 [a] | ±0.53 | 2.06 [a] | ±0.64 | 0.909 |
| *trans*-Linalooloxide | 2.17 [a] | ±0.72 | 2.74 [a] | ±0.88 | 2.49 [a] | ±0.61 | 0.666 |
| Linalool | 5.71 [a] | ±1.40 | 6.82 [a] | ±0.82 | 6.91 [a] | ±0.94 | 0.378 |
| Terpinen-4-ol | 0.068 [a] | ±0.020 | 0.070 [a] | ±0.024 | 0.069 [a] | ±0.032 | 0.999 |
| α-Terpineol | 2.00 [a] | ±0.52 | 2.19 [a] | ±0.02 | 2.42 [a] | ±0.20 | 0.343 |
| β-Citronellol | 0.163 [a] | ±0.048 | 0.747 [b] | ±0.051 | 0.130 [a] | ±0.030 | 0.051 |
| Nerol | 15.1 [a] | ±0.6 | 16.4 [a] | ±1.8 | 11.3 [a] | ±4.0 | 0.120 |
| Geraniol | 20.7 [a] | ±1.7 | 24.1 [b] | ±1.4 | 19.0 [a] | ±1.1 | 0.013 |
| **Norisoprenoids** |  |  |  |  |  |  |  |
| 3-oxo-α-Ionol | 5.44 [b] | ±0.43 | 2.80 [a] | ±1.41 | 4.40 [ab] | ±1.56 | 0.102 |
| α-Ionol | 0.035 [a] | ±0.017 | 0.015 [a] | ±0.010 | 0.030 [a] | ±0.018 | 0.329 |
| **Benzenoids** |  |  |  |  |  |  |  |
| Vanillin | 0.175 [a] | ±0.095 | 9.41 [c] | ±1.36 | 1.60 [b] | ±0.22 | 0.001 |
| Methyl vanillate | 4.01 [a] | ±0.21 | 5.17 [b] | ±0.29 | 4.92 [b] | ±0.02 | 0.001 |
| Ethyl vanillate | 0.365 [a] | ±0.129 | 0.737 [b] | ±0.209 | 0.747 [b] | ±0.188 | 0.065 |
| Benzyl Alcohol | 303.7 [b] | ±58.2 | 112.4 [a] | ±10.7 | 163.9 [a] | ±21.5 | 0.002 |

Values in the same row with different letters indicate statistically significant differences ($p < 0.05$). <LOQ: Value below the limit of quantification.

## 2.2. Volatile Composition of Aged Wine

After model aging, wines were different for 28 volatile compounds (Table 3) and seven bound compounds ($p < 0.05$) (Table 4). The fruttaio and the on-vine samples showed significant differences ($p < 0.05$) for 17 compounds, three of which were glycosidically bound precursors. The PCA analysis (Figure 2) after wine model aging showed a total variance of 63.1%. Three clusters were formed corresponding to the three conditions studied: control, fruttaio and on-vine. The compounds that most characterized these three groups were basically the same that were obtained in wine samples before model aging: the class of esters, acids and alcohols for control samples, terpenes and benzenoids for withered samples.

**Table 3.** Concentration (µg/L) of free compounds in aged wine samples. Mean, standard deviation (SD) and ANOVA.

|  | Control |  | Fruttaio |  | On-Vine |  |  |
| --- | --- | --- | --- | --- | --- | --- | --- |
| **Compounds** | **Mean** | **SD** | **Mean** | **SD** | **Mean** | **SD** | **_p_-Value** |
| **Alcohols** |  |  |  |  |  |  |  |
| 2-Butanol | 3019.8 [b] | ±237.4 | 1980.9 [a] | ±291.9 | 3276.2 [b] | ±112.6 | 0.001 |
| 1-Butanol | 247.4 [a] | ±37.2 | 452.1 [b] | ±16.7 | 206.9 [a] | ±50.3 | 0.000 |
| 1-Pentanol | 17.9 [a] | ±0.6 | 15.0 [a] | ±5.2 | 14.2 [a] | ±1.9 | 0.398 |
| Isoamyl alcool | 60,167.0 [a,b] | ±1467.0 | 55,052.5 [a] | ±5134.7 | 63,751.1 [b] | ±2291.3 | 0.051 |
| Phenylethyl Alcohol | 20,612.4 [a] | ±2721.8 | 25,084.0 [ab] | ±2609.2 | 28,513.4 [b] | ±4720.3 | 0.083 |
| **C6 Alcohols** |  |  |  |  |  |  |  |
| 1-Hexanol | 3199.2 [c] | ±22.3 | 2231.4 [b] | ±208.7 | 1787.8 [a] | ±102.2 | <0.0001 |
| *trans*-3-Hexen-1-ol | 51.9 [c] | ±2.6 | 40.4 [b] | ±1.3 | 26.2 [a] | ±7.4 | 0.001 |
| *cis*-3-Hexen-1-ol | 609.6 [b] | ±7.7 | 110.0 [a] | ±39.5 | 74.0 [a] | ±11.1 | <0.0001 |

**Table 3.** *Cont.*

| Compounds | Control Mean | SD | Fruttaio Mean | SD | On-Vine Mean | SD | $p$-Value |
|---|---|---|---|---|---|---|---|
| **Acetate Esters** | | | | | | | |
| Ethyl acetate | 19,178.22 [a] | ±7696.5 | 32,697.7 [b] | ±2026.1 | 38,187.8 [b] | ±5575.2 | 0.015 |
| Isoamyl acetate | 3732.0 [b] | ±188.2 | 1435.2 [a] | ±852.1 | 760.8 [a] | ±300.4 | 0.001 |
| n-Hexyl acetate | 82.8 [b] | ±3.1 | 8.9 [a] | ±2.1 | 8.0 [a] | ±2.8 | 0.000 |
| Phenylethyl acetate | 55.2 [b] | ±1.9 | 29.0 [a] | ±10.2 | 23.8 [a] | ±2.2 | 0.002 |
| **Ethyl Esters** | | | | | | | |
| Ethyl butanoate | 255.2 [a] | ±12.2 | 296.2 [b] | ±20.1 | 231.7 [a] | ±24.2 | 0.018 |
| Ethyl 3-methylbutanoate | 56.6 [a] | ±2. 9 | 56.2 [a] | ±3.5 | 53.2 [a] | ±7.5 | 0.685 |
| Ethyl hexanoate | 491.2 [c] | ±18.2 | 357.1 [b] | ±32.5 | 300.4 [a] | ±29.1 | 0.000 |
| Ethyl octanoate | 245.1 [b] | ±7.7 | 153.0 [a] | ±37.2 | 138.9 [a] | ±14.4 | 0.003 |
| Ethyl decanoate | 18.8 [a] | ±1.7 | 16.0 [a] | ±6.6 | 12.1 [a] | ±3.1 | 0.246 |
| Ethyl lactate | 3704.2 [b] | ±158.1 | 3656.8 [b] | ±99.3 | 3194.4 [a] | ±352.9 | 0.065 |
| **Fatty Acids** | | | | | | | |
| 3-Methylbutanoic acid | 706.1 [b] | ±115.7 | 306.1 [a] | ±8.8 | 448.0 [ab] | ±194.8 | 0.025 |
| Hexanoic acid | 3753.0 [b] | ±77.1 | 1833.3 [a] | ±156.2 | 1878.8 [a] | ±235.0 | <0.0001 |
| Octanoic acid | 2901.9 [b] | ±140.7 | 1288.8 [a] | ±183.5 | 1405.4 [a] | ±166.7 | <0.0001 |
| **Terpenes** | | | | | | | |
| cis-Linalool oxide | 7.85 [a] | ±0.53 | 8.57 [ab] | ±0.59 | 9.60 [b] | ±0.66 | 0.031 |
| trans-Linalool oxide | 4.84 [a] | ±0.30 | 6.07 [b] | ±0.18 | 6.06 [b] | ±0.28 | 0.002 |
| Linalool | 15.8 [a] | ±3.17 | 23.1 [a] | ±3.1 | 18.80 a | ±3.3 | 0.142 |
| Terpinen-1-ol | <LOQ | | <LOQ | | <LOQ | | |
| Terpinen-4-ol | 11.7 [a] | ±2.7 | 17.5 [b] | ±2.0 | 13.4 [a,b] | ±1.8 | 0.044 |
| Ho-trienol | 0.013 [a] | ±0.006 | 0.018 [a] | ±0.003 | 0.022 [a] | ±0.003 | 0.115 |
| α-Terpineol | 34.2 [a] | ±3.3 | 37.5 [a] | ±3.9 | 34.6 [a] | ±3.3 | 0.511 |
| Nerol | 1.47 [a] | ±0.57 | 1.33 [a] | ±0.75 | 2.29 [a] | ±0.06 | 0.147 |
| Geraniol | 2.97 [a] | ±0.14 | 4.31 [a] | ±1.99 | 2.64 [a] | ±1.02 | 0.319 |
| β-Citronellol | 2.71 [a] | ±2.37 | 7.92 [b] | ±1.60 | 4.38 [a] | ±0.99 | 0.027 |
| p-Menthane-1.8-diol | 0.417 [b] | ±0.060 | 0.214 [a] | ±0.069 | 0.404 [b] | ±0.054 | 0.011 |
| α-Phellandrene | 0.802 [a] | ±0.023 | 0.518 [a] | ±0.237 | 0.620 [a] | ±0.076 | 0.218 |
| 1,4-Cineole | 0.202 [a] | ±0.169 | 0.286 [a] | ±0.041 | 0.323 [a] | ±0.045 | 0.398 |
| 1,8-Cineole | 0.188 [a] | ±0.032 | 0.307 [a] | ±0.024 | 0.362 [a] | ±0.018 | 0.341 |
| Limonene | 0.408 [a] | ±0.178 | 0.497 [a] | ±0.026 | 0.430 [a] | ±0.038 | 0.596 |
| γ-Terpinen | 1.01 [a] | ±0.13 | 0.294 [a] | ±0.131 | 0.992 [a] | ±0.040 | 0.664 |
| p-Cymene | 0.162 [a] | ±0.061 | 0.252 [b] | ±0.016 | 0.155 [a] | ±0.018 | 0.034 |
| Terpinolene | 0.115 [a] | ±0.036 | 0.175 [b] | ±0.015 | 0.098 [a] | ±0.013 | 0.017 |
| **Norisoprenoids** | | | | | | | |
| β-Damascenone | 3.49 [a] | ±0.05 | 3.08 [a] | ±0.18 | 3.58 [a] | ±0.10 | 0.418 |
| α-Ionone | 0.282 [a] | ±0.128 | 0.448 [a] | ±0.099 | 0.480 [a] | ±0.287 | 0.441 |
| α-Ionol | 0.130 [a] | ±0.0317 | 0.140 [a] | ±0.094 | 0.295 [a] | ±0.096 | 0.554 |
| Vitispirane | 5.07 [a] | ±4.66 | 8.33 [a] | ±0.51 | 7.72 [a] | ±0.89 | 0.367 |
| TPB | 0.080 [b] | ±0.013 | 0.049 [a] | ±0.006 | 0.055 [a] | ±0.008 | 0.015 |
| TDN | 4.25 [a] | ±1.34 | 4.20 [a] | ±0.68 | 4.50 [a] | ±0.35 | 0.906 |
| 3-oxo-α-Ionol | 17.6 [a,b] | ±4.2 | 11.6 [a] | ±2.8 | 20.7 [b] | ±3.4 | 0.080 |
| **Benzenoids** | | | | | | | |
| 4-Ethyl guaiacol | 0.190 [a] | ±0.049 | 0.295 [a] | ±0.152 | 0.153 [a] | ±0.193 | 0.500 |
| 4-Vinyl guaiacol | 15.5 [a] | ±4.4 | 21.3 [a] | ±0.7 | 20.2 [a] | ±5.1 | 0.235 |
| 2,6-Dimethoxyphenol | 0.078 [a] | ±0.014 | 0.110 [a] | ±0.029 | 0.208 [a] | ±0.043 | 0.220 |
| Benzyl Alcohol | 153.8 [a] | ±27.3 | 275.0 [c] | ±12.1 | 215.2 [b] | ±16.3 | 0.001 |
| Vanillin | 5.36 [a] | ±1.04 | 7.26 [b] | ±0.37 | 6.39 [ab] | ±0.61 | 0.050 |
| Methyl vanillate | 4.83 [a] | ±0.50 | 5.96 [b] | ±0.37 | 5.79 [b] | ±0.34 | 0.030 |
| Ethyl vanillate | 34.4 [a] | ±6.0 | 56.7 [b] | ±3.7 | 69.7 [b] | ±3.4 | 0.012 |

Values in the same row with different letters indicate statistically significant differences ($p < 0.05$). <LOQ: Value below the limit of quantification.

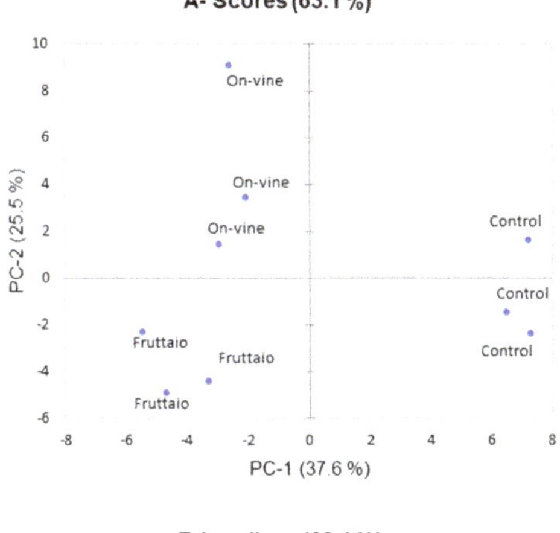


**Figure 2.** Principal component analysis showing aged wine samples scores (**A**) and loadings (**B**), only loadings with score value >0.75 were shown. Loadings plot number correspond to: (F1) 2-Butanol, (F2) 1-Butanol, (F5) Phenyl ethyl alcohol, (F6) 1-Hexanol, (F8) *cis*-3-Hexenol, (F10) Isoamyl acetate, (F11) n-Hexyl acetate, (F12) Phenylethyl acetate, (F13) Ethyl butanoate, (F15) Ethyl hexanoate, (F16) Ethyl octanoate, (F19) 3-Methylbutanoic acid, (F20) Hexanoic acid, (F21)Octanoic acid, (F22) *cis*-Linalool oxide, (F23) *trans*-Linalool oxide, (F25)Terpinen-1-ol, (F26) Terpinen-4-ol, (F31) β-Citronellol, (F32) p-Menthane-1,8-diol, (F38) p-Cymene, (F39) Terpinolene, (F50) Benzyl alcohol, (B3) bound *cis*-3-Hexenol, (B8) bound α-Terpineol, (B9) bound β-Citronellol, (B11) bound Geraniol, (B12) bound Benzyl alcohol, (B14) bound Phenylethyl alcohol, B16) bound 3-oxo-α-Ionol, (B17) bound Methyl vanillate.

**Table 4.** Concentration (μg/L) of bound compounds in aged wine samples. Mean, standard deviation (SD) and ANOVA.

| | Control | | Fruttaio | | On-Vine | | |
|---|---|---|---|---|---|---|---|
| Compounds | Mean | SD | Mean | SD | Mean | SD | *p*-Value |
| **Alcohols** | | | | | | | |
| Phenylethyl alcohol | 121.4 [ab] | ±7.8 | 96.2 [a] | ±21.9 | 146.9 [b] | ±9.4 | 0.015 |
| **C6 alcohols** | | | | | | | |
| 1-Hexanol | 84.6 [a] | ±6.7 | 84.8 [a] | ±14.0 | 82.8 [a] | ±7.4 | 0.965 |
| *trans*-3-Hexen-1-ol | 0.588 [a] | ±0.035 | 0.852 [b] | ±0.163 | 0.750 [ab] | ±0.065 | 0.054 |
| *cis*-3-Hexen-1-ol | 22.6 [b] | ±1.6 | 11.2 [a] | ±5.2 | 13.8 [a] | ±0.8 | 0.011 |

**Table 4.** *Cont.*

| Compounds | Control Mean | SD | Fruttaio Mean | SD | On-Vine Mean | SD | *p*-Value |
|---|---|---|---|---|---|---|---|
| **Terpenes** | | | | | | | |
| *cis*-Linalooloxide | 2.23 [a] | ±0.26 | 2.41 [a] | ±0.46 | 2.79 [a] | ±0.32 | 0.229 |
| *trans*-Linalooloxide | 2.83 [a] | ±0.74 | 3.05 [a] | ±0.71 | 3.68 [a] | ±0.64 | 0.366 |
| Linalool | <LOQ | | <LOQ | | <LOQ | | |
| Terpinen-4-ol | <LOQ | | <LOQ | | <LOQ | | |
| α-Terpineol | 0.182 [a] | ±0.029 | 0.977 [b] | ±0.253 | 0.977 [b] | ±0.140 | 0.009 |
| β-Citronellol | 0.175 [a] | ±0.004 | 1.19 [b] | ±0.20 | 0.000 [a] | ±0.096 | <0.0001 |
| Nerol | 3.60 [a] | ±0.60 | 4.68 [a] | ±0.20 | 3.95 [a] | ±0.73 | 0.128 |
| Geraniol | 4.08 [a] | ±1.03 | 6.95 [b] | ±0.56 | 5.20 [ab] | ±1.63 | 0.060 |
| **Norisoprenoids** | | | | | | | |
| 3-oxo-α-Ionol | 2.59 [a] | ±0.22 | 2.11 [a] | ±0.37 | 3.17 [b] | ±0.24 | 0.011 |
| α-Ionol | <LOQ | | <LOQ | | <LOQ | | |
| **Benzenoids** | | | | | | | |
| Vanillin | 3.76 [a] | ±0.07 | 3.61 [a] | ±0.30 | 3.76 [a] | ±0.16 | 0.595 |
| Methyl vanillate | 3.94 [a] | ±0.09 | 5.22 [b] | ±0.31 | 5.02 [b] | ±0.28 | 0.001 |
| Ethyl vanillate | 0.317 [a] | ±0.132 | 0.737 [a] | ±0.135 | 0.668 [a] | ±0.209 | 0.134 |
| Benzyl Alcohol | 130.5 [b] | ±6.2 | 65.7 [a] | ±3.4 | 105.9 [ab] | ±8.5 | 0.023 |

Values in the same row with different letters indicate statistically significant differences ($p < 0.05$). <LOQ: Value below the limit of quantification.

## 3. Discussion

Post-harvest withering plays a central role in determining the compositional and sensory characteristics of Valpolicella red wines [1,3]. From a quantitative point of view, the main physiological change associated with this traditional practice is water loss, that is carried out up to an average of 30% weight loss depending on wine style. This has major implication for grape composition, most notably increased concentration of metabolites such as sugars, phenolics, and certain aroma compounds, directly influencing composition of the resulting wine. Additional important consequences of increased sugar levels are related to changes in yeast metabolism, which can further impact wine composition. However, it has been recently shown that post-harvest withering is not simply a dehydration process, with many complex metabolic transformations beyond simple concentration taking place inside the berry, inducing important modifications in the pool of grape secondary metabolites, including volatile compounds [17]. In consideration of this complex scenario, one of the purposes of the present study was to investigate how and to what extent withering of the grapes affects the volatile composition of the resulting wine. Second, and most important, this study had the objective to assess the potential of an alternative withering approach to modulate Corvina wine volatile composition. Although post-harvest withering is traditionally carried out in warehouses (locally called "fruttaio"), there is an ongoing interest towards the exploration of alternative strategies that can be applied to obtain a suitable degree of over-ripening or withering, also with the aim of producing alternative wine types and styles [2]. Among these, cane-cut on-vine has been shown to positively influence wine aroma and phenolic composition [6–12]. In the present study, an alternative approach to on-vine withering, still based on blocking xylem flow but not involving cane cutting, was investigated in comparison with conventional fruttaio withering. As sugar levels at grape crush were similar for both withering modalities, any difference is expected to result from differences in grape composition in terms of secondary metabolites or interaction with yeast.

### 3.1. Influence of Grape Withering on Volatile Composition of Corvina Wines

Analysis of free and glycosidically-bound volatile compounds of the wines at bottling showed that withering of the grapes significantly affected wine aroma compounds, influencing the concentrations of various classes of volatiles. Free compounds can have a direct influence on wine aroma while the bound compounds can act as an aroma reservoir that is released during aging. At a general level, fermentation-derived volatiles such as esters, higher alcohols, and acids, as well as grape-related compounds such as certain norisoprenoids, were mostly associated with non-withered grapes, whereas withering resulted in higher wine content in terpenes and benzenoids (Figure 1 and Table 1). Among compounds known to impact red wine aroma, acetate esters (i.e., isoamyl acetate) and ethyl fatty acid esters (i.e., ethyl hexanoate and octanoate) were strongly influenced by withering, which resulted in a significant decrease in the concentration of nearly all the analyzed esters. Esters are related to red wine's fruity character [23] and are formed during alcoholic fermentation involving amino acid metabolism in the case of acetates, and fatty acid metabolism for ethyl esters [24]. The production of esters by yeast is influenced by several factors, and different studies have reported an influence of grape maturity [25] and levels of withering [4,9,26] on wine ester content, suggesting and influence of must sugar content on ester production. In agreement with these reports, wines from withered Corvina grapes, which at harvest displayed additional 2 Brix compared to control grapes, showed lower ester content. In particular, acetate esters were more impacted, in spite of the fact that higher alcohol content, a precursor of acetates, was not so different across treatments and in some cases was even greater in withered samples. Ester/alcohol ratios were calculated to establish esterification rates of the different esters, and in the case of acetates it appeared clear that acetylation was much higher in fermentation of non-withered grapes (Figure 3). Likewise, a higher acetylation rate was also observed for the control wine in the case of the ethyl ester of the branched chain fatty acid 3-methylbutanoic acid, also derived from amino acid metabolism. Conversely, although concertation of ethyl esters was higher in control samples, esterification of the corresponding fatty acid was similar in all treatments, so that it can be inferred that wine ester levels were determined by concentration of the corresponding fatty acid. It can be therefore assumed that, under our experimental conditions, withering impact on esters was due on one hand to reduced acetylation and on the other hand to reduced production of fatty acids, which would be in agreement with the observations of Saerens et al. (2008) [27]. In the case of ethyl fatty acid esters, the reduced availability of short chain fatty acid precursors could be due to the greater availability of unsaturated fatty acids in musts from withered grapes [28], which would result in reduced medium chain fatty acids biosynthesis [29]. An increase in available lipids can also reduce the expression of the *ATF1* gene and therefore lower acetyl transferase activity catalyzing the acetylation reaction [30]. Interestingly, ethyl acetate showed a completely different trend, its concentration increasing significantly in wines from withered grapes. Although acetyl transferase activities are expected to play a role in ethyl acetate formation by *S. cerevisiae*, recent observations indicated that acetyl transferases other than Atf1 and Atf2 contribute significantly to production of this ester [31], which could explain its different response to withering.

$C_6$ alcohols were also found to discriminate, with a high level of significance, control wines from the two withering modalities. $C_6$ alcohols contribute to the "leafy" and "herbaceous" odors of wines [32]. In control samples *cis*-3-hexenol showed and odor active value (OAV, calculated as ration between concentration and odor threshold) higher than one therefore potentially contributed to wine aroma (OAV = 1.3). Instead, in withered samples, the $C_6$ alcohols had OAV values lower than one. $C_6$ alcohols are formed during berry crushing by enzymatic oxidation of grape unsaturated fatty acids, initiated by grape lipoxygenase enzymes [33]. Zenoni et al. (2016) [17] reported a decrease in the expression of lipoxygenase genes during withering of Corvina, which could explain the decrease in $C_6$ alcohols observed here. However, other studies indicated an opposite trend [8,16], an increase in $C_6$ aldehydes and alcohols during postharvest grape dehydration of Malvasia grape was reported [8], suggesting that more complex patterns could occur.

Various terpenes were affected by withering, although trends varied depending on the specific molecule. The importance of monoterpene alcohols and cyclic terpenes to Corvina wines aroma was recently described, in particular for linalool [34,35]. In the present study, linalool was the main monoterpene alcohol detected and its concentration was significantly decreased by withering, in agreement with previous findings [17,36]. Considering that in control wines linalool had an OAV

= 1.1, a possible contribution to wine aroma characteristics can be expected, whereas in wines from withered grapes this was not the case (OAV = 0.59 and 0.66 for fruttaio and on-vine respectively). Terpenes are produced in grapes through both the 1-deoxy-D-xylulose-5-phosphate/methylerythritol phosphate (DOXP/MEP) pathway and the mevalonic acid (MVA) pathway. In Corvina the influence of withering on these pathways is complex, with upper steps of the pathway being downregulated but late biosynthetic steps upregulated [16]. In addition to free forms of terpenes, grapes also contain non-volatile glycosylated forms of these compounds, which in Corvina can contribute significantly to terpenes level in finished through enzymatic and chemical hydrolysis during vinification [35,37]. Although in the present study differences in glycosidically-bound terpenes in the finished wines were relatively small, we observed generally higher concentrations of bound terpenes in withered wines (Table 2). Contrary to linalool, citronellol, the second most abundant monoterpene alcohols detected, increased with withering. In non-aromatic grapes such as Corvina, formation of citronellol is connected to the ability of yeast to reduce available geraniol including the portion derived from hydrolysis of geraniol glyosidic precursors [37,38]. Bound geraniol in finished wines increased in one of the withering modalities, supporting a possible contribution of bound geraniol to free citronellol levels. In wines from withered Pinot noir, Moreno et al. (2008) [9] also observed increased wine citronellol content. Among other terpenes, withering was consistently associated with increased contents of linalool oxides, limonene and 1,4-cineole. Small increases in the content of terpinen-4-ol in wine were also observed with withering, in agreement with the observations of Zenoni et al. (2016) [17]. The contribution of linalool oxides, limonene, 1,4-cineole and terpinen-4-ol to wine aroma seemed to be limited because their concentrations were found to be lower than the respective odor thresholds.

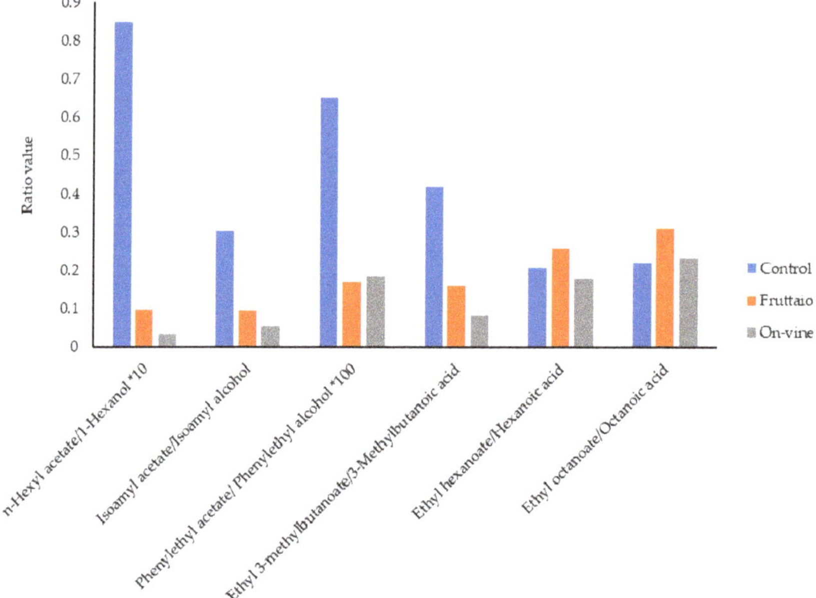


**Figure 3.** Ratio between esters and related precursors.

Comparted to terpenes, norisoprenoids, were affected to a smaller extent by withering, with concentration of the potent odorant β-damascenone decreasing. The formation of this compound during winemaking is associated with multiple pathways involving acid- or yeast-mediated hydrolysis of different precursors [39,40]. The negative influence of withering on damascenone wine content could be due to complex factors and requires further investigation, also considering that other norisoprenoids such as 3-oxo-α-ionol had an opposite behavior and were found in higher concentrations in withered samples.

The benzenoids ethyl vanillate, and benzyl alcohol were found in higher concentrations in withered samples, unlike previously reports by Bellincontro et al. (2016) [4], albeit with more intense withering.

Considering that wines made from withered grapes are generally destined to age, the volatile profile evolution of withered and non-withered samples was investigated by means of a model aging protocol [34]. Data showed that after model aging, the three sample modalities showed differences for a smaller number of compounds compared to young wines. Variations in compound concentrations are reported in Table 5. Esters remained one major factor discriminating wines from withered grapes after aging (Table 3 and Figure 2), with control wines exhibiting higher levels of ethyl fatty acids and acetate esters. Trends during aging were, however, different, with withered wines typically showing reduced losses of even increases in some cases compared to control. Esters can be formed or degraded according to wine pH and the ester/acid ratio. As a consequence, esters produced in a higher amount by yeast during fermentation, such as isoamyl acetate, tend to decrease during aging while branched-chain fatty acid esters increase [41–43]. Considering that the pH of the different samples was similar, we can conclude that by reducing esters formation during fermentation, withering resulted in reduced ester losses. One exception to this was observed for ethyl hexanoate, for which the rate of hydrolysis was similar in all treatments.

**Table 5.** Composition of Corvina wines made from not withered grapes, withered on-vine and withered in "fruttaio".

|          | Brix | pH            | Alcohol % vol |
|----------|------|---------------|---------------|
| Control  | 21   | 2.96 ± 0.05   | 13.6 ± 0.2    |
| On-vine  | 23.2 | 2.92 ± 0.08   | 15.0 ± 0.2    |
| Fruttaio | 23.2 | 3.03 ± 0.06   | 15.2 ± 0.2    |

Aging patterns of certain terpenes also showed differences that could be associated with withering. For example, linalool in control wines during aging decreased from being sensorially active (OAV = 1.1) to an OAV < 1. Instead, in withered wines, linalool concentration tended to increase with aging. Particularly in the Fruttaio samples, after aging linalool had an increase of 8.3 µg/L, 1.5 times higher than young Fruttaio wines. This increase could be due to a higher content of glycosylated precursors in withered samples. However, the analysis of bound compounds in young wines did not show significant differences between samples that could explain the observed differences. It may be that different precursors forms of linalool exist in our samples that were not quantified with the employed method. *cis*-Linalool oxide increased more markedly in wines from withered grapes, and this could be attributed to acid hydrolysis of glycosidic precursors (diendiol) [44], and 3,7-dimethyloct-1-ene-3,6,7-triol (triol) [45]. This last pathway seemed more consistent in this sample set, as both *cis*- and *trans*-linalool oxide bound precursors did not decrease with aging. 1,8-Cineole also displayed substantially different behaviors during aging between wines from withered and non-withered grapes, with concentration increasing during aging only in withered wines. This could be due to the fact that young wines from withered grapes exhibited significantly higher content of tepinen-4-ol, which we have recently shown to be a precursor to 1,8-cineole in Corvina wines [34].

β-Damascenone evolution with aging also highlighted a major difference associated with withering. Evolution of β-damascenone in Corvina wines during aging is characterized by a complex trend with an initial increase followed by a decline [34], reflecting simultaneous release from precursors (until available) followed by degradation through various reactions [46,47]. In the present study, the concentration of β-damascenone after model aging remained stable in control wines, and increased in withered samples reaching, in the case of on-vine samples, the level of control (Figure 4). Samples withered in fruttaio showed the most important increase of β-damascenone.

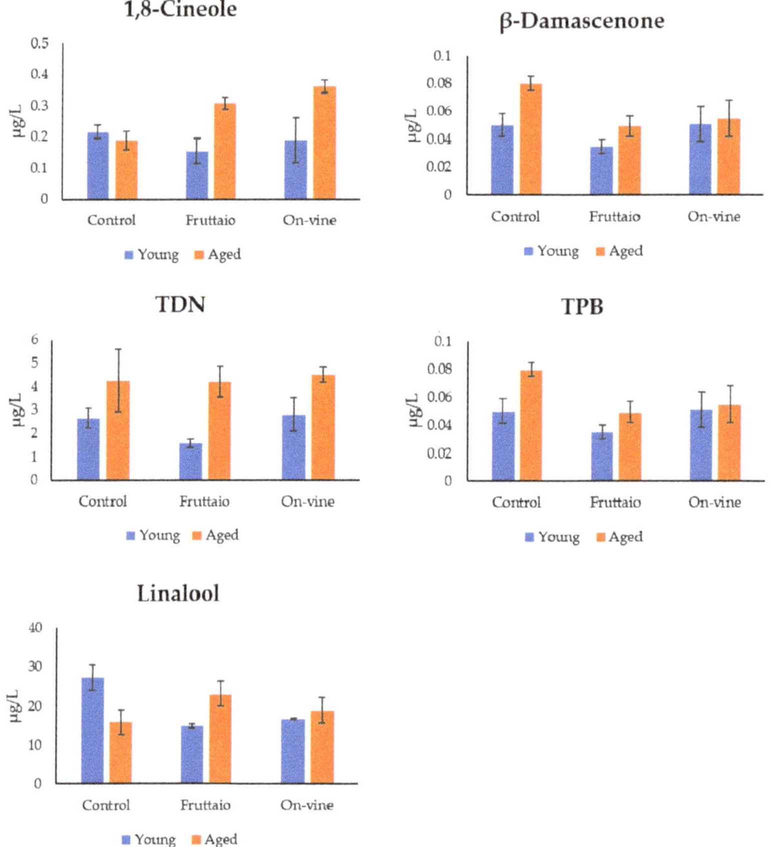

**Figure 4.** 1,8-Cineole, β-damascenone, (*E*)-1-(2,3,6-trimethylphenyl)buta-1,3-diene (TPB),1,1,6-Trimethyl-1,2-dihydronapthalene (TDN) and linalool content in young and model aged wines.

The (*E*)-1-(2,3,6-trimethylphenyl)buta-1,3-diene (TPB) also showed interesting differences in aged wines. A significant important increase has been observed after aging only in control samples. A slight increase occurred in samples withered in fruttaio while on-vine samples did not show any changes with aging. TPB has a tobacco aroma at low concentrations and geranium like odor at higher concentrations [48]. Its concentration in Corvina wine has been correlated with wine ageing [34]. It has been suggested that in red wine rich in tannins, TPB could react with polyphenols, resulting in a lower concentration like in Shiraz and Cabernet Sauvignon (Janusz 2003). It is reported that withered wines have a higher polyphenol content [10,12] while Corvina is known to be poor in polyphenols, so a TPB–tannin reaction may explain the lower content found in aged withered samples.

TDN variation was higher in fruttaio samples; however, it should be noticed that at the end of model aging samples of all the three modalities, it reached the same concentration level. TDN has a kerosene-like aroma, the concentration in wine was reported to be influenced by grape sun exposure, wine age, pH, and storage temperature [49,50]. Our data suggested that TDN concentration in wine was not affected by grape withering, the level of TDN formed in withered wine could depend on the TDN precursors accumulated just before the start of the withering process, harvest or peduncle twist.

The occurrence of aroma notes related to TDN is often associated with aged wines; however, in the control and on-vine samples, an OAV of 1.32 and 1.39, respectively, was observed already in the young wine, indicating a possible sensory contribution.

## 3.2. Influence of Withering Modality on Volatile Composition of Corvina Wine

In comparison with the differences due to withering, those associated with the withering modality were quantitatively less important and restricted to a small number of volatiles. In young wines, small but statistically significant increases in the concentration of different wine esters and citronellol were observed in fruttaio withering compared to on-vine withering, whereas linalool, β-damascenone, and α-ionol were mostly associated with on-vine withering. The trends observed for glycosylated volatiles were different, as in this case fruttaio wines exhibited a higher content of bound terpenes such as geraniol and nerol, whereas benzenoids and certain norisoprenoids were more abundant in on-vine withering. Several studies have investigated the influence of on-vine over-ripening or even on-vine withering on grape composition, but only in a few cases these were compared with other methods of withering. Zamboni et al. (2008) [16] provided interesting insights in differences existing at a transcriptomic level between on-vine and off-vine (fruttaio) withering of Corvina, indicating that differences in transcripts associated with secondary metabolites were minor [16]. However, on-vine withering did not involve any blockage of vascular tissues, so results are hard to compare with the present study.

Interestingly, some differences between the two withering modalities could be observed after aging. For example, the above-mentioned trend of reduced ester loss was lower in the case of on-vine withering, to the point that some esters actually increased during aging of wines from on-vine withering. Increases in certain grape-derived compounds were also dependent on withering, as in the case of 3-oxo-α-ionol, p-menthane-1,8-diol. Overall, it appeared that the two different withering conditions induced similar types of changes, mostly modulating the extent of such changes.

## 4. Materials and Methods

### 4.1. Chemicals

Octan-2-ol (97%), 1-hexanol (99%), *cis*-3-hexenol (98%), *trans*-3-hexenol (97%), vanillin (99%), 2,6-dimethoxyphenol (99%), linalool (97%), terpinen-4-ol (≥95%), α-terpineol (90%), nerol (≥97%), geraniol (98%), linalool oxide (≥97%), β-citronellol (95%), p-cymene (99%), terpinolene (≥85%), γ-terpinene (≥97%), limonene (97%), 1,8-cineole (99%), 1,4-cineole(≥98.5%), β-damascenone (≥98%), isoamyl alcohol (98%), benzyl alcohol (≥99%), 2-phenylethanol (≥99%), ethyl acetate (99%), ethyl butanoate (99%), ethyl 3-methyl butanoate ((≥98%), isoamyl acetate (≥95%), ethyl hexanoate (≥95%), phenylethyl acetate (99%), n-hexyl acetate (≥98%), ethyl lactate (≥98%), ethyl octanoate (≥98%), ethyl decanoate (≥98%), hexanoic acid (≥99%), octanoic acid (≥98%), α-phellandrene (95%), p-menthane-1,8-diol (97%), 3-methylbutanoic acid (99%), α-ionone (90%), 1-pentanol (99%), 1-butanol (≥99%), 2-butanol (≥99%), ethyl guaiacol (≥99%), vinyl guaiacol (≥98%), methyl-vanillate (99%) and ethyl vanillate (99%), were supplied by Sigma Aldrich (Milan, Italy). Dichloromethane (≥99.8%) and methanol (≥99.8%), were provided by Honeywell (Seelze, Germany). Sodium chloride (≥99.5%) was supplied by Sigma Aldrich (Milan, Italy).

### 4.2. Wine Samples

Wine samples were produced in the experimental facility of Masi Agricola. Corvina grapes from the 2017 vintage were obtained from a single 3 ha vineyard (45°29'22.9" N 10°46'20.5" E) located in the town of Lazise, 25 km west of Verona. The vineyard site was flat, with an altitude of 70 m asl. Vines had 12 years of age and were trained with a double arch cane system, with an average of 60,000 gems/ha and a yearly production of 11–12 tons/ha. Upon achievement of a sugar level of 21 Brix (27 of September), three experimental modalities were applied. Control grapes were hand harvested, placed in 7 kg harvest bins and transferred to the experimental winery where they were directly vinified as described later. A second batch, labelled "fruttaio", was harvested on the same day and the harvest bins were placed in a non-conditioned withering warehouse until November 4, when the berries had achieved a sugar content of 23.2 °Brix. Average conditions in the warehouse over the same period for the previous 10 years indicated a gradual temperature decrease (from 16 °C to 7 °C) and a progressive increase in relative humidity (from 55% to 80%). A third modality, labelled "on-vine", was obtained by applying a peduncle twist in order to block vascular tissue and induce grape dehydration (Figure 5). Weather conditions in the vineyard area during the on-vine withering period were obtained from the Arpa Veneto meteorological database (http://www.arpa.veneto.it/). These conditions were in line with the typical conditions of the area, and were as follows: average daily minimum temperature of 7.5 °C), average daily maximum temperature of 20.6 °C, average daily mean temperature of 13.1 °C, average daily minimum relative humidity 44%, average daily maximum relative humidity of 97%, total precipitations 19 mm (1 rainy day). Upon achievement of a sugar content of 23.2 Brix (25 October), grapes were hand harvested as for the other modalities and were vinified. All vinifications were carried out in triplicate. For each vinification, 100 kg of grapes were destemmed and crushed, and the obtained musts were added with 100 mg/L of potassium metabilsulfite. Fermentations were conducted 75 L steel tanks by inoculation with the the proprietary *S. cerevisiae* yeast MASY03 (Microbion, Castel d'Azzano, Italy). At the end of fermentation, potassium metabisulphite was added in order to reach 30 mg/L of free $SO_2$, wines were then filtrated at 1 micron and bottled. Sample bottles were stored at 16 °C until analysis. Data concerning grapes at harvest and wine at bottling are summarized in Table 5.

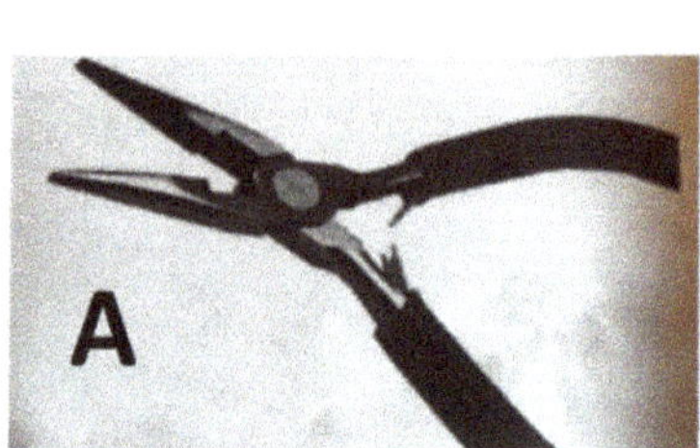

**Figure 5.** Plier for the peduncle twist (**A**), and grape after peduncle twist (**B**). Photo courtesy of Masi.

### 4.3. Wines Model Aging

Model aging was carried out as described by Slaghenaufi et al. (2019) [35], by placing 115 mL of wine in glass vial and crimped leaving 0.8 mL of headspace corresponding to 2 mg/L of oxygen. Vials were then crimped and sealed with Araldite glue and stored at 40 °C for 12 weeks.

### 4.4. Volatile and Glycosidically-Bound Compound Analysis

Volatile and glycosidically-bound compounds have been analyzed as described by Slaghenaufi et al. (2019) [35] with minor modification. In total, 50 mL of sample was added with 20 μL of internal standard solution (2-octanol at 42 mg/L in ethanol) and diluted with 50 mL of distilled water. The solution was then loaded on a BOND ELUT-ENV, SPE cartridge, containing 1 g of sorbent (Agilent Technologies, Santa Clara, CA, USA), previously activated with 20 mL of methanol and equilibrated with 20 mL of water. The cartridge was then washed with 15 mL of water. Free volatile compounds were eluted with 10 mL of dichloromethane, and then concentrated under gentle nitrogen stream to 200 μL prior to GC injection. Bound compounds were recovered with 20 mL of methanol. Methanol was then evaporated under vacuum. Bound compounds were then dissolved in 5 mL of citrate buffer (pH 5). were added to dissolve bound compounds to that 200 μL of an enzyme preparation AR2000 (DSM, Brussels, Belgium, prepared at 70 mg/mL in citrate buffer) were added and incubated at 37 °C for 24 h under shaking (150 rpm).

A calibration curve was prepared for each analyte using seven concentration points and three replicate solutions per point in model wine (12% *v/v* ethanol, 3.5 gr/L tartaric acid, pH 3.5) [51]. A total of 20 μL of internal standards 2-octanol (42 mg/L in ethanol), was added to the solution. SPE extraction and GC-MS analysis were performed as described above for the samples. Calibration curves were obtained using Chemstation software (Agilent Technologies) by linear regression, plotting the response ratio (analyte peak area/internal standard peak area) against concentration ratio (analyte added concentration/internal standard concentration). Method characteristics are reported in Table 6. The 3-oxo-α-ionol analysis was semi-quantitative and they were expressed as µg/L of 2-octanol equivalent (internal standard) as for this compound no commercial standard was available.

**Table 6.** Retention indices, quantification ions of studied compounds.

| | Method [1] | LRI [1] | Identification [2] | Quantitation Ion *m/z* | Qualifier Ions *m/z* | LOD (µg/L) | LOQ (µg/L) |
|---|---|---|---|---|---|---|---|
| 1-Butanol | a | 1159 | RS | 56 | 55 | 0.02 | 0.06 |
| 2-Butanol | a | 1020 | RS | 59 | | 0.20 | 0.6 |
| 1-Pentanol | a | 1256 | RS | 55 | 56, 57, 70 | 0.04 | 0.11 |
| Isoamyl alcohol | a | 1220 | RS | 57 | 55, 56, 70 | 0.02 | 0.06 |
| Phenylethyl Alcohols | a | 1920 | RS | 91 | 65, 92, 122 | 1.95 | 5.84 |
| 1-Hexanol | a | 1316 | RS | 56 | 55, 69 | 0.76 | 2.27 |
| *trans*-3-Hexen-1-ol | a | 1379 | RS | 67 | 55, 69, 82 | 0.40 | 1.21 |
| *cis*-3-Hexen-1-ol | a | 1391 | RS | 68 | 55, 69, 83 | 1.23 | 3.68 |
| Ethyl acetate | b | 895 | RS | 88 | 61, 70 | 0.5 | 1.58 |
| Isoamyl acetate | a | 1125 | RS | 70 | 55, 60, 87 | 0.03 | 0.1 |
| n-Hexyl acetate | a | 1271 | RS | 56 | 55, 61, 84 | 0.03 | 0.1 |
| Ethyl 3-methyl butanoate | a | 1069 | RS | 88 | 57, 60, 85 | 0.30 | 0.9 |
| Ethyl butanoate | a | 1032 | RS | 71 | 88 | 0.01 | 0.04 |
| Ethyl hexanoate | a | 1240 | RS | 88 | 60, 99 | 5.82 | 17.47 |
| Ethyl octanoate | a | 1430 | RS | 88 | 57, 100, 127 | 0.54 | 1.63 |
| Ethyl decanoate | a | 1640 | RS | 88 | 71, 101, 155 | 0.16 | 0.49 |
| Ethyl lactate | a | 1340 | RS | 75 | 88, 90 | 2.1 | 6.3 |
| 3-Methylbutanoic acid | a | 1667 | RS | 60 | 87 | 0.17 | 0.52 |
| Hexanoic acid | a | 1839 | RS | 60 | 73, 87 | 0.15 | 0.46 |
| Octanoic acid | a | 2071 | RS | 60 | 73, 101, 115 | 0.00 | 0.01 |
| *cis*-Linalooloxide | b | 1437 | RS | 59 | 111, 94 | 0.02 | 0.07 |
| *trans*-Linalooloxide | b | 1469 | RS | 59 | 111, 94 | 0.02 | 0.07 |
| Linalool | b | 1547 | RS | 71 | 121, 93 | 0.08 | 0.25 |
| Geraniol | b | 1860 | RS | 93 | 123, 121, 105 | 0.06 | 0.2 |
| β-Citronellol | b | 1771 | RS | 69 | 82, 81, 67 | 0.07 | 0.21 |
| α-Terpineol | b | 1701 | RS | 136 | 121, 93, 59 | 0.23 | 0.7 |
| α-Phellandrene | b | 1180 | RS | 93 | 136, 91 | 0.001 | 0.003 |
| γ-Terpinen | b | 1188 | RS | 121 | 93, 126 | 0.03 | 0.1 |
| Limonene | b | 1198 | RS | 136 | 139, 125, 111 | 0.03 | 0.1 |
| 1,4-Cineole | b | 1186 | RS | 154 | 139, 111, 108 | 0.003 | 0.011 |
| 1,8-Cineole | b | 1217 | RS | 154 | 139, 111, 108 | 0.003 | 0.011 |
| p-Cymene | b | 1271 | RS | 119 | 134, 91 | 0.02 | 0.06 |
| Terpinolene | b | 1283 | RS | 121 | 136, 93 | 0.03 | 0.09 |
| Terpinen-1-ol | b | 1581 | LRI MS | 136 | 121, 81 | - | - |
| Terpinen-4-ol | b | 1614 | RS | 71 | 111, 93, 86 | 0.02 | 0.05 |
| p-Menthane-1,8-diol | a | 2250 | RS | 96 | 88, 139 | 0.03 | 0.09 |
| Ho-trienol | b | 1585 | LRI MS | 82 | 67, 71 | - | - |
| Nerol | b | 1812 | RS | 93 | 121, 84, 69 | 0.04 | 0.12 |
| β-Damascenone | b | 1825 | RS | 69 | 190, 121, 105 | 0.01 | 0.03 |
| α-Ionone | b | 1853 | RS | 121 | 136, 192 | 0.02 | 0.06 |
| α-Ionol | b | 1925 | RS | 95 | 123, 138 | 0.04 | 0.12 |
| 3-Oxo-α-ionol | a | 2555 | LRI MS | 108 | 152 | - | - |
| Vitispirane | b | 1523 | LRI MS | 192 | 177, 93 | - | - |
| TPB | b | 1828 | LRI MS | 172 | 157, 142 | - | - |
| TDN | b | 1745 | LRI MS | 157 | 172, 142 | - | - |
| Benzyl Alcohols | a | 1874 | RS | 106 | 105, 77, 51 | 0.03 | 0.1 |
| Vanillin | a | 2572 | RS | 151 | 81, 152, 109 | 0.01 | 0.02 |
| 4-Ethyl guaiacol | a | 1988 | RS | 137 | 122, 152 | 0.03 | 0.09 |
| 4-Vinyl guaiacol | a | 2212 | RS | 150 | 107, 135 | 0.07 | 0.21 |
| Ethyl vanillate | a | 2665 | RS | 151 | 168, 196 | 2.36 | 7.09 |
| Methyl vanillate | a | 2630 | RS | 151 | 123, 182 | 0.97 | 2.91 |
| 2,6-Dimethoxyphenol | a | 2270 | RS | 154 | 95, 111, 139 | 0.01 | 0.03 |

[1] Extraction method: a (SPE) and b (SPME) [2] Linear Retention Index (LRI) were determined on DB-WAX polar column, as described by van Den Dool and Kratz (1963) [52]. RS identified using reference standard; LRI MS tentatively identified by comparing the Linear Retention Index and mass spectra with those of literature.

Terpenoids have been analyzed by SPME-GC-MS as described by Slaghenaufi and Ugliano (2018) [34]. In total, 5 mL of wine added with 5 µL of internal standard solution (octen-2-ol at 420 mg/L in ethanol) was placed into a 20 mL vial, together with 5 mL of mQ water (18.2 MΩ-cm) and 3 g of NaCl. The sample was equilibrated for 1 min at 40 °C. Subsequently SPME extraction was performed using a 50/30 µm divinylbenzene–carboxen–polydimethylsiloxane (DVB/CAR/PDMS) fiber (Supelco, Bellafonte, PA, USA) exposed to sample headspace for 60 min at 40 °C. The fiber was then desorbed into the injector port of a HP 7890A (Agilent Technologies) gas chromatographer coupled to a 5977B mass spectrometer. Injection was performed at 250 °C for 5 min in splitless mode. Chromatographic separation was done using a DB-WAX capillary column (30 m × 0.25, 0.25 µm film thickness, Agilent Technologies). Helium was used as carrier gas at 1.2 mL/min of constant flow rate. The temperature of the GC oven was initially kept at 40 °C for 3 min, and then programmed to raise at 230 °C at 4 °C/min, maintained for 20 min. Mass spectrometer operated in electron ionization (EI) at 70 eV with ion source temperature at 250 °C and quadrupole temperature at 150 °C. Acquisition was done in Selected Ion Monitoring (SIM). Quantification was performed using calibration curve obtained by standards addition at 7 different concentration levels in Corvina wine. A total of 5 µL of internal standards 2-octanol (420 mg/L in Ethanol), 5 mL of water and 3 g of NaCl were added to 5 mL of standard solutions. GC-MS analysis was performed as described above for the samples. Linear term for calibration curves were obtained using Chemstation software (Agilent Technologies) by linear regression, plotting the response ratio (analyte peak area/internal standard peak area) against concentration ratio (analyte added concentration/internal standard concentration). The analysis of vitispirane, terpinen-1-ol, TPB, TDN, and ho-trienol was semiquantitative as no standards was available. Results for these molecules were expressed as µg/L of 2-octanol equivalent (internal standard) (Table 6).

*4.5. Statistic*

Data treatment, ANOVA, Tukey post-hoc test and PCA were performed using XLSTAT 2017 (Addinsoft SARL, Paris, France).

## 5. Conclusions

The present study allowed to characterize the influence of post-harvest withering of Corvina grapes on the aroma profile of wines. The aromatic contribution given by the withering on-vine or in a traditional withering warehouse (fruttaio) was also been evaluated.

Withering resulted in a lower content in fermentation-derived volatiles such as esters, higher alcohols, and acids, as well as grape-related compounds such as $C_6$ alcohols, and certain norisoprenoids like β-damascenone. Terpenes showed different behaviors according to the compound. Linalool, the major terpene found in the sample wines analyzed, and *cis*-linalool oxide were negatively influenced by the withering process while β-citronellol and 1,4-cineole showed a different trend and it was found in higher concentrations in withered samples. The same trend was observed for ethyl acetate, ethyl vanillate, benzyl alcohol and vanillin

The aroma profile of wines obtained by whitering in fruttaio was characterized by higher concentrations of esters such as ethyl acetate compared to on-vine withering. Wines withered in fruttaio were also distinguished by higher concentrations of β-citronellol and 3-oxo-α-ionol, while on-vine withering showed higher content of β-damascenone. The withering process as well as the technique employed also influenced the behavior of compounds during aging, showing different variation.

Overall, the results of the present study indicate that on-vine withering with blocked xylem is an interesting alternative to conventional fruttaio withering for the production of wines where a mild withering is requested. Although on-vine withering can only be carried out in years where climatic conditions are suitable, the possibility to explore this kind of withering technique is of interest to reduce the workload of fruttaio facilities and the energy cost associated with their functioning, reducing the environmental impact of the winemaking process.

**Author Contributions:** A.B. and M.U. conceived and designed the experiments. V.Z. and A.D.C. produced samples and data analysis. D.S., A.P., and G.L. performed all the experimental work and data analysis, D.S. and M.U. interpreted the data and drafted the manuscript. All authors reviewed and edited the manuscript. All authors have read and agreed to the published version of the manuscript.

**Funding:** This research was funded by Masi Agricola winery.

**Acknowledgments:** We warmly thank Masi Agricola, for providing wine samples used in this research, and sharing grape data and information.

**Conflicts of Interest:** Davide Slaghenaufi, Alessandro Prandi, Giovanni Luzzini and Maurizio Ugliano declare no conflict of interest. Anita Boscaini, Vittorio Zandonà and Anfrea Dal Cin work at Masi Agricola, funder of the project. They contributed to design of the study and sample collection.

## References

1. Paronetto, L.; Dellaglio, F. Chapter 9 Amarone: A modern wine coming from an ancient production technology. In *Advances in Food and Nutrition Research, Speciality Wine*; Ronald, S., Jackson, Eds.; Academic Press: Waltham, MA, USA, 2011; Volume 63, pp. 285–306.

2. Mencarelli, F.; Bellincontro, A. Recent advances in postharvest technology of the wine grape to improve the wine aroma. *J. Sci. Food Agric.* **2018**. [CrossRef]

3. Consorzio Valpolicella. Available online: http://www.consorziovalpolicella.it/uploads/files/Attachment/Disciplinare_Amarone.pdf (accessed on 30 March 2020).

4. Bellincontro, A.; Matarese, F.; D'Onofrio, C.; Accordini, D.; Tosi, E.; Mencarelli, F. Management of postharvest grape withering to optimise the aroma of the final wine: A case study on Amarone. *Food Chem.* **2016**, *213*, 378–387. [CrossRef] [PubMed]

5. Fedrizzi, B.; Tosi, E.; Simonato, B.; Finato, F.; Cipriani, M.; Caramia, G.; Zapparoli, G. Changes in wine aroma composition according to botrytized berry percentage: A preliminary study on amarone wine. *Food Technol. Biotechnol.* **2011**, *49*, 529–535.

6. Chou, H.-C.; Suklje, K.; Antalick, G.; Schmidtke, L.M.; Blackman, J. Late-season shiraz berry dehydration that alters composition and sensory traits of wine. *J. Agric. Food Chem.* **2018**, *66*, 7750–7757. [CrossRef] [PubMed]

7. Bellincontro, A.; De Santis, D.; Botondi, R.; Villa, I.; Mencarelli, F. Different post- harvest dehydration rates affect quality characteristics and volatile compounds of Malvasia, Trebbiano and Sangiovese grapes for wine production. *J. Sci. Food Agric.* **2004**, *84*, 1791–1800. [CrossRef]

8. Costantini, V.; Bellincontro, A.; De Santis, D.; Botondi, R.; Mencarelli, F. Metabolic changes of malvasia grapes for wine production during postharvest drying. *J. Agric. Food Chem.* **2006**, *54*, 3334–3340. [CrossRef]

9. Moreno, J.; Cerpa-Calderón, F.; Cohen, S.D.; Fang, Y.; Qian, M.; Kennedy, J.A. Effect of postharvest dehydration on the composition of pinot noir grapes (Vitis vinifera L.) and wine. *Food Chem.* **2008**, *109*, 755–762. [CrossRef]

10. Mencarelli, F.; Bellincontro, A.; Nicoletti, I.; Cirilli, M.; Muleo, R.; Corradini, D. Chemical and biochemical change of healthy phenolic fractions in winegrape by means of postharvest dehydration. *J. Agric. Food Chem.* **2010**, *58*, 7557–7564. [CrossRef]

11. Noguerol-Pato, R.; González-Álvarez, M.; González-Barreiro, C.; Cancho-Grande, B.; Simal-Gándara, J. Evolution of the aromatic profile in garnacha tintorera grapes during raisining and comparison with that of the naturally sweet wine obtained. *Food Chem.* **2013**, *139*, 1052–1061. [CrossRef]

12. Ossola, C.; Giacosa, S.; Torchio, F.; Segade, S.R.; Caudana, A.; Cagnasso, E.; Gerbi, V.; Rolle, L. Comparison of fortified, sfursat, and passito wines produced from fresh and dehydrated grapes of aromatic black cv. Moscato nero ( Vitis vinifera L.). *Food Res. Int.* **2017**, *98*, 59–67. [CrossRef]

13. Zoccatelli, G.; Zenoni, S.; Savoi, S.; Santo, S.D.; Tononi, P.; Zandonà, V.; Cin, A.D.; Guantieri, V.; Pezzotti, M.; Tornielli, G.B. Skin pectin metabolism during the postharvest dehydration of berries from three distinct grapevine cultivars. *Aust. J. Grape Wine Res.* **2013**, *19*, 171–179. [CrossRef]

14. Rolle, L.; Giacosa, S.; Gerbi, V.; Bertolino, M.; Novello, V. Varietal comparison of the chemical, physical, and mechanical properties of five colored table grapes. *Int. J. Food Prop.* **2013**, *16*, 598–612. [CrossRef]

15. Versari, A.; Parpinello, G.P.; Tornielli, G.B.; Ferrarini, R.; Giulivo, C. Stilbene compounds and stilbene synthase expression during ripening, wilting, and UV treatment in grape cv. Corvina. *J. Agric. Food Chem.* **2001**, *49*, 5531–5536. [CrossRef] [PubMed]

16.   Zamboni, A.; Minoia, L.; Ferrarini, A.; Tornielli, G.B.; Zago, E.; Delledonne, M.; Pezzotti, M. Molecular analysis of post-harvest withering in grape by AFLP transcriptional profiling. *J. Exp. Bot.* **2008**, *59*, 4145–4159. [CrossRef] [PubMed]

17.   Zenoni, S.; Fasoli, M.; Guzzo, F.; Santo, S.D.; Amato, A.; Anesi, A.; Commisso, M.; Herderich, M.J.; Ceoldo, S.; Avesani, L.; et al. Disclosing the molecular basis of the postharvest life of berry in different grapevine genotypes1. *Plant Physiol.* **2016**, *172*, 1821–1843. [CrossRef] [PubMed]

18.   Mencarelli, F.; Tonutti, P. *Sweet, Reinforced and Fortified Wines*; John Wiley & Sons, Ltd.: Chichester, UK, 2013.

19.   Ferreira, V.; López, R.; Cacho, J.F. Quantitative determination of the odorants of young red wines from different grape varieties. *J. Sci. Food Agric.* **2000**, *80*, 1659–1667. [CrossRef]

20.   Francis, L.; Newton, J. Determining wine aroma from compositional data. *Aust. J. Grape Wine Res.* **2005**, *11*, 114–126. [CrossRef]

21.   Sacks, G.L.; Gates, M.J.; Ferry, F.X.; Lavin, E.H.; Kurtz, A.J.; Acree, T.E. Sensory threshold of 1,1,6-trimethyl-1,2-dihydronaphthalene (TDN) and concentrations in young riesling and non-riesling wines. *J. Agric. Food Chem.* **2012**, *60*, 2998–3004. [CrossRef]

22.   Antalick, G.; Tempère, S.; Suklje, K.; Blackman, J.; Deloire, A.; De Revel, G.; Schmidtke, L.M. Investigation and sensory characterization of 1,4-cineole: A potential aromatic marker of australian cabernet sauvignon wine. *J. Agric. Food Chem.* **2015**, *63*, 9103–9111. [CrossRef]

23.   Escudero, A.; Campo, E.; Fariña, L.; Cacho, J.; Ferreira, V. Analytical characterization of the aroma of five premium red wines. insights into the role of odor families and the concept of fruitiness of wines. *J. Agric. Food Chem.* **2007**, *55*, 4501–4510. [CrossRef]

24.   Swiegers, J.H.; Bartowsky, E.; Henschke, P.; Pretorius, I.S. Yeast and bacterial modulation of wine aroma and flavour. *Aust. J. Grape Wine Res.* **2005**, *11*, 139–173. [CrossRef]

25.   Antalick, G.; Suklje, K.; Blackman, J.; Meeks, C.; Deloire, A.; Schmidtke, L.M. Influence of grape composition on red wine ester profile: Comparison between cabernet sauvignon and shiraz cultivars from australian warm climate. *J. Agric. Food Chem.* **2015**, *63*, 4664–4672. [CrossRef] [PubMed]

26.   Genovese, A.; Gambuti, A.; Piombino, P.; Moio, L. Sensory properties and aroma compounds of sweet Fiano wine. *Food Chem.* **2007**, *103*, 1228–1236. [CrossRef]

27.   Saerens, S.M.G.; Delvaux, F.R.; Verstrepen, K.J.; Van Dijck, P.; Thevelein, J.M. Parameters affecting ethyl ester production by saccharomyces cerevisiae during fermentation. *Appl. Environ. Microbiol.* **2007**, *74*, 454–461. [CrossRef]

28.   Le Fur, Y.; Hory, C.; Bard, M.-H.; Olson, A. Evolution of phytosterols in Chardonnay grape berry skins during last stages of ripening. *Vitis* **1994**, *33*, 127–131.

29.   Saerens, S.M.G.; Delvaux, F.R.; Verstrepen, K.J.; Thevelein, J.M. Production and biological function of volatile esters in Saccharomyces cerevisiae. *Microb. Biotechnol.* **2010**, *3*, 165–177. [CrossRef]

30.   Mason, A.B.; Dufour, J.-P. Alcohol acetyltransferases and the significance of ester synthesis in yeast. *Yeast* **2000**, *16*, 1287–1298. [CrossRef]

31.   Kruis, A.; Levisson, M.; Mars, A.E.; Van Der Ploeg, M.; Daza, F.G.; Ellena, V.; Kengen, S.; Van Der Oost, J.; Weusthuis, R.A. Ethyl acetate production by the elusive alcohol acetyltransferase from yeast. *Metab. Eng.* **2017**, *41*, 92–101. [CrossRef]

32.   Benkwitz, F.; Tominaga, T.; Kilmartin, P.; Lund, C.; Wohlers, M.; Nicolau, L. Identifying the chemical composition related to the distict aroma characterisitics of New Zealand Sauvignon blanc wines. *Am. J. Enol. Vitic.* **2012**, *63*, 62–72. [CrossRef]

33.   Ugliano, M. Enzymes in Winemaking. In *Wine Chemistry and Biochemistry*; Springer: New York, NY, USA, 2008; pp. 103–126.

34.   Slaghenaufi, D.; Ugliano, M. Norisoprenoids, Sesquiterpenes and Terpenoids Content of Valpolicella Wines During Aging: Investigating Aroma Potential in Relationship to Evolution of Tobacco and Balsamic Aroma in Aged Wine. *Front. Chem.* **2018**, *6*, 66. [CrossRef]

35.   Slaghenaufi, D.; Guardini, S.; Tedeschi, R.; Ugliano, M. Volatile terpenoids, norisoprenoids and benzenoids as markers of fine scale vineyard segmentation for Corvina grapes and wines. *Food Res. Int.* **2019**, *125*, 108507. [CrossRef] [PubMed]

36.   Giacosa, S.; Giordano, M.; Vilanova, M.; Cagnasso, E.; Segade, S.R.; Rolle, L. On-vine withering process of 'Moscato bianco' grapes: Effect of cane-cut system on volatile composition. *J. Sci. Food Agric.* **2018**, *99*, 1135–1144. [CrossRef] [PubMed]

37. Ugliano, M.; Bartowsky, E.J.; McCarthy, J.; Moio, L.; Henschke, P.A. Hydrolysis and Transformation of Grape Glycosidically Bound Volatile Compounds during Fermentation with ThreeSaccharomycesYeast Strains. *J. Agric. Food Chem.* **2006**, *54*, 6322–6331. [CrossRef] [PubMed]

38. Slaghenaufi, D.; Indorato, C.; Troiano, E.; Luzzini, G.; Felis, G.E.; Ugliano, M. Fate of Grape-Derived Terpenoids in Model Systems Containing Active Yeast Cells. *J. Agric. Food Chem.* **2020**. [CrossRef] [PubMed]

39. Sefton, M.A.; Skouroumounis, G.K.; Elsey, G.M.; Taylor, D.K. Occurrence, sensory impact, formation, and fate of damascenone in grapes, wines, and other foods and beverages. *J. Agric. Food Chem.* **2011**, *59*, 9717–9746. [CrossRef] [PubMed]

40. Lloyd, N.D.R.; Capone, D.L.; Ugliano, M.; Taylor, D.K.; Skouroumounis, G.K.; Sefton, M.A.; Elsey, G.M. Formation of damascenone under both commercial and model fermentation conditions. *J. Agric. Food Chem.* **2011**, *59*, 1338–1343. [CrossRef]

41. Díaz-Maroto, M.C.; Schneider, R.; Baumes, R. Formation pathways of ethyl esters of branched short-chain fatty acids during wine aging. *J. Agric. Food Chem.* **2005**, *53*, 3503–3509. [CrossRef]

42. Moio, L.; Ugliano, M.; Genovese, A.; Gambuti, A.; Pessina, R.; Piombino, P. Effect of antioxidant protection of must on volatile compounds and aroma shelf life of falanghina (vitis viniferal.) wine. *J. Agric. Food Chem.* **2004**, *52*, 891–897. [CrossRef]

43. Antalick, G.; Perello, M.-C.; De Revel, G. Esters in Wines: New insight through the establishment of a database of french wines. *Am. J. Enol. Vitic.* **2014**, *65*, 293–304. [CrossRef]

44. Wilson, B.; Strauss, C.R.; Williams, P.J. Changes in free and glycosidically bound monoterpenes in developing muscat grapes. *J. Agric. Food Chem.* **1984**, *32*, 919–924. [CrossRef]

45. Williams, P.J.; Strauss, C.R.; Wilson, B. Hydroxylated linalool derivatives as precursors of volatile monoterpenes of muscat grapes. *J. Agric. Food Chem.* **1980**, *28*, 766–771. [CrossRef]

46. Skouroumounis, G.K.; Massy-Westropp, R.A.; Sefton, M.A.; Williams, P.J. Precursors of damascenone in fruit juices. *Tetrahedron Lett.* **1992**, *33*, 3533–3536. [CrossRef]

47. Daniel, M.A.; Puglisi, C.J.; Capone, D.L.; Sefton, M.A.; Elsey, G.M. Rationalising the formation of damascenone: Synthesis and hydrolysis of damascenone precursors and their analogues, in both aglycone and glycoconjugate form. *J. Agric. Food Chem.* **2008**, *56*, 9183–9189. [CrossRef] [PubMed]

48. Janusz, A.; Capone, D.L.; Puglisi, C.J.; Perkins, M.; Elsey, G.M.; Sefton, M.A. (*E*)-1-(2,3,6-Trimethylphenyl)buta-1,3-diene: A potent grape-derived odorant in wine. *J. Agric. Food Chem.* **2003**, *51*, 7759–7763. [CrossRef] [PubMed]

49. Pinto, M.M.M. Carotenoid breakdown products the—Norisoprenoids—In wine aroma. *Arch. Biochem. Biophys.* **2009**, *483*, 236–245. [CrossRef] [PubMed]

50. Winterhalter, P.; Gök, R. TDN and β-damascenone: Two important carotenoid metabolites in wine. *Chem. Stud. Success Field-Test. Evid.-Based Guide* **2013**, *1134*, 125–137. [CrossRef]

51. Andújar-Ortiz, I.; Moreno-Arribas, M.V.; Martin-Alvarez, P.; Pozo-Bayon, M.A. Analytical performance of three commonly used extraction methods for the gas chromatography–mass spectrometry analysis of wine volatile compounds. *J. Chromatogr. A* **2009**, *1216*, 7351–7357. [CrossRef]

52. Dool, H.V.D.; Kratz, P.D. A generalization of the retention index system including linear temperature programmed gas—Liquid partition chromatography. *J. Chromatogr. A* **1963**, *11*, 463–471. [CrossRef]

**Sample Availability:** Samples are not available from the authors.

*Article*

# Characterization of Volatile Component Changes in Jujube Fruits during Cold Storage by Using Headspace-Gas Chromatography-Ion Mobility Spectrometry

Lvzhu Yang [1,2], Jie Liu [1,2], Xinyu Wang [1,2], Rongrong Wang [3,*], Fang Ren [4], Qun Zhang [2], Yang Shan [2] and Shenghua Ding [1,2,*]

[1]  Longping Branch Graduate School, Hunan University, Changsha 410125, China; ylzhu0115@163.com (L.Y.); jie195712@163.com (J.L.); wxy25994@163.com (X.W.)

[2]  Provincial Key Laboratory for Fruits and Vegetables Storage Processing and Quality Safety, Agricultural Product Processing Institute, Hunan Academy of Agricultural Sciences, Changsha 410125, China; zqun208@163.com (Q.Z.); sy6302@sohu.com (Y.S.)

[3]  College of Food Science and Technology, Hunan Agricultural University, Changsha 410128, China

[4]  G.A.S. Department of Shandong Hanon Science Instrument Co., Ltd., Jinan 253000, China; amanda@hanon.cc

*  Correspondence: sdauwrr@163.com (R.W.); shhding@hotmail.com (S.D.); Tel.: +86-731-84617093 (R.W.); +86-731-84691006 (S.D.)

Received: 20 September 2019; Accepted: 28 October 2019; Published: 30 October 2019

**Abstract:** Volatile components in jujube fruits from *Zizyphus jujuba* Mill. *cv.* Dongzao (DZ) and *Zizyphus jujuba* Mill. *cv.* Jinsixiaozao (JS) were analyzed under different cold storage periods via headspace-gas chromatography-ion mobility spectrometry (HS-GC-IMS). Results identified 53 peaks that corresponded to 47 compounds and were mostly alcohols, aldehydes, esters, and ketones. Differences in the volatile components of jujube fruits were revealed in topographic plots and fingerprints. For DZ, 3-pentanone was the characteristic component of fresh fruits. After storage for 15 days, dipropyl disulfide became the most special substance. Moreover, when stored for 30 and 45 days, the fruits had some same volatile components, like 2-pentyl furan and diallyl sulfide. However, for DZ stored for 60 days, esters were the prominent constituent of the volatile components, simultaneously, some new alcohols appeared. For JS, 2-ethyl furan was the representative of fresh fruits, and 2-butoxyethanol content was the most abundant after 15 and 30 days of storage. Different from that in DZ, the content of ester in JS increased after storage for 45 days. Substances such as amyl acetate dimer, methyl salicylate, and linalool greatly contributed to the jujube flavor during the late storage period. Principal component analysis (PCA) showed that fresh samples and refrigerated fruits were effectively distinguished. Heat map clustering analysis displayed the similarity of volatile components in different samples and was in accordance with PCA results. Hence, the volatile components of jujube fruits can be readily identified via HS-GC-IMS, and jujube fruits can be classified at different periods based on the difference of volatile components.

**Keywords:** jujube fruits; volatile components; headspace-gas chromatography-ion mobility spectrometry (HS-GC-IMS); cold storage; principal component analysis (PCA)

---

## 1. Introduction

Jujube (*Zizyphus jujuba* Mill) tree belongs to the Rhamnaceae family, which is indigenous to China and is distributed worldwide, in places such as Asia, northern Africa, southern Europe, the Middle East, and the southwestern USA. This tree has a history of more than 4000 years, with over

700 cultivars found in China [1–3]. Phytochemical analytical studies showed that jujube fruits are rich in nutrients, including fiber, sugars, organic acids, amino acids, vitamins, and trace minerals [4]. They also contain high levels of functional components, such as polysaccharides, triterpene acids, phenolics, and cyclic nucleotides, which exhibit multiple health-promoting properties, such as antioxidant and anti-inflammatory properties and liver protection [5–9]. Recognized for its delicious taste and health beneficial properties, jujube fruits have been consumed for thousands of years as ordinary fruits and Chinese traditional medicine.

In addition to their nutritional and biological activity, jujube fruits are favored as food by consumers due to their unique flavor. Fresh jujube fruits show an extraordinary flavor, but they are highly perishable when not handled properly due to their high moisture content, which leads to the loss of their commercial value. Cold storage is a common means of delaying the deterioration of quality and prolonging shelf-life effectively, which can affect the physicochemical properties of jujube fruits. Liu et al. found that the contents of chlorophyll, ascorbic acid, and soluble solids in fresh-cut jujubes without any treatment were reduced after storage at 4 °C for a certain period [10]. Kou et al. [11] reported that when jujube fruits were stored at 0 °C, the contents of total soluble solids, ascorbic acid, and chlorophylls continuously declined, and the anthocyanin content firstly increased and then decreased. Furthermore, Günther et al. [12] studied kiwifruit and found that cold storage also affected flavor. Simultaneously, changes in post-harvest metabolic and anaerobic environment greatly influence the fruit flavor, anaerobic conditions can enhance the flavor quality of fruits by producing certain aromatic compounds during storage [13]. Aroma profiles are important characteristics to evaluate the quality of fruits. Several studies have reported that the volatile compounds of jujube fruits are affected by many factors, such as growth period [14], extraction methods [15], the load [16], and processing methods [17–20]. However, few reports have been associated with changes in volatile compounds of jujube fruits during cold storage, which should be considered. The difference in volatile compounds among different periods of jujube fruit cold storage is unclear. Therefore, the purpose of this study was to monitor the changes in flavor and identify its characteristics at different times.

Ion mobility spectrometry (IMS) is an instrumental analytical technique of separating the ions of detected substances based on their ion mobility velocity under atmospheric pressure [21]. It is a convenient and efficient instrument with the advantages of simple sample preparation, easy operation, high sensitivity, and quick analytical speed. Even trace volatile compounds can also be detected in a short time [22,23]. Furthermore, ion mobility notably allows the separation of isomers and isobaric compounds, which cannot be separated even with ultra-high resolution instruments [24]. At first, compounds extracted from the sample enter the ionization chamber directed by a carrier gas, and the analyte is charged after being ionized. Then, a series of reactions occur and reactant ions $[H^+(H_2O)_n]$ are generated. In an IMS instrument, if the proton affinity of the analyte is higher than the proton affinity of water, it will react with the reactant ions. Based on the content of the analyte, their chemical nature or the drift tube temperature, product ions such as protonated monomers or proton-bound dimers were produced [25]. Subsequently, analyte ions enter the drift region through the Bradbury–Nielsen–Shutter. Analyte ions react against the reverse drift gas under the action of the electric field and migrate to the right end to reach the right detector, the drift velocity of ions depends on their charge, mass, and shape [26]. Finally, the Faraday–Plate detects ions and outputs electrical signals. The results are expressed in terms of voltage units. IMS has been used for chemical warfare agents [27], illicit drug detection [28], analysis of explosives [29], and environmental monitoring [30]. The instrument is highly sensitive to high electronegativity and high proton affinity compounds, which can detect a large number of compounds from different chemical families, such as alcohols, aldehydes, aromatics, esters, and ketones [30]. Combining IMS with other instruments is a good way to increase its advantages and produce a good analysis result. In recent years, HS-GC-IMS has been extensively applied to investigate volatile compounds in food science, such as *Tricholoma matsutake* Singer [22], jujube fruits [31], eggs [26], Iberian ham [32,33], and honey [25]. As a consequence, HS-GC-IMS could be used to identify the volatile components of jujube fruits at different periods.

In this study, HS-GC-IMS was used to analyze the variations in the volatile compositions of jujube fruits at different storage periods, and the fingerprints were established to confirm the characteristic substance of each period. The results will provide a new method for studying the flavor of jujube fruits, which will help to rapidly select the best storage time of jujube fruits.

## 2. Results and Discussion

### 2.1. Volatile Components Identification of All Samples at Different Storage Periods

The aromatic components of fruits undergo many complicated changes during storage, such as vitamin degradation and phenol oxidation, resulting in changes in fruit flavor. Rodrigo et al. [34] found that the aroma of peaches was related to storage conditions and fruit quality characteristics. As the refrigeration progressed, the aroma of peaches dropped considerably, and the flavor and fruit sweetness, juiciness, and texture were strongly correlated. In this study, the volatile components of jujube fruits at different storage periods were determined by HS-GC-IMS. The samples were ionized in the column and then identified using ion mobility systems based on retention and drift times. The qualitative analysis of volatile components in jujube fruits is shown by numbers in Figure 1, where the ordinate represented the retention time, and the abscissa represented the drift time. A total of fifty-three peaks, forty-seven components were identified from the GC×IMS library (Figure 1 and Table 1), including fourteen alcohols, six aldehydes, nine esters, six ketones, two organic acids, two furans, three pyrazines, four sulfur-containing compounds, linalool oxide, and 2-methoxy-4-cresol. Among them, sulfur-containing compounds, linalool oxide, and 2-methoxy-4-cresol were detected in jujube fruits for the first time. This finding may be ascribed to the differences in detection methods and essential differences of raw materials. The identified components are listed in Table 1, which includes the compound name, CAS number, molecular formula, molecular weight, retention index, retention time, and drift time. Moreover, other substances with signals were detected but could not be determined were not listed. When moving through the drift region, due to the formation of adducts between the analyzed ions and neutral molecules (such as dimers and trimers), multiple signals were observed for a single compound [35]. The compounds of 1-octen-3-one, 3-hydroxy-2-butanone, 1-pentanol, heptanal, amyl acetate, and ethyl propanoate exhibited two peaks due to the presence of both monomer and dimer.

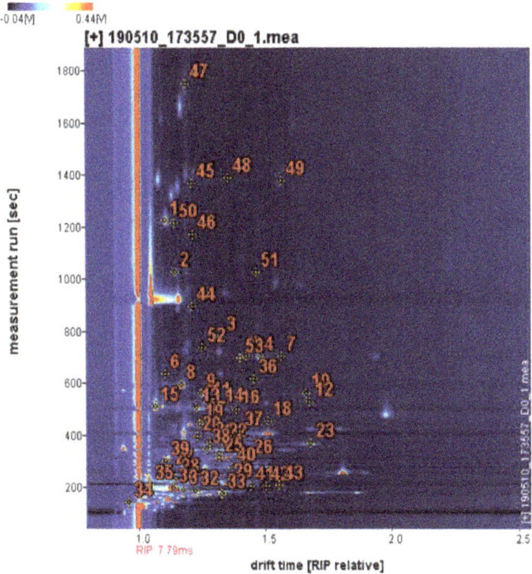

**Figure 1.** Ion migration spectra of jujube fruits stored for different times at 4 °C. The numbers are identified volatile components.

Table 1. The information on identified compounds of jujube fruits.

| NO. | Compound | CAS# | Formula | MW | RI | Rt (Sec) | Dt (RIP Relative) | Comment | Identification Approach |
|---|---|---|---|---|---|---|---|---|---|
| 1 | Furfuryl alcohol | C98000 | $C_5H_6O_2$ | 98.1 | 1661.4 | 1226.886 | 1.1116 | | RI, Dt |
| 2 | Octan-2-ol | C123966 | $C_8H_{18}O$ | 130.2 | 1401.0 | 704.688 | 1.4313 | | RI, Dt |
| 3 | (E)-3-Hexen-1-ol | C928972 | $C_6H_{12}O$ | 100.2 | 1327.8 | 562.241 | 1.2529 | | RI, Dt |
| 4 | 2-Butoxyethanol | C111762 | $C_6H_{14}O_2$ | 118.2 | 1400.9 | 704.400 | 1.5715 | | RI, Dt |
| 5 | 2-Heptanol | C543497 | $C_7H_{16}O$ | 116.2 | 1291.4 | 500.190 | 1.3920 | | RI, Dt |
| 6 | 1-Pentanol | C71410 | $C_5H_{12}O$ | 88.1 | 1262.0 | 456.411 | 1.2576 | monomer | RI, Dt |
| 7 | 1-Pentanol | C71410 | $C_5H_{12}O$ | 114.2 | 1262.0 | 456.411 | 1.5164 | dimer | RI, Dt |
| 8 | 2-Methyl-1-butanol | C137326 | $C_5H_{12}O$ | 88.1 | 1215.7 | 397.445 | 1.2367 | | RI, Dt |
| 9 | 2-Hexanol | C626937 | $C_6H_{14}O$ | 102.2 | 1194.9 | 374.160 | 1.2880 | | RI, Dt |
| 10 | 3-Methyl-2-butanol | C598754 | $C_5H_{12}O$ | 88.1 | 1124.2 | 306.504 | 1.4417 | | RI, Dt |
| 11 | 2-Propanol | C67630 | $C_3H_8O$ | 60.1 | 908.5 | 188.339 | 1.2264 | | RI, Dt |
| 12 | 2-Methyl-1-propanol | C78831 | $C_4H_{10}O$ | 74.1 | 1090.2 | 279.704 | 1.3727 | | RI, Dt |
| 13 | Linalool | C78706 | $C_{10}H_{18}O$ | 154.3 | 1496.8 | 896.913 | 1.2161 | | RI, Dt |
| 14 | Citronellol | C106229 | $C_{10}H_{20}O$ | 156.3 | 1742.8 | 1390.307 | 1.3610 | | RI, Dt |
| 15 | 3-Octanol | C589980 | $C_8H_{18}O$ | 130.2 | 1397.5 | 697.657 | 1.4073 | | RI, Dt |
| 16 | (E)-2-Octenal | C2548870 | $C_8H_{14}O$ | 126.2 | 1438.5 | 779.930 | 1.3345 | | RI, Dt |
| 17 | (E)-2-Heptenal | C18829555 | $C_7H_{12}O$ | 112.2 | 1326.9 | 560.684 | 1.6667 | | RI, Dt |
| 18 | Heptanal | C111717 | $C_7H_{14}O$ | 114.2 | 1191.5 | 370.381 | 1.3412 | monomer | RI, Dt |
| 19 | Heptanal | C111717 | $C_7H_{14}O$ | 114.2 | 1193.2 | 372.270 | 1.6841 | dimer | RI, Dt |
| 20 | 3-Methylbutanal | C590863 | $C_5H_{10}O$ | 86.1 | 917.4 | 190.741 | 1.1796 | | RI, Dt |
| 21 | (E)-2-undecenal | C53448070 | $C_{11}H_{20}O$ | 168.3 | 1736.9 | 1378.477 | 1.5700 | | RI, Dt |
| 22 | 5-Methylfurfural | C620020 | $C_6H_6O_2$ | 110.1 | 1562.0 | 1027.52 | 1.4726 | | RI, Dt |
| 23 | 1-Octen-3-one | C4312996 | $C_8H_{14}O$ | 126.2 | 1308.2 | 527.716 | 1.2773 | monomer | RI, Dt |
| 24 | 1-Octen-3-one | C4312996 | $C_8H_{14}O$ | 126.2 | 1308.2 | 527.716 | 1.6826 | dimer | RI, Dt |
| 25 | 1-Hydroxypropan-2-one | C116096 | $C_3H_6O_2$ | 74.1 | 1295.3 | 506.307 | 1.2352 | | RI, Dt |
| 26 | 3-Hydroxy-2-butanone | C513860 | $C_4H_8O_2$ | 88.1 | 1298.6 | 511.779 | 1.0710 | monomer | RI, Dt |
| 27 | 3-Hydroxy-2-butanone | C513860 | $C_4H_8O_2$ | 88.1 | 1295.5 | 506.629 | 1.3273 | dimer | RI, Dt |
| 28 | 2,3-Butanedione | C431038 | $C_4H_6O_2$ | 86.1 | 1021.6 | 233.375 | 1.1507 | | RI, Dt |
| 29 | 3-Pentanone | C96220 | $C_5H_{10}O$ | 86.1 | 989.2 | 215.961 | 1.3600 | | RI, Dt |
| 30 | 2-Methyl-3-heptanone | C13019200 | $C_8H_{16}O$ | 128.2 | 1169.0 | 347.292 | 1.2730 | | RI, Dt |
| 31 | Ethyl octanoate | C106321 | $C_{10}H_{20}O_2$ | 172.3 | 1401.0 | 704.688 | 1.4841 | | RI, Dt |

**Table 1.** *Cont.*

| NO. | Compound | CAS# | Formula | MW | RI | Rt (Sec) | Dt (RIP Relative) | Comment | Identification Approach |
|---|---|---|---|---|---|---|---|---|---|
| 32 | Amyl acetate | C628637 | $C_7H_{14}O_2$ | 130.2 | 1150.4 | 329.460 | 1.3196 | monomer | RI, Dt |
| 33 | Amyl acetate | C628637 | $C_7H_{14}O_2$ | 130.2 | 1134.6 | 315.334 | 1.3240 | dimer | RI, Dt |
| 34 | Ethyl propanoate | C105373 | $C_5H_{10}O_2$ | 102.1 | 942.6 | 197.947 | 1.1418 | monomer | RI, Dt |
| 35 | Ethyl acetate | C141786 | $C_4H_8O_2$ | 88.1 | 866.3 | 176.930 | 1.3333 | | RI, Dt |
| 36 | Hexyl acetate | C142927 | $C_8H_{16}O_2$ | 144.2 | 1230.8 | 415.502 | 1.4044 | | RI, Dt |
| 37 | Ethyl propanoate | C105373 | $C_5H_{10}O_2$ | 102.1 | 964.6 | 205.502 | 1.4486 | dimer | RI, Dt |
| 38 | Propyl acetate | C109604 | $C_5H_{10}O_2$ | 102.1 | 968.8 | 207.090 | 1.5065 | | RI, Dt |
| 39 | Ethyl isobutanoate | C97621 | $C_6H_{12}O_2$ | 116.2 | 970.1 | 207.619 | 1.5604 | | RI, Dt |
| 40 | Methyl salicylate | C119368 | $C_8H_8O_3$ | 152.1 | 1732.1 | 1368.733 | 1.2162 | | RI, Dt |
| 41 | 2-Methylpropanoic acid | C79312 | $C_4H_8O_2$ | 88.1 | 1562.1 | 1027.802 | 1.1494 | | RI, Dt |
| 42 | 2-Methylbutanoic acid | C116530 | $C_5H_{10}O_2$ | 102.1 | 1632.9 | 1169.846 | 1.2188 | | RI, Dt |
| 43 | 2-Ethylpyrazine | C13925003 | $C_6H_8N_2$ | 108.1 | 1365.8 | 634.530 | 1.1116 | | RI, Dt |
| 44 | 2-Ethyl-6-methylpyrazine | C13925036 | $C_7H_{10}N_2$ | 122.2 | 1345.4 | 594.936 | 1.1737 | | RI, Dt |
| 45 | Acetylpyrazine | C22047252 | $C_6H_6N_2O$ | 122.1 | 1654.4 | 1212.857 | 1.1463 | | RI, Dt |
| 46 | 2-Pentyl furan | C3777693 | $C_9H_{14}O$ | 130.2 | 1253.1 | 444.179 | 1.2477 | | RI, Dt |
| 47 | 2-Ethylfuran | C3208160 | $C_6H_8O$ | 96.1 | 975.6 | 209.900 | 1.0454 | | RI, Dt |
| 48 | Dimethyldisulphide | C624920 | $C_2H_6S_2$ | 94.2 | 1074.6 | 268.202 | 1.1262 | | RI, Dt |
| 49 | Dimethyl sulfide | C75183 | $C_2H_6S$ | 62.1 | 737.5 | 142.102 | 0.9636 | | RI, Dt |
| 50 | Dipropyl disulfide | C629196 | $C_6H_{14}S_2$ | 150.3 | 1356.5 | 616.330 | 1.4603 | | RI, Dt |
| 51 | Diallyl sulfide | C592881 | $C_6H_{10}S$ | 114.2 | 1118.4 | 301.687 | 1.1088 | | RI, Dt |
| 52 | Linalool oxide | C60047178 | $C_{10}H_{18}O_2$ | 170.3 | 1417.2 | 737.207 | 1.2585 | | RI, Dt |
| 53 | 2-Methoxy-4-cresol | C93516 | $C_8H_{10}O_2$ | 138.2 | 1921.7 | 1749.150 | 1.1921 | | RI, Dt |

MW: molecular mass; RI: retention index; Rt: retention time; Dt: drift time.

## 2.2. Differential Analysis of the Topographic Plots of Volatile Components in Jujube Fruits at Different Storage Periods

For an intuitive observation and comparison, topographic plots were used to characterize the substances of different jujube fruits during storage. From the 3D topographic plot (Figure 2), it can be clearly observed that with prolonged storage periods, the content of some compounds decreased, and new substances were formed. For DZ, some of the volatile components in the yellow circles disappeared after 15 days of storage but new peaks appeared after 60 days. Moreover, new peaks with high retention time were found after 45 days of storage, as indicated by the black circles (Figure 2A). For JS, more new peaks appeared when the storage time was extended to 45 days (Figure 2B). Hence, compared with fresh jujube fruits, volatile components were formed during the storage process [36]. During fruit ripening, the production of aroma volatiles (especially esters) was regulated by ethylene signaling pathways. Moreover, the production of volatile aroma components was strongly hampered in ethylene-suppressed fruits [37]. It could be inferred that after a certain time of refrigeration, the increase in ethylene synthesis led to more aroma components.

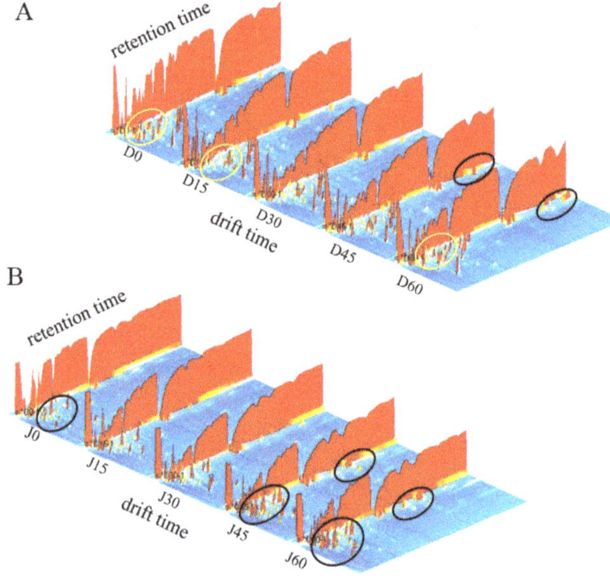


**Figure 2.** 3D-topographic plots of jujube fruits at different times. Jujube fruits are Dongzao shown in (**A**), D0, D15, D30, D45, and D60 represent refrigeration for 0, 15, 30, 45, and 60 days, respectively. Jujube fruits are Jinsixiaozao shown in (**B**), J0, J15, J30, J45, and J60 represent refrigeration for 0, 15, 30, 45, and 60 days, respectively.

Considering that the 3D spectrum was rough, the overhead view was used for comparison, as represented in Figure 3. Different volatile organic components (VOCs) were different points in the picture, which was highly convenient for observation and analysis. The red vertical line at the abscissa was the reactive ion peak (RIP) at normalized drift times for DZ and JS of 7.79 and 7.81, respectively. The differential contrast model was used to compare the differences among the samples. The fresh sample was selected as the reference, and the spectrum of the other samples deducted the reference. The background after deduction was white showed that the VOCs were the same. Red spots indicate that the content of the substance was higher than the reference, and blue spots indicate that the content was lower than that of the reference. Compared with that of untreated fruits, more red spots are located in the retention time range of 900–1400 s of DZ after storage, and the VOCs changed inconspicuously in the retention time range of 100–600 s at the topographic plot (Figure 3A). For JS, many red spots appeared, and the entire retention time was covered. Especially after 45 days, most of the signals were much higher than that of fresh fruits. This finding indicated that the VOCs of two species varied at different refrigeration times and were affected by variety when the fruits were under similar conditions. Further ripening of the jujube fruits during storage promoted the synthesis of certain volatile components. The formation of fruit flavor was a dynamic process in which the fruit continuously synthesized volatile aroma substances [38]. In the post-harvest ripening stage of bananas, the ester content in the yellow ripening period increased, and the alcohol content in the over-ripe period increased [39].

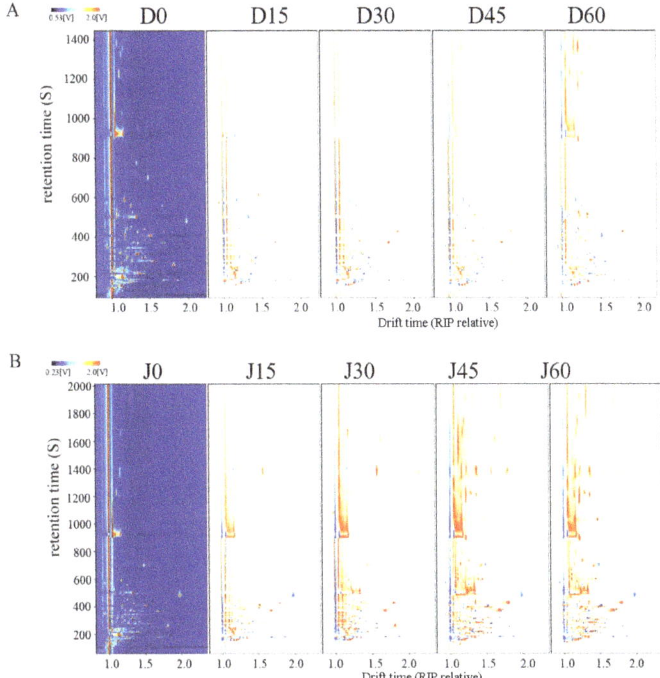

**Figure 3.** 2D-topographic plots of jujube fruits at different times. Jujube fruits are Dongzao shown in (**A**), D0, D15, D30, D45, and D60 represent refrigeration for 0, 15, 30, 45, and 60 days, respectively. Jujube fruits are Jinsixiaozao shown in (**B**), J0, J15, J30, J45, and J60 represent refrigeration for 0, 15, 30, 45, and 60 days, respectively.

### 2.3. Fingerprints of VOCs in Jujube Fruits at Different Storage Periods

Although the topographic plots showed the tendency of volatile components, it is difficult to make an accurate judgment for the dense material on the map. The use of fingerprint was a good way to solve this problem. According to the peak signal of the topographic plots, the fingerprints of jujube fruits were formed (Figures 4 and 5). In the fingerprints, each row represents the entire signal peak of one sample, and each column represents the same substance in different samples. Each cell represents the content of a substance at different times. Colors represent the content of volatile compounds. The brighter the color, the higher the content. Two compounds with the same name in the fingerprints were the monomers and their dimers. The drift time of dimers was increased due to their proton affinity and higher content [40]. By utilizing the fingerprints, the VOCs between different samples can be compared intuitively, moreover, the dynamic changes of each substance can be revealed. The unidentified substances are represented by numbers in the fingerprints. During the whole storage period, the volatile components detected in JS were more than those in DZ, which may be caused by the differences in the varieties and genetic factors of the two jujube fruits.

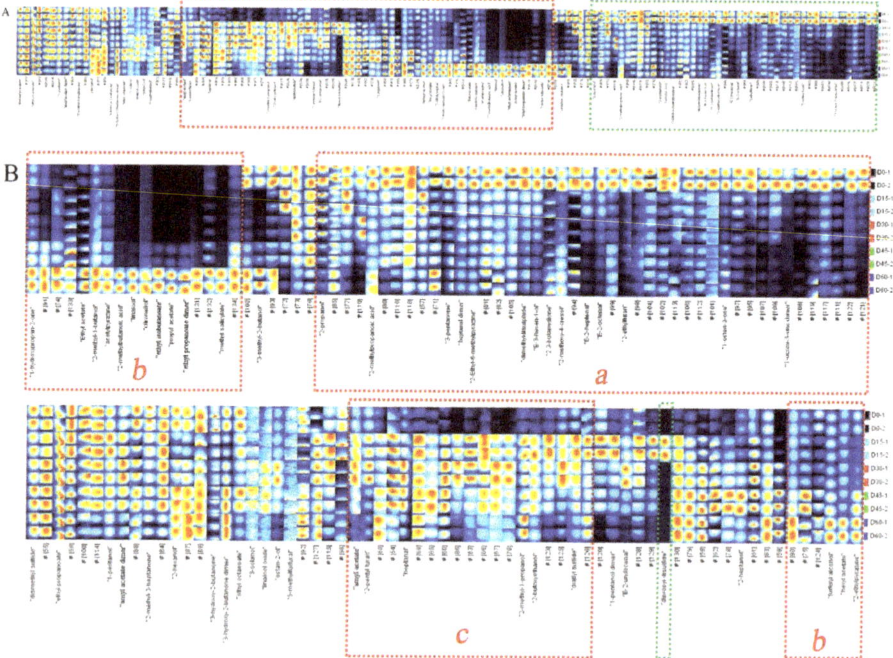

**Figure 4.** Fingerprint of volatile compounds of Dongzao samples. (**A**) Fingerprint of all Dongzao, (**B**) two parts of the fingerprint.

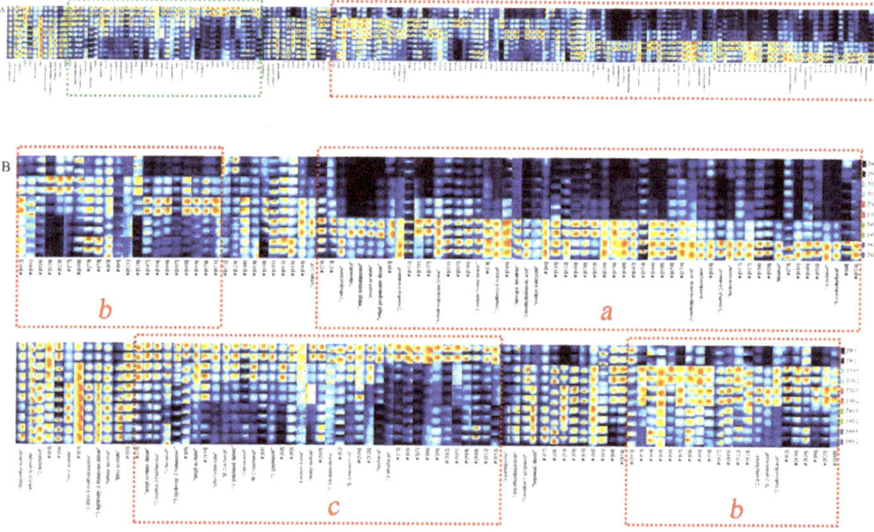

**Figure 5.** Fingerprint of volatile compounds of Jinsixiaozao samples. (**A**) Fingerprint of all Jinsixiaozao, (**B**) two parts of the fingerprint.

The VOCs of the jujube fruits constantly changed during storage. By comparing the intensity of spot for the profiles of VOCs at different stages, changes in the substances during storage (increased, decreased, disappeared or fluctuated) can be determined. Substances in the green-framed areas of Figures 4A and 5A were the most abundant in fresh jujube fruits but dramatically decreased or even disappeared during the later period of storage. For DZ, 2-propanol, 3-methyl-2-butanol, 3-pentanone, and heptanal dimer were detected, while amyl acetate, ethyl octanoate, (*E*)-2-octenal, and 1-pentanol dimers were detected in JS. Simultaneously, it can be clearly seen that, compared to other periods, the relative amount percentage of these substances was the highest in fresh fruits during the entire storage period (Table S1). This finding may be due to the degradation of these substances during refrigeration.

By contrast, with prolonged storage time, some new compounds appeared, which showed strong signal intensity and bright color (the red-framed areas of Figures 4A and 5A). These substances included linalool, ethyl isobutanoate, propyl acetate, ethyl ester, and methyl salicylate. This finding may be attributed to the complex physiological metabolism in fruits during storage, which mainly were a series of reactions, such as fatty acid, amino acid, and carbohydrate metabolism. Esters, which are mainly derived from the lipoxygenase pathway and amino acid metabolism are associated with the "fruity" attributes of fruit flavor, and its levels typically increase in the later periods of the ripening process [41]. Moreover, peaches stored after pre-storage were sweeter and had higher levels of propyl acetate, amyl acetate, and 2-methyl-1-butanol than control fruits [42]. The VOCs of the unframed part in the fingerprints presented few changes during the whole storage period, indicating that these substances were relatively stable during the cold storage of jujube fruits.

*2.4. Changes of VOCs during Different Storage Periods*

Changes in volatile components during different storage periods have been observed from the fingerprints. To compare the changes of each substance clearly, the fingerprint of each jujube fruit was divided into two parts based on the characteristic volatiles presented in different parts (Figures 4B and 5B). The representative VOCs detected in fresh DZ and JS were found in the region *a* of Figure 4B and region *c* of Figure 5B, respectively. The same ingredients included 1-octen-3-one, 2-ethylfuran, (*E*)-2-octenal, (*E*)-2-heptenal, 2,3-butanedione, (*E*)-3-hexen-1-ol, and heptanal. Chen et al. [43] studied ten different varieties of Chinese jujube fruits and found that aldehydes had the highest content and contributed to the aroma of fresh jujube fruits, moreover, (*E*)-2-hexenal, hexanal, (*Z*)-2-heptenal, benzaldehyde, and (*E*)-2-nonenal were the common volatile components in fresh jujube fruits. This finding differed from our results, presumably due to differences in varieties and detection methods. Heredity can determine precursors, enzymes, and their activity in the formation of flavor components in jujube fruits. Compared with fresh jujube fruits, these parameters decreased significantly after 15 days of storage in DZ and 30 days in JS. Aldehydes significantly affected the flavor of fresh jujube fruits, which generated from the oxidation of fatty acids and the metabolism of amino acids [44]. They can affect the overall aroma of samples at low content for their low odor threshold values [45]. (*E*)-2-Heptenal was characterized by soap and almond flavor, (*E*)-2-octenal was associated with roasted, cucumber, nutty, and fatty characteristics, and heptanal has a fishy, nutty and sweet apricot note flavor [46].

Many different changes of VOCs have been observed during storage. As shown in Figure 4B, the contents of 3-methyl-2-butanol, heptanal dimer, 2-ethyl-6-methylpyrazine, dimethyldisulphide, (*E*)-3-hexen-1-ol, (*E*)-2-octenal, (*E*)-2-heptenal, 2-ethylfuran, and 1-octen-3-one dimer in fresh DZ were much higher than that after storage. During storage, their contents continuously declined, (*E*)-2-octenal, (*E*)-2-heptenal, 2-ethylfuran, and 1-octen-3-one dimer vanished after storage. It showed that 2-methylpropanoic acid, 3-pentanone, 2,3-butanedione, 2-methoxy-4-cresol, and 1-octen-3-one were important for the flavor of fresh DZ. The contents of 3-pentanone and 2,3-butanedione were initially the highest in DZ, which decreased significantly after a certain period of storage and increased slightly after 60 days, while in JS, 3-pentanone with low content in fresh fruits. Low-carbon saturated ketones have a special aroma, and many ketones have been found in cheese, which are the main volatile constituents of cheese with a unique flavor [47]. In addition, it can be observed that the change of 2-methylpropanoic acid was consistent with 3-pentanone and 2,3-butanedione. However, 2-methylbutanoic acid was different from them, it appeared with a strong signal at 60 days, which was not detected during the storage time of 0–45 days. Additionally, its relative amount percentage was up to 100% when stored for 60 days, before this period, this value was basically less than 10% (Table S1). Interestingly, only these two short-chain organic acids were identified in the two cultivars of jujube fruits. In JS, they appeared in storage for 45 days and then enhanced (Figure 5B). Organic acid is derived from the oxidation of fatty acids, which affects the flavor of jujube fruits [48]. During ripening, fruits undergo an esterification reaction, consuming a large amount of acid. When the jujube fruits are ripe, the amount of acids becomes low [49]. However, long-term storage resulted in the oxidation and decomposition of fatty acids. Besides, it should be noted that high content of organic acids can damage fruit quality and deteriorate the flavor of jujube fruits. In addition, processing methods also affect the content of organic acids. Chen et al. reported that acids were the major group among all the volatile chemicals in dried jujube fruits. Moreover, pre-treatment of jujube fruits with 5% $CO_2$ can induce a decrease in acid content during drying, which is mainly caused by a decrease in lauric acid and nutmeg acid content [18]. 2-Methoxy-4-cresol was a newly detected component in the VOCs of jujube fruits, which appeared with high content in fresh DZ but not found in fresh JS and showed a pleasant clove-like flavor [50]. For DZ, its signal intensity was almost invisible after storage for 15–30 days, then gradually strengthened. In JS, it appeared after 45 days with a bright color in the fingerprint plot.

Some volatile compounds, mostly including esters, were rare in fresh samples of two jujube fruits but abundant in the later storage period, which were the most important volatile components at the end of storage (Figures 4B and 5B), including hexyl acetate, ethyl acetate, ethyl propanoate, propyl acetate, ethyl isobutanoate, and methyl salicylate. Most of these compounds show a typical fruity or floral flavor, which may greatly affect the aroma of fruits. Zhou et al. found that low-temperature treatment could prevent the loss of aromatic esters during the ripening of 'Nanguo' pears at room temperature [51]. During the cold storage period, although complete ripening of the jujube fruits was inhibited, ripening proceeded at a slower pace, thus generating more esters. It can be found that the relative amount percentage of ethyl propanoate, propyl acetate, ethyl isobutanoate, and methyl salicylate was lower in the two cultivars of jujube at 0–30 days of storage and increased after 45 days. Moreover, it is worth noting that long-term storage at low temperatures inhibits the activity of enzymes, such as lipoxygenase and alcohol acyltransferase, which are the key enzymes in the aroma metabolism of fruits esters, resulting in a reduction in fruit flavor [52]. The jujube fruits in this investigation were refrigerated for 45–60 days, showed a good flavor. However, some substances had an adverse effect on the flavor of jujube fruits when the content was too high. The odor of high content of ethyl acetate was not described as "fruity" but rather as an off-flavor. Its elevated levels are usually associated with the over-ripeness and/or anaerobic metabolism of the fruit [53].

Two alcohols, including linalool and citronellol, showed the same variation tendency of the whole storage period, which was not detected at the initial stages of storage but appeared at a later time. For DZ they were detected with high signal intensity when the storage time was 60 days, but they appeared earlier in JS, showing a high content after 45 days of storage. This condition might have occurred because, with the prolongation of storage time, the glycoside precursors and other precursors for linalool and citronellol synthesis were formed in the jujube fruits. He et al. found that the content of linalool was closely related to the temperature of lemon-flavored hard tea during storage. At room temperature, its content increased slightly but decreased significantly when the temperature increased [54]. An interesting phenomenon was that the monoterpene alcohol of transgenic citrus peels could induce resistance against fungal invasion [55]. Therefore, whether jujube fruits were infected by fungi in the late storage period was uncertain, and the increased in linalool and citronellol content may be related to the self-defensive mechanism of fruits.

Meanwhile, 2-ethylpyrazine and acetylpyrazine with weak signal in fresh jujube fruits, and eye-catching signals have emerged in the fingerprints until storage for 45 days. Pyrazines were the characteristic flavor of the Maillard reaction and were often found in dried jujube fruits, which show a roasted or nutty flavor [56]. During drying, heat treatment could accelerate the progress of the Maillard reaction, resulting in more furans and pyrazines were formed. In this study, jujube fruits were stored at low temperature, which did not provide a good condition for the formation of pyrazines compared with drying.

Some substances with the highest content during the middle of storage. Amyl acetate, 2-pentyl furan, 2-butoxyethanol, diallyl sulfide, heptanal, 1-pentanol dimer, and (*E*)-2-undecenal were the characteristic VOCs when DZ was stored at 4 °C for 15–30 days. Among these compounds, 2-pentyl furan, 2-butoxyethanol, and (*E*)-2-undecenal were the characteristic VOCs in JS in the same period. Heptanal, 2-methyl-1-propanol, and 2-butoxyethanol remained in DZ and were maintained at a high content during the subsequent storage period. In addition, the content of furfuryl alcohol continued to increase during storage, reaching a maximum of 60 days. Linalool and citronellol were synthesized in large quantities at the late storage stage. Most alcohols showed increased contents in the later stages, indicating that aldehydes were reduced and converted into corresponding alcohols at this stage, thus promoting the esterification of alcohols with acids produced by anaerobic respiration during storage. However, the signal intensity of 2-butoxyethanol disappeared later in JS, but the heptanal dimer was largely accumulated.

Dipropyl disulfide was only detected when the storage period was 15 days in DZ samples, but its content increased in JS after storage for 45 days. Three other sulfide compounds, namely, dimethyl sulfide, diallyl sulfide, and dimethyldisulphide were identified. The signal of dimethyl sulfide was observed in the whole storage process, and minor changes in signal intensity were observed in all samples. Diallyl sulfide and dimethyldisulphide were the characteristic VOCs of fresh JS and DZ, respectively. Sulfur-containing compounds are widely found in vegetables [57]. Diallyl sulfide was the characteristic flavor of garlic, dipropyl disulfide was associated with onion, and dimethyldisulphide was abundant in the volatile components of cabbage. Sulfur-containing compounds commonly arise from sulfur-bearing precursors, and in jujube fruits, sulfur-containing amino acids primarily exist, such as methionine. Low content of sulfur compounds can enhance the aromatic flavor of jujube fruits. Pyrazines and sulfur-containing compounds had low odor thresholds, and they acted with other compounds to enhance the overall aroma in jujube fruits [58,59]. Some substances, such as ethyl propanoate, amyl acetate dimer, 2-methyl-3-heptanone, and 3-hydroxy-2-butanone, existed in all periods of two jujube fruits, and the signal had minor changes.

### 2.5. PCA of Jujube Fruits at Different Storage Periods

PCA is a multivariate statistical analysis method that uses multiple variables to linearly transform to select a few significant variables. The main features are extracted for linear analysis by reducing the dimensionality of the data, and the main information is retained in several unrelated principal components [60]. Generally, when the cumulative contribution rate reaches 60%, PCA model is selected as the separation model [61]. Li et al. found the volatile components of different tissue parts of tomato showed disparate distributions among four varieties by PCA [62]. In this study, PCA was performed to analyzed the variation of 53 identified volatile compounds in the two cultivars of jujube. Firstly, all data were normalized to calculate the covariance matrix and its eigenvalues and eigenvectors, which derived from the corresponding peak area of each volatile component of jujube fruits. Then, the principal component was determined, and the corresponding contribution rate was calculated. Finally, a classification program was adopted in a smaller space to illustrate the relationship of jujube fruits during different storage periods. The results were shown in Figures 6 and 7, a clear separation trend of jujube fruits at different storage periods in two principal components can be observed.

As shown in Figures 6 and 7, the first principal component (PC1) and the second principal component (PC2) explained 82% and 84.6% of the total variables of the model in DZ and JS, respectively. A remarkable difference was found between fresh jujube fruits and refrigerated ones in volatile components. On the axis, a large distance is found between fresh DZ and the other samples, the four other samples in different storage periods can be distinguished easily. Among them, the ones stored for 15 days and 30 days are closer (Figure 6A), which indicates that the volatile components of DZ in these two periods are similar. In JS, changes of the sample are distributed from right to left in the PCA, as the storage time was extended (Figure 7A). Jujube fruits stored for 0–30 days are distributed in the right quadrant, while the ones stored for 45–60 days are in the left quadrant with a higher similarity. Based on the PCA, the jujube fruits at different periods of two cultivars are separated well. Additionally, some information on the volatile compounds was lost during the statistical re-modeling. As shown in the loading plots (Figures 6B and 7B), the length of the arrow reflects the extent of information loss, the shorter the arrow, the more information lost [20]. For example, linalool oxide and 5-methyfurfural suffered the most with the loss of their information of DZ, but in JS, dimethyl sulfide loss was the highest.

To get more details, the biplots were used (Figures 6C and 7C). From the biplots, it can be clearly seen that 2-methoxy-4-cresol, heptanal and some ketones (1-octen-3-one and 2,3-butanedione) were positively related to the fresh DZ. However, for JS, 3-hydroxy-2-butanone contributed a lot to the flavor of the fresh fruits. When storage for 15 days, (*E*)-undecenal and 2-penty furanl were closely related to the DZ, 2-butoxyethanol was positively related to JS. In the biplots, the relationship between specific volatile components and jujube fruits in a certain period was demonstrated. While at the end of the storage, most esters were positively related to the jujube fruits, while 2-methylbutanoic acid and 2-methylpropanoic acid were found to be positively related to JS. These results were in accordance with the above fingerprints.

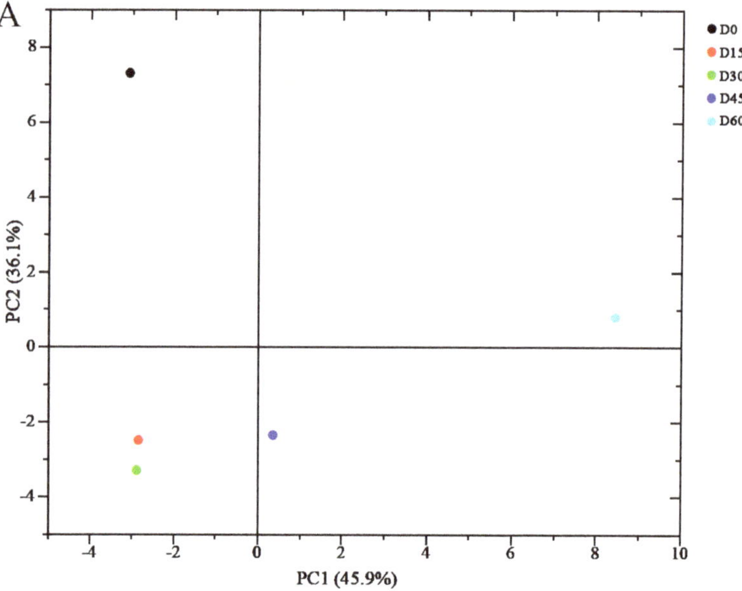


**Figure 6.** *Cont.*

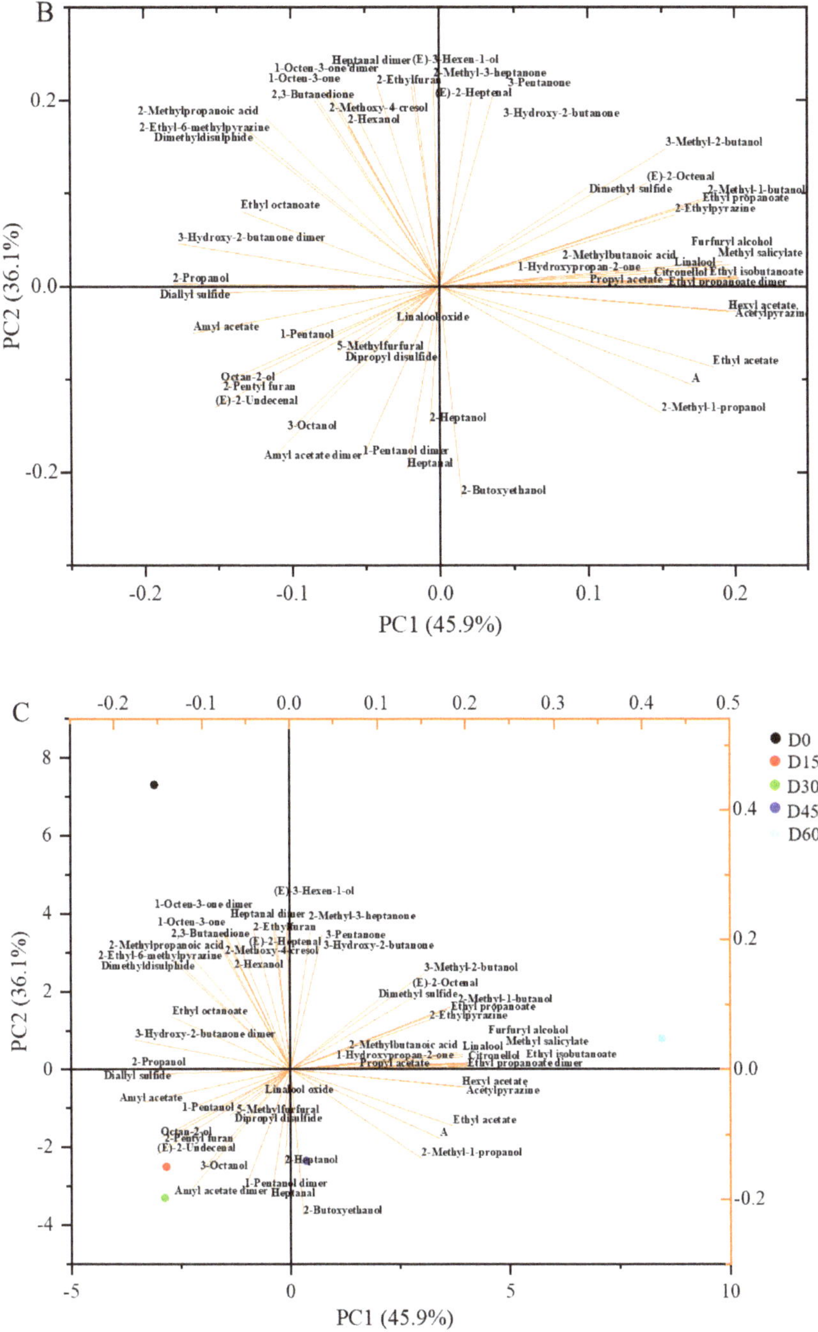

**Figure 6.** PCA analysis of Dongzao. (**A**) score plot of the first two principal components, D0, D15, D30, D45, and D60 represent refrigeration for 0, 15, 30, 45, and 60 days, respectively. (**B**) loading plot of different variances, (**C**) biplot of PCA.

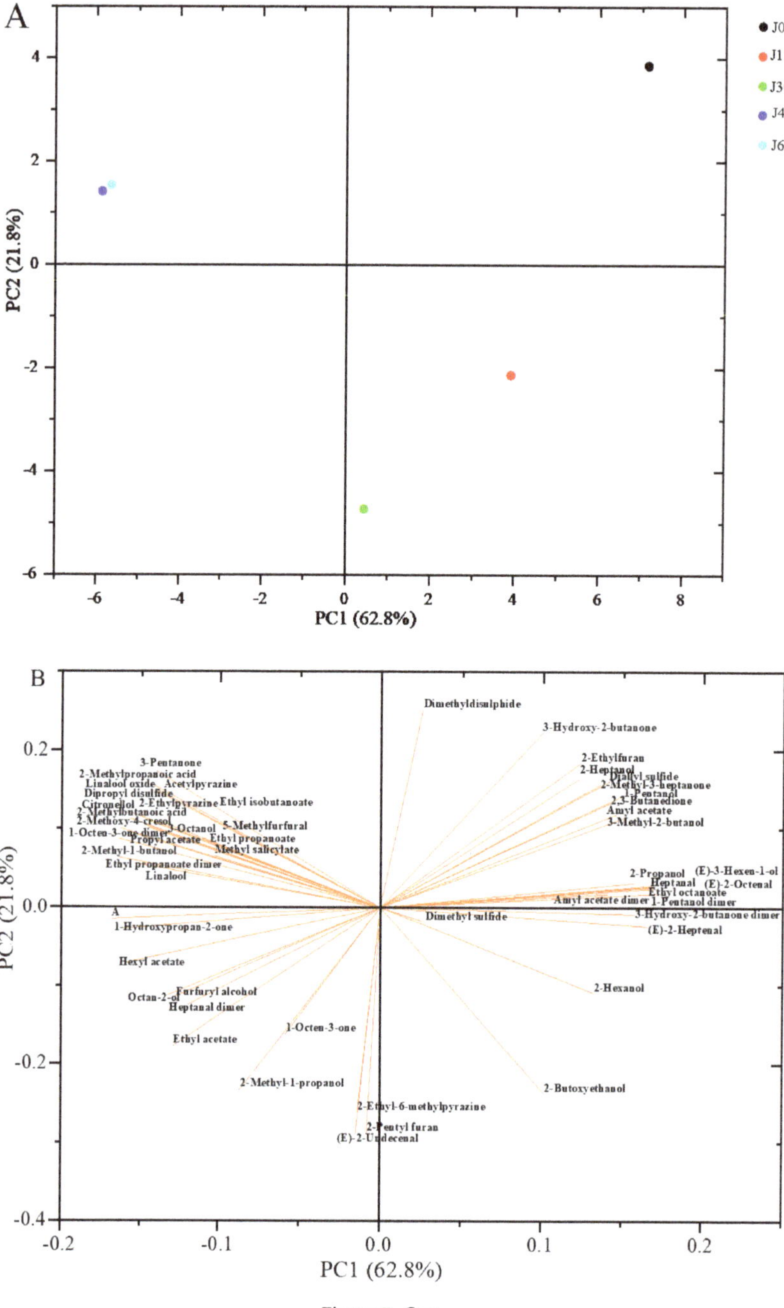



**Figure 7.** *Cont.*

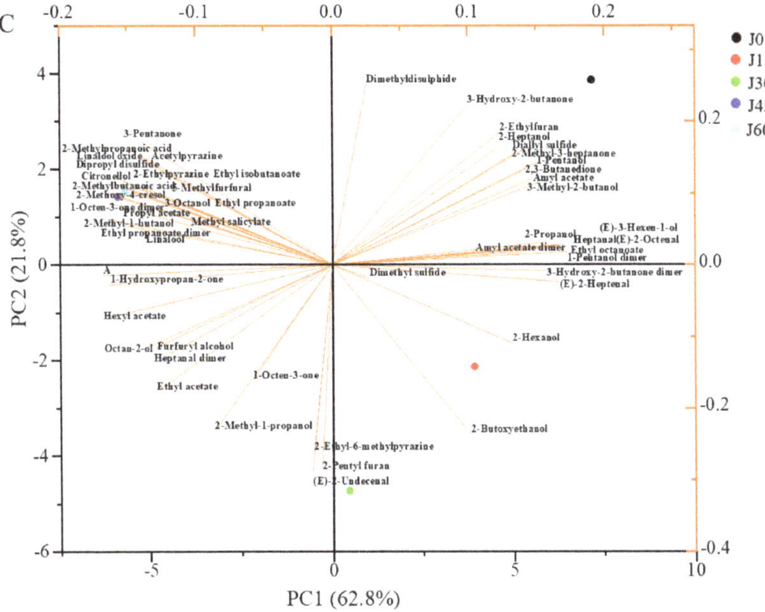


**Figure 7.** PCA analysis of Jinsixiaozao. (**A**) Score plot of the first two principal components, J0, J15, J30, J45, and J60 represent refrigeration for 0, 15, 30, 45, and 60 days, respectively. (**B**) Loading plot of different variances, (**C**) biplot of PCA.

## 2.6. Cluster Analysis of VOCs of Jujube Fruits from Different Periods Based on the Heat Map

To further understand the differences in VOCs of two jujube fruits at different storage periods, cluster analysis was performed using a heat map (Figure 8). According to the vertical direction of the heat map, all samples were classified into two main categories. Jujube fruits from JS that were stored for 45 and 60 days and those from DZ that were stored for 60 days were clustered together to form a group (J45, J60, and D60 in Figure 8), the rest of the samples were clustered together to form another group (J0 to J30, and D0 to D45 in Figure 8). At the late period of storage, the volatile components of DZ were very similar to those in JS and were quite different from those in the early storage period. The second group can also be divided into four categories. Among them, fresh jujube fruits from different cultivars were clustered into two different groups. Jujube fruits from JS, which were stored for 15 and 30 days, were grouped together (J15 and J30 in Figure 8), and those from DZ were stored for 15, 30 and 45 days, which were from the same class (D15, D30, and D45 in Figure 8).

According to the above results, we can infer that the volatile components of DZ and JS largely differ due to the differences in their varieties. The content of 1-octen-3-one and 2-methoxy-4-cresol was higher in fresh DZ than that in JS. Samples from D60, J45, and J60 were similar in 2-pentyl furan, 1-octen-3-one, ethyl acetate and propyl acetate content. Cluster analysis show that the storage time greatly influenced the volatile components of jujube fruits, and the flavor of jujube fruits varied during different periods. This finding was consistent with the fingerprint and PCA.

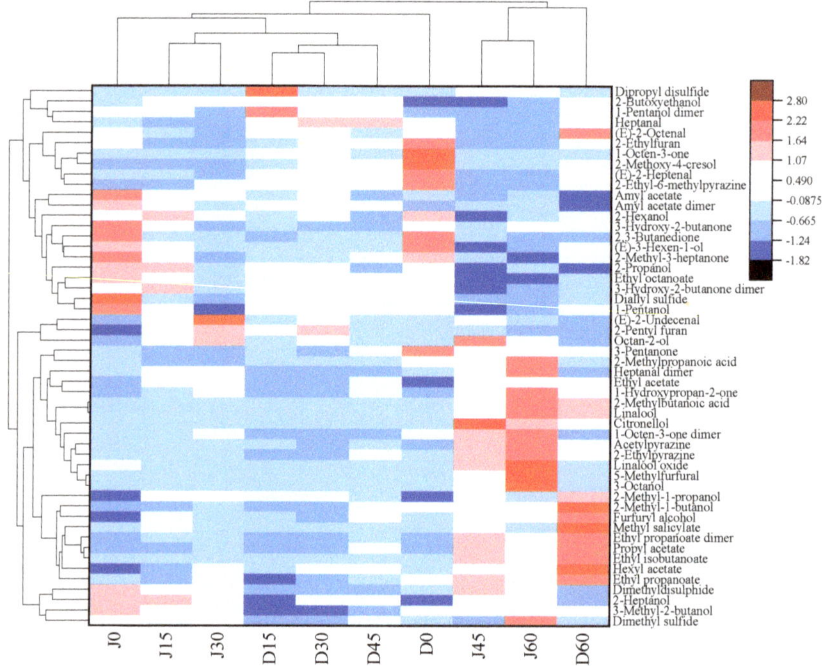

**Figure 8.** Heat map and cluster analysis of jujube fruits at different periods. D0, D15, D30, D45, and D60 represent DZ refrigerated for 0, 15, 30, 45, and 60 days, respectively; J0, J15, J30, J45, and J60 represent JS refrigerated for 0, 15, 30, 45, and 60 days, respectively.

## 3. Materials and Methods

### 3.1. Experimental Materials

Two cultivars of jujube fruits were used for this experiment, namely, *Zizyphus jujuba* Mill. *cv.* 'Dongzao' and *Zizyphus jujuba* Mill. *cv.* 'Jinsixiaozao'. Fresh DZ were harvested in September 2018, from a garden in Zhanhua County, Shandong province, China (118°7′56″ E, 37°41′53″ N). JS were harvested from an experimental field station in Qidong County, Hengyang City, Hunan Province, China (111°59′14″ E, 26°48′3″ N). After harvest, they were transported to the laboratory within 24 h with a cold chain. Jujube fruits with intact appearance, similar morphological properties, and free from visible blemish, disease, and mechanical injury were selected as raw materials. Samples were stored at 4 °C with a humidity of 90% and obtained every 15 days until storage for 60 days. The collected samples (500 g) were de-nucleated and freeze-dried, then the dried jujube fruits were crushed into powder with a grinder (RoyalstarRS-FS1401, Zhongshan Rongshida Kitchen & Bathroom Appliance Co., Ltd., Zhongshan, China), after that, collected samples were passed through a 60-mesh sieve, sealed, and stored at −80 °C for subsequent analyses.

### 3.2. Apparatuses

Vacuum freeze dryer LGJ-25C (Beijing Sihuan Instrument Factory Co., Ltd. Beijing, China), HS-GC-IMS instrument: the GC-IMS FlavourSpec® Gesellschaft für Analytische Sensorsysteme mbH (G.A.S., Dortmund, Germany). The device was equipped with an autosampler unit (CTC Analytics AG, Zwingen, Switzerland).

### 3.3. HS-GC-IMS Analysis

Analyses of jujube fruits were performed using the HS-GC-IMS instrument as described by Xin et al. with slight modifications [31]. Specifically, 1.0 g of fine powder was weighed and placed into a 20 mL headspace-glass sampling vial. The samples were incubated at 60 °C for 10 min. After incubation, 500 μL of headspace was automatically injected into the injector under splitless injection mode with a syringe at 85 °C. The GC was performed with an RTX-WAX (30 m, 0.53 mm ID, 1 μm film thickness, Restek Co., Bellefonte, PA, USA) capillary column to separate volatile components and coupled to IMS at 45 °C. Nitrogen (99.999% purity) was used as the carrier gas under the following programmed flow: 2 mL/min for 2 min and maintenance for 8 min, 20 mL/min for 10 min and 50 mL/min for 15 min, and flow then stopped. The instrument was performed under ambient pressure. The analytes were separated at 40 °C in the column and then ionized in the IMS ionization chamber of 45 °C. The drift gas (nitrogen gas) was set at 150 mL/min.

In this experiment, the instrument was standardized with n-ketones whose retention index was linear, because IMS had no response to alkanes. The retention index (RI) of volatile compounds was calculated by using n-ketones C4-C9 (Sinopharm Chemical Reagent Beijing Co., Ltd., Beijing, China) as external references. Volatile compounds were identified by comparing RI and the drift time (the time it takes for ions to reach the collector through drift tube, in milliseconds) of standard in the GC-IMS library (Gesellschaft für Analytische Sensorsysteme mbH, Dortmund, Germany).

### 3.4. Data Analysis

The instrumental analysis software included laboratory analytical viewer (LAV, G.A.S., Dortmund, Germany), three plug-ins (G.A.S., Dortmund, Germany), and GC × IMS library search, which can be used for sample analysis from different perspectives. The spectra were analyzed using the LAV software, and the difference profiles and fingerprints of volatile components were constructed using the Reporter and Gallery plug-ins. The NIST and IMS databases were built into the software for qualitative analysis of the materials. The principal component analysis (PCA) and heat map were used for clustering analysis of samples. The heat map and PCA were generated using Origin 2018 software (OriginLab, Northampton, MA, USA).

## 4. Conclusions

In this study, 47 volatile compounds were identified in jujube fruits at different storage periods at 4 °C via HS-GC-IMS. Among the identified substances, alcohols, esters, ketones and aldehydes were predominant, and a few organic acids were found. At the same time, some new compounds were found, namely, four sulfur-containing compounds (dimethyldisulphide, dimethyl sulfide, dipropyl disulfide, diallyl sulfide), linalool oxide, and 2-methoxy-4-cresol. The differences of the volatile compounds at different cold storage times between varieties and individuals were revealed by using topographic plots and fingerprints. Aldehydes and ketones were the main volatile components in fresh fruits and the early storage, while esters changed obviously and became the main VOCs at the end of the storage, which synthesized in large quantities at 45 days (JS) and 60 days (DZ), respectively. According to the PCA and heat map, the samples from different periods were well separated, the stored jujube fruits kept a long distance from the fresh one on the PCA map. The results showed that HS-GC-IMS could identify the characteristic volatile compounds of jujube fruits. It could provide an efficient method for determining the storage period of jujube fruits due to simple sample preparation steps and shorter analysis time.

**Supplementary Materials:** The following are available online, Table S1: Relative amount percentage (%) of the same compound at different refrigerated times.

**Author Contributions:** L.Y. performed the experiments, conducted the analyses, and wrote the original draft. S.D. and R.W. conceived the experiment. J.L., X.W., Q.Z., F.R., and Y.S. critically revised the manuscript from format to content.

**Funding:** The study was supported by the National Natural Science Foundation of China (31501543), Key Research Project of Hunan Province, China (2017NK2112), and Agricultural Science and Technology Innovation Project of Hunan Province, China (2019JG01, 2019TD04).

**Conflicts of Interest:** The authors declare that there is no conflict of interest.

## Abbreviations

| | |
|---|---|
| HS-GC-IMS | headspace-gas chromatography-ion mobility spectrometry |
| DZ | *Zizyphus jujuba* Mill. *cv.* Dongzao |
| JS | *Zizyphus jujuba* Mill. *cv.* Jinsixiaozao |
| VOCs | volatile organic compounds |
| PCA | PCA principal component analysis |

## References


1. Li, H.M.; Li, F.; Wang, L.; Sheng, J.C.; Xin, Z.H.; Zhao, L.Y.; Xiao, H.M.; Zheng, Y.H.; Hu, Q.H. Effect of nano-packing on preservation quality of Chinese jujube (*Ziziphus jujuba* Mill. var. *inermis* (Bunge) Rehd). *Food Chem.* **2009**, *114*, 547–552. [CrossRef]
2. Hernández, F.; Legua, P.; Melgarejo, P.; Martínez, R.; Martínez, J.J. Phenological growth stages of jujube tree (*Ziziphus jujube*): Codification and description according to the BBCH scale. *Ann. Appl. Biol.* **2015**, *166*, 136–142. [CrossRef]
3. Wojdyło, A.; Carbonell-Barrachina, Á.A.; Legua, P.; Hernández, F. Phenolic composition, ascorbic acid content, and antioxidant capacity of Spanish jujube (*Ziziphus jujube* Mill.) fruits. *Food Chem.* **2016**, *201*, 307–314. [CrossRef] [PubMed]
4. Li, J.W.; Fan, L.P.; Ding, S.D.; Ding, X.L. Nutritional composition of five cultivars of Chinese jujube. *Food Chem.* **2007**, *103*, 454–460. [CrossRef]
5. Choi, S.H.; Ahn, J.B.; Kozukue, N.; Levin, C.E.; Friedman, M. Distribution of free amino acids, flavonoids, total phenolics, and antioxidative activities of jujube (*Ziziphus jujuba*) fruits and seeds harvested from plants grown in Korea. *J. Agric. Food Chem.* **2011**, *59*, 6594–6604. [CrossRef]
6. Guo, S.; Duan, J.A.; Zhang, Y.; Qian, D.W.; Tang, Y.P.; Zhu, Z.H.; Wang, H.Q. Contents changes of triterpenic acids, nucleosides, nucleobases, and saccharides in jujube (*Ziziphus jujuba*) fruit during the drying and steaming process. *Molecules* **2015**, *20*, 22329–22340. [CrossRef]
7. Liu, G.P.; Liu, X.Q.; Zhang, Y.C.; Zhang, F.; Wei, T.; Yang, M.; Wang, K.M.; Wang, Y.J.; Liu, N.; Cheng, H.L.; et al. Hepatoprotective effects of polysaccharides extracted from *Zizyphus jujube* cv. Huanghetanzao. *Int. J. Biol. Macromol.* **2015**, *76*, 169–175. [CrossRef]
8. González-Gallego, J.; Sánchez-Campos, S.; Tuñón, M.J. Anti-inflammatory properties of dietary flavonoids. *Nutr. Hosp.* **2007**, *22*, 287–293.
9. Ma, Q.; Wang, L.H.; Jiang, J.G. Hepatoprotective effect of flavonoids from *Cirsium japonicum* DC on hepatotoxicity in comparison with silymarin. *Food Funct.* **2016**, *7*, 2179–2184. [CrossRef]
10. Liu, H.C.; Xu, M.D. Changes in quality characteristics of fresh-cut jujubes as affected by pressurized nitrogen treatment. *Innov. Food Sci. Emerg. Technol.* **2015**, *30*, 43–50. [CrossRef]
11. Kou, X.H.; He, Y.L.; Li, Y.F.; Chen, X.Y.; Feng, Y.C.; Xue, Z.H. Effect of abscisic acid (ABA) and chitosan/nano-silica/sodium alginate composite film on the colour development and quality of postharvest Chinese winter jujube (*Zizyphus jujuba* Mill. cv. Dongzao). *Food Chem.* **2019**, *270*, 385–394. [CrossRef] [PubMed]
12. Günther, C.S.; Marsh, K.B.; Winz, R.A.; Harker, R.F.; Wohlers, M.W.; White, A.; Goddard, M.R. The impact of cold storage and ethylene on volatile ester production and aroma perception in 'Hort16A' kiwifruit. *Food Chem.* **2015**, *169*, 5–12. [CrossRef]
13. Both, V.; Brackmann, A.; Thewes, F.R.; de Freitas Ferreira, D.; Wagner, R. Effect of storage under extremely low oxygen on the volatile composition of 'Royal Gala' apples. *Food Chem.* **2014**, *156*, 50–57. [CrossRef] [PubMed]
14. Bandeira Reidel, R.V.; Melai, B.; Cioni, P.L.; Pistelli, L. Chemical composition of volatiles emitted by *Ziziphus jujuba* during different growth stages. *Plant Biosyst.* **2018**, *152*, 825–830. [CrossRef]

15. Wang, H.; Li, P.; Sun, S.H.; Zhang, Q.D.; Su, Y.; Zong, Y.L.; Xie, J.P. Comparison of liquid-liquid extraction, simultaneous distillation extraction, ultrasound-assisted solvent extraction, and headspace solid-phase microextraction for the determination of volatile compounds in jujube extract by gas chromatography/mass spectrometry. *Anal. Lett.* **2014**, *47*, 654–674.

16. Galindo, A.; Noguera-Artiaga, L.; Cruz, Z.N.; Burló, F.; Hernández, F.; Torrecillas, A.; Carbonell-Barrachina, Á.A. Sensory and physico-chemical quality attributes of jujube fruits as affected by crop load. *LWT Food Sci. Technol.* **2015**, *63*, 899–905. [CrossRef]

17. Wang, R.R.; Ding, S.H.; Zhao, D.D.; Wang, Z.F.; Wu, J.H.; Hu, X.S. Effect of dehydration methods on antioxidant activities, phenolic contents, cyclic nucleotides, and volatiles of jujube fruits. *Food Sci. Biotechnol.* **2016**, *25*, 137–143. [CrossRef]

18. Chen, K.; Gao, L.; Li, Q.; Li, H.R.; Zhang, Y. Effects of $CO_2$ pretreatment on the volatile compounds of dried Chinese jujube (*Zizyphus jujuba* Miller). *Food Sci. Technol.* **2017**, *37*, 578–584. [CrossRef]

19. Pu, Y.H.; Ding, T.; Lv, R.L.; Cheng, H.; Liu, D.H. Effect of drying and storage on the volatile compounds of jujube fruit detected by electronic nose and GC-MS. *Food Sci. Technol. Res.* **2018**, *24*, 1039–1047. [CrossRef]

20. Wang, L.N.; Wang, Y.Q.; Wang, W.Z.; Zheng, F.P.; Chen, F. Comparison of volatile compositions of 15 different varieties of Chinese jujube (*Ziziphus jujuba* Mill.). *J. Food Sci. Technol.* **2019**, *56*, 1631–1640. [CrossRef]

21. Zhang, L.X.; Shuai, Q.; Li, P.W.; Zhang, Q.; Ma, F.; Zhang, W.; Ding, X.X. Ion mobility spectrometry fingerprints: A rapid detection technology for adulteration of sesame oil. *Food Chem.* **2016**, *192*, 60–66. [CrossRef] [PubMed]

22. Li, M.Q.; Yang, R.W.; Zhang, H.; Wang, S.L.; Chen, D.; Lin, S.Y. Development of a flavor fingerprint by HS-GC-IMS with PCA for volatile compounds of *Tricholoma matsutake* Singer. *Food Chem.* **2019**, *290*, 32–39. [CrossRef] [PubMed]

23. Gerhardt, N.; Birkenmeier, M.; Sanders, D.; Rohn, S.; Weller, P. Resolution-optimized headspace gas chromatography-ion mobility spectrometry (HS-GC-IMS) for non-targeted olive oil profiling. *Anal. Bioanal. Chem.* **2017**, *409*, 3933–3942. [CrossRef] [PubMed]

24. Arce, L.; Gallegos, J.; Garrido, R.; Medina, L.M.; Sielemann, S.; Wortelmann, T. Ion mobility spectrometry a versatile analytical tool for metabolomics application in food science. *Curr. Metab.* **2014**, *2*, 264–271. [CrossRef]

25. Wang, X.R.; Yang, S.P.; He, J.N.; Chen, L.Z.; Zhang, J.Z.; Jin, Y.; Zhou, J.H.; Zhang, Y.X. A green triple-locked strategy based on volatile-compound imaging, chemometrics, and markers to discriminate winter honey and *sapium* honey using headspace gas chromatography-ion mobility spectrometry. *Food Res. Int.* **2019**, *119*, 960–967. [CrossRef] [PubMed]

26. Cavanna, D.; Zanardi, S.; Dall'Asta, C.; Suman, M. Ion mobility spectrometry coupled to gas chromatography: A rapid tool to assess eggs freshness. *Food Chem.* **2019**, *271*, 691–696. [CrossRef]

27. Makinen, M.A.; Anttalainen, O.A.; Sillanpää, M.E. Ion mobility spectrometry and its applications in detection of chemical warfare agents. *Anal. Chem.* **2010**, *82*, 9594–9600. [CrossRef]

28. Verkouteren, J.R.; Staymates, J.L. Reliability of ion mobility spectrometry for qualitative analysis of complex, multicomponent illicit drug samples. *Forensic Sci. Int.* **2011**, *206*, 190–196. [CrossRef]

29. De las Nieves Peiró, M.; Armenta, S.; Garrigues, S.; de la Guardia, M. Determination of 3, 4-methylenedioxypyrovalerone (MDPV) in oral and nasal fluids by ion mobility spectrometry. *Anal. Bioanal. Chem.* **2016**, *408*, 3265–3273. [CrossRef]

30. Márquez-Sillero, I.; Aguilera-Herrador, E.; Cárdenas, S.; Valcárcel, M. Ion-mobility spectrometry for environmental analysis. *TrAC Trends Anal. Chem.* **2011**, *30*, 677–690. [CrossRef]

31. Sun, X.; Gu, D.Y.; Fu, Q.B.; Gao, L.; Shi, C.H.; Zhang, R.T.; Qiao, X.G. Content variations in compositions and volatile component in jujube fruits during the blacking process. *Food Sci. Nutr.* **2019**, *7*, 1387–1395. [CrossRef] [PubMed]

32. Martín-Gómez, A.; Arroyo-Manzanares, N.; Rodríguez-Estévez, V.; Arce, L. Use of a non-destructive sampling method for characterization of Iberian cured ham breed and feeding regime using GC-IMS. *Meat Sci.* **2019**, *152*, 146–154. [CrossRef] [PubMed]

33. Arroyo-Manzanares, N.; Martín-Gómez, A.; Jurado-Campos, N.; Garrido-Delgado, R.; Arce, C.; Arce, L. Target vs. spectral fingerprint data analysis of Iberian ham samples for avoiding labelling fraud using headspace-gas chromatography-ion mobility spectrometry. *Food Chem.* **2018**, *246*, 65–73. [CrossRef]

34. Infante, R.; Farcuh, M.; Meneses, C. Monitoring the sensorial quality and aroma through an electronic nose in peaches during cold storage. *J. Sci. Food Agric.* **2008**, *88*, 2073–2078. [CrossRef]

35. Rodríguez-Maecker, R.; Vyhmeister, E.; Meisen, S.; Martinez, A.R.; Kuklya, A.; Telgheder, U. Identification of terpenes and essential oils by means of static headspace gas chromatography-ion mobility spectrometry. *Anal. Bioanal. Chem.* **2017**, *409*, 6595–6603. [CrossRef] [PubMed]

36. Osorio, S.; Scossa, F.; Fernie, A. Molecular regulation of fruit ripening. *Front. Plant Sci.* **2013**, *4*, 198. [CrossRef]

37. Atkinson, R.G.; Gunaseelan, K.; Wang, M.Y.; Luo, L.K.; Wang, T.C.; Norling, C.L.; Johnston, S.L.; Maddumage, R.; Schröder, R.; Schaffer, R.J. Dissecting the role of climacteric ethylene in kiwifruit (*Actinidia chinensis*) ripening using a 1-aminocyclopropane-1-carboxylic acid oxidase knockdown line. *J. Exp. Bot.* **2011**, *62*, 3821–3835. [CrossRef]

38. Xi, W.P.; Zheng, H.W.; Zhang, Q.Y.; Li, W.H. Profiling taste and aroma compound metabolism during apricot fruit development and ripening. *Int. J. Mol. Sci.* **2016**, *17*, 998. [CrossRef]

39. Liu, T.T.; Yang, T.S. Optimization of solid-phase microextraction analysis for studying change of headspace flavor compounds of banana during ripening. *J. Agric. Food Chem.* **2002**, *50*, 653–657. [CrossRef]

40. Lantsuzskaya, E.V.; Krisilov, A.V.; Levina, A.M. Structure of the cluster ions of ketones in the gas phase according to ion mobility spectrometry and ab initio calculations. *Russ. J. Phys. Chem. A* **2015**, *89*, 1838–1842. [CrossRef]

41. Bengtsson, M.; Bäckman, A.C.; Liblikas, I.; Ramirez, M.I.; Borg-Karlson, A.K.; Ansebo, L.; Anderson, P.; Lofqvist, J.; Witzgall, P.; Bäckman, A.C.; et al. Plant odour analysis of apple: Antennal response of codling moth females to apple volatiles during phenological development. *J. Agric. Food Chem.* **2001**, *49*, 3736–3741. [CrossRef] [PubMed]

42. Cano-Salazar, J.; López, M.L.; Crisosto, C.H.; Echeverría, G. Volatile compound emissions and sensory attributes of 'Big Top' nectarine and 'Early Rich' peach fruit in response to a pre-storage treatment before cold storage and subsequent shelf-life. *Postharvest Biol. Technol.* **2013**, *76*, 152–162. [CrossRef]

43. Chen, Q.Q.; Song, J.X.; Bi, J.F.; Meng, X.J.; Wu, X.Y. Characterization of volatile profile from ten different varieties of Chinese jujubes by HS-SPME/GC-MS coupled with E-nose. *Food Res. Int.* **2018**, *105*, 605–615. [CrossRef] [PubMed]

44. Wang, D.; Cai, J.; Zhu, B.Q.; Wu, G.F.; Duan, C.Q.; Chen, G.; Shi, Y. Study of free and glycosidically bound volatile compounds in air-dried raisins from three seedless grape varieties using HS-SPME with GC-MS. *Food Chem.* **2015**, *177*, 346–353. [CrossRef]

45. Wang, L.N.; Zhu, J.C.; Wang, Y.Q.; Wang, X.; Chen, F.; Wang, X.W. Characterization of aroma-impact compounds in dry jujubes (*Ziziphus jujube* Mill.) by aroma extract dilution analysis (AEDA) and gas chromatography-mass spectrometer (GC-MS). *Int. J. Food Prop.* **2018**, *21*, 1844–1853. [CrossRef]

46. Varlet, V.; Prost, C.; Serot, T. Volatile aldehydes in smoked fish: Analysis methods, occurrence and mechanisms of formation. *Food Chem.* **2007**, *105*, 1536–1556. [CrossRef]

47. Bertolino, M.; Dolci, P.; Giordano, M.; Rolle, L.; Zeppa, G. Evolution of chemico-physical characteristics during manufacture and ripening of Castelmagno PDO cheese in wintertime. *Food Chem.* **2011**, *129*, 1001–1011. [CrossRef]

48. Reineccius, G. *Flavor Chemistry and Technology*, 2nd ed.; Taylor and Francis Routledge: Boca Raton, FL, USA, 2006.

49. Xiao, Z.B.; Liu, J.H.; Chen, F.; Wang, L.Y.; Niu, Y.W.; Feng, T.; Zhu, J.C. Comparison of aroma-active volatiles and their sensory characteristics of mangosteen wines prepared by *Saccharomyces cerevisiae* with GC-olfactometry and principal component analysis. *Nat. Prod. Res.* **2015**, *29*, 656–662. [CrossRef]

50. Söllner, K.; Schieberle, P. Decoding the Key Aroma compounds of a Hungarian-type salami by molecular sensory science approaches. *J. Agric. Food Chem.* **2009**, *57*, 4319–4327. [CrossRef]

51. Zhou, X.; Dong, L.; Li, R.; Zhou, Q.; Wang, J.W.; Ji, S.J. Low temperature conditioning prevents loss of aroma-related esters from 'Nanguo' pears during ripening at room temperature. *Postharvest Biol. Technol.* **2015**, *100*, 23–32. [CrossRef]

52. Zhu, X.Y.; Luo, J.; Li, Q.M.; Li, J.; Liu, T.X.; Wang, R.; Chen, W.X.; Li, X.P. Low temperature storage reduces aroma-related volatiles production during shelf-life of banana fruit mainly by regulating key genes involved in volatile biosynthetic pathways. *Postharvest Biol. Technol.* **2018**, *146*, 68–78. [CrossRef]

53. Zlatić, E.; Zadnik, V.; Fellman, J.; Demšar, L.; Hribar, J.; Čejić, Ž.; Vidrih, R. Comparative analysis of aroma compounds in 'Bartlett' pear in relation to harvest date, storage conditions, and shelf-life. *Postharvest Biol. Technol.* **2016**, *117*, 71–80. [CrossRef]

54. He, F.; Qian, Y.L.; Qian, M.C. Flavour and chiral stability of lemon-flavored hard tea during storage. *Food Chem.* **2018**, *239*, 622–630. [CrossRef] [PubMed]

55. Rodríguez, A.; Kava, V.; Latorre-García, L.; da Silva, G.J., Jr.; Pereira, R.G.; Glienke, C.; Ferreira-Maba, L.S.; Vicent, A.; Shimada, T.; Peña, L. Engineering d-limonene synthase down-regulation in orange fruit induces resistance against the fungus *Phyllosticta citricarpa* through enhanced accumulation of monoterpene alcohols and activation of defence. *Mol. Plant Pathol.* **2018**, *19*, 2077–2093. [CrossRef] [PubMed]

56. Wang, D.; Duan, C.Q.; Shi, Y.; Zhu, B.Q.; Javed, H.U.; Wang, J. Free and glycosidically bound volatile compounds in sun-dried raisins made from different fragrance intensities grape varieties using a validated HS-SPME with GC-MS method. *Food Chem.* **2017**, *228*, 125–135. [CrossRef] [PubMed]

57. Kremr, D.; Bajerová, P.; Bajer, T.; Eisner, A.; Adam, M.; Ventura, K. Using headspace solid-phase microextraction for comparison of volatile sulphur compounds of fresh plants belonging to families *Alliaceae* and *Brassicaceae*. *J. Food Sci. Technol.* **2015**, *52*, 5727–5735. [CrossRef]

58. Mishra, P.K.; Tripathi, J.; Gupta, S.; Variyar, P.S. Effect of cooking on aroma profile of red kidney beans (*Phaseolus vulgaris*) and correlation with sensory quality. *Food Chem.* **2017**, *215*, 401–409. [CrossRef]

59. Müller, R.; Rappert, S. Pyrazines: Occurrence, formation and biodegradation. *Appl. Microbiol. Biotechnol.* **2010**, *85*, 1315–1320. [CrossRef]

60. Jo, D.; Kim, G.R.; Yeo, S.H.; Jeong, Y.J.; Noh, B.S.; Kwon, J.H. Analysis of aroma compounds of commercial cider vinegars with different acidities using SPME/GC-MS, electronic nose, and sensory evaluation. *Food Sci. Biotechnol.* **2013**, *22*, 1559–1565. [CrossRef]

61. Wu, Z.B.; Chen, L.Z.; Wu, L.M.; Xue, X.F.; Zhao, J.; Li, Y.; Ye, Z.H.; Lin, G.H. Classification of Chinese honeys according to their floral origins using elemental and stable isotopic compositions. *J. Agric. Food Chem.* **2015**, *63*, 5388–5394. [CrossRef]

62. Li, J.; Di, T.J.; Bai, J.H. Distribution of volatile compounds in different fruit structures in four tomato cultivars. *Molecules* **2019**, *24*, 2594. [CrossRef] [PubMed]

**Sample Availability:** Samples of the compounds are not available from the authors.

*Article*

# Tracking Sensory Characteristics of Virgin Olive Oils During Storage: Interpretation of Their Changes from a Multiparametric Perspective

Ana Lobo-Prieto [1], Noelia Tena [2], Ramón Aparicio-Ruiz [2], María T. Morales [2] and Diego L. García-González [1,*]

[1]   Instituto de la Grasa (CSIC), Ctra. de Utrera, km. 1, Campus Universitario Pablo de Olavide-building 46, 41013 Sevilla, Spain; ana.lobo@ig.csic.es
[2]   Department of Analytical Chemistry, Faculty of Pharmacy, University of Seville, Prof. García González, 2, 41012 Seville, Spain; noelia.tena@ig.csic.es (N.T.); aparicioruiz@cica.es (R.A.-R.); tmorales@us.es (M.T.M.)
*   Correspondence: dlgarcia@ig.csic.es; Tel.: +34-954-611-550

Academic Editor: Eugenio Aprea
Received: 2 March 2020; Accepted: 3 April 2020; Published: 7 April 2020

**Abstract:** Virgin olive oil is inevitably subject to an oxidation process during storage that can affect its stability and quality due to off-flavors that develop before the oil surpasses its 'best before' date. Many parameters are involved in the oxidation process at moderate conditions. Therefore, a multiparametric study is necessary to establish a link between physico-chemical changes and sensory quality degradation in a real storage experiment. In this context, a storage experiment of 27 months was performed for four monovarietal virgin olive oils, bottled in transparent 500-mL PET bottles and subjected to conditions close to a supermarket scenario. Volatile composition, quality parameters and phenolic compounds were determined monthly. Simultaneously, an accredited sensory panel assessed their sensory characteristics. The stability of the fresh samples was also studied with the oxidative stability index (OSI) and mesh cell-FTIR. (*E*)-2-hexenal, (*Z*)-3-hexen-1-ol and (*E*)-2-hexen-1-ol were identified as markers of the fruity attribute. Hexanal and nonanal were also identified as compounds that were associated with the rise of median of defect during storage. Some disagreements were observed between the sensory assessment and the OSI analyzed by Rancimat. However, the increase of concentration of rancid markers agreed with the increase of aldehyde band measured with mesh cell-FTIR.

**Keywords:** virgin olive oil; volatile compounds; sensory assessment; storage; SPME-GC; oxidation; oxidative stability; mesh cell-FTIR

---

## 1. Introduction

The production of virgin olive oil (VOO) is limited to several months per year; this leads to the necessity of storing the oil to ensure a continuous supply for consumers. This storage is carried out at various levels in the food chain, for example, in tanks during trading, or in bottles in retailers. During storage, VOO is exposed to external variables that cause changes to its composition that lead to a loss of its nutritional quality and changes of its sensory characteristics [1]. It is well known that VOO is more resistant to oxidation than other edible oils because of its composition. In spite of this, numerous studies [2–5] have demonstrated that storage conditions have a strong influence in the degradations of oils that may cause some problems in retailing. Thus, light and temperature—even in mild conditions—can affect considerably the stability and quality of VOO [6–11] during the shelf life, leading to a loss of its nutritional properties and ultimately, a development of off-flavors that result in rancidity. This problem of stability has sometimes led to discrepancies between the results of control-testing and information declared on the label. This is the case of some extra virgin olive oils that—after a storage period—may be unexpectedly downgraded to virgin olive oil category. For this reason, regulatory bodies have stablished some specific requirements concerning the storage conditions of olive oil [12,13]. Additionally, International Olive Council has recently approved a document with the best practice guidelines for the storage of VOO [14].

The need of a major control of VOO stability is not new, although the increasing importance of sensory quality of this product and the stricter standards today [15] have encouraged producers and standardization bodies to find new methods to evaluate and understand VOO stability. Some methods are based on the study of the quality parameters during oil storage under conditions that are similar to real ones. These methods require a long period of study (several months), so they are not useful when a rapid answer is required. Other methods use accelerated conditions to obtain results in a shorter time (i.e., several hours/days), such as Rancimat, but they accelerate the oxidation process by means of drastic conditions and the results do not correlate well with the real oxidation process and do not take photooxidation into account [7].

Several studies have proposed rapid methods that assess VOO stability under conditions that are close to the real ones. Thus, Schaal oven test allows determining stability of the oils at 63 °C [16]. Other techniques, such as electron spin resonance spectroscopy [17,18] and differential scanning calorimetry [19], have been also used to assess oil stability at 70 °C or lower temperatures. Recently, our group developed a new procedure based on mesh cell-FTIR spectroscopy [6] to assess VOO stability at room temperature considering the effect of photooxidation [7,20]. The methods that use moderate conditions provide information that is easily correlated with the real degradation taking place in VOO under real storage. However, the interpretation of results is hindered by the fact that many parameters (chemical, physico-chemical, sensory) are evolving at the same time with different kinetics. Thus, it is difficult to establish a simple rule based on these parameters to assess if an oil is clearly out of the 'best before' date for its consumption. A comprehensive study of all these parameters and their inter-relationships among them is necessary to understand VOO shelf-life.

In the last decade, numerous authors have focused on tracking different compounds and parameters of VOO during its shelf life. The majority of these studies have aimed to monitor the changes in the quality parameters of VOO during the storage under different conditions and with different types of containers [5,21]. Other authors have centered their studies on the loss of VOO healthy compounds, such as phenols, during the shelf life under different storage conditions [22–24]. Due to the changes in sensory characteristics that take place during shelf-life, some of the studies were based on volatile analysis during storage at different conditions [3,8,25,26]. Some of these studies do not use conditions that are commonly used in a supermarket, where oils are exposed to room temperature and light/dark cycles of 12 h. Therefore, currently a disparity of results is observed when the volatile compounds and other parameters are evaluated during the storage of VOOs, with a resulting difficulty in the interpretation of oxidation processes at moderate conditions.

Aroma has a strong influence on the consumer's rejection or acceptability of VOOs that have been stored for several months [7]. For this reason, stablishing a link between the chemical changes taking place in the oils during the storage and the sensory changes during this time is needed to predict better the 'best before' date of the oil. The aim of this study was to evaluate the VOO quality changes generated by the oxidation process to which the oil is inevitably subjected during its storage. Thus, a long storage experiment of 27 months was performed, period during which VOOs from three different cultivars were exposed under moderate conditions simulating a supermarket scenario. When moderate conditions of temperature and light intensity are used, the changes on sensory properties can be subtle and difficult to interpret. In order to facilitate this interpretation, in addition to the volatile composition and sensory assessment, chemical parameters directly or indirectly related to virgin olive oil quality were analyzed month by month. Furthermore, the oxidative stabilities of fresh oils evaluated with accelerated procedures (Rancimat and mesh cell-FTIR) were also considered. With all this information, the moments in which a remarkable sensory change take place was identified and explained by changes produced in the volatile composition and the other indexes of quality. Finally, the results of stability assessment on the fresh samples carried out by Rancimat and mesh-cell FTIR were analyzed according to the development of the off-flavors during the storage of the four monocultivar samples.

## 2. Results and Discussion

### 2.1. Characterization of Fresh VOOs

Four VOOs were selected from three different cultivars (Hojiblanca, Picual and two from Arbequina) for this study. These cultivars were selected to cover different chemical composition and for their genuine sensory characteristics [27,28].

The quality parameters were determined for the fresh four VOOs before starting the storage ("time zero") with the aim of characterizing their actual quality at the moment of bottling. Free acidity, extinction coefficients ($K_{270}$ and $K_{232}$), peroxide value (PV), fatty acid composition, total phenol content, and results from Rancimat and sensory assessment (medians of the defect and the fruity attribute) are shown in Table 1, together with date of extraction and cultivar of each VOO. According to the results obtained for PV, $K_{270}$ and $K_{232}$, all the values were below the limits stated in EC regulation [13] for the classification as "extra virgin olive oil" category. However, the sample VOO2 was pointed out as the most oxidized sample before starting the experiment despite all the samples were collected from the vertical centrifuge and the storage experiment started shortly after. Thus, in all these parameters, VOO2 showed the highest values, although far from the maximum limits for the "extra virgin olive oil" category. However, the sensory quality parameters (medians of defect and the fruity attribute showed in Table 1) revealed that VOO4 was initially categorized within the "virgin olive oil" category instead of "extra virgin olive oil" category. In this sample, panelists detected a winey-vinegary defect (median of defect = 2.1) before starting the storage. The organoleptic assessment before the storage experiment reported the sensory differences associated to the cultivars [27,28]. Thus, assessors identified an intense fruity and green odor in the fresh sample of VOO1 (VOO1-0m), which explained the highest median value for the fruity attribute (Table 1). VOO3-0m was characterized by a high median of the fruity attribute and the panelists described it as a fruity, bitter and pungent oil with some fig and wood notes, typical from Picual cultivar [29]. Whereas, VOO2-0m and VOO4-0m, from Arbequina cultivar, showed a delicate fruitiness with slight bitter and pungent notes, showing the lowest median of the fruity attribute.

**Table 1.** Quality parameters, together with the quality criteria of extra virgin olive oil (EVOO) according to EC regulation No 2568/91 and updates; stability index, fatty acid composition and total phenol content in the fresh virgin olive oils, before of the bottling. The virgin olive oil codes, dates of the extraction and the cultivars are shown.

| Chemical Parameters | | Virgin Olive Oil Codes | | | | Quality Criteria (EVOO) |
|---|---|---|---|---|---|---|
| | | VOO1 | VOO2 | VOO3 | VOO4 | |
| Cultivar | | Hojiblanca | Arbequina | Picual | Arbequina | |
| Extraction Date (dd/mm/yy) | | 31/10/15 | 3/12/2015 | 12/11/2015 | 20/11/2015 | |
| Organoleptic | Mf [a] | 4.7 | 3.5 | 3.8 | 3 | >0 |
| characteristics | Md [b] | 0 | 0 | 0 | 2.1 | =0 |
| Free fatty acid (% m/m of oleic acid) | | 0.15 | 0.21 | 0.20 | 0.22 | $\leq$0.8 |
| Peroxide value (PV) (meq $O_2$/kg) | | 4.30 | 5.13 | 3.63 | 4.80 | $\leq$20 |
| $K_{232}$ ($K_{1cm}^{1\%}$) | | 1.52 | 1.87 | 1.72 | 1.84 | $\leq$2.50 |
| $K_{270}$ ($K_{1cm}^{1\%}$) | | 0.06 | 0.19 | 0.04 | 0.18 | $\leq$0.22 |
| Total phenols (mg/kg) | | 226.71 | 338.90 | 534.82 | 468.10 | |
| Oil stability index (OSI) at 100 °C (h) | | 38.71 | 22.95 | 82.80 | 53.60 | |
| C16:0 (%) | | 14.30 | 15.46 | 10.79 | 16.02 | |
| C16:1 (%) | | 0.93 | 1.77 | 1.19 | 1.87 | |
| C17:0 (%) | | 0.18 | 0.11 | 0.04 | 0.15 | |
| C17:1 (%) | | 0.29 | 0.24 | 0.08 | 0.27 | |
| C18:0 (%) | | 2.56 | 1.95 | 2.49 | 1.91 | |
| C18:1 (%) | | 72.38 | 66.71 | 80.28 | 65.93 | |
| C18:2 (%) | | 7.29 | 12.22 | 3.45 | 12.36 | |
| C18:3 (%) | | 1.08 | 0.66 | 0.82 | 0.65 | |
| C20:0 (%) | | 0.47 | 0.40 | 0.41 | 0.39 | |
| C20:1 (%) | | 0.31 | 0.30 | 0.28 | 0.28 | |
| C22:0 (%) | | 0.14 | 0.13 | 0.12 | 0.12 | |
| C24:0 (%) | | 0.09 | 0.07 | 0.065 | 0.06 | |
| C12:0, C14:0, tC18:1, tC18:2, tC18:3 (%) | | <0.01 | <0.01 | <0.01 | <0.01 | |
| $\Sigma$SFA (%) | | 17.73 | 18.11 | 13.91 | 18.64 | |
| $\Sigma$ MUFA (%) | | 73.90 | 69.01 | 81.82 | 68.35 | |
| $\Sigma$ PUFA (%) | | 8.37 | 12.88 | 4.27 | 13.01 | |
| $\Sigma$ UFA (%) | | 82.27 | 81.89 | 86.09 | 81.36 | |

[a] Median of the fruity attribute. [b] Median of defect.

In order to assess the oil susceptibility to oxidation, the content of total phenols and the fatty acids composition were determined (Table 1). Furthermore, the oil stability index (OSI) was determined by Rancimat method, which values are shown in Table 1. The results show the following stability order of the oils (from more to less stable): VOO3, VOO4, VOO1, VOO2. Thus, VOO3 showed the highest oxidative stability (82.80 h), which can be explained by its high concentration of phenols (534.82 mg/kg) and monounsaturated fatty acids (81.82%). VOO2 and VOO4, which were characterized by a medium phenol content and the highest percentage of polyunsaturated fatty acids (Table 1), showed totally different oxidative stability values between them. Thus, VOO4 showed better stability (53.60 h) than VOO2 (22.95 h), the former showing a higher phenol concentration (468.10 mg/kg) than the latter (338.90 mg/kg). Sample VOO1 showed an intermediate situation because its oxidative stability value was 38.71 h. Although this sample had a high monounsaturated fatty acid percentage, it showed the lowest total phenol content (226.71 mg/kg).

Regarding the volatile composition, Table 2 shows the concentration of the identified volatile compounds of the fresh oils before starting the storage experiment, their odor thresholds and their sensory attributes. These volatile concentrations provide useful information about their oxidation state or the presence of some oxidative/fermentative defects [1,3,30,31] before the storage. Table 2 shows the high content of C6 aliphatic compounds, such as hexanal, (*E*)-2-hexenal, hexyl acetate, hexanol, (*E*)-3-hexen-1-ol, (*Z*)-3-hexen-1-ol, (*E*)-2-hexen-1-ol, (*Z*)-2-hexen-1-ol and (*Z*)-3-hexenyl acetate, which derived from linoleic and linolenic acids through the lipoxygenase (LOX) pathway [1,32,33]. The total concentration for this group of compounds, which provide pleasant notes to the oil, represented more than 20% of the total concentration of volatiles in all samples. Thus, the highest concentration for C6 lipoxygenase products was found in VOO1-0m, representing 33% of its total volatile compounds, with a value of 18.26 mg/kg. Whereas, the percentages and concentrations values for the rest of samples were 25.33% and 11.13 mg/kg for VOO2-0m, 24.41% and 15.45 mg/kg for VOO3-0m and 25.90% and 11.09 mg/kg for VOO4-0m. (*E*)-2-hexanal is one of the most abundant compounds in all fresh samples, with a concentration value that ranged from 4.53 to 5.81 mg/kg. Hexanal and hexanol showed high concentration values as well, in the range of 1.08–3.83 mg/kg. These three compounds strongly contributed to the aroma of all fresh samples since their concentrations exceeded their odor threshold value (Table 2). The high amount of (*Z*)-3-hexen-1-ol in VOO1 (1.10 mg/kg) and (*Z*)-3-hexenyl acetate in VOO3 (1.73 mg/kg) are also remarkable. These compounds are characterized by ripe fruity, bitter and green sensory attributes.

**Table 2.** Volatile compounds identified in the studied virgin olive oils, with their codes, their concentrations (mg/kg) at two different moments (before and after the storage experiment), the odor threshold and the sensory attributes of each volatile compound.

| | | VOO1 | | VOO2 | | VOO3 | | VOO4 | | Odor Threshold (mg/kg) | Aroma Sensory Descriptor |
| | | Hojiblanca | | Arbequina | | Picual | | Arbequina | | | |
| CODE | Months of Storage | 0 | 27 | 0 | 27 | 0 | 27 | 0 | 27 | | |
|---|---|---|---|---|---|---|---|---|---|---|---|
| 1 | Octane | 0.33 [a] | 4.24 [a,b] | 1.06 [a,b] | 12.87 [a,b] | 0.65 [a] | 13.95 [a,b] | 2.012 [a,b] | 14.13 [a,b] | 0.94 | Sweet, alkane |
| 2 | Methyl acetate | 0.51 [a,b] | 0.59 [a,b] | 1.28 [a,b] | 0.89 [a,b] | 0.13 [a] | 0.18 [a] | 1.03 [a,b] | 0.49 [a,b] | 0.20 | Solvent, fruit |
| 3 | Butanal | 0.05 [a,b] | 0.03 [a,b] | 0.05 [a,b] | 0.02 [a,b] | 0.07 [a,b] | 0.03 [a,b] | 0.04 [a,b] | 0.03 [a,b] | 0.08 | Green, pungent |
| 4 | Ethyl acetate | 0.70 [a] | 0.61 [a] | 1.11 [a,b] | 1.01 [a,b] | 0.16 [a] | 0.19 [a] | 1.59 [a,b] | 1.00 [a,b] | 0.94 | Sticky, sweet |
| 5 | Butan-2-one | 0.48 [a] | 0.37 [a] | 0.29 [a] | 0.28 [a] | 0.55 [a] | 0.41 [a] | 0.12 [a] | 0.10 [a] | 40.00 | Ethereal, fruit |
| 6 | 2-methylbutanal | 0.04 [a,b] | 0.03 [a,b] | 0.15 [a,b] | 0.11 [a,b] | 0.05 [a,b] | 0.08 [a,b] | 0.16 [a,b] | 0.12 [a,b] | 0.005 | Malty |
| 7 | 3-methylbutanal | 0.02 [a,b] | 0.03 [a,b] | 0.06 [a,b] | 0.11 [a,b] | 0.04 [a,b] | 0.08 [a,b] | 0.09 [a,b] | 0.05 [a,b] | 0.005 | Malty |
| 8 | Ethanol | 19.63 [a] | 15.59 [a] | 12.17 [a] | 5.96 [a] | 21.09 [a] | 15.67 [a] | 15.88 [a] | 9.45 [a] | 30.00 | Alcohol |
| 9 | Ethyl propanoate | 0.76 [a,b] | 0.64 [a,b] | 0.32 [a,b] | 0.26 [a,b] | 0.24 [a,b] | 0.29 [a,b] | 0.10 [a] | 0.05 [a] | 0.10 | Fruit, strong |
| 10 | 3-pentanone | 4.83 [a,b] | 4.04 [a,b] | 2.41 [b] | 2.20 [b] | 3.46 [a,b] | 2.82 [a,b] | 2.23 [a,b] | 1.60 [a,b] | 70.00 | Sweet, fruit |
| 11 | Butan-2-ol | 0.07 [a] | 0.04 [a] | 0.09 [a] | 0.05 [a] | 0.09 [a] | 0.04 [a] | 0.06 [a] | 0.02 [a] | 0.15 | Winey |
| 12 | Hexanal [c] | 3.83 [a,b] | 2.33 [a,b] | 2.08 [a,b] | 3.75 [a,b] | 2.62 [a,b] | 2.19 [a,b] | 2.41 [a,b] | 2.72 [a,b] | 0.08 | Green-sweet |
| 13 | 2-methylpropan-1-ol | 0.12 [a] | 0.02 [a] | 0.04 | 0.01 | 0.02 [a] | 0.01 [a] | 0.01 [a] | 0.02 [a] | 1.00 | Wine, solvent |
| 14 | 1-penten-3-ol | 0.96 [a,b] | 0.65 [a,b] | 0.44 [a,b] | 0.31 [a] | 0.51 [b] | 0.52 [b] | 0.40 [a] | 0.31 [a] | 0.40 | Pungent, butter |
| 15 | (E)-2-pentenal | 0.46 [a,b] | 0.23 [a] | 0.16 | 0.16 | 0.26 | 0.30 [b] | 0.12 [a] | 0.31 [a,b] | 0.30 | Green, apple |
| 16 | Butan-1-ol | 0.30 [a] | 0.07 [a] | 0.01 [a] | 0.01 [a] | 0.10 | 0.01 | 0.04 [a] | 0.03 [a] | 0.40 | Fatty–medicine |
| 17 | Heptanal | 0.10 | 0.09 | 0.06 [a] | 0.47 [a] | 0.05 [a] | 0.30 [a] | 0.08 [a] | 0.42 [a] | 0.50 | Oily, fatty, Woody |
| 18 | 2-methylbutan-1-ol | 0.01 | 0.01 | Nd [d] | Nd [d] | 0.01 [a] | 0.05 [a] | 0.02 | 0.03 | 0.30 | Winey, spicy |
| 19 | 3-methylbutan-1-ol | 0.03 | 0.03 | 0.04 [a] | 0.04 [a] | 0.02 [a] | 0.01 [a] | 0.08 | 0.06 | 0.10 | Woody, whiskey |
| 20 | (E)-2-hexenal [c] | 4.53 [a,b] | 1.21 [a,b] | 5.03 [a,b] | 3.87 [a,b] | 5.81 [a,b] | 3.31 [a,b] | 5.71 [a,b] | 2.57 [a,b] | 0.42 | Green, apple-like |
| 21 | Octan-3-one | 0.08 [a] | 0.03 [a] | 0.04 [a] | 0.01 [a] | 0.15 | 0.09 | 0.02 [a] | 0.08 [a] | 0.75 | Herb, butter |
| 22 | Pentanol | 0.03 [a] | 0.01 [a] | 0.01 [a] | 0.06 [a] | 0.06 | 0.03 | 0.01 [a] | 0.01 [a] | 0.47 | Fruity |
| 23 | 1-octen-3-one | 0.17 [a] | 0.07 [a] | 0.16 | 0.43 | 0.11 | 0.11 | 0.10 [a] | 0.20 [a] | 0.01 | Mushroom, mold |
| 24 | Hexyl acetate [c] | 1.87 [b] | 1.96 [b] | 0.85 | 0.75 | 1.69 [a,b] | 1.52 [a,b] | 0.89 [a] | 0.70 [a] | 1.04 | Green, fruity, sweet |
| 25 | Octan-2-one | 0.08 [a] | 0.03 [a] | 0.02 [a] | 0.01 [a] | 0.12 [b] | 0.04 | 0.02 [a] | 0.01 [a] | 0.51 | Mold, green |
| 26 | Octanal | 0.07 | 1.38 [b] | 0.42 [b] | 0.98 [b] | 1.01 [b] | 1.00 [b] | 0.55 [a,b] | 0.31 [a,b] | 0.32 | Fatty, sharp, citrus-like |
| 27 | (Z)-3-hexenyl acetate [c] | 0.63 [a] | 0.31 [a] | 0.56 [a] | 0.50 [a] | 1.73 [b] | 0.98 | 0.48 [a] | 0.31 [a] | 0.20 | Green |
| 28 | (E)-2-heptenal | 0.08 [a,b] | 0.05 [a,b] | 0.02 [a,b] | 0.10 [a,b] | 0.02 [a,b] | 0.08 [a,b] | 0.02 [a,b] | 0.09 [a,b] | 0.005 | Oxidized, tallow |
| 29 | 6-methyl-5-hepten-2-one | 0.01 [a] | 0.01 [a] | 0.01 | 0.02 | 0.02 [a] | 0.02 [a] | 0.01 [a] | 0.01 [a] | 1.00 | Pungent, green |
| 30 | Hexanol [c] | 3.56 [a,b] | 3.54 [a,b] | 1.57 [a,b] | 1.41 [a,b] | 1.65 [a,b] | 1.43 [a,b] | 1.08 [a,b] | 0.88 [a,b] | 0.40 | Fruit, banana, soft |
| 31 | (E)-3-hexen-1-ol [c] | 0.46 [a] | 0.06 [a] | 0.10 [a] | 0.11 [a] | 0.21 [a] | 0.09 [a] | 0.05 [a] | 0.03 [a] | 1.00 | Green |
| 32 | (Z)-3-hexen-1-ol [c] | 1.10 [a] | 0.45 [a] | 0.16 [a] | 0.12 [a] | 0.29 | 0.16 | 0.05 [a] | 0.04 [a] | 1.10 | Green |
| 33 | Nonanal | 0.22 [a,b] | 0.29 [a,b] | 0.19 [a,b] | 0.80 [a,b] | 0.24 [a,b] | 0.39 [a,b] | 0.22 [a,b] | 0.61 [a,b] | 0.15 | Fatty, waxy, pungent |
| 34 | 1-octen-3-ol | 0.14 [a,b] | 0.04 [a,b] | 0.05 [a,b] | 0.08 [a,b] | 0.14 [a,b] | 0.04 [a,b] | 0.05 [a,b] | 0.09 [a,b] | 0.001 | Moldy, earthy |
| 35 | (E)-2-hexen-1-ol [c] | 0.83 [a] | 0.23 [a] | 0.50 [a] | 0.37 [a] | 0.85 [a] | 0.44 [a] | 0.25 [a] | 0.19 [a] | 5.00 | Green grass, leaves |
| 36 | (Z)-2-hexen-1-ol [c] | 1.43 [a] | 0.43 [a] | 0.29 [a] | 0.30 [a] | 0.59 [a] | 0.23 [a] | 0.16 [a] | 0.13 [a] | 1.00 | Green |

**Table 2.** *Cont.*

| | | VOO1 | | VOO2 | | VOO3 | | VOO4 | | Odor Threshold (mg/kg) | Aroma Sensory Descriptor |
| | | Hojiblanca | | Arbequina | | Picual | | Arbequina | | | |
| CODE | Months of Storage | 0 | 27 | 0 | 27 | 0 | 27 | 0 | 27 | | |
|---|---|---|---|---|---|---|---|---|---|---|---|
| 37 | Acetic acid | 1.63 [a,b] | 1.56 [a,b] | 5.16 [b] | 6.36 [b] | 2.94 [a,b] | 3.78 [a,b] | 0.63 [b] | 0.79 [b] | 0.50 | Sour, vinegary |
| 38 | Propanoic acid | 0.14 [a] | 0.24 [a] | 0.18 [a] | 0.19 [a] | 0.13 [a] | 0.20 [a] | 0.10 | 0.13 | 0.72 | Pungent, sour |
| 39 | Butanoic acid | 0.47 [a] | 0.23 [a] | 0.13 [a] | 0.06 [a] | 0.25 | 0.24 | 0.25 | 0.15 | 0.65 | Rancid, cheese |
| 40 | 2-methylpropanoic acid | 0.08 [a] | 0.04 [a] | 0.07 [a] | 0.05 [a] | 0.10 [a] | 0.06 [a] | 0.06 [a] | 0.04 [a] | - | Butter, cheese, rancid |
| 41 | (E)-2-decenal | 0.13 [b] | 0.07 [b] | 0.12 [a,b] | 0.80 [a,b] | 2.75 [a,b] | 3.37 [a,b] | 2.55 [a,b] | 3.47 [a,b] | 0.01 | Painty, fishy, fatty |
| 42 | Pentanoic acid | 0.24 | 0.25 | 0.13 [a] | 0.24 [a] | 0.29 | 0.15 | 0.11 [a] | 0.20 [a] | 0.60 | Unpleasant, pungent |
| 43 | Hexanoic acid | 1.50 [b] | 1.78 [b] | 0.57 [a] | 1.22 [a,b] | 2.51 [a,b] | 0.57 [a] | 0.63 [a] | 1.23 [a,b] | 0.70 | Pungent, rancid |
| 44 | Heptanoic acid | 1.27 [a,b] | 2.97 [a,b] | 3.99 [a,b] | 1.52 [a,b] | 4.96 [a,b] | 1.16 [a,b] | 1.15 [a,b] | 4.00 [a,b] | 0.10 | Rancid, fatty |
| 45 | Octanoic acid | 1.07 [a] | 2.62 [a] | 1.77 [a] | 1.57 [a] | 4.47 [a] | 1.33 [a] | 1.03 [a] | 2.14 [a] | 3.00 | Oily, fatty |
| 46 | Nonanoic acid | 0.02 [a] | 0.08 [a] | 0.02 [a] | 0.05 [a] | 0.12 [a] | 0.02 [a] | 0.02 [a] | 0.05 [a] | - | Fat, must, sweat, sour |
| | Total volatiles | 55.00 | 49.54 | 43.94 | 50.47 | 63.29 | 57.91 | 42.65 | 49.36 | | |

[a] Significant difference ($p > 0.05$) in the concentration of the compound at the beginning and end of the storage; [b] the concentration of the compound exceeds its odor threshold (Odor Activity Value, OAV >1). [c] Compounds derived from the lipoxygenase (LOX) pathway. [d] Not detected.

On the other hand, the analysis of the fresh samples pointed out that the compounds responsible for rancid defect [31], such as heptanal, octanal, (*E*)-2-heptanal, nonanal and (*E*)-2-decenal were found at low concentrations before the storage. The Hojiblanca oil (VOO1-0m) was characterized by the lowest concentration of octanal (0.07 mg/kg for VOO1-0m vs. 0.42 mg/kg, 1.01 mg/kg and 0.55 mg/kg for VOO2-0m, VOO3-0m and VOO4-0m, respectively). The total amount of carboxylic acids also pointed out a higher degradation of VOO2-0m (12.02 mg/kg) and VOO3-0m (15.75 mg/kg) oils compared to VOO1-0m (6.41 mg/kg) and VOO4-0m (3.97 mg/kg). Furthermore, ethanol and acetic acid, which are typically found at high concentrations in oils with fermentative defects (e.g., winey-vinegary defect) [31,34], were identified in the fresh samples, although their concentrations were not enough to produce a remarkable sensory impact. Thus, the concentrations of ethanol were lower than its odor threshold in all cases. On the contrary, in the case of acetic acid, the concentrations were higher than the odor threshold in all the oils (Table 2). Only VOO2-0m showed a particularly high concentration of acetic acid (5.16 mg/kg), which was at least 2 times the concentration found in the other oils (Table 2).

### 2.2. Chemical Changes during the Storage Experiment

In order to assess the chemical changes that take place during the VOO shelf life, a storage experiment was carried out for 27 months under controlled conditions (see Section 3.1). Quality parameters, volatile composition and sensory characteristics (panel test) were monthly determined during this period of storage.

The indicators of the VOO quality alteration showed an increase during the storage experiment at moderate conditions. Figure 1 shows the evolution of quality parameters and total phenol concentration per each sample during the long-term storage.

In all cases, the final values were inside the "extra virgin olive oil" category according to the limits stated in EC regulation [13], except for $K_{270}$. Thus, this parameter exceeded the legislative limit stablished for the "extra virgin olive oil" category in the first months of storage (1–5 month) for the four VOOs. Furthermore, the increase of the $K_{270}$ is faster in VOO1 and VOO3 than in the other two VOOs in the first five months of storage. On the other hand, the total phenol concentration decreased during the storage experiment in all cases. Their concentrations showed a higher decrease during the first twenty months of storage compared to the seven last months in all VOO. These results agree with the results found by other authors [22,23]. VOO3 and VOO4 underwent the highest concentration decrease, which were respectively 297.21 and 329.88 mg/kg. The other two samples showed a slighter decrease of their concentrations, with a reduction value of 133.24 mg/kg for VOO1 and 155.09 mg/kg for VOO2.

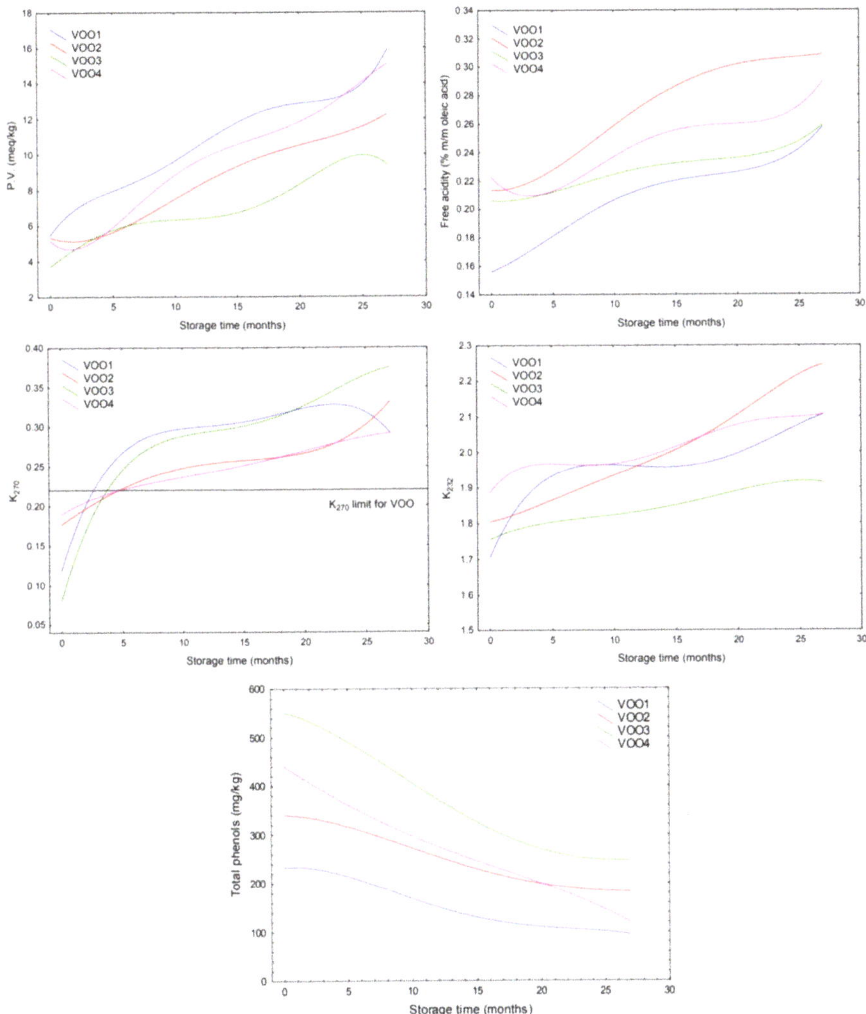

**Figure 1.** Time-course plots of peroxide value (PV), free acidity (FFA), ultraviolet absorbance at 270 nm ($K_{270}$) and 232 nm ($K_{232}$) and total phenols concentration per each virgin olive oil (VOO) (polynomial fitting). According to [13]: Limits for extra virgin olive oil (EVOO): PV $\leq$ 20 meq $O_2$/kg, FFA $\leq$ 0.8%, $K_{270} \leq 0.22$, $K_{232} \leq 2.50$. Limits for virgin olive oil (VOO): PV $\leq$ 20 meq $O_2$/kg, FFA $\leq$ 2.0%, $K_{270} \leq 0.25$, $K_{232} \leq 2.60$.

In the course of the storage experiment, panelists identified some flavor changes in the samples, which resulted in a variation in their medians of the fruity attribute and the defect. Figure 2 shows the evolution of the medians of the defect and the fruity attributes for each sample during the storage. This figure also shows the variations of VOO category of the oils during the storage, according to the limits established in European regulation [13]. The sensory assessment results revealed that the pleasant odor attributes decreased during the storage. In all cases, a reduction of the median of the fruity attributes was observed. VOO2 showed the fastest decrease, so displaying a drop from 3.5 to 2.0 units in the median of the fruity attribute during the first five months of storage, while the rest of the samples kept their initial values at this time. On the other hand, the median of defect showed that VOO2 was downgraded to "virgin olive oil" category rapidly, in the fifth month of storage (VOO2-5m) because assessors detected a winey-vinegary defect at this time. The time-trend changes of $K_{232}$ and FFA also pointed out that this sample was the least stable, since the final values of these parameters (2.45% and 0.32% respectively) were the highest compared with the rest of the VOOs (Figure 1). The next VOO undergoing a downgrade of category was VOO3. Thus, this oil changed to "virgin olive oil" category in the tenth month of storage (VOO3-10m) when assessors detected a winey-vinegary defect as well. Although the evolution of $K_{270}$ for this oil pointed out that this oil was unstable (the final value after 27 months of storage was 0.38), the time-trend changes of the other parameters pointed out its stability during the storage. Thus, this oil showed the lowest values of $K_{232}$, PV and FFA compared with the rest of the stored VOOs at the end of the storage period (Figure 1). The off-flavors detected by the panelist after 10 months may have been in the oil from the beginning of the storage and been masked by the high intensity of green and fruity attributes (Figure 2). Finally, VOO1 changed to "virgin olive oil" category in the fifteenth month of storage (VOO1-15m), because an incipient rancid defect was detected, this sample being the one that needed more time to undergo a change of category. These results do not match with the low phenol content of this VOO and its rapid evolution of PV and $K_{270}$ (Figure 1). Furthermore, despite the changes in the medians of the fruity attribute and the defect were more moderate during the last months (months 15–27) than in the first ones (months 0–14) in VOO1 (Figure 2), this sample underwent another change of category to "lampante virgin olive oil" at month 27 (VOO1-27m). This oil was identified as the least stable according to the evolution of PV, which is associated with the first step of the oxidation process and reached its maximum value (15.27 meq $O_2$/kg) at the end of the storage period. With respect to VOO4, which was initially categorized as "virgin olive oil", this oil showed an incipient rancidity in the fourth month of storage and the median of defect raised above 3.5 in month 18 (VOO4-18m) and consequently the oil was downgraded to "lampante virgin olive oil" category at this time.

The total concentration of volatile compounds is showed in Table 2, where the initial (month 0) and final values (month 27) per sample are displayed. Furthermore, Tables S1–S4 (Supplementary Material) show the concentrations of the volatile compounds in the VOOs during the storage experiment. These values revealed a moderate change in the total concentration of volatiles during the storage time. Thus, a maximum of 15% of variation was observed when comparing the total concentration of volatile compounds between the beginning and the end of the storage experiment (Table 2). However, the panelists detected important changes in the sensory characteristics of the samples that led to a change in the category. In order to extract more information about the changes taking place during the storage, the total concentration of volatiles was studied regarding 5 different chemical series: aldehydes, alcohols, esters, ketones and carboxylic acids. Table S5 shows the concentration of these chemical series at the beginning and the end of the storage. In this case, the maximum variation of concentration was found in carboxylic acids. Thus, the concentration of this chemical series was up to 54.56% higher at the end of the storage experiment in VOO4 (Table S5). The maximum percentages of variation for the other chemical series were 39.40% for aldehydes (VOO1), 43.18% for alcohols (VOO2), 37.26% for esters (VOO4) and 21.41% for ketones (VOO3).

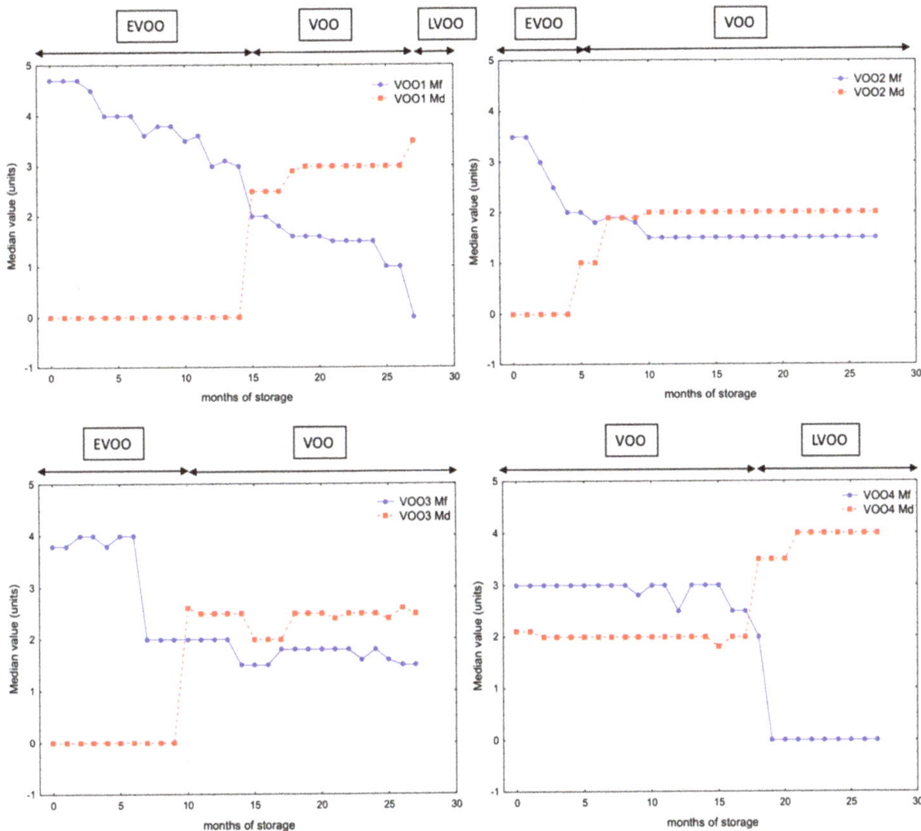



**Figure 2.** Changes of the oil category, the median of the fruity attribute (Mf) and the median of defect (Md) for the four virgin olive oils (VOO1–VOO4) during the storage.

The interpretation of the sensory changes during storage by means of volatile compounds requires the study of the individual compounds, in particular of those that are odor-active. In a first stage, the content of all the individual compounds (Table 2) was studied during the storage experiment in order to identify which compounds underwent significant changes ($p$-value $< 0.05$) during the storage experiment and an ANOVA analysis was performed comparing concentration values from the months 0–3 vs. 24–27 (fresh vs. aged oils). Most of the volatile compounds (at least 75%) showed statistically significant changes in all the oils (Table 2). Overall, the storage time showed a great effect on the volatile compound's concentration in all cases except for some alcohols, such as 2-methylbutan-1-ol. The relative standard deviation (RSD%) of the concentration values was calculated for those compounds presenting significant changes ($p$-value $< 0.05$) comparing the initial and the last sample stored in the experiment (0 and 27 months). The objective was to identify the compounds whose concentration varied in a greater extent. Table 3 shows the compounds whose concentration changed with a RSD% higher than 50% in the four VOOs. These compounds were different in each oil and the total number of compounds with RSD% $> 50$% per oil points out the oxidation state of them at the end of the experiment. These compounds show that VOO1 and VOO3 underwent a decrease in the concentration of the majority of the selected compounds, which have pleasant attributes, such as (*E*)-3-hexen-1-ol, which explains the loss of their initial positive attributes (median of the fruity attribute) during the storage (Figure 2). Sample VOO3 showed a large increase of the concentration of heptanal and

(*E*)-2-heptenal, which are related to oily and oxidized aroma descriptors [1,31,35]. Finally, VOO2 and VOO4 were in an advanced oxidation state at the end of the storage, as it is pointed out by the increase of the content of the volatiles related to the rancid defect [36], such as nonanal, heptanal and hexanoic acid.

In a next step—and in order to gain a better understanding about which compounds have more influence on the virgin olive oil aroma—the odor activity value (OAV) of each volatile compound was determined at every month during the entire storage time for the different oils. This value results from the ratio of the concentration of the compound to its odor threshold [37]. Many of the compounds derived from the lipoxygenase (LOX) pathway, which contribute with a green and fruit aroma sensory descriptors [1], showed an OAV > 1 in all samples during the entire storage time, such as hexanal, (*E*)-2-hexenal, hexyl acetate, hexanol and (*Z*)-3-hexenyl acetate. The OAV of hexanal and (*E*)-2-hexenal were particularly high: they were in the ranges of 51.10–27.69 and 13.69–2.84, respectively.

Other compounds with unpleasant sensory descriptors showed a high OAV from the beginning of the storage, such as 1-octen-3-ol in VOO1 (135.94), VOO2 (47.57) and VOO4 (47.99), heptanoic acid in VOO3 (49.49) and (*E*)-2-decenal in VOO4 (255.44). 1-octen-3-ol provides moldy odor to the oil, whereas heptanoic acid and (*E*)-2-decenal are volatile markers of the rancid defect [31,38]. Other compounds related to the rancid defect also showed OAV > 1 but at lower extent, such as nonanal and (*E*)-2-heptenal. They were found at low concentrations although their odor thresholds were low enough (0.005 mg/kg and 0.15 mg/kg for (*E*)-2-heptenal and nonanal respectively) to have some impact on the sensory characteristics of the oil. Thus, their OAV in the fresh samples were 2.96–3.49 and 1.26–1.63 for (*E*)-2-heptenal and nonanal, respectively.

Once the compounds that underwent the most significant changes were identified (Table 3), all volatile compounds from Table 2 were studied to identify those whose concentration changes were better correlate to sensory changes. For this purpose, a correlation matrix was performed between the OAV of the volatile compounds (Table 2) and the results from the sensory assessment (medians of the fruity attribute and the defect) monthly obtained. Only (*E*)-2-hexenal, which exceeded its odor threshold in all samples during the storage experiment, showed a high correlation coefficient (0.70–0.96) with the median of the fruity attribute in the studied oils. Particularly, a strong correlation was found between these two variables in VOO1 (R = 0.96). This compound, which is considered as a freshness marker in vegetables oils [31,39], contributes with green and fruity attributes (Table 2). Figure 3 shows a double-y graph where the median of the fruity attribute and the OAV for (*E*)-2-hexenal are plotted per each VOO. Although (*E*)-2-hexenal concentration was reduced in a range of 1.16–3.33 mg/kg (Table 2), this compound with pleasant sensory descriptor showed an OAV higher than 1 during the entire storage time in all samples (OAV > 2.84 in all cases). Other compounds showing high correlation coefficients (>0.70) with median of the fruity attribute in some particular oils were (*Z*)-3-hexen-1-ol (R = 0.93 and 0.70 in VOO1 and VOO4, respectively) and (*E*)-2-hexen-1-ol (R = 0.96 in VOO1). The concentration of these two compounds is highly influenced by the stage of ripeness [1].

**Table 3.** Selected volatile compounds that showed statistically significant changes (*p*-value < 0.05) and RSD% higher than 50% during the storage at moderate conditions.

| VOO1 | | VOO2 | |
| --- | --- | --- | --- |
| Compound | RSD% | Compound | RSD% |
| Octane [a] | 120.82 | Octane | 119.81 |
| Butan-2-ol | 50.00 | Butanal | 58.59 |
| 2-methylpropan-1-ol | 93.90 | 2-methylpropan-1-ol | 75.00 |
| Butan-1-ol | 90.08 | Butan-1-ol | 69.05 |
| (*E*)-2-hexenal | 82.03 | Heptanal [a] | 107.53 |
| Octan-3-one | 70.06 | Octan-3-one | 92.07 |
| Pentanol | 78.22 | Octan-2-one | 68.64 |
| Octan-2-one | 74.39 | Octanal [a] | 56.58 |
| Octanal [a] | 128.77 | 1-octen-3-one [a] | 64.17 |
| 1-octen-3-one | 59.92 | (*E*)-2-heptenal [a] | 99.97 |
| (*E*)-2-heptenal [a] | 71.08 | 6-methyl-5-hepten-2-one [a] | 64.13 |
| 6-methyl-5-hepten-2-one | 61.66 | Nonanal [a] | 87.46 |
| (*E*)-3-hexen-1-ol | 107.40 | Butanoic acid | 51.36 |
| (*Z*)-3-hexen-1-ol | 59.21 | (*E*)-2-decenal [a] | 104.44 |
| 1-octen-3-ol | 72.03 | Hexanoic acid [a] | 50.98 |
| (*E*)-2-hexen-1-ol | 79.88 | Heptanoic acid | 63.61 |
| (*Z*)-2-hexen-1-ol | 76.50 | Nonanoic acid [a] | 78.26 |
| Heptanoic acid [a] | 56.85 | | |
| Octanoic acid [a] | 59.30 | | |
| Nonanoic acid [a] | 94.12 | | |
| VOO3 | | VOO4 | |
| Compound | RSD% | Compound | RSD% |
| Octane [a] | 128.80 | Octane [a] | 106.09 |
| Butanal | 56.41 | Methyl acetate | 50.03 |
| Butan-2-ol | 64.18 | Butan-2-ol | 80.69 |
| 2-methylpropan-1-ol | 71.54 | 2-methylpropan-1-ol [a] | 64.37 |
| Butan-1-ol | 118.65 | (*E*)-2-pentenal [a] | 62.07 |
| Heptanal [a] | 100.26 | Heptanal [a] | 94.48 |
| 2-methylbutan-1-ol [a] | 99.86 | (*E*)-2-hexenal | 53.76 |
| Octan-2-one | 77.44 | Octan-3-one [a] | 79.59 |
| (*E*)-2-heptenal [a] | 95.01 | Pentanol [a] | 53.96 |
| 6-methyl-5-hepten-2-one | 97.30 | (*E*)-2-heptenal [a] | 95.28 |
| (*E*)-3-hexen-1-ol | 58.40 | 6-methyl-5-hepten-2-one | 80.14 |
| 1-octen-3-ol | 74.51 | Nonanal [a] | 66.11 |
| (*Z*)-2-hexen-1-ol | 63.51 | Heptanoic acid [a] | 78.27 |
| Hexanoic acid | 88.83 | Octanoic acid [a] | 49.46 |
| Heptanoic acid | 87.90 | Nonanoic acid [a] | 80.15 |
| Octanoic acid | 76.55 | | |
| Nonanoic acid | 105.60 | | |

[a] Compounds that underwent an increase of their concentrations during the storage.

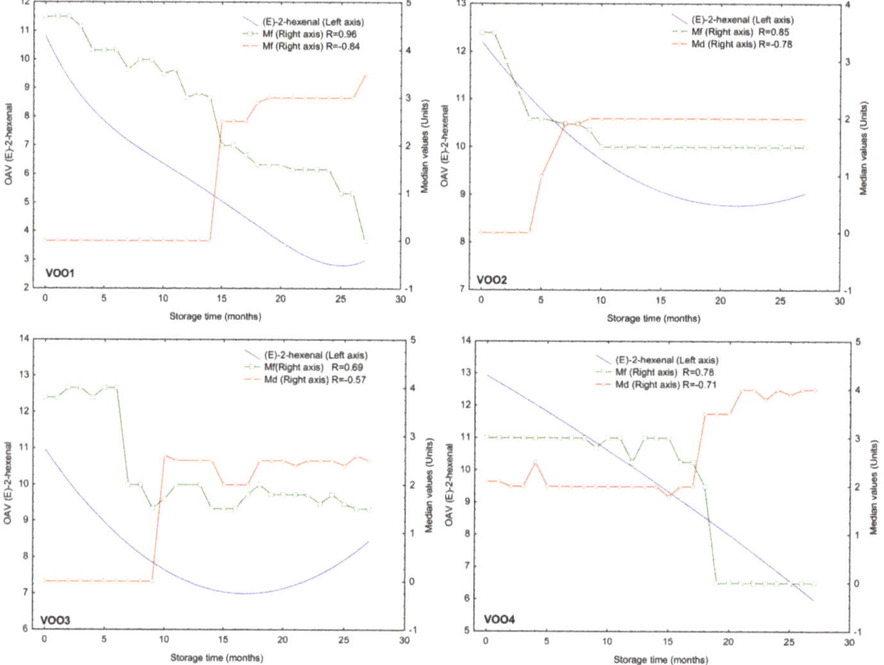

**Figure 3.** Median of the fruity attribute (Mf), median of defect (Md) and (*E*)-2-hexenal OAV (odor activity value) of the four virgin olive oils (VOO) during the storage experiment. The regression coefficients (R) between Mf and Md in relation to the OAV of (*E*)-2-hexenal are shown.

The OAV values of hexanal was also highly correlated (>0.70) with medians of the fruity attribute, although only in two oils. Thus, the correlation coefficients were 0.91 and 0.76 for VOO1 and VOO3, respectively, while a negative value (−0.66) was obtained for VOO2 and VOO4. Hexanal was not selected by ANOVA when comparing its concentration in the initial and last months of storage (Table 2) and it did not show a change in its concentration with a RSD% > 50% (Table 3). However, the study of its concentration and OAV during the storage under moderate conditions may provide useful information about the oxidation state of the samples since this compound is also produced during oxidation and it has an evident implication in virgin olive oil rancidity [3,40]. Figure 4 shows the OAV of hexanal and the medians of the fruity attribute and the defect represented on a double-y graph with respect to the storage time. Two kinds of trends of its OAV during the storage was observed. Thus, the hexanal OAV decreased in VOO1 and VOO3 during the storage; it shows the opposite trend in both Arbequina samples, VOO2 and VOO4, in which its concentration increased from the beginning of the storage. The decrease of OAV in VOO1 and VOO3 can be explained by a loss of hexanal present in the fresh oil originated from the LOX pathway while hexanal was later produced during oxidation at lower extent [1]. Figure 4 shows a similar trend in the decrease of median of the fruity attribute for VOO1—and in lesser degree—for VOO3, which explains the high correlation coefficients in these two cases. On the contrary, the opposite time-trends of OAV and median of the fruity attribute in VOO2 and VOO4 explain the negative value of the correlation coefficient (−0.66). However, the correlation coefficients when comparing OAV and median of defect in these two oils were positive although always below 0.8 (0.56 and 0.61 for VOO2 and VOO4, respectively) (Figure 4).

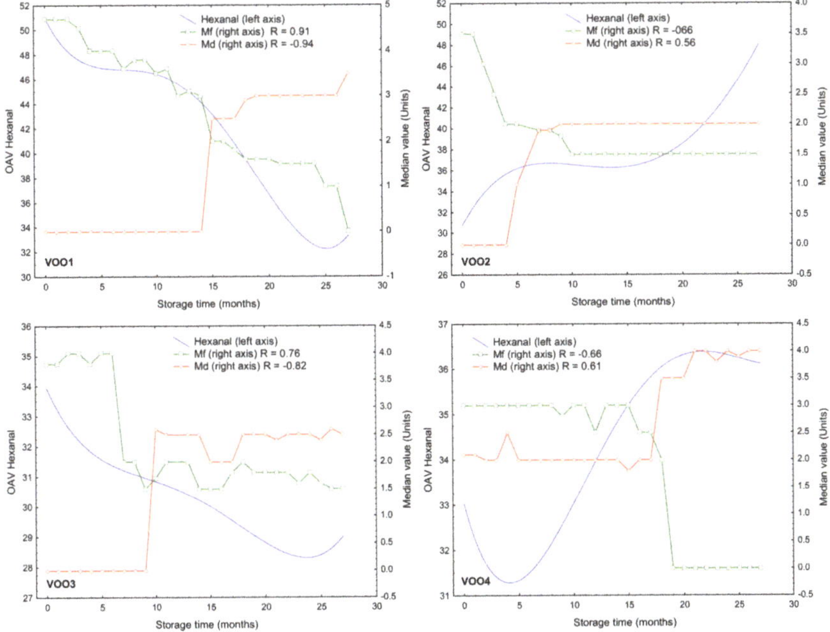

**Figure 4.** Double-y graph of the median of the fruity attribute (Mf), median of defect (Md) and the OAV (odor activity value) of hexanal for each virgin olive oil (VOO) during the storage experiment. The regression coefficients (R) between Mf and Md in relation to the OAV of hexanal are shown.

The increase of hexanal OAV and concentrations in VOO2 and VOO4 is explained by the decomposition reactions of hydroperoxides formed from the fatty acids [3,41], contributing to the off flavor of the sample with an intense greasy odor [42]. In fact, these off-flavors associated with rancidity is detected by the assessors at month 5 in VOO2 (sample VOO2-5m) and at month 19 in VOO4 (sample VOO4-19m) (Figure 2). These results indicate that VOO2 and VOO4 were in a more advanced oxidation state than the rest of the studied oils in the course of the storage experiment. This finding was not in agreement with the oxidative stability index from Rancimat method in the particular case of VOO4. Thus, VOO4 showed a high stability for Rancimat (53.60 h, the second most stable VOO) (Table 1). However, in the storage experiment, this VOO was undergraded to lampante category in the 19th month, while the rest of oils never were classified as lampante (VOO2, VOO3) or were classified as lampante later (VOO1 in 25th month). The evolution of PV values in VOO4 also shows a faster oxidation compared with the others, except for VOO1 (Figure 1). These results agree with the already reported relationship of hexanal with rancidity in aged oils [31,34,42].

In order to study the changes in the sensory characteristics and the volatile composition of the oils from a multivariate perspective, a principal component analysis (PCA) was carried out with the concentration values of the selected volatile compounds that showed a RSD% > 50% in the storage experiment (Table 3) and the medians of the fruity attribute and the defect. Figure 5 shows the resulting PCA plots for the 4 VOOs. The PCA plots show that the median of the fruity attribute and the median of defect were well separated by factor 1 in all VOOs, which were located in opposite quadrants. On the other hand, the plotted compounds were clustered by factor 1 according to their link to the median of the fruity attribute and the median of defect during the storage period. The median of the fruity attribute and the compounds associated to it were found in the left quadrant while the median of defect and the compounds related to it were located in the right quadrant. These results revealed than some compounds show a higher correlation with median of defect; this relationship was different depending on the oxidative state of the oil in the storage experiment. The PCA results show that the selected compounds that originated from the lipoxygenase pathway (Table 2), such as (*E*)-2-hexenal and (*Z*)-2-hexen-1-ol and the median of fruity attribute were plotted in the same quadrant in VOO1, VOO3 and VOO4. In the case of VOO2, however, the median of the fruity attribute appears to be only associated to (*Z*)-2-hexen-1-ol and (*Z*)-3-hexen-1-ol, but not with the rest of these compounds, which can be explained by the slight change in their concentrations in this oil during the storage time instead of a reduction in their concentration as in the other oils (Table 2). Furthermore, other compounds, such as butanal and butan-1-ol, were associated to the median of the fruity attribute in all VOOs, due to their concentration decrease during the storage, except for butan-1-ol in VOO2. This compound contributes with an aroma that is closer to the negative attributes (fatty and medicine for butan-1-ol) [43]. Other compounds contributing with negative attributes were plotted near the median of the fruity attribute, which is explained by the fact that their concentrations also decreased over time: methyl acetate in both VOO2 and VOO4, octan-2-one and 1-octen-3-one in VOO1 and nonanoic acid in VOO3.

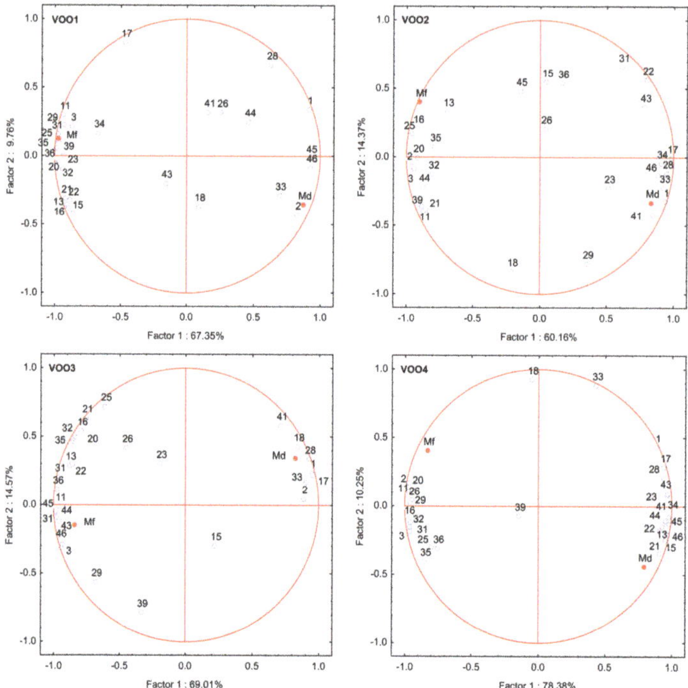

**Figure 5.** Principal component analysis plots (PCA) of volatile compounds that showed significant changes (*p* < 0.05) and RSD% > 50% in the storage experiment, the median of the fruity attribute (Mf) and the median of defect (Md) for the storage time. An individual plot was performed for each virgin olive oil (VOO). Note. Codes are indicated in Table 2.

Regarding the median of defect value, the PCA plots show an association of this median with the concentrations of octane, (*E*)-2-heptenal and (*E*)-2-decenal in all cases, these compounds being related with sensory defects and the two latter contributing with oxidized and fatty notes [1]. Moreover, nonanal and the median of defect were plotted in the same quadrant in VOO1, VOO2 and VOO3, which points out that the autooxidation process takes place during the storage experiment. However, this association was not observed in the PCA for VOO4, in which nonanal and median of defect were plotted in different quadrants despite this oil was the first one that downgraded to lampante at the end of the storage. These results may point out that the relationship of nonanal concentration and median of defect is more evident at earlier stages of oxidation as it is the case in VOO1, VOO2 and VOO3. On the other hand, a strong association was found for heptanal and the median of defect in VOO2, VOO3 and VOO4. In VOO1, this association was not found in the PCA plot, probably due to the fact that it was the oil that underwent the oxidation at lower extent in the first half of the storage experiment and the downgrading of category (from "extra virgin olive oil "to "virgin olive oil") took place the latest (month 15) (Figure 2). Finally, the PCA plots showed that the hexanoic acid is related to the median of defect in the cases of VOO2 and VOO4.

In a second PCA, only the concentrations of the same volatile compounds (Table 3) were studied without including the medians of the fruity attribute and the defect in the data set. The objective was to check the score plot (samples) to study the changes of the volatile profile of the oils during the storage experiment with a multivariate perspective. Thus, Figure S1 shows a score plot per each VOO stored, in which the samples collected every month are represented against the factors. The score plot of VOO1 shows a change in the distribution of the samples from month 15 (VOO1-15m), which is the moment when this sample downgraded to "virgin olive oil" category (Figure 2). The score plot for VOO2 pointed out a change in the distribution of the monthly collected samples in the month 8 (VOO2-8m) when the assessors identified an incipient rancidity in this oil for the first time (Figure 2). Furthermore, in VOO3, the change of the trend of the samples was identified at month 10 (VOO3-10m) when assessors detected a winey-vinegary defect in this samples that caused an increase of 2.6 in its median of defect and, consequently, its downgrading to "virgin olive oil" category (Figure 2). Finally, in sample VOO4 two changes in the trend of the samples were observed in the score plot. Thus, a change in the distribution of the samples was identified at month 8 (VOO4-8m) and another change was detected at month 19 (VOO4-19m). In this last month, the oil downgraded to "lampante olive oil" according to the panel test (Figure 2). These results evidence that the selected compounds were able to explain the sensory changes identified by the assessor during the storage period.

The results obtained by Rancimat method (Table 1) before the storage experiment and the sensory assessment carried out during the storage (Figure 2) showed a different order in the stability of VOO. Thus, the order of oils from more to less stable according to Rancimat was VOO3, VOO4, VOO1 and VOO2, while this order was different according to the month in which a change of category takes place (from later to sooner): VOO1, VOO3, VOO2 and VOO4. That means that the results from Rancimat tests do not necessarily correlate with the actual stability in sensory terms. The evolution of PV values also pointed out VOO4 as one of the samples that faster was oxidized (Figure 1).

Finally, the results from the mesh cell-FTIR experiment, an innovative method to evaluate stability of the oils under moderate conditions, was examined with the aim of comparing these results with the actual stability of the oils according to the sensory changes. Mesh cell-FTIR allowed monitoring the chemical changes of the oils during an incubation time of 576 h under conditions of light and temperature (400 lx and 35 °C) that were closer to the real storage conditions compared with other accelerated tests that use high temperatures (>100 °C) [41]. The mesh cell-FTIR experiment was carried out with the fresh oils before starting the storage experiment. The spectral band assigned to the C=O stretching of unsaturated aldehydes (1685 cm$^{-1}$) was monitored during the incubation time since it is related with secondary oxidation products and it rises during oxidation [6,20]. The maximum intensity for the aldehydes band was found in sample VOO2 and VOO4 with a value of 0.35 and 0.33 respectively. The other two oils showed lower values, these values being 0.28 for VOO3 and 0.25 for VOO1. These results revealed that both Arbequina VOOs were more susceptible to oxidation at moderate condition than the other two oils (VOO1 and VOO3). Considering this measurement, the order of stability (from high to low) was VOO1, VOO3, VOO4 and VOO2, which is close to the stability order according to the sensory testing (VOO1, VOO3, VOO2 and VOO4). Thus, mesh cell-FTIR pointed out that VOO1 was the most stable sample and, in fact, this sample changed of quality category the latest. On the contrary, Rancimat test pointed out that this sample was the second most unstable. These differences in results can be explained by the different mechanisms of oxidation that are involved depending on the conditions [44]. It illustrates the necessity of studying the VOO oxidative stability at moderate conditions and including light as a relevant variable.

In a farther study, with the aim of studying the ability of the mesh cell-FTIR band assigned to aldehydes to represent the changes of volatile composition of VOO during the storage, the results obtained by mesh cell-FTIR were compared with the concentration increment of the compounds associated to the rancid defect and that showed significant changes during the storage. The concentration increment was calculated using the final and initial concentrations of these compounds (showed in Table 2). Figure 6 shows a double-y column graph in which the mesh cell-FTIR absorbance of the band assigned to aldehydes and the concentration increment (mg/kg) of nonanal, heptanal and (E)-2-heptenal of each VOO during the storage period are shown. Although the concentration increment of the selected volatiles revealed more differences between the samples than the results obtained by mesh cell, they revealed the same order of oxidative stability of the samples.

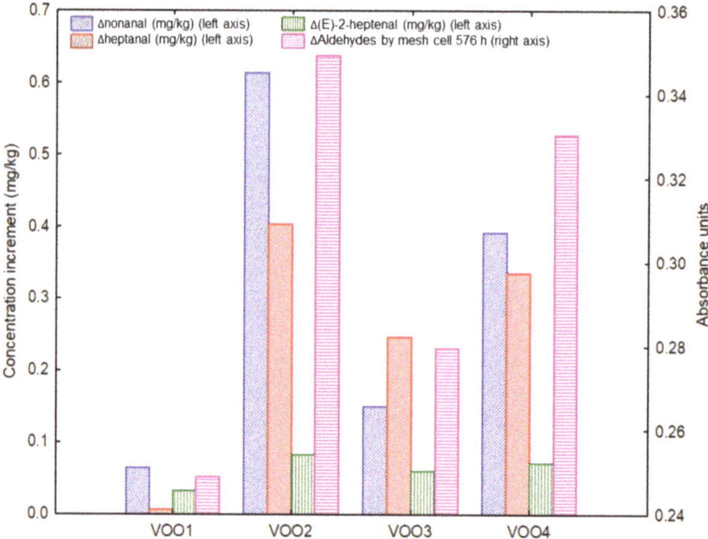


**Figure 6.** Double-y column graph in which the absorbance of the band assigned to aldehydes and the concentration increment (Δ) of nonanal, heptanal and (E)-2-heptenal are represented for the all studied virgin olive oils (VOOs) during the storage period.

## 3. Materials and Methods

### 3.1. Samples

Four monovarietal VOOs from the cultivars Picual, Hojiblanca and Arbequina (2 different samples of the last cultivar) provided by different producers were selected. These cultivars were chosen to cover different chemical compositions. The codes used to identify the VOOs and their respective cultivars were: VOO1, Hojiblanca; VOO2, Arbequina-1; VOO3, Picual; and VOO4, Arbequina-2. In order to guarantee that the samples VOO1–VOO4 were fresh at the beginning of the experiment, they were directly collected from the vertical centrifuge at the oil mills and then filtered to remove water. This time was considered as "time zero" and the storage experiment started just after collecting the samples.

VOOs was packaged in 27 transparent PET bottles of 500 mL—which are commonly used to bottle VOO—and they were hermetically sealed. The oil bottles were stored during 27 months in a compartment specially designed for it, where samples were exposed to light intensity of 1000 lx in 12 h light/dark cycles, simulating the conditions of a supermarket shelf under controlled conditions of temperature and humidity. The maximum and minimum of temperature and humidity were measured daily, being 29.7 °C–16.3 °C and 70%–21%, respectively. A bottle per oil was taken from the compartment monthly, analyzed and discarded afterwards. Thus, the analyses were carried out on VOOs from bottles newly opened. In order to identify the samples corresponding to each month, the number of each month and the letter "m" was added to the initial code (e.g., VOO1-5m means VOO of Hojiblanca cultivar after 5 months of storage).

### 3.2. Quality Parameters

The fatty acid composition and the trans fatty acid content were determined in the fresh samples (month 0) following the standard method (COI/T.20/Doc. No 33) [45].

Quality parameters were determined to confirm the quality category before starting the storage and to track their change during the experiment. These parameters were peroxide value (PV), free acidity (free fatty acids or FFA), which were determined by titration according to their respective standard methods (COI/T.20/Doc. No 35 and COI/T.20/Doc. No 34) [46,47]; and ultra-violet absorbance measured through the extinction coefficients ($K_{270}$ and $K_{232}$), which were determined according to the standard method (COI/T.20/Doc. No 19) [48].

### 3.3. Oil Stability Index (OSI)

The VOOs were analyzed to determine their oxidative stability by Rancimat method, also called oil stability index (OSI), before starting the storage experiment. OSI was determined according to the standard method AOCS Cd 12b-9 [49]. This method consists in heating samples at 100 °C while a continuous stream of air (20 L/h) is passing through the samples. The air is bubbled through a vessel with 60 mL of deionized water. The reported result is increase of the conductivity of the water due to the formation of volatile organic acids.

### 3.4. Mesh Cell-FTIR Analysis

Mesh cell-FTIR analyses were carried out with the fresh VOOs, following the method proposed by Tena et al. [6]. Aliquots of the fresh oils (20 μL) were deposited onto the mesh and they were stored at 35 °C and 400 lx. The spectral changes during the experiment were monitored daily during an incubation time of 576 h using a Bruker vertex 70 FTIR spectrometer (Bruker, Optics, Germany) equipped with a deuterated triglycine sulfate (DTGS) detector. The spectra were collected and manipulated with OPUS software version 7.2 (Bruker Optics, Ettlingen, Germany). The band assigned to the C=O stretching of unsaturated aldehydes ($1685$ cm$^{-1}$) was selected to track the changes during incubation. The peak heights of this band were measured relative to a selected single-point baseline at $1576$ cm$^{-1}$ by implementing a macro programed on Omnic 7.3 (Thermo Electron Inc., Madison, WI, USA).

### 3.5. Determination of Volatile Compounds

The volatile compounds were determined by solid phase microextraction-gas chromatography (SPME-GC) including a preconcentration step carried out on a multipurpose sample autosampler (Gerstel, Mülheim an der Ruhr, Germany) in which temperature and time of the process were automatically controlled by Gerstel Maestro (v1.4) software (Gerstel GmbH et Co.KG, Mülheim an der Ruhr, Germany).

The oil sample (2 g) was placed in a 20 mL glass vial, tightly capped with polytetrafluoroethylene (PTFE) septum and left for 10 min at 40 °C to allow for the equilibration of the volatiles in the headspace. After the equilibration time, the septum covering each vial was pierced with a SPME needle and the fiber was exposed to the headspace for 40 min. A 1-cm StableFlex divinylbenzene/carboxen/polydimethylsiloxane (DVB/CAR/PDMS) composite SPME fiber (50/30 μm film thickness) was used (Supelco, Bellefonte, PA, USA). The fiber was previously conditioned following the instructions of the supplier.

The volatiles adsorbed by the fiber were thermally desorbed in the hot injection port of a 7820A gas chromatograph (Agilent Technologies, Madrid, Spain) with a flame ionization detector for 5 min at 260 °C, with the purge valve off (splitless mode). An Agilent J&W GC DB-WAX capillary column (60 m × 0.25 mm internal diameter, 0.25 μm coating) (Agilent Technologies, Santa Clara, CA, USA) was used. The carrier gas was hydrogen, at a flow rate of 1.5 mL/min. The oven temperature was held at 40 °C for 10 min and then programmed to rise 3 °C/min to a final temperature of 200 °C.

The identification of the volatiles was carried out with standards [29]. Additionally, the identification was checked by analyzing the samples with mass spectrometry (GC7820-MSD5975, Agilent Technology, Santa Clara, CA, USA) following the strategy described in previous works [38,50]. Thus, the information from mass spectra, and their comparison with standards in the FID chromatograms, and linear retention indexes (LRIs) were considered for a full identification. The quantification of volatiles was carried out by using 4-methyl-2-pentanol as internal standard and correcting the concentrations by the relative response factors. These factors were calculated following the procedure described by Oliver-Pozo et al. [51] using two ranges of concentration, 0.05–5.0 and 0.5–30 mg/kg.

The odor activity values (OAVs) [52,53] of the volatile compounds were calculated in all the monthly collected samples, in order to study the sensory influence of each compound on the total aroma of VOO and its evolution during the storage.

### 3.6. Sensory Assessment

The organoleptic assessment of the olive oil samples was carried out monthly by the accredited panel of Instituto de la Grasa (UNE-EN-ISO/IEC 17025) (Seville, Spain) [54] using the standard method COI/T.20/Doc. No 15/Rev.10 [55]. Eight-twelve trained assessors qualified the samples by odor descriptors and established if the samples had any defects, to determine the progression of off-flavors during the storage under moderate conditions.

The results monthly generated from the sensory assessment were based on the calculation of the medians of the fruity attribute and the defect for the four stored VOOs. This test provided a sequential information about the sensory characteristics of the samples and allowed identifying changes in the category of the oils and in their sensory profile.

### 3.7. Determination of Phenol Content

The method for the determination of phenol composition was based on the method described by Aparicio-Ruiz et al. [56]. The sample (2.5 g) was solved in 6 mL of hexane together with *p*-hydroxyphenylacetic (0.12 mg/mL) and *o*-coumaric (0.01 mg/mL) as internal standards. The phenolic fraction was extracted with methanol by solid phase extraction using diol-bonded phase cartridges. The extracted phenolic fraction was concentrated and injected in the HPLC system (Agilent Technologies 1200, Waghaeusel–Wiesental, Germany), equipped with a diode array detector. The column was a LiChrospher 100RP-18 column (4.0 mm i.d. × 250 mm; 5 μm, particle size) (Merck KGaA, Darmstadt, Germany) maintained at 30 °C. The gradient elution, at a flow rate of 1.0 mL/min, was achieved by using a mixture of water/ortho-phosphoric acid (99.5:0.5 *v/v*) (solvent A) and methanol/acetonitrile (50:50 *v/v*) (solvent B). The change in solvent gradient was programed as follows: from 95% (A)-5% (B) to 70% (A)-30% (B) in 25 min; 65% (A)-35% (B) in 10 min; 60% (A)-40% (B) in 5 min; 30% (A)-70%(B) in 10 min and 100% (B) in 5 min, followed by 5 min of maintenance. The chromatographic signals were obtained at 235, 280 and 335 nm. The quantification of the phenols, cinnamic acid and lignans was carried out at 280 nm using p-hydroxyphenylacetic acid as internal standard. The quantification of flavones was done at 335 nm by using *o*-coumaric acid as internal standard. The response factors and recoveries were based on the procedure developed by Mateo et al. (2001) [57].

### 3.8. Statistical Analysis

The STATISTICA 8 package (Statsoft, Tulsa, OK, USA) was used to carry out the statistical analysis. A one-way ANOVA analysis was performed comparing concentration values from the months 0–3 vs. 24–27 (fresh vs. aged oils). Significance was accepted when $p < 0.05$.

Principal component analysis (PCA) was carried out on the concentration values of volatile compounds and the medians of defect and fruity attribute to explore the data from a multivariate perspective and to support the relationship between compounds and the observed changes in the sensory assessment.

## 4. Conclusions

This study shows the complexity of predicting the shelf-life of VOO overall when the sensory quality is considered as a relevant criterion in addition to physico-chemical parameters. Thus, the sensory defects that can appear during storage can be detected by consumers with a resulting refusal of the product. On the other hand, this change of quality can be marked enough to result in a downgrading of category (from "extra virgin olive oil" to "virgin olive oil" and even to "lampante olive oil"), with the consequent non-conformity result when the oil category is checked by a panel test. This change of quality is explained by the changes in the concentrations of volatile compounds during storage. Thus, while the compounds contributing with positive sensory attributes (C6 and C5 compounds) reduce their concentration, those compounds coming from oxidation (mainly aldehydes and acids) increase their concentration. The panel tests revealed that the reduction of the compounds contributing to positive notes can result in the detection of fermentative sensory defects (e.g., winey-vinegary) that the oil already contained but they were masked by the intense green aroma. The oil stability studied by Rancimat at the beginning of the storage produced results that disagreed with the actual sensory changes detected in the storage experiment. The stability studied by mesh-cell FTIR revealed that the band assigned to aldehydes permitted to establish a stability order of the oils that was closer to the order of stability according to the sensory changes (determined by panel test). This result underscores the importance of considering light and moderate temperatures when studying the oil stability in order to avoid unexpected quality-category downgrading. The results also showed that the changes on sensory characteristics and volatile profile followed a different trend depending on the studied oil and there is not a uniform change rate that could be stablished for all the oils.

**Supplementary Materials:** The following are available online: Figure S1. PCA score plots of cases (monthly collected samples) calculated using the volatile compounds which showed significant changes ($p < 0.05$) and RSD% > 50% during the storage time. The numbers indicate the month when the sample was collected and analyzed in the storage experiment. Table S1. Concentration of volatile compounds (mg/kg) in VOO1 during the storage experiment. Table S2. Concentration of volatile compounds (mg/kg) in VOO2 during the storage experiment. Table S3. Concentration of volatile compounds (mg/kg) in VOO3 during the storage experiment. Table S4. Concentration of volatile compounds (mg/kg) in VOO4 during the storage experiment. Table S5. Concentration of the chemical series of the volatile compounds identified in the oils (mg/kg) in two different moments (before and after the storage experiment).

**Author Contributions:** Conceptualization, D.L.G.-G. and N.T.; data curation, A.L.-P. and R.A.-R.; investigation, D.L.G.-G., M.T.M., R.A.-R., N.T. and A.L.-P.; methodology, D.L.G.-G., M.T.M., R.A.-R., N.T. and A.L.-P.; project administration, D.L.G.-G.; resources, A.L.-P. and M.T.M; supervision, D.L.G.-G., M.T.M., R.A.-R., N.T. and A.L.-P.; writing—original draft, A.L.-P. and D.L.G.-G.; writing—review & editing, N.T., R.A.-R. and M.T.M. All authors have read and agreed to the published version of the manuscript.

**Funding:** This research was funded by the Spanish Research State Agency (research projects AGL2015-69320-R and RTI2018-101546-B-C21).

**Acknowledgments:** The authors would like to thank the collaboration of the accredited panel of the Instituto de la Grasa (CSIC), in particular to Mr. Fernando Martínez Román for the sensory assessment carried out of this study and the fruitful discussion about the sensory changes of the oils.

**Conflicts of Interest:** The authors declare no conflict of interest.

## References

1.  Morales, M.T.; Aparicio-Ruiz, R.; Aparicio, R. Chromatographic methodologies: Compounds for olive oil odor issues. In *Handbook of Olive Oil: Analysis and Properties*; Aparicio, R., Harwood, J., Eds.; Springer US: Boston, MA, USA, 2013; pp. 261–309.

2. Tena, N.; Aparicio, R.; García-González, D.L. Chemical changes of thermoxidized virgin olive oil determined by excitation-emission fluorescence spectroscopy (EEFS). *Food Res. Int.* **2012**, *45*, 103–108. [CrossRef]

3. Morales, M.T.; Ríos, J.J.; Aparicio, R. Changes in the volatile composition of virgin olive oil during oxidation: Flavors and off-flavors. *J. Agric. Food Chem.* **1997**, *45*, 2666–2673. [CrossRef]

4. Sikorska, E.; Khmelinskii, I.V.; Sikorski, M.; Caponio, F.; Bilancia, M.T.; Pasqualone, A.; Gomes, T. Fluorescence spectroscopy in monitoring of extra virgin olive oil during storage. *Int. J. Food Sci. Technol.* **2008**, *43*, 52–61. [CrossRef]

5. Méndez, A.I.; Falqué, E. Effect of storage time and container type on the quality of extra-virgin olive oil. *Food Control* **2007**, *18*, 521–529. [CrossRef]

6. Tena, N.; Aparicio, R.; García-González, D.L. Virgin olive oil stability study by mesh cell-FTIR spectroscopy. *Talanta* **2017**, *167*, 453–461. [CrossRef]

7. Tena, N.; Aparicio, R.; García-González, D.L. Photooxidation effect in liquid lipid matrices: Answers from an innovative FTIR spectroscopy strategy with "mesh Cell" incubation. *J. Agric. Food Chem.* **2018**, *66*, 3541–3549. [CrossRef]

8. Genovese, A.; Caporaso, N.; Sacchi, R. Temporal changes of virgin olive oil volatile compounds in a model system simulating domestic consumption: The role of biophenols. *Food Res. Int.* **2015**, *77*, 670–674. [CrossRef]

9. Hernández-Sánchez, N.; Lleó, L.; Ammari, F.; Cuadrado, T.R.; Roger, J.M. Fast fluorescence spectroscopy methodology to monitor the evolution of extra virgin olive oils under illumination. *Food Bioprocess Technol.* **2017**, *10*, 949–961. [CrossRef]

10. Iqdiam, B.M.; Welt, B.A.; Goodrich-Schneider, R.; Sims, C.A.; Baker, G.L.; Marshall, M.R. Influence of headspace oxygen on quality and shelf life of extra virgin olive oil during storage. *Food Packag. Shelf Life* **2020**, *23*, 100433. [CrossRef]

11. Esposto, S.; Selvaggini, R.; Taticchi, A.; Veneziani, G.; Sordini, B.; Servili, M. Quality evolution of extra-virgin olive oils according to their chemical composition during 22 months of storage under dark conditions. *Food Chem.* **2020**, *311*, 126044. [CrossRef]

12. International Olive Council. Trade Standard Applying to Olive Oils and Olive-Pomace Oils. COI/T.15/Nc No 3/Rev. 14 2019, Madrid, Spain. 2019. Available online: https://www.internationaloliveoil.org/wp-content/uploads/2019/12/trade-standard-REV-14-Eng.pdf (accessed on 6 April 2020).

13. European Commission. Commission regulation (EEC) No 2568/91 of 11 July 1991 on the characteristics of olive oil and olive-residue oil and on the relevant methods of analysis (and updates). 1991. Available online: https://eur-lex.europa.eu/legal-content/EN/TXT/PDF/?uri=CELEX:01991R2568-20191020&qid=1586170284196&from=ES (accessed on 6 April 2020).

14. International Olive Council. Best Practice Quidelines for the Storage of Olive Oils and Olive-Pomace Oils for Human Consumption. COI/Bps/Doc. No1 2018, Madrid, Spain. 2018. Available online: https://www.internationaloliveoil.org/wp-content/uploads/2019/11/COI-BPS-Doc.-No-1-2018-Eng.pdf (accessed on 6 April 2020).

15. Izquierdo, M.; Lastra-Mejías, M.; González-Flores, E.; Cancilla, J.C.; Aroca-Santos, R.; Torrecilla, J.S. Deep thermal imaging to compute the adulteration state of extra virgin olive oil. *Comput. Electron. Agric.* **2020**, *171*, 105290. [CrossRef]

16. Maszewska, M.; Florowska, A.; Dłużewska, E.; Wroniak, M.; Marciniak-Lukasiak, K.; Żbikowska, A. Oxidative stability of selected edible oils. *Molecules* **2018**, *23*, 15–17. [CrossRef]

17. Velasco, J.; Andersen, M.L.; Skibsted, L.H. Evaluation of oxidative stability of vegetable oils by monitoring the tendency to radical formation. A comparison of electron spin resonance spectroscopy with the Rancimat method and differential scanning calorimetry. *Food Chem.* **2004**, *85*, 623–632. [CrossRef]

18. Jiang, S.; Xie, Y.; Li, M.; Guo, Y.; Cheng, Y.; Qian, H.; Yao, W. Evaluation on the oxidative stability of edible oil by electron spin resonance spectroscopy. *Food Chem.* **2020**, *309*, 125714. [CrossRef]

19. Qi, B.; Zhang, Q.; Sui, X.; Wang, Z.; Li, Y.; Jiang, L. Differential scanning calorimetry study - Assessing the influence of composition of vegetable oils on oxidation. *Food Chem.* **2016**, *194*, 601–607. [CrossRef]

20. Trypidis, D.; García-González, D.L.; Lobo-Prieto, A.; Nenadis, N.; Tsimidou, M.Z.; Tena, N. Real time monitoring of the combined effect of chlorophyll content and light filtering packaging on virgin olive oil photo-stability using mesh cell-FTIR spectroscopy. *Food Chem.* **2019**, *295*, 94–100. [CrossRef]

21. Mishra, P.; Lleó, L.; Cuadrado, T.; Ruiz-Altisent, M.; Hernández-Sánchez, N. Monitoring oxidation changes in commercial extra virgin olive oils with fluorescence spectroscopy-based prototype. *Eur. Food Res. Technol.* **2018**, *244*, 565–575. [CrossRef]

22. Di Stefano, V.; Melilli, M.G. Effect of storage on quality parameters and phenolic content of Italian extra-virgin olive oils. *Nat. Prod. Res.* **2020**, *34*, 78–86. [CrossRef]

23. Gómez-Alonso, S.; Mancebo-Campos, V.; Salvador, M.D.; Fregapane, G. Evolution of major and minor components and oxidation indices of virgin olive oil during 21 months storage at room temperature. *Food Chem.* **2007**, *100*, 36–42. [CrossRef]

24. Krichene, D.; Salvador, M.D.; Fregapane, G. Stability of virgin olive oil phenolic compounds during long-term storage (18 months) at temperatures of 5–50 °C. *J. Agric. Food Chem.* **2015**, *63*, 6779–6786. [CrossRef]

25. Vichi, S.; Pizzale, L.; Conte, L.S.; Buxaderas, S.; López-Tamames, E. Solid-phase microextraction in the analysis of virgin olive oil volatile fraction: Modifications induced by oxidation and suitable markers of oxidative status. *J. Agric. Food Chem.* **2003**, *51*, 6564–6571. [CrossRef]

26. Kotsiou, K.; Tasioula-Margari, M. Changes occurring in the volatile composition of Greek virgin olive oils during storage: Oil variety influences stability. *Eur. J. Lipid Sci. Technol.* **2015**, *117*, 514–522. [CrossRef]

27. Taticchi, A.; Esposto, S.; Servili, M. The basis of the sensory properties of virgin olive oil. In *Olive Oil Sensory Science*; Monteleone, E., Langstaff, S., Eds.; John Wiley and sons, Ltd.: Chichester, UK, 2014; pp. 33–54.

28. Uceda, M.; Aguilera, M.; De la, P.; Jiménez, A.; Beltrán, G. Variedades de olivo y aceituna. Tipos de aceites. In *El Aceite de Oliva Virgen: Tesoro de Andalucía*; Fernández Gutiérrez, A., Segura Carretero, A., Eds.; Fundación Unicaja: Málaga, Spain, 2009; pp. 109–137.

29. Luna, G.; Morales, M.T.; Aparicio, R. Characterisation of 39 varietal virgin olive oils by their volatile compositions. *Food Chem.* **2006**, *98*, 243–252. [CrossRef]

30. Quintanilla-Casas, B.; Bustamante, J.; Guardiola, F.; García-González, D.L.; Barbieri, S.; Bendini, A.; Toschi, T.G.; Vichi, S.; Tres, A. Virgin olive oil volatile fingerprint and chemometrics: Towards an instrumental screening tool to grade the sensory quality. *LWT* **2020**, *121*, 108936. [CrossRef]

31. Morales, M.T.; Luna, G.; Aparicio, R. Comparative study of virgin olive oil sensory defects. *Food Chem.* **2005**, *91*, 293–301. [CrossRef]

32. Angerosa, F.; Servili, M.; Selvaggini, R.; Taticchi, A.; Esposto, S.; Montedoro, G. Volatile compounds in virgin olive oil: Occurrence and their relationship with the quality. *J. Chromatogr. A* **2004**, *1054*, 17–31. [CrossRef]

33. Salas, J.J.; García-González, D.L.; Aparicio, R. Volatile compound biosynthesis by green leaves from an Arabidopsis thaliana hydroperoxide lyase knockout mutant. *J. Agric. Food Chem.* **2006**, *54*, 8199–8205. [CrossRef]

34. García-González, D.L.; Aparicio, R. Detection of vinegary defect in virgin olive oils by metal oxide sensors. *J. Agric. Food Chem.* **2002**, *50*, 1809–1814. [CrossRef]

35. Cecchi, L.; Migliorini, M.; Giambanelli, E.; Rossetti, A.; Cane, A.; Melani, F.; Mulinacci, N. Headspace solid-phase microextraction–gas chromatography–mass spectrometry quantification of the volatile profile of more than 1200 virgin olive oils for supporting the panel test in their classification: Comparison of different chemometric approaches. *J. Agric. Food Chem.* **2019**, *67*, 9112–9120. [CrossRef]

36. Aparicio, R.; Morales, M.T.; García-González, D.L. Towards new analyses of aroma and volatiles to understand sensory perception of olive oil. *Eur. J. Lipid Sci. Technol.* **2012**, *114*, 1114–1125. [CrossRef]

37. Grosch, W. Determination of potent odourants in foods by aroma extract dilution analysis (AEDA) and calculation of odour activity values (OAVs). *Flavour Fragr. J.* **1994**, *9*, 147–158. [CrossRef]

38. Romero, I.; García-González, D.L.; Aparicio-Ruiz, R.; Morales, M.T. Validation of SPME-GCMS method for the analysis of virgin olive oil volatiles responsible for sensory defects. *Talanta* **2015**, *134*, 394–401. [CrossRef] [PubMed]

39. Cavalli, J.F.; Fernandez, X.; Lizzani-Cuvelier, L.; Loiseau, A.M. Characterization of volatile compounds of French and Spanish virgin olive oils by HS-SPME: Identification of quality-freshness markers. *Food Chem.* **2004**, *88*, 151–157. [CrossRef]

40. Morales, M.T.; Przybylski, R. Olive oil oxidation. In *Handbook of Olive Oil: Analysis and Properties*; Aparicio, R., Harwood, J., Eds.; Springer US: Boston, MA, USA, 2013; pp. 479–522.

41. Tena, N.; Lobo-Prieto, A.; Aparicio, R.; García-González, D.L. Storage and preservation of fats and oils. In *Encyclopedia of Food Security and Sustainability*; Ferranti, P., Berry, E., Jock, A., Eds.; Elsevier: Amsterdam, The Netherlands, 2018; pp. 605–618.

42. García-González, D.L.; Vivancos, J.; Aparicio, R. Mapping brain activity induced by olfaction of virgin olive oil aroma. *J. Agric. Food Chem.* **2011**, *59*, 10200–10210. [CrossRef]

43. Vivancos, J.; Tena, N.; Morales, M.T.; Aparicio, R.; García-González, D.L. A neuroimaging study of pleasant and unpleasant olfactory perceptions of virgin olive oil. *Grasas Aceites* **2016**, *67*, e157. [CrossRef]

44. Velasco, J.; Dobarganes, C. Oxidative stability of virgin olive oil. *Eur. J. Lipid Sci. Technol.* **2002**, *104*, 661–676. [CrossRef]

45. International Olive Council. Determination of fatty acid methyl esters by gas chromatography. COI/T.20/Doc. No 33/Rev.1 2017, Madrid, Spain. 2017. Available online: https://www.internationaloliveoil.org/wp-content/uploads/2019/11/COI-T.20-Doc.-No-33-Rev.-1-2017.pdf (accessed on 6 April 2020).

46. International Olive Council. Determination of peroxide value. COI/T.20/Doc. No 35/Rev.1 2017, Madrid, Spain. 2017. Available online: https://www.internationaloliveoil.org/wp-content/uploads/2019/11/Method-COI-T.20-Doc.-No-35-Rev.-1-2017.pdf (accessed on 6 April 2020).

47. International Olive Council. Determination of free fatty acids, cold method. COI/T.20/Doc. No 34/Rev. 1 2017, Madrid, Spain. 2017. Available online: https://www.internationaloliveoil.org/wp-content/uploads/2019/11/COI-T.20-Doc.-No-34-Rev.-1-2017.pdf (accessed on 6 April 2020).

48. International Olive Council. Spectrophotometric investigation in the ultraviolet. COI/T.20/Doc. No 19/Rev. 5 2019, Madrid, Spain. 2019. Available online: https://www.internationaloliveoil.org/wp-content/uploads/2019/11/Method-COI-T.20-Doc.-No-19-Rev.-5-2019-2.pdf (accessed on 6 April 2020).

49. AOCS. Oil Stability Index (OSI). Cd 12b-92 Champaign, IL. 1993. Available online: https://www.aocs.org/attain-lab-services/methods/methods/method-detail?productId=111524 (accessed on 6 April 2020).

50. Oliver-Pozo, C.; Trypidis, D.; Aparicio, R.; García-González, D.L.; Aparicio-Ruiz, R. Implementing dynamic headspace with SPME sampling of virgin olive oil volatiles: Optimization, quality analytical study, and performance testing. *J. Agric. Food Chem.* **2019**, *67*, 2086–2097. [CrossRef]

51. Oliver-Pozo, C.; Aparicio-Ruiz, R.; Romero, I.; García-González, D.L. Analysis of volatile markers for virgin olive oil aroma defects by SPME-GC/FID: Possible sources of incorrect data. *J. Agric. Food Chem.* **2015**, *63*, 10477–10483. [CrossRef]

52. Aparicio, R.; Morales, M.T. Characterization of olive ripeness by green aroma compounds of virgin olive oil. *J. Agric. Food Chem.* **1998**, *46*, 1116–1122. [CrossRef]

53. Buettner, A.; Schieberle, P. Influence of mastication on the concentrations of aroma volatiles - Some aspects of flavour release and flavour perception. *Food Chem.* **2000**, *71*, 347–354. [CrossRef]

54. Normalización, A.E. de Requisitos generales para la competencia de los laboratorios de ensayo y calibración. (ISO/IEC 17025:2017). Asoc. Española Norm. 2017, Madrid, Spain. Available online: https://www.une.org/encuentra-tu-norma/busca-tu-norma/norma/?c=N0059467 (accessed on 6 April 2020).

55. International Olive Council. Sensory analysis of olive oil. Method for the organoleptic assessment of virgin olive oil. COI/T.20/Doc. No 15/Rev. 10 2018, Madrid, Spain. 2018. Available online: https://www.internationaloliveoil.org/wp-content/uploads/2019/11/COI-T20-Doc.-15-REV-10-2018-Eng.pdf (accessed on 6 April 2020).

56. Aparicio-Ruiz, R.; García-González, D.L.; Oliver-Pozo, C.; Tena, N.; Morales, M.T.; Aparicio, R. Phenolic profile of virgin olive oils with and without sensory defects: Oils with non-oxidative defects exhibit a considerable concentration of phenols. *Eur. J. Lipid Sci. Technol.* **2016**, *118*, 299–307. [CrossRef]

57. Mateos, R.; Espartero, J.L.; Trujillo, M.; Ríos, J.J.; León-Camacho, M.; Alcudia, F.; Cert, A. Determination of phenols, flavones, and lignans in virgin olive oils by solid-phase extraction and high-performance liquid chromatography with diode array ultraviolet detection. *J. Agric. Food Chem.* **2001**, *49*, 2185–2192. [CrossRef] [PubMed]

**Sample Availability:** Samples of the compounds (virgin olive oils collected every month during the storage experiment) are available from the authors.

 *molecules*

*Article*

# Validation of a LLME/GC-MS Methodology for Quantification of Volatile Compounds in Fermented Beverages

Eduardo Coelho [1], Margarida Lemos [1], Zlatina Genisheva [1], Lucília Domingues [1], Mar Vilanova [2,]* and José M. Oliveira [1,]*

[1]  CEB—Centre of Biological Engineering, University of Minho, 4710-057 Braga, Portugal; e.coelho@ceb.uminho.pt (E.C.); margarida.lemos.20@gmail.com (M.L.); zlatina@deb.uminho.pt (Z.G.); luciliad@deb.uminho.pt (L.D.)
[2]  Spanish National Research Council (MBG-CSIC), Salcedo, 36143 Pontevedra, Spain
*  Correspondence: mvilanova@mbg.csic.es (M.V.); jmoliveira@deb.uminho.pt (J.M.O.); Tel.: +34-986-854-800 (M.V.); +351-253-604-400 (J.M.O.)

Academic Editor: Eugenio Aprea
Received: 8 January 2020; Accepted: 29 January 2020; Published: 31 January 2020

**Abstract:** Knowledge of composition of beverages volatile fraction is essential for understanding their sensory attributes. Analysis of volatile compounds predominantly resorts to gas chromatography coupled with mass spectrometry (GC-MS). Often a previous concentration step is required to quantify compounds found at low concentrations. This work presents a liquid-liquid microextraction method combined with GC-MS (LLME/GC-MS) for the analysis of compounds in fermented beverages and spirits. The method was validated for a set of compounds typically found in fermented beverages comprising alcohols, esters, volatile phenols, and monoterpenic alcohols. The key requirements for validity were observed, namely linearity, sensitivity in the studied range, accuracy, and precision within the required parameters. Robustness of the method was also evaluated with satisfactory results. Thus, the proposed LLME/GC-MS method may be a useful tool for the analysis of several fermented beverages, which is easily implementable in a laboratory equipped with a GC-MS.

**Keywords:** fermented beverages; volatile compounds; analytical method; liquid-liquid microextraction; GC-MS

---

## 1. Introduction

The flavor, which is one of the most important sensory attributes of fermented, alcoholic, and distilled beverages (cider, wine, beer, vinegar, spirits, vodka, whiskey, among others), is determined by a vast and diverse number of volatile compounds, arising either from raw material (e.g., grapes, barley, hops), yeast/bacteria fermentations, which are secondary metabolites [1–3], or from ageing when applied (e.g., in oak wood) [4]. These volatile compounds belong to diverse chemical families like alcohols, esters, aldehydes and carbonyls, volatile fatty acids, volatile phenols, sulphur compounds, terpenes, norisoprenoids, lactones, furans, and more [3,5–7], which are often found in very low concentrations.

Since volatile compounds of fermented/alcoholic beverages are highly correlated with the sensory characteristics of the products, its identification and quantification acquires crucial significance for understanding beverages organoleptic properties and further develop product quality. In addition, the presence/absence or the amount of each individual component may be a marker of the used technology or the indication of a product defect. The analysis of individual volatile compounds must comprise a chromatographic separation, which is followed by a generic or a selective identification (e.g., flame ionization detector, electron capture detector, flame photometric detector, mass spectrometric detector) [8]. Recently, some authors have correlated FTIR spectra with some specific compounds or groups of compounds [9,10].

Apart from the major volatile compounds present in amounts of mg/L (e.g., 3-methyl-1-butanol), which may be analyzed by direct injection, those presented in lower amounts ranging from a few µg/L (e.g., linalool) or even scarce ng/L (e.g., 4-methyl-4-mercapto-2-pentanone) must be concentrated before the chromatographic separation. This step could be achieved by mixing a solvent with the sample, as in liquid-liquid extraction (LLE) [11,12] and liquid-liquid microextraction (LLME) [13,14]. Several adaptations/modifications of LLE/LLME methods can be envisioned, e.g., the evaporation of solvent for increasing concentration, adsorption of volatiles in a solvent drop (single-drop microextraction – SDME) [15], or even adsorption/desorption of the compounds using a polymeric phase (sorbent-phase extraction – SPE) [16]. Solvent-free techniques include solid-phase micro-extraction (SPME) [17,18], usually in the headspace of the sample (HS-SPME) [19,20], and stir-bar sorptive extraction (SBSE) [21,22]. Some of these methods, developed to analyze volatiles in alcoholic/fermented beverages, are generic considering that they allow the identification and quantification of the majority of compounds, where their range of application depends on the solvents and/or the sorbents' polarity. Specific methodologies based on polymeric materials, sometimes applying derivatization procedures, were developed to quantify specific compounds or classes of compounds [23].

For a method to be applied in the laboratory, it must be validated to ensure its reliability and the quality of the obtained results. Several points must be addressed for a method to be valid, namely its linearity, specificity, quantification range, limits of detection and quantification, sensitivity, precision, and accuracy. Optionally, robustness and reproducibility studies can be performed to reinforce the methods applicability and efficiency [24–26].

This work aims to validate a liquid-liquid microextraction method (LLME) first published by Oliveira and collaborators [27], which only reported its use for the analysis of three C6-alcohols (1-hexanol, *E*-3-hexenol and *Z*-3-hexenol), exclusively in wine. As the method provided satisfactory performance and results, its feasibility for the analysis of a broader range of compounds and a wider variety of matrices remained to be validated. The presented LLME method combined with GC-MS poses as an additional alternative to analyze volatile compounds in alcoholic/fermented beverages. This procedure can be applied in any laboratory equipped with a GC-MS by any technician, using only ordinary glassware and low amounts of sample and solvents. High throughput applications are envisioned as the procedure enables handling a substantial number of samples and screening a large number of volatiles in a short period of time.

## 2. Results and Discussion

### *2.1. Linearity and Sensitivity*

Linearity and sensitivity of the proposed LLME method were evaluated by outlining calibration curves for each analyte, using a solution of pure standards. Compounds selected for calibration of the method were chosen on the basis of their contribution for the volatile and aromatic fraction of fermented products, which are considered to be representative of the analytes generally found in beer, wine, spirits, and vinegar. Acids were left out of the validation study by considering the difficulties of maintaining them in a standard solution due to their reaction with some other components in the mixture.

Regressions were performed from the obtained data with the corresponding coefficients presented in Table 1. Good linear regressions were obtained for extraction and quantification using the LLME method, with values of $R^2 > 0.995$ for all of the studied analytes. The $R^2$ value is a useful indicator of the regression quality. However, according to Kruve and collaborators [26] and Araújo [28], it cannot be considered as a standalone measure to validate a method linearity, which must be further validated by a statistical lack-of-fit *F*-test. Lack–of-fit tests were performed for the regression curves obtained for each analyte, according to the recommendations of Araujo [28], since all regressions were demonstrated to be linear, with the $F$ obtained being lower than the tabulated one for the corresponding degrees of freedom. This linearity reflects not only the directly proportional response of the MS detector, but also the direct proportionality in the extraction of analytes by LLME.

**Table 1.** Reference, purity ($P$), and concentration range ($C$) for each analyte, and Pearson correlation coefficient ($R^2$), limit of quantification ($LOQ$), and response factor of the method ($R_f$), with respective confidence limits ($p$ = 0.05), obtained from the calibration curves.

| Compound | Reference | $P$/% | Range $C$/(μg/L) | $R^2$ | $LOQ$/(μg/L) | $R_f$ |
|---|---|---|---|---|---|---|
| 4-methyl-2-pentanone | Fluka 02474 | $\geq$ 99.7 | 24.8 to 248 | 0.9991 | 6.9 | 1.32 ± 0.05 |
| Ethyl butyrate | Aldrich E15701 | 99 | 5.76 to 576 | 0.9995 | 4.7 | 1.58 ± 0.04 |
| Ethyl 2-methylbutyrate | Aldrich 306886 | 99 | 2.48 to 248 | 0.9997 | 1.8 | 0.87 ± 0.02 |
| Ethyl 3-methylbutyrate | Aldrich 112283 | 98 | 3.12 to 312 | 0.9993 | 2.2 | 0.91 ± 0.03 |
| 3-methyl-1-butyl acetate | Aldrich 306967 | $\geq$ 99 | 21.32 to 2132 | 0.9990 | 3.9 | 2.00 ± 0.07 |
| Ethyl hexanoate | Aldrich 148962 | $\geq$ 99 | 9.64 to 964 | 0.9978 | 2.2 | 1.32 ± 0.07 |
| Hexyl acetate | Aldrich 108154 | 99 | 2.76 to 276 | 0.9983 | 2.9 | 1.57 ± 0.08 |
| 3-methyl-1-pentanol | Aldrich 111112 | 99 | 25.6 to 256 | 0.9968 | 14.2 | 4.63 ± 0.30 |
| Ethyl lactate | Aldrich E34102 | 98 | 113.2 to 1132 | 0.9978 | 107.4 | 44.90 ± 2.45 |
| 1-hexanol | Fluka 73117 | > 99.9 | 14.72 to 1472 | 0.9976 | 6.7 | 3.63 ± 0.20 |
| E-3-hexen-1-ol | Aldrich 224715 | 97 | 6.32 to 632 | 0.9971 | 5.1 | 5.11 ± 0.32 |
| Z-3-hexen-1-ol | Fluka 53056 | $\geq$ 98 | 7.20 to 720 | 0.9968 | 5.9 | 5.23 ± 0.34 |
| Linalool | Aldrich L2602 | 97 | 4.76 to 476 | 0.9998 | 3.2 | 1.71 ± 0.03 |
| Diethyl succinate | Aldrich 112402 | 99 | 6.12 to 612 | 0.9977 | 2.4 | 1.25 ± 0.07 |
| α-terpineol | Merck 8.21078 | $\geq$ 98 | 2.60 to 260 | 0.9979 | 2.6 | 1.37 ± 0.07 |
| Citronellol | Aldrich C83201 | 95 | 2.72 to 272 | 0.9999 | 2.2 | 1.43 ± 0.02 |
| Nerol | Aldrich 268909 | 97 | 3.04 to 304 | 0.9988 | 3.1 | 1.83 ± 0.07 |
| 2-phenylethyl acetate | Fluka 46030 | > 99 | 10.32 to 1032 | 0.9995 | 2.6 | 1.39 ± 0.03 |
| Geraniol | Aldrich 163333 | 98 | 3.08 to 308 | 0.9994 | 2.4 | 1.26 ± 0.04 |
| Guaiacol | Aldrich G10903 | 98 | 2.92 to 292 | 0.9984 | 5.1 | 2.65 ± 0.12 |
| 4-ethylphenol | Aldrich E44205 | 99 | 4.88 to 488 | 0.9983 | 4.2 | 2.03 ± 0.10 |

Extraction selectivity was maintained throughout the tested concentrations, which enabled proper quantification of the analytes in the mixture. All regressions presented intercept values not significantly different from zero ($p$ > 0.05) and, therefore, equations are only based on the slope, similarly to the previously reported works for other LLE methods [12]. Moreover, the baseline value is subtracted for the integration of peaks in the chromatogram using background correction in the software, which also justifies the absence of the intercept value.

Sensitivity is defined as the change in the method response, which corresponds to a change in the measured quantity and is intrinsically related to the slope of the calibration curve [29]. In this case, $R_f$ is the inverse of the slope. This factor is, therefore, a measure of the method's sensitivity in terms of the relative response of each compound in relation to the response of the internal standard. A higher response factor means a higher variation of the compound's concentration for a given variation of the signal, which, therefore, accounts for a lower sensitivity. Overall, response factors obtained for esters ranged between 1 and 2, with the exception of ethyl lactate for which the response factor was highly superior while attaining the value of 44.9 and accounting for a lower sensitivity of the method toward this compound. Similar to esters, monoterpenic alcohols as well as 4-methyl-2-pentanone presented response factors between 1.2 and 1.8, which was followed by volatile phenols that presented slightly higher $R_f$ values of about 2. With higher response factors, and, therefore, lower sensitivity alcohols, presented $R_f$ values between 3.6 and 5.1. This variation in the response factor is a combination of different extraction selectivity by the LLME method and differences in ionization and detection in the MS. Response factors seem to be similar within groups of compounds, which, despite not excluding the need for determining a specific analyte response for a proper quantification, can aid in the prediction of the response for compounds within the same group. With response factors between 1 and 5 for the majority of compounds, it is believed that the method has good sensitivity. Therefore, the method complies with the first base requirements for validation, being that the LLME method in the study presents proper sensitivity and linearity.

### 2.2. Limits of Detection and Quantification

Limits of detection (*LOD*) and quantification (*LOQ*) deal with the minimum amount of compound possible to detect and quantify, respectively. As stated by Brettell and Lester [30], two strategies can be used for determining *LOD*: a statistical approach, which is more likely to generate artificially low *LOD* values, and an experimental approach, which is attained by decreasing the analyte concentration until the identification criterion is no longer met. This generates higher and closer values to reality *LOD* values [30]. Since the mass spectrum of a given compound/peak can be compared with spectra of private or commercial spectrum libraries, the occurrence of a match at a given retention time ensures reliability of compound detection and identification. Additionally, considering the GC-MS method used, the *LOD* value will be related with the *LOQ* value as, if a given compound identification is reliable, its quantification from the chromatogram is possible. Hence, the statistical calculation of the *LOD* value has no practical application, and the experimental approach was performed for determining the *LOQ* value, which is of greater use. Several recommendations can be found for determining the *LOQ* value but considering the focus of the method. A conservative approach was chosen for its establishment by following the recommendations of Kruve and collaborators [25]. Therefore, the minimum amount of analyte detected and quantified was taken into account for determining *LOQ*, with the obtained values presented in Table 1. As demonstrated, *LOQ* values ranged from 2 µg/L to 7 µg/L for most compounds, with the exception of 3-methyl-1-pentanol (14.2 µg/L) for which the minimum tested concentration was higher, and ethyl lactate (107.4 µg/L) due to the lower sensitivity obtained. The obtained values are, in their majority, about 2 to 10 folds lower when compared with the values reported by Ferreira and collaborators [13], which worked with similar compounds and concentrations. As a cross validation for acceptance of this value, the measurements are within the 20% of the relative standard deviation (*RSD*), as stipulated by Brettel and Lester [30].

### 2.3. Precision

As reported by Kruve and collaborators [26], precision can be quantified as the relative standard deviation/coefficient of variation of replicate analysis. In this work, we evaluated two types of precision, the repeatability (a single operator in the same run conditions), and the intermediate precision (different operators, different run conditions but the same laboratory). The *RSD* values were calculated for the two scenarios which were presented in Table 2. For evaluating the intermediate precision, analyses using the proposed LLME method were performed by two operators with one experienced in its execution and one with reduced experience in the laboratory and with the method. Five replicates were measured by each operator using an independent equipment and apparatus, where the GC-MS is the only equipment in common for the analysis of extracts.

As visible in Table 2, *RSD* observed for evaluation of repeatability was considerably low, ranging from 3.3% to 9.0%. When analysing the *RSD* values obtained for intermediate precision (involving two different operators), a higher variation can be observed ranging from about 6.0% to 19.7%. This higher dispersion of the measurements can be justified by the differences in the experience of the operators, where deviations in the addition of an internal standard or differences in the interpretation and integration of chromatograms can lead to a higher dispersion of results. Establishment of critical *RSD* values for a method to be precise depends strongly on the application intended. Several limits have been proposed, which are the most common considered *RSD* < 15% of the nominal value [26,31] or as high as 20% for environmental or food samples [32]. As seen in the results, values of *RSD* regarding repeatability were all below the minimum level accepted. In addition, despite the higher *RSD* values obtained for intermediate precision, the majority of compounds were still below the acceptable limit of 15% with the exception of 3-methyl-1-pentanol, ethyl lactate, *E*-3-hexen-1-ol, and *Z*-3-hexen-1-ol, which still fall below the limit of 20% proposed by Huber [32]. Thus, the method is considered precise and can be performed with satisfactory outputs.

**Table 2.** Values obtained for evaluation of precision, measured as relative standard deviation (*RSD*), accuracy, expressed as relative error (*RE*), and robustness, quantified by compound recovery (*Rec*).

| Compound | Repeatability | Intermediate Precision | Accuracy | Robustness | | |
|---|---|---|---|---|---|---|
| | *RSD*/% | *RSD*/% | *RE*/% | *Rec*/% ($t$ = 30 min) | *Rec*/% (Synthetic Wine) | *Rec*/% (Synthetic Vinegar) |
| 4-methyl-2-pentanone | 6.5 | 9.3 | 11.2 | 103.2 | 100.6 | 103.1 |
| Ethyl butyrate | 5.3 | 7.4 | 10.8 | 99.9 | 91.0 | 95.2 |
| Ethyl 2-methylbutyrate | 9.0 | 9.3 | 13.2 | 96.6 | 86.0 | 83.0 |
| Ethyl 3-methylbutyrate | 5.3 | 6.7 | 20.5 | 98.0 | 88.8 | 91.8 |
| 3-methyl-1-butyl acetate | 4.5 | 5.7 | 1.6 | 104.7 | 93.9 | 96.2 |
| Ethyl hexanoate | 3.3 | 6.0 | 2.9 | 100.6 | 101.9 | 100.3 |
| Hexyl acetate | 3.8 | 11.2 | 15.5 | 98.0 | 92.3 | 97.8 |
| 3-methyl-1-pentanol | 6.3 | 18.0 | 2.3 | 96.1 | 148.6 | 113.6 |
| Ethyl lactate | 8.4 | 18.9 | 2.8 | 88.5 | 161.6 | 106.7 |
| 1-hexanol | 5.0 | 12.5 | 15.8 | 91.3 | 115.3 | 95.9 |
| E-3-hexen-1-ol | 7.0 | 18.0 | 14.2 | 86.0 | 116.3 | 96.2 |
| Z-3-hexen-1-ol | 6.8 | 19.7 | 15.9 | 85.8 | 116.4 | 90.5 |
| Linalool | 4.0 | 10.8 | 9.5 | 103.8 | 98.6 | 93.4 |
| Diethyl succinate | 3.3 | 10.3 | 13.3 | 113.0 | 117.7 | 113.4 |
| $\alpha$-terpineol | 4.9 | 10.2 | 10.8 | 109.7 | 112.7 | 105.6 |
| Citronellol | 4.4 | 12.5 | 6.6 | 87.8 | 90.6 | 86.1 |
| Nerol | 5.9 | 13.8 | 13.7 | 108.6 | 100.0 | 93.2 |
| 2-phenylethyl acetate | 3.4 | 7.9 | 0.7 | 108.6 | 109.6 | 109.9 |
| Geraniol | 2.6 | 9.6 | 16.7 | 107.2 | 97.3 | 93.9 |
| Guaiacol | 6.3 | 14.0 | 1.6 | 97.3 | 120.5 | 105.5 |
| 4-ethylphenol | 4.5 | 9.9 | 5.0 | 72.9 | 112.5 | 97.1 |

## 2.4. Accuracy

Accuracy was determined by the addition of a known amount of the analytes in the study to a real sample (spiking) and quantification of the analytes in the spiked sample. For this purpose, a commercial beer was analyzed using the proposed LLME method both in its original state and after spiking, as recommended by the guidelines for method validation [26]. To better assess accuracy, the theoretical expected concentration ($C_{expect}$ = $C_{beer}$ + $C_{spik}$) was compared with the concentration measured using the LLME method ($C_{determin}$). According to multiple *t*-tests for comparisons ($p \leq 0.05$), no differences were found between the expected and the measured concentrations. As a more appropriate measure of accuracy, the deviation of the measured concentrations regarding the expected values was calculated, and expressed as a relative error (*RE*) [31]. In agreement with the results reported by González and collaborators [31], this value cannot exceed 15% for the method to be accurate (except for determinations at the *LOQ* where 20% is accepted). As shown in Table 2, *RE* values were within the 15% limit established for the studied compounds. Thus, the method is considered to be accurate when complying with another key requirement for validity.

## 2.5. Robustness

As stated, robustness can be defined as the ability of the method to endure slight variations and maintain its result [25]. To assess the robustness of the method, two criteria were evaluated including variation of contact time and the matrix effect, which were identified as the main variables affecting the LLME method. The effect of an increased stirring and extraction time was tested to evaluate the possible occurrence of differences in compound extraction. Again, as performed for accuracy, possible differences in compound recovery and quantification were statistically determined by the *t*-test, comparing the measured concentration with the expected concentration of the compounds, and evaluating the recovery of target analytes by taking into account the known dilution and concentration of the solution of standards.

Regarding the increase of stirring time, no statistically significant differences were observed for the measurements performed with 30 min of stirring ($p > 0.05$). Extraction of the compounds using 15 min stipulated in the method is shown to be sufficient for the total recuperation of analytes, which is maintained independently of the longer contact time. For a better assessment of robustness, recovery was calculated for each compound in accordance with Kruve and collaborators [26], with the values presented in Table 2. Similarly, when observed for accuracy, critical recovery values can be established for the acceptance of recovery in determining robustness. In this sense, values of recovery between 70% and 110% for measurements ranging from 10 µg/L to 100 µg/L, or 80% to 110% for measurements above 100 µg/L, are considered acceptable [33]. Therefore, the recovery values obtained with increased stirring time were within the acceptable range.

Regarding the matrix effect, the main focus was to evaluate if the recovery and quantification of the analyses would be affected by other components in the mixture. For control purposes, two synthetic matrixes were tested including a solution mimicking wine composition and another mimicking vinegar composition. For the majority of compounds, recovery values were also within the acceptable ranges previously referred, with the exception of those obtained for 3-methyl-1-pentanol and ethyl lactate in the synthetic wine matrix. The higher recovery observed for these compounds can be caused by a higher efficiency and selectivity in their extraction and, therefore, an accuracy test or validation in wine would be advised for the specific quantification of these compounds. Nevertheless, only two compounds in one matrix showed recovery values outside the acceptable range. The remaining compounds were properly quantified in the synthetic wine as well as all compounds in the synthetic vinegar matrix. Considering the overall results obtained under multiple conditions, global robustness of the method can be considered satisfactory.

## 3. Materials and Methods

### 3.1. LLME-GC/MS Method

#### 3.1.1. Liquid-Liquid Microextraction of Volatile Compounds

In a 10 mL culture tube (Pyrex, ref. 1636/26MP), 8 mL of sample, clarified by centrifugation if necessary, 2.46 µg of internal standard (4-nonanol, Merck ref. 818773), and a magnetic stir bar were added. Extraction was done by stirring samples with 400 µL of dichloromethane (Merck, ref. 106054), at room temperature for 15 min, using a magnetic stirrer. Tubes were placed vertically and agitation was regulated in order to maintain dispersion of solvent micro-drops without reaching the sample surface. After cooling at 0 °C for 10 min, the magnetic stir bar was removed and the organic phase was detached by centrifugation (5118 $g$, 5 min, 4 °C). Using a glass Pasteur pipette, the extract was recovered into a vial, dried with anhydrous sodium sulphate (Merck, ref. 1.06649), and transferred to a new vial for storage at –20 °C before analysis.

#### 3.1.2. Chromatographic Analysis

Gas chromatographic analysis of volatile compounds was performed using a GC-MS Varian Saturn 2000 (Varian, Walnut Creek, CA, USA) equipped with a 1079 injector, an ion-trap mass spectrometer, and a Sapiens-Wax MS capillary column (30 m × 0.15 mm, 0.15 µm film thickness, Teknokroma, Barcelona, Spain). The temperature of the injector and the MS transfer line were both set to 250 °C. The oven temperature was held at 60 °C, for 2 min, then programmed to rise from 60 °C to 234 °C, at 3 °C/min, and from 234 °C to 260 °C at 5 °C/min. Lastly, it was held for 10 min at 260 °C. The carrier gas was helium GHE4× (Praxair, Maia, Portugal), at a constant flow rate of 1.3 mL/min. A 1 µL injection was made in the split-less mode, for 30 s (split vent of 30 mL/min). The detector was set to an electronic impact mode (70 eV) with an acquisition range ($m/z$) from 35 to 300 at an acquisition rate of 610 ms.

#### 3.1.3. Identification of Volatile Compounds

Identification of volatile compounds was preformed using the software Star – Chromatography Workstation version 6.9.3 (Varian), by comparing mass spectra and retention indices with those of pure standard compounds.

### 3.2. Method Validation

#### 3.2.1. Base Standard Solution

To perform the method validation, a hydroalcoholic solution (7%, by volume; ethanol Fisher, 99.8%), using Milli-Q water, was initially prepared, which was the solvent used for compound dilution. First, a concentrated solution of the volatile compounds was prepared by adding each compound, by weighing using an analytical scale (Kern ABJ), to the hydroalcoholic solution, at a concentration of 1000× the highest concentration presented in Table 1. The base standard solution was then prepared by diluting the concentrated solution by a factor of 1000 with the hydroalcoholic solution to attain the highest concentrations specified in Table 1 (maximum value of the cited range). Compounds were chosen as being representative of the chemical groups with the higher impact in the volatile fraction and sensory properties of fermented beverages, such as wine, beer, and vinegar. These were purchased as pure standards with the purity and suppliers indicated in Table 1.

#### 3.2.2. Linearity

Calibration curves were constructed by using six points, corresponding to different concentrations obtained by the dilution of the base standard solution in the hydroalcoholic solution. Each solution was analyzed in triplicate by the proposed method. The average area ratios (*i.e.*, peak area of the compound $x$, $A_x$, to the peak area of the internal standard, $A_{IS}$) were plotted against the concentration ratios (i.e., concentration of the compound $x$, $C_x$, to the concentration of the internal standard, $C_{IS}$) to obtain the calibration curves in accordance with Equation (1).

$$\frac{A_x}{A_{IS}} = b \times \frac{C_x}{C_{IS}} \tag{1}$$

From each curve, slope ($b$) and regression coefficient ($R^2$) were calculated, and linearity was evaluated by a lack-of-fit $F$-test. Response factors ($R_f$) were also calculated for each compound as the inverse of the slope $\left(R_f = \frac{1}{b}\right)$. The limit of quantification (*LOQ*) was determined as the minimum concentration of the compound that could be trustily quantified.

#### 3.2.3. Precision

Two different measures of precision were evaluated for validation of the LLME method such as repeatability and intermediate precision. Repeatability was evaluated by the analysis of five replicate samples in the same conditions of the proposed method. As a measure of repeatability, the relative standard deviation (*RSD*) was calculated according to Equation (2).

$$\frac{RSD}{\%} = \frac{s}{\bar{x}} \times 100 \tag{2}$$

where $s$ stands for standard deviation and $\bar{x}$ represents the average of the measured values. To evaluate intermediate precision, independent measurements of dissimilar samples were performed at different times by independent operators, where the *RSD* was also calculated as stated in Equation (2).

#### 3.2.4. Accuracy

In the absence of reference materials, accuracy was investigated by spiking and recovery. A commercial beer was used for analysis by the proposed method in its original state and after the addition (spiking) of a known mass of the analyte to the sample. The relative error (*RE*) of the determined concentration was calculated based on Equation (3), i.e., calculating the concentration of each compound in the spiked beer ($C_{\text{determ}}$) against its expected concentration ($C_{\text{expect}}$).

$$\frac{RE}{\%} = \frac{C_{\text{determ}} - C_{\text{expect}}}{C_{\text{expect}}} \times 100 \qquad (3)$$

### 3.2.5. Robustness

Other parameters were studied to evaluate the susceptibility of the method to changes that might occur during routine analysis (use a different matrix or an extended extraction time). The matrix effect was evaluated using two different matrices mimicking a wine and a vinegar, respectively, by adding the same volatile compounds under evaluation at an intermediate concentration. Apart from volatile compounds, the synthetic wine comprises ethanol (12% by volume, Fisher, 99.8%), tartaric acid (5 g/L, Sigma, 99.5%), glycerol (7.5 g/L, Himedia, 99.5%), and malic acid (2 g/L, Acros Organics, 99%). Synthetic vinegar was prepared using 10 g/L of citric acid (Panreac, 99.5%) and 50 g/L of acetic acid (Sigma). Three replicates were carried out for each matrix. The effect of the change of the duration of the extraction time was also evaluated. Accordingly, three replicates of the extraction procedure were done by stirring the sample for 30 min instead of 15 min of the proposed method. Recovery (*Rec*) of target compounds, expressed as a percentage, was evaluated by calculating the measured concentration ($C_{\text{measur}}$) vs. the expected concentration ($C_{\text{expect}}$), as stated in Equation (4).

$$\frac{Rec}{\%} = \frac{C_{\text{measur}}}{C_{\text{expect}}} \times 100 \qquad (4)$$

## 4. Conclusions

The LLME method presented in this work is a reliable alternative for the analysis of compounds participating in the volatile fraction of fermented beverages. The method is linear for the studied ranges and has good sensitivity, which varies depending on the chemical group of compounds. The method is precise and has shown good repeatability and intermediate precision. Variations were performed for the analytical matrix and for protocol execution. The LLME method was also demonstrated to be robust. Lastly, the method is accurate and adequate for application in real samples. Having complied with all the parameters needed, the LLME method presented in this work is, therefore, valid for application in the analysis of fermented beverages and, certainly, to distilled beverages/spirits, after a convenient dilution with water to reach an alcoholic strength, by volume, of about 15%.

**Author Contributions:** Conceptualization, E.C., Z.G., M.V., and J.M.O. Methodology, E.C., Z.G., M.V., and J.M.O. Investigation, E.C., M.L., and Z.G. Validation, E.C., M.L., Z.G., and J.M.O. Formal analysis, E.C., M.L., and J.M.O. Writing- original draft, E.C. and J.M.O. Writing- review and editing, E.C., Z.G., and M.V. Visualization E.C. and J.M.O. Supervision L.D. and J.M.O. Project administration, L.D. and J.M.O. Funding acquisition, L.D. and J.M.O. All authors have read and agreed to the published version of the manuscript.

**Funding:** This study was supported by the Portuguese Foundation for Science and Technology(FCT) under the scope of the strategic funding of UIDB/04469/2020 unit and BioTecNorte operation (NORTE-01-0145-FEDER-000004) was funded by the European Regional Development Fund under the scope of Norte2020—Programa Operacional Regional do Norte, and also by Xunta de Galicia (Plan Galego de Investigación, Desenvolvemento e Innovación Tecnolóxica—INCITE).



## References

1. Briggs, D.E.; Boulton, C.A.; Brookes, P.A.; Stevens, R. *Brewing Science and Practice*; Woodhead Publishing Limited: Cambridge, UK; CRC Press LLC: Boca Raton, FL, USA, 2004; p. 9062. ISBN 978185573.
2. Robinson, A.L.; Boss, P.K.; Solomon, P.S.; Trengove, R.D.; Heymann, H.; Ebeler, S.E. Origins of Grape and Wine Aroma. Part 1. Chemical Components and Viticultural Impacts. *Am. J. Enol. Vitic.* **2013**, *65*, 1–24. [CrossRef]

3.  Hirst, M.B.; Richter, C.L. Review of aroma formation through metabolic pathways of saccharomyces cerevisiae in beverage fermentations. *Am. J. Enol. Vitic.* **2016**, *67*, 361–370. [CrossRef]

4.  Le Floch, A.; Jourdes, M.; Teissedre, P.-L. Polysaccharides and lignin from oak wood used in cooperage: Composition, interest, assays: A review. *Carbohydr. Res.* **2015**, *417*, 94–102. [CrossRef] [PubMed]

5.  Ugliano, M.; Henschke, P.A. Yeasts and wine flavour. In *Wine Chemistry and Biochemistry*; Moreno-Arribas, M.V., Polo, M.C., Eds.; Springer: New York, NY, USA, 2009; pp. 313–392. ISBN 978-0-387-74116-1.

6.  Coelho, E.; Vilanova, M.; Genisheva, Z.; Oliveira, J.M.; Teixeira, J.A.; Domingues, L. Systematic approach for the development of fruit wines from industrially processed fruit concentrates, including optimization of fermentation parameters, chemical characterization and sensory evaluation. *LWT* **2015**, *62*, 1043–1052. [CrossRef]

7.  Coelho, E.; Genisheva, Z.; Oliveira, J.M.; Teixeira, J.A.; Domingues, L. Vinegar production from fruit concentrates: Effect on volatile composition and antioxidant activity. *J. Food Sci. Technol.* **2017**, *54*, 4112–4122. [CrossRef]

8.  Robinson, A.; Boss, P.K.; Solomon, P.S.; Trengove, R.D.; Heymann, H.; Ebeler, S.E. Origins of Grape and Wine Aroma. Part 2. Chemical and Sensory Analysis. *Am. J. Enol. Vitic.* **2014**, *65*, 25–42. [CrossRef]

9.  Dambergs, R.; Gishen, M.; Cozzolino, D. A review of the state of the art, limitations, and perspectives of infrared spectroscopy for the analysis of wine grapes, must, and grapevine tissue. *Appl. Spectrosc. Rev.* **2015**, *50*, 261–278. [CrossRef]

10. Genisheva, Z.; Quintelas, C.; Mesquita, D.; Ferreira, E.; Oliveira, J.; Amaral, A. New PLS analysis approach to wine volatile compounds characterization by near infrared spectroscopy (NIR). *Food Chem.* **2018**, *246*, 172–178. [CrossRef]

11. Perestrelo, R.; Fernandes, A.; Albuquerque, F.; Marques, J.C.; Câmara, J. Analytical characterization of the aroma of Tinta Negra Mole red wine: Identification of the main odorants compounds. *Anal. Chim. Acta* **2006**, *563*, 154–164. [CrossRef]

12. Andújar-Ortiz, I.; Moreno-Arribas, M.; Martin-Alvarez, P.; Pozo-Bayon, M.A. Analytical performance of three commonly used extraction methods for the gas chromatography–mass spectrometry analysis of wine volatile compounds. *J. Chromatogr. A* **2009**, *1216*, 7351–7357. [CrossRef] [PubMed]

13. Ferreira, V.; Lopez, R.; Escudero, A.; Cacho, J.F. Quantitative determination of trace and ultratrace flavour active compounds in red wines through gas chromatographic–ion trap mass spectrometric analysis of microextracts. *J. Chromatogr. A* **1998**, *806*, 349–354. [CrossRef]

14. Ortega, C.; López, R.; Cacho, J.; Ferreira, V. Fast analysis of important wine volatile compounds development and validation of a new method based on gas chromatographic-flame ionisation detection analysis of dichloromethane microextracts. *J. Chromatogr. A* **2001**, *923*, 205–214. [CrossRef]

15. Jain, A.; Verma, K.K. Recent advances in applications of single-drop microextraction: A review. *Anal. Chim. Acta* **2011**, *706*, 37–65. [CrossRef] [PubMed]

16. Oliveira, J.M.; Oliveira, P.; Baumes, R.L.; Maia, O. Changes in aromatic characteristics of Loureiro and Alvarinho wines during maturation. *J. Food Compos. Anal.* **2008**, *21*, 695–707. [CrossRef]

17. Zhu, H.; Zhu, J.; Wang, L.; Li, Z. Development of a SPME-GC-MS method for the determination of volatile compounds in Shanxi aged vinegar and its analytical characterization by aroma wheel. *J. Food Sci. Technol.* **2016**, *53*, 171–183. [CrossRef]

18. Souza-Silva, É.A.; Gionfriddo, E.; Pawliszyn, J. A critical review of the state of the art of solid-phase microextraction of complex matrices II. Food analysis. *TrAC Trends Anal. Chem.* **2015**, *71*, 236–248. [CrossRef]

19. Cirlini, M.; Caligiani, A.; Palla, L.; Palla, G. HS-SPME/GC–MS and chemometrics for the classification of Balsamic Vinegars of Modena of different maturation and ageing. *Food Chem.* **2011**, *124*, 1678–1683. [CrossRef]

20. Martín, S.G.; Herrero, C.; Peña, R.; Barciela, J.; Crecente, R.M.P. Solid-phase microextraction gas chromatography–mass spectrometry (HS-SPME-GC–MS) determination of volatile compounds in orujo spirits: Multivariate chemometric characterisation. *Food Chem.* **2010**, *118*, 456–461. [CrossRef]

21. Castro, L.F.; Ross, C.F. Determination of flavour compounds in beer using stir-bar sorptive extraction and solid-phase microextraction. *J. Inst. Brew.* **2015**, *121*, 197–203. [CrossRef]

22. Horak, T.; Čulík, J.; Kellner, V.; Jurková, M.; Čejka, P.; Hašková, D.; Dvořák, J. Analysis of Selected Esters in Beer: Comparison of Solid-Phase Microextraction and Stir Bar Sorptive Extraction. *J. Inst. Brew.* **2010**, *116*, 81–85. [CrossRef]

23. Rodríguez-Bencomo, J.J.; Schneider, R.; Lepoutre, J.P.; Rigou, P. Improved method to quantitatively determine powerful odorant volatile thiols in wine by headspace solid-phase microextraction after derivatization. *J. Chromatogr. A* **2009**, *1216*, 5640–5646. [CrossRef]

24. Hartmann, C.; Smeyers-Verbeke, J.; Massart, D.L.; McDowall, R.D. Validation of bioanalytical chromatographic methods. *J. Pharm. Biomed. Anal.* **1998**, *17*, 193–218. [CrossRef]

25. Kruve, A.; Rebane, R.; Kipper, K.; Oldekop, M.-L.; Evard, H.; Herodes, K.; Ravio, P.; Leito, I. Tutorial review on validation of liquid chromatography–mass spectrometry methods: Part I. *Anal. Chim. Acta* **2015**, *870*, 29–44. [CrossRef] [PubMed]

26. Kruve, A.; Rebane, R.; Kipper, K.; Oldekop, M.-L.; Evard, H.; Herodes, K.; Ravio, P.; Leito, I. Tutorial review on validation of liquid chromatography–mass spectrometry methods: Part II. *Anal. Chim. Acta* **2015**, *870*, 8–28. [CrossRef]

27. Oliveira, J.M.; Faria, M.; Sá, F.; Barros, F.; Araújo, I.M. C6-alcohols as varietal markers for assessment of wine origin. *Anal. Chim. Acta* **2006**, *563*, 300–309. [CrossRef]

28. Araujo, P. Key aspects of analytical method validation and linearity evaluation. *J. Chromatogr. B* **2009**, *877*, 2224–2234. [CrossRef]

29. Magnusson, B.; Örnemark, U. *Eurachem Guide: The Fitness for Purpose of Analytical Methods—A Laboratory Guide to Method Validation and Related Topics*, 2nd ed.; Eurachem: Torino, Italy, 2014; Available online: http://eurachem.org (accessed on 2 January 2020).

30. Brettell, T.A.; Lester, R.E. Validation and QA/QC of Gas Chromatographic Methods. *Mod. Pract. Gas Chromatogr.* **2004**, 969–988.

31. Gonzalez, O.; Blanco, M.E.; Iriarte, G.; Bartolomé, L.; Maguregui, M.I.; Alonso, R.M.; Rojas, R.M.A. Bioanalytical chromatographic method validation according to current regulations, with a special focus on the non-well defined parameters limit of quantification, robustness and matrix effect. *J. Chromatogr. A* **2014**, *1353*, 10–27. [CrossRef]

32. Huber, L. Validation of analytical methods: review and strategy. *LC-CG Int.* **1998**, *11*, 96–105.

33. AOAC International AOAC Official Methods of Analysis—Appendix K: Guidelines for Dietary Supplements and Botanicals. Available online: http://www.eoma.aoac.org/app_k.pdf (accessed on 2 January 2020).

**Sample Availability:** Samples of the compounds are not available from the authors.

*Article*

# Unveiling the Molecular Basis of Mascarpone Cheese Aroma: VOCs analysis by SPME-GC/MS and PTR-ToF-MS

Vittorio Capozzi [1], Valentina Lonzarich [2], Iuliia Khomenko [3], Luca Cappellin [4], Luciano Navarini [2,*] and Franco Biasioli [3]

[1] Institute of Sciences of Food Production, National Research Council (CNR), URT c/o CS-DAT, Via Michele Protano, 71121 Foggia, Italy; vittorio.capozzi@ispa.cnr.it
[2] Aromalab, illycaffè s.p.a., Area di Ricerca, Padriciano 99, 34149 Trieste, Italy; valentina.lonzarich@illy.com
[3] Department of Food Quality and Nutrition, Research and Innovation Centre, Fondazione Edmund Mach (FEM), via E. Mach 1, 38010 San Michele all'Adige, Italy; iuliia.khomenko@fmach.it (I.K.); franco.biasioli@fmach.it (F.B.)
[4] Department of Chemical Sciences, University of Padua, Via F. Marzolo 1, 35131 Padova, Italy; luca.cappellin@unipd.it
* Correspondence: luciano.navarini@illy.com

Academic Editor: Eugenio Aprea
Received: 31 January 2020; Accepted: 2 March 2020; Published: 10 March 2020

**Abstract:** Mascarpone, a soft-spread cheese, is an unripened dairy product manufactured by the thermal-acidic coagulation of milk cream. Due to the mild flavor and creamy consistency, it is a base ingredient in industrial, culinary, and homemade preparations (e.g., it is a key constituent of a widely appreciated Italian dessert 'Tiramisù'). Probably due to this relevance as an ingredient rather than as directly consumed foodstuff, mascarpone has not been often the subject of detailed studies. To the best of our knowledge, no investigation has been carried out on the volatile compounds contributing to the mascarpone cheese aroma profile. In this study, we analyzed the Volatile Organic Compounds (VOCs) in the headspace of different commercial mascarpone cheeses by two different techniques: Headspace-Solid Phase Microextraction-Gas Chromatography-Mass Spectrometry (HS-SPME GC-MS) and Proton-Transfer Reaction-Mass Spectrometry coupled to a Time of Flight mass analyzer (PTR-ToF-MS). We coupled these two approaches due to the complementarity of the analytical potential—efficient separation and identification of the analytes on the one side (HS-SPME GC-MS), and effective, fast quantitative analysis without any sample preparation on the other (PTR-ToF-MS). A total of 27 VOCs belonging to different chemical classes (9 ketones, 5 alcohols, 4 organic acids, 3 hydrocarbons, 2 furans, 1 ester, 1 lactone, 1 aldehyde, and 1 oxime) have been identified by HS-SPME GC-MS, while PTR-ToF-MS allowed a rapid snapshot of volatile diversity confirming the aptitude to rapid noninvasive quality control and the potential in commercial sample differentiation. Ketones (2-heptanone and 2-pentanone, in particular) are the most abundant compounds in mascarpone headspace, followed by 2-propanone, 2-nonanone, 2-butanone, 1-pentanol, 2-ethyl-1-hexanol, furfural and 2-furanmethanol. The study also provides preliminary information on the differentiation of the aroma of different brands and product types.

**Keywords:** mascarpone cheese; dairy product; VOCs; PTR-ToF-MS; HS-SPME GC-MS; aroma; ketones; alcohols; Tiramisù; milk cream

---

## 1. Introduction

Mascarpone cheese is a soft-spread dairy unripened product manufactured by the thermal-acidic coagulation of milk cream [1]. Mascarpone represents an interesting cheese processing method, in which direct acidification is applied. The raw materials for its production are milk cream (containing 80% dry weight lipids, 2.8% to 6% protein) and acidifying substances (single or mixed), such as acetic, citric, tartaric, or lactic acids, vinegar or lemon juice, with a final pH ranging from 5.7 to 6.6 [2]. The cream is heated up to 85–95 °C and, while stirring, acid is added in order to force matrix coagulation [3,4]. During the intensive heating, the whey protein denatures and aggregates or sticks to the casein micelles and the fat globule membrane [3]. As a result of this reaction, whey proteins partly remain in the cheese matrix during the draining step (about 20 h), obtaining the typical texture and flavor of mascarpone cheese [3]. This typical Italian cheese was once produced domestically by the farmers of some northern regions and consumed immediately after production. Due to its traditional importance, mascarpone is included in the list of traditional agro-food products (*Prodotto Agroalimentare Tradizionale*) [5], a list of Italian traditional regional food products. More recently, it has been industrially produced to satisfy the increasing demand driven by two main sensory characteristics—the mild flavor and the creamy consistency. In fact, due to these attributes, mascarpone cheese is a base ingredient in industrial, culinary, and homemade preparations. The best example is its use in the preparation of one of the most widely appreciated Italian desserts—the Tiramisù. In spite of its popularity and its increasing economic relevance, the scientific literature does not report a characterization of Volatile Organic Compounds (VOCs) released by this peculiar dairy matrix. In order to characterize for the first time the VOCs associated with the headspace of mascarpone cheese, among various analytical techniques, we exploit the complementarity of Gas Chromatography-Mass Spectrometry (GC-MS) and Proton Transfer Reaction-Mass Spectrometry coupled to a Time of Flight mass analyzer (PTR-ToF-MS) [6]. Gas Chromatography-Mass Spectrometry (GC-MS) is the reference method in the analysis of VOCs in the field of environmental, food, flavour and fragrance, medical and forensic sciences [7]. Solid-Phase Microextraction (SPME) combined with static headspaces (HS-SPME), in particular, offers relatively high-throughput performance and does not require extended sample preparation [8]. Moreover, it is reproducible, simple, and effective, and eliminates interference compounds from the sample matrix with improvement in the selectivity of the analysis. PTR-ToF-MS uses proton transfer to induce chemical ionization of the sample headspace directly introduced into a drift tube, where volatile organic compounds can react with $H_3O^+$ ions formed in a hollow cathode ion source. The protonated particles are analyzed according to their mass/charge ratio (*m/z*) using a quadrupole or Time-of-Flight (ToF) mass analyzer and eventually detected as ion counts/second (cps) by a secondary electron multiplier or multichannel plates [9]. The outcome is a rapid (< 1 s) mass resolved fingerprint of the total volatile profile of the sample, measuring most VOCs at ultralow concentrations (a few pptv) and high mass resolution [10]. These analytical approaches are complementary. In fact, PTR-MS provides analytical information that is mostly limited to concentration and m/z ratios, i.e., sum formula, while isobar separation and compound identification needs usually the support of GC analysis [6]. PTR-MS, however, guarantees rapid and direct analysis and high sensitivity [11].

Using this integrated approach, the present study represents a first step towards the comprehension of the molecular basis of sensory perceptions associated with the consumption of mascarpone cheese and, more relevantly, of products that use mascarpone as raw material. Furthermore, within the panel of tested samples, we preliminary explored variables such as different manufacturers and delactosed mascarpone productions.

## 2. Results and Discussion

### 2.1. HS-SPME GC-MS Results

Solid-phase microextraction (SPME) is a very popular analytical extraction technique used before GC-MS headspace (HS) analysis thanks to its ease-to-use, the possibility of automation, and good sensitivity. SPME utilizes a short, thin, solid rod of fused silica coated with an absorbent/adsorbent polymer. The coated fused silica (the SPME fiber) is attached to a metal rod, and both are protected by a metal sheath that covers the fiber when not in use. SPME is particularly well suited to the analysis of dairy products being capable of extracting a broader range of analytes than most other sample preparation methods [12]. Moreover, thanks to the relatively low temperatures and short times at which headspace SPME extraction is performed, the risks to induce thermal artifacts are extremely low if compared with other techniques such as simultaneous distillation-extraction (SDE) [13].

Methods developed for the analysis of organic compounds from aqueous samples by SPME coupled to GC have been used to analyze VOCs in fresh and ripened dairy productions [8,14]. A wide range of fibers with varying affinities for specific classes of volatile organic compounds is available. After a preliminary screening of seven different types of SPME fibers (100 μm PDMS (polydimethylsiloxane), 60 μm PEG (Carbowax-Polyethylene Glycol), 85 μm PA (Polyacrylate), 75 μm CAR/PDMS (Carboxen/Polydimethylsiloxane), 85 μm CAR/PDMS, 50/30 μm DVB/CAR/PDMS (Divinylbenzene/Carboxen/Polydimethylsiloxane), and 65 μm PDMS/DVB (Polydimethylsiloxane/Divinylbenzene)) 75 μm CAR/PDMS was chosen, because it provides the higher number of extracted volatiles. This fiber has been suggested to work particularly well for the analysis of volatiles in dairy products [12]. Extraction conditions have also been preliminarily explored by checking the effect of different times (10 min up to 4 h) and temperature (40 and 60 °C). A good compromise between the multiplicity of extracted volatiles and peak intensity was found by headspace exposing the fiber for 60 min at 60 °C and this experimental condition was selected for the present characterization.

A total of 27 compounds belonging to different chemical classes (nine ketones, five alcohols, four acids, three hydrocarbons, two furans, one ester, one lactone, one aldehyde, and one oxime) have been identified. Ketones, which might induce fruity and floral sensory notes, are common constituents of most dairy products [15–17] and by far the most important class of compounds contributing to the mascarpone cheese aroma. In particular, 5 different ketones (2-heptanone > 2-pentanone > 2-propanone > 2-nonanone ≈ 2-butanone) represent almost 75–80% of the sample headspace. The compounds 2-heptanone and 2-pentanone characterized by odor descriptors including sweet, fruity, orange peel, and herbaceous [15] are the dominating volatile organic compounds in all samples. Several alcohols have been detected, but differently from ketones, are not present in all samples—1-pentanol and 2-ethyl-1-hexanol, both common primary alcohols detected in dairy products [15], are ubiquitous, and ethanol and 1,2 propandiol have been detected only in one sample (Manufacturer B), suggesting a possible technological origin. Other minor compounds, including short- and moderate-chain even-numbered fatty acids ($C_4$–$C_{12}$), ethyl acetate, δ-hexalactone, toluene, benzaldehyde and methoxyphenyl oxime have been already found in cheese products [16,18,19]. The two hydrocarbons 2,4-dimethylheptene and 2,2,4,6,6-pentamethylheptane have been detected only in one sample (Manufacturer A) and the latter has been detected in the volatile fraction of butter [20]. Furfural and 2-furanmethanol, identified in all mascarpone cheese samples, have been found to contribute to the nutty and roasted aroma of Parmigiano-Reggiano cheese [21].

In order to provide a general overview of volatile composition of three different samples (M1-M3) of Mascarpone cheese analyzed by HS-SPME GC-MS, we performed multivariate data analysis using Principal Component Analysis (PCA), reporting the graphical result in Figure 1.

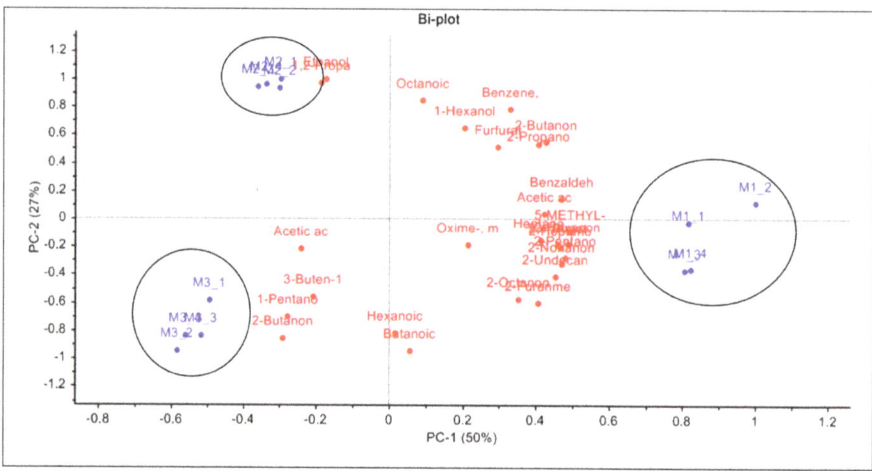

**Figure 1.** Principal Component Analysis (PCA) biplot of the 3 different commercial samples of mascarpone (M1, M2, and M3). For each sample, the mean (n = 4) is represented by the sample name. Score plot was given by the Volatile Organic Compound (VOC) content for each sample and loading plot of the single volatile organic compounds. The codes correspond to the samples indicated in Table 1.

In the figure, it is possible to observe the samples (scores) and variables (loadings) plots related to the first two principal components, which (cumulated) explain the 77% of the total variance (PC1, 50.0%; PC2, 27.0%) associated with the data set. A clear separation of mascarpone cheese M1 from the other samples is observable along the PC1. PC2 explains the parting between M2 and M3 mascarpone samples. The replicates belong to the same commercial mascarpone batch that is well-clustered together, while is possible to highlight a clear separation of the three samples on the biplot. Observing the loadings (i.e., the involvement of the single volatiles), it is possible to have an idea of the different influence of the diverse volatiles in justifying variance observed trends.

**Table 1.** List of commercial 'Mascarpone' samples analyzed in the present study. All samples (M1–M12) were investigated by PTR-ToF-MS analysis. Underlined samples (M1–M3) were evaluated also by HS-SPME GC-MS.

| Sample | Claimed Characteristics | Manufacturer |
|---|---|---|
| **M1** | Mascarpone | A |
| **M2** | Mascarpone | B |
| **M3** | Mascarpone | C |
| **M4** | Mascarpone | B |
| **M5** | Mascarpone | C |
| **M6** | Mascarpone | C |
| **M7** | Mascarpone | M |
| **M8** | Mascarpone | M |
| **M9** | Mascarpone without lactose | M |
| **M10** | Mascarpone without lactose | M |
| **M11** | Mascarpone without lactose | C |
| **M12** | Mascarpone without lactose | C |

## 2.2. PTR-ToF-MS Results and Comparison with HS-SPME GC-MS Findings

As other Direct-Injection Mass Spectrometric (DIMS) technologies, PTR-MS finds application in many sectors, from environmental sciences to food chemistry, and from biological studies to medical applications. With this regard, we recently described a tailored system, that found application in this study, achieved connecting PTR-ToF-MS with an automated sampler, and associated custom-made data analysis applications that improve the versatility of the analytical approach in the determination of VOCs in association with i) huge numbers of samples, ii) bioprocesses monitoring, and iii) high numbers of variables to be considered [11].

PTR-MS has been already exploited to study VOCs associated to dairy products such as mozzarella cheese [22], Grana Padano, Parmigiano Reggiano, and Grana Trentino cheeses [23], liquid whey [24], butter and butter oils (by means of quadrupole-based PTR-MS analyses, sensory analyses and classical chemical analyses) [25], milk and whey powders [26], anhydrous milk fat [26], and fermented milk-based beverages (yogurt and kefir) [27,28].

All samples included in this study have been analyzed by PTR-ToF-MS. A total of 411 mass peaks were detected and extracted. Upon comparison with the blanks, 92 peaks were kept that are significantly different between various manufacturers ($p < 0.01$ with Bonferroni correction) and tentatively identified on the basis of exact mass, isotopic ration, and literature [29]. PTR-MS allowed the detection and characterization of a larger number of VOCs/VOC fragments, which was larger than the number of volatiles identified by GC. For the PTR analysis, all vials were incubated alternatively at 40 °C or at 60 °C (data not shown) for 30 min before PTR-MS analysis. The last one was the temperature at which good results were obtained by HS-SPME GC-MS. However, with PTR, even at 40 °C, the analysis was successfully performed and results were reliable. For this reason, we report the data performed at 40 °C, a temperature closer to the real mascarpone cheese testing conditions. One-way ANOVA followed by Tukey HSD test was carried out to compare and underline significant differences among the assessed mascarpone samples. For each peak, we obtained the concentration of the corresponding VOC ion in the headspace of all explored samples. Boxplots reported in Figure 2 illustrate the observed trends for 6 ions among the tested samples, as illustrative cases. In detail, the figure proposes the behaviors corresponding to the peaks at *m/z* 73.065 (tentatively identified as 2-butanone), *m/z* 75.044 (tentatively identified as propionic acid), m/z 83.086 (tentatively identified as hexanol fragment), *m/z* 87.080 (tentatively identified as 2-pentanone/isoprenol), *m/z* 98.105, and *m/z* 101.096 (tentatively identified as 2-hexanone). The intensity corresponding to the mass peak *m/z* 73.065 reaches the highest values in the delactosed samples produced by Manufacturer C, while the standard productions belonging to the same manufacturer registered the lowest values (as all mascarpone batches of Producer M) (Figure 2a). Samples from Manufacturers A and B present intermediate intensities for this peak (Figure 2a). In accordance with these results, 2-butanone was found to be variable in different types of whey [30]. In only the M2 batch did we detected a relevant intensity for the mass peak *m/z* 75.044 (Figure 2b), tentatively identified as propionic acid, a compound that can be responsible for a dairy taste/odor with a pronounced fruity lift [31].

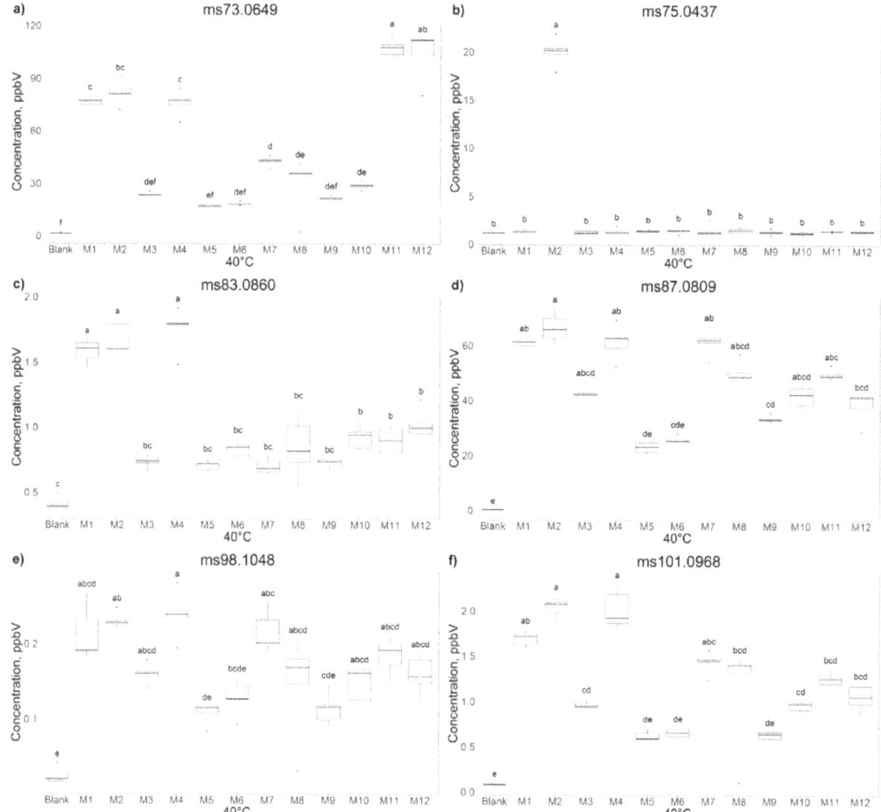

**Figure 2.** The boxplots indicated by letters (**a–f**) represent selected volatiles found in association with the different commercial mascarpone samples such as m/z 73.0649—$C_4H_9O^+$—t.i. 2-Butanone, 75.0437—$C_3H_7O_2^+$—t.i. Methyl acetate, 83.0860—$C_6H_{11}^+$—t.i. fragment of Hexanal/Hexenol, 87.0809—$C_5H_{11}O^+$—t.i. 2-Pentanone/3-Buten-1-ol, 3-methyl-, 98.1048—isotope of $C_7H_{13}^+$—t.i. Heptanal, 101.0968—$C_6H_{13}O^+$—t.i. 2-Hexanone. Different letters indicate a significant difference between different samples ($p < 0.05$, one-way ANOVA, Tukey HSD).

The mass peak *m/z* 83.086 has been found with pronounced intensities in the samples produced by the Manufacturers A and B (Figure 2c). Hexanal was included among the high-content compounds identified in samples belonging to dairy products [32] and described as having a fatty, green, grassy, powerful, penetrating characteristic fruity odor and taste [31]. A similar trend can be underlined for the intensities of mass peak *m/z* 101.096 (Figure 2f). Finally, a considerable variability can be highlighted for the intensities corresponding to the mass peaks m/z 87.080 and 98.105 (Figure 2d,e).

Other than this kind of punctual analysis, PTR analysis offers also the opportunity to depict a global analysis of molecular fingerprinting associated with the headspaces of the different samples. Considering that the present work deals with an integrated analytical approach, we propose a PTR data set selected in light of the comparison with GC data. In fact, we defined a new subset of the PTR-ToF-MS data including only the mass peaks that were found also using the HS-SPME GC-MS technique. As a result, we have a new matrix (Table 2) of twenty peaks corresponding to the masses of protonated molecular ions of compounds such as acetic acid (sour pungent, cider vinegar, slightly malty with a brown nuance; naturally occurring in various dairy products, it has a role in butter and cheese flavors), acetoin (acidic, sour, cheesy, dairy, creamy with a fruity nuance; normally occurs in butter, milk, and cheeses), acetone (characteristic aromatic odor, pungent, somewhat sweet taste; naturally occurring in fermented dairy products), ethanol (slight, characteristic odor and a burning taste; naturally occurring in blue cheese, cheddar cheese, Swiss cheese), furfural (characteristic penetrating odor typical of cyclic aldehydes; naturally occurring in cheeses), hexanoic acid (sickening, sweaty, rancid, sour, sharp, pungent, cheesy, fatty, unpleasant odor reminiscent of copra oil; naturally occurring in cheeses, butter, milk), and octanoic acid (mildly unpleasant odor and a burning, rancid taste, also reported as having a faint, fruity-acid odor and slightly sour taste; natural component of butter fat, occurring in cheeses) [31,33].

**Table 2.** Volatile compounds detected by both Proton Transfer Reaction-Mass Spectrometry coupled to a Time of Flight mass analyzer (PTR-ToF-MS) and SPME/GC-MS in association with mascarpone samples.

| Compound | Chemical Class | Protonated Ion | |
|---|---|---|---|
| | | *m/z* | Sum Formula |
| Ethanol | Alcohols | 47.049 | $C_2H_7O^+$ |
| 2-Propanone | Ketones | 59.049 | $C_3H_7O^+$ |
| Acetic acid | Organic acids | 61.028 | $C_2H_5O_2^+$ |
| 2-Butanone | Ketones | 73.065 | $C_4H_9O^+$ |
| 1,2-Propanediol = Propylene glycol | Alcohols | 77.060 | $C_3H_9O_2^+$ |
| 2-Pentanone/3-Buten-1-ol, 3-methyl- | Ketones/Alcohols | 87.080 | $C_5H_{11}O^+$ |
| 2-Butanone, 3-hydroxy- (B) / Butanoic acid/Acetic acid ethyl ester | Ketones/Organic acids/Esters | 89.060 | $C_4H_9O_2^+$ |
| Toluene | Hydrocarbons | 93.070 | $C_7H_9^+$ |
| Furfural | Furans | 97.028 | $C_5H_5O_2^+$ |
| 2-Hexanone | Ketones | 101.096 | $C_6H_{13}O^+$ |
| Benzaldehyde | Aldehyde | 107.049 | $C_7H_7O^+$ |
| 5-Methyl-delta-valerolactone | Lactones | 115.075 | $C_6H_{11}O_2^+$ |
| 2-Heptanone | Ketones | 115.112 | $C_7H_{15}O^+$ |
| Hexanoic acid | Organic acids | 117.091 | $C_6H_{13}O_2^+$ |
| 2,4-Dimethyl-1-heptene | Hydrocarbons | 127.148 | $C_9H_{19}^+$ |
| 2-Octanone | Ketones | 129.127 | $C_8H_{17}O^+$ |
| 1-Hexanol, 2-ethyl- | Alcohols | 131.143 | $C_8H_{19}O^+$ |
| 2-Nonanone | Ketones | 143.143 | $C_9H_{19}O^+$ |
| Octanoic acid | Organic acids | 145.122 | $C_8H_{17}O_2^+$ |
| Oxime-, methoxy-phenyl- | Oxime | 152.071 | $C_8H_{10}NO_2^+$ |
| 2-Undecanone | Ketones | 171.174 | $C_{11}H_{23}O^+$ |
| Heptane, 2,2,4,6,6-pentamethyl | Hydrocarbons | 171.211 | $C_{12}H_{27}^+$ |

Statistical tests were performed on the new matrix in an attempt at understanding the impact of these VOCs on the characterization of the different mascarpone cheese samples. The results obtained for the twelve experimental modes were visualized by means of principal component analysis (PCA), with each point representing a distinct sample (Figure 3), maximizing explained variability in two dimensions.

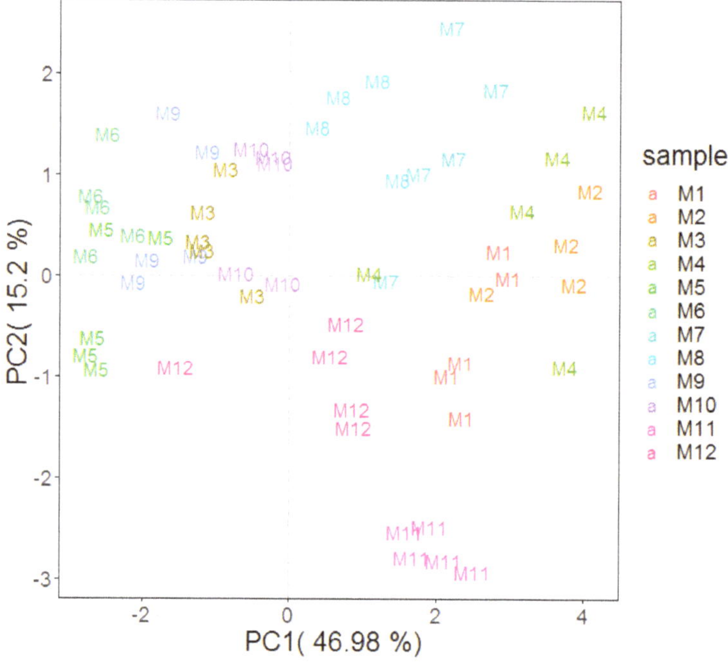

**Figure 3.** Analysis of mascarpone VOCs profile assessed by PTR-ToF-MS. Plot depicts the VOC profile distribution of the twelve Mascarpone over the PCA score plot defined by the first two principal components. The codes correspond to the samples indicated in Table 1.

Separation among Mascarpone samples according to the first two components accounted for about 62% of the total variance. It is possible to highlight how the replicates belonging to the same sample generally clustered together. In addition, a good separation among the different samples is also depicted. Considering all variables connoting the panel of different mascarpone cheese analyzed, it is mandatory to underline that the studied diversity in terms of different producers and classic versus delactosed was not selected in order to delve into the effect of these parameters. In fact, it was just a heterogeneous panel selected in order to provide a broad description of the overall VOCs associated with this traditional dairy production. However, it is possible to foresee some preliminary differences, such as clear groups among mascarpone cheese samples belonging to the same manufacturer and a general (more or less pronounced) separation between classic and delactosed samples within the same producer (Figure 3). These pieces of evidence suggest the need for further studies with tailored sampling in order to test the potential of a PTR-based approach as a discriminatory tool to monitor these variables. Considering the sensory changes among mascarpone cheese samples, our study confirmed the presence of a diversification comparing different batches and different producers already described in terms of spreadability [34]. In fact, Cattaneo et al. [34], studying eighteen batches from six different manufacturers, noticed differences in four viscometric parameters they selected to assess changes of rheological aptitude of mascarpone cheeses. This sensory variability calls attention to the need for versatile tools for the industrial quality control also in the case of mascarpone cheese, a topic of generally significant interest in the food industry [35,36].

In Table 3, it is possible to delve into the results for a more representative number of mass peaks, underlining significant differences among concentrations reported for 22 protonated ions out of the 92 selected after comparison with the blanks. From this analysis, it is possible to notice how the trends for selected mass peak intensities follow a certain producer-dependent behavior. It is also clear how the probabilities to find selected mass peaks associated to given experimental variables considerably increase using the PTR-based technique, due to the potential of an untargeted approach. The opportunity to have a wide (untargeted analysis) and fast (rapid time of analysis without any sample preparation/extraction/destruction) view of the VOCs associated with mascarpone headspaces confirmed the aptitude of this analytical approach to allow rapid noninvasive quality control for the food industry (e.g., [37,38]), already explored in the dairy industry but on other matrices (e.g., [25,26]). An approach that can i) simplify the selection of mascarpone as an ingredient in the food industry and ii) boost the quality improvements in the production of this fresh cheese.

**Table 3.** Organic compounds associated to mascarpone headspace detected by PTR-ToF-MS. Black color indicates compounds identified also by SPME/GC-MS. For each compound, different letters indicate a significant difference between different samples according to ANOVA and Tukey HSD ($p < 0.05$). The codes correspond to the samples indicated in Table 1. In the parenthesis, the different producers.

| MM | TM | SF | M1 (A) | M2 (B) | M3 (C) | M4 (B) | M5 (C) | *p*-Value |
|---|---|---|---|---|---|---|---|---|
| 41.039 | 41.039 | $C_3H_5^+$ | 21.2 ± 0.9 [b] | 27 ± 3 [c] | 11.4 ± 0.6 [a] | 21 ± 1 [b] | 10 ± 1 [a] | $1 \times 10^{-13}$ |
| 43.018 | 43.018 | $C_2H_3O^+$ | 30.0 ± 0.7 [b] | 44 ± 3 [c] | 24 ± 2 [a] | 33 ± 2 [b] | 24 ± 5 [a] | $3 \times 10^{-9}$ |
| 43.054 | 43.054 | $C_3H_7^+$ | 11.5 ± 0.6 [c] | 16 ± 1 [d] | 3.5 ± 0.5 [a] | 6.7 ± 0.4 [b] | 3.1 ± 0.4 [a] | $4 \times 10^{-17}$ |
| 45.033 | 45.033 | $C_2H_5O^+$ | 113 ± 6 [b] | 175 ± 12 [c] | 81 ± 3 [a] | 114 ± 10 [b] | 88 ± 23 [a] | $2 \times 10^{-9}$ |
| 47.049 | 47.049 | $C_2H_7O^+$ | 8 ± 3 [a] | 52 ± 39 [c] | 10 ± 8 [a] | 16 ± 1 [ab] | 44 ± 5 [bc] | $2 \times 10^{-3}$ |
| 55.054 | 55.054 | $C_4H_7^+$ | 13.3 ± 0.4 [c] | 14.9 ± 0.8 [d] | 8.1 ± 0.3 [b] | 15 ± 1 [d] | 6.0 ± 0.4 [a] | $7 \times 10^{-16}$ |
| 57.070 | 57.070 | $C_4H_9^+$ | 6.0 ± 0.1 [c] | 5.6 ± 0.6 [c] | 3.9 ± 0.2 [b] | 12 ± 1 [d] | 2.7 ± 0.1 [a] | $7 \times 10^{-17}$ |
| 59.049 | 59.049 | $C_3H_7O^+$ | 1062 ± 35 [c] | 976 ± 72 [c] | 563 ± 29 [b] | 1230 ± 116 [d] | 355 ± 26 [a] | $9 \times 10^{-15}$ |
| 61.029 | 61.028 | $C_2H_5O_2^+$ | 10 ± 3 [a] | 26 ± 6 [b] | 18 ± 5 [ab] | 11 ± 2 [a] | 24 ± 10 [b] | $4 \times 10^{-4}$ |
| 63.026 | 63.026 | $C_2H_7S^+$ | 15.1 ± 0.3 [c] | 16 ± 1 [c] | 7.2 ± 0.4 [b] | 20 ± 2 [d] | 3.3 ± 0.5 [a] | $4 \times 10^{-16}$ |
| 69.070 | 69.07 | $C_5H_9^+$ | 6.4 ± 0.2 [a] | 8.3 ± 0.6 [b] | 6.0 ± 0.4 [a] | 9.0 ± 0.8 [b] | 6.2 ± 0.6 [a] | $2 \times 10^{-8}$ |
| 71.086 | 71.086 | $C_5H_{11}^+$ | 1.2 ± 0.1 [ab] | 1.7 ± 0.8 [b] | 0.7 ± 0.1 [a] | 1.3 ± 0.1 [ab] | 0.72 ± 0.05 [a] | $1 \times 10^{-3}$ |
| 73.065 | 73.065 | $C_4H_9O^+$ | 77 ± 2 [b] | 82 ± 7 [b] | 24 ± 1 [a] | 76 ± 7 [b] | 17.2 ± 0.9 [a] | $1 \times 10^{-16}$ |
| 75.044 | 75.044 | $C_3H_7O_2^+$ | 1.4 ± 0.2 [a] | 20 ± 1 [b] | 1.2 ± 0.2 [a] | 1.4 ± 0.3 [a] | 1.4 ± 0.2 [a] | $1 \times 10^{-21}$ |
| 83.086 | 83.086 | $C_6H_{11}^+$ | 1.6 ± 0.1 [b] | 1.7 ± 0.1 [b] | 0.7 ± 0.0 [a] | 1.8 ± 0.2 [b] | 0.71 ± 0.04 [a] | $4 \times 10^{-14}$ |
| 87.044 | 87.044 | $C_4H_7O_2^+$ | 3.7 ± 0.5 [bc] | 3.8 ± 0.9 [bc] | 3.0 ± 0.4 [ab] | 4.3 ± 0.4 [c] | 2.0 ± 0.3 [a] | $2 \times 10^{-5}$ |
| 87.081 | 87.08 | $C_5H_{11}O^+$ | 61 ± 2 [c] | 66 ± 5 [c] | 43 ± 2 [b] | 61±6 [c] | 23 ± 2 [a] | $6 \times 10^{-13}$ |
| 89.060 | 89.06 | $C_4H_9O_2^+$ | 2.2 ± 0.6 [a] | 5.2 ± 0.6 [c] | 2.9 ± 0.3 [ab] | 2±1 [ab] | 3.5 ± 0.3 [b] | $2 \times 10^{-6}$ |
| 97.102 | 97.101 | $C_7H_{13}^+$ | 2.3 ± 0.1 [c] | 2.7 ± 0.2 [d] | 1.8 ± 0.0 [b] | 2.6±0.1 [d] | 1.1 ± 0.1 [a] | $4 \times 10^{-14}$ |
| 101.097 | 101.096 | $C_7H_7O^+$ | 1.7 ± 0.1 [c] | 2.0 ± 0.1 [d] | 1.0 ± 0.0 [b] | 2.0±0.2 [d] | 0.6 ± 0.1 [a] | $5 \times 10^{-14}$ |
| 115.113 | 115.112 | $C_7H_{15}O^+$ | 28 ± 1 [c] | 30 ± 2 [c] | 20.5 ± 0.6 [b] | 30±2 [c] | 12 ± 1 [a] | $7 \times 10^{-14}$ |
| 143.145 | 143.143 | $C_9H_{19}O^+$ | 2.1 ± 0.1 [c] | 2.5 ± 0.2 [d] | 1.7 ± 0.1 [b] | 2.4±0.2 [d] | 1.1 ± 0.1 [a] | $2 \times 10^{-12}$ |

This panel of 22 peaks includes only 9 masses detected also by the GC analysis, thus providing a broader overview of the diversity among samples associated with VOCs content. Comparing these findings with a recent PTR headspace analysis of other dairy product of industrial interest (milk powder, whey powder and anhydrous milk fat), the mass peaks 47.049, 63.026, 73.065, 87.081, 89.060, 101.097, 115.113, 143.145 seem to be peculiar of mascarpone headspace [26], indicating a potential role of the corresponding volatiles in shaping perceptions associated to Mascarpone consumption. Additionally, on the other hand, we found variable trends in mass peaks already detected in association with the headspaces of skim milk powder (43.018, 61.029, 87.044, 97.102), whole milk powder (41.039, 43.018, 45.033, 55.054, 61.029, 71.086, 75.044, 83.086, 87.044), whey powder (43.018, 59.049, 61.029, 75.044), and anhydrous milk fat (43.018, 43.054, 57.070, 69.070) [26]. This partial and specific overlapping, in terms of volatiles content, with the headspaces of other dairy ingredients/products, can be probably of help in the understanding of the unique sensory properties of mascarpone matrix.

Finally, in order to provide more complete information about the preliminary potential that arises from the PTR data in terms of separation of delactosed products, we propose two PCA representations, analyzing samples with or without lactose for the Manufacturers C and M, respectively (Figure 4).

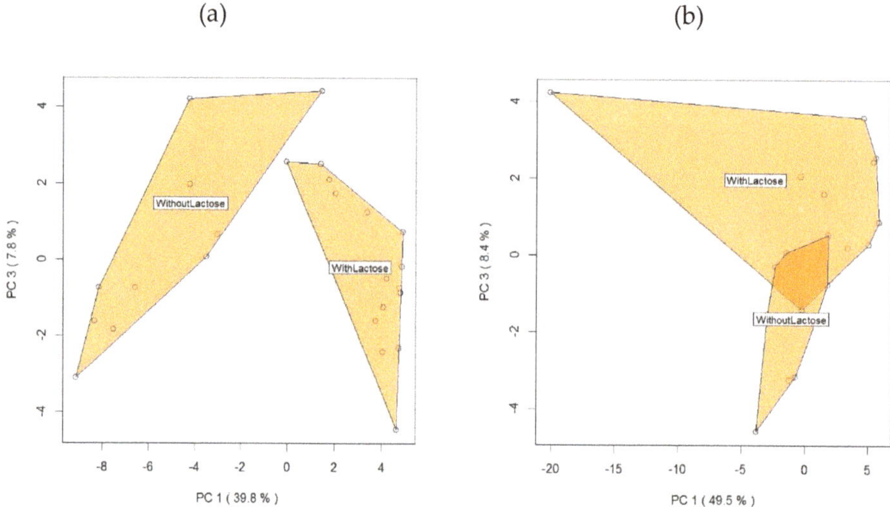



**Figure 4.** Analysis of mascarpone VOCs profile assessed by PTR-ToF-MS for the Manufacturers C (**a**) and M (**b**) plotted by the first and the third principal components. The labels and the selected areas indicate the separation between samples with or without lactose.

Figure 4a (Manufacturer C) and 4b (Manufacturer M) show that mascarpone samples classic and delactosed in this subset (different producers) are separated along the first and third PC (explaining 47.6% and 57.9% of the total variance, respectively). Even if preliminary, these results confirm the potential of PTR-TOF-MS analysis for the quality evaluation of lactose-free dairy products. In fact, recently, this analytical approach found application to monitor VOC variability in ultrahigh temperature lactose-free milk samples (assessing the impact of storage time and the of the use of different lactase preparations) [39]

## 3. Materials and Methods

### 3.1. Sample Selection and Preparation

A total of 12 different mascarpone batches were studied in this project that are listed in Table 1. The corresponding chemicophysical characteristics are reported in Table S1.

We obtained the samples from different local markets and stored them at 4 °C. The samples represent different manufacturers, all analyzed within the expiration date, and both plain and delactosed Mascarpone.

### 3.2. HS-SPME GC-MS Measurements

Aliquots of 8 mL of sample were placed in a 20 mL vials that were immediately sealed with a silicone rubber Teflon cap and crimped with aluminium seal. Then samples were heated at 60 °C and kept at the same temperature for 30 min while a polydimethylsiloxane/divinylbenzene SPME fibre (Supelco, Bellefonte, PA, USA) was exposed to the headspace over the surface of each sample in order to collect the compounds in the vapour phase. The exposure time was optimized in preliminary experimental trials. The SPME coating containing the headspace volatile compounds was inserted into the GC injection port and then thermally desorbed at 250 °C for 10 min in a 6890 GC (Agilent Technologies, Santa Clara, CA, United States). Compounds were eluted by a He gas flow of 1,4 mL/min in split mode (split 1:4) and separated using a 60 m Varian FactorFour WAXms capillary column (film thickness 0.25 mm, 0.25 mm internal diameter) (Varian, Middelburg, The Netherlands). The oven temperature, initially set to 35 °C, was increased to 210 °C at 4 °C/min, then to 240 °C at a rate of 20 °C/min, and then this final temperature was held for 5 min. The mass spectrometer was set to electron ionization mode (MS-EI) generated at 70 eV, and mass spectra were collected in full scan mode, collecting ions from 39 to 250 *m/z*. The volatile compounds studied were identified by comparing their mass spectra and their retention times to those of reference standards analyzed at the same conditions and by comparison with spectra recorded in the Wiley 6 N mass spectral library (Wiley, Hoboken, NJ, USA) and, when needed, to literature references. Due to the lengthy HS-SPME GC/MS analysis, only four samples have been analysed by this method. For each sample, four replicates were analyzed.

### 3.3. PTR-ToF-MS Measurements

A commercial PTR-ToF-MS 8000 instrument (Ionicon Analytik GmbH, Innsbruck, Austria) was used for the headspace measurements. The instrumental conditions in the drift tube were as following—drift voltage 550 V, drift temperature 110 °C, drift pressure 2.30 mbar affording an E/N value of 140 Townsend (1 Td = $10^{-17}$ V.cm$^2$). Sampling was performed with a flow rate of 40 sccm. The mass resolution (m/$\Delta$m) was at least 3800. Measurements were performed in an automated way by using a multipurpose GC automatic sampler (Gerstel GmbH, Mulheim am Ruhr, Germany) as previously described [11]. The measurement order, both samples and replicates, was randomized to avoid memory effects. All vials were incubated at 40 °C for 30 min before PTR-MS analysis. Each sample was measured for 30 s, at an acquisition rate of 1 spectrum per second with an overall throughput of one sample every 5 min. The experiment was repeated at 60 °C, the temperature at which HS-SPME GC-MS provided better results. The entire experiment was repeated three times and empty vials, containing lab air, were measured together with the sample set and considered as "blanks". Data processing of PTR-ToF-MS spectra included dead time correction, external calibration and peak extraction steps performed according to a procedure described elsewhere [40]. The baseline of the mass spectra was removed after averaging the whole measurement and peak detection and peak area extraction was performed by using modified Gaussian to fit the data [41]. To determine the concentrations of volatile compounds in ppbv (part per billion by volume) the formulas described by Lindinger et al. were used by assuming a constant reaction rate coefficient ($k_R$=2 × $10^{-9}$ cm$^3$/s) for $H_3O^+$ as primary ion [42].

### 3.4. Statistical Analyses

Data exploration was based on Principal Component Analysis (PCA) of centered and scaled data. Analysis of variance (ANOVA) with Bonferroni correction was performed for selection of mass peaks in the sample headspace which are significantly higher than blanks. After this step, one-way ANOVA followed by Tukey's HSD ($p < 0.05$) was applied to evaluate the significant differences among mascarpone samples. All analyses were performed with core functions of R programming language (R Development Core Team, R Foundation for Statistical Computing, Vienna, Austria, 2014) and its external packages (ChemometricsWithR, DiscriMiner, prospectr). In some cases, in order to interpret the results of the experiment, the entire dataset was divided into smaller subsets based on different criteria (e.g., producer, lactose content).

## 4. Conclusions

Using two complementary analytical approaches, Headspace-Solid Phase Microextraction-Gas Chromatography-Mass Spectrometry (HS-SPME GC-MS) and Proton-Transfer Reaction-Mass Spectrometry coupled to a Time of Flight mass analyzer (PTR-ToF-MS), the present work provides a first description of Volatile Organic Compounds (VOCs). In addition, we underline the differences in VOC content susceptible to characterize the aroma of different brands and product types (classic and lactose-free). On the whole, the dominance of volatiles generally associated to floral, fruity, sweet, and nutty notes might contribute to explain the delicate sensory impression perceived by smelling this fresh dairy product. Unfortunately, the aroma profile of the present investigation cannot be discussed in light of previous literature that is, as mentioned, very scarce. Considering the wide number of products that use mascarpone as raw material, such as the popular Tiramisù and coffee mascarpone cream, this study provides information to design future studies conceived to assess the contribution of this unripened cheese to the sensory characteristics of final products.

**Supplementary Materials:** The following are available online, Table S1: Monitored chemico-physical characteristics for the list of 'Mascarpone' samples analyzed in the present study.

**Author Contributions:** Conceptualization, F.B., and L.N.; methodology, V.L., I.K., L.C., F.B., and L.N.; software, L.C.; validation, L.C., F.B., and L.N.; formal analysis, V.L., and I.K.; investigation, V.C., V.L., and I.K.; resources, F.B. and L.N.; data curation, V.C., V.L., and I.K.; writing—original draft preparation, V.C., V.L., and I.K.; writing—review and editing, L.C., F.B., and L.N.; visualization, V.C., V.L., I.K., F.B. and L.N.; supervision, F.B., and L.N.; project administration, F.B., and L.N.; funding acquisition, F.B., and L.N. All authors have read and agreed to the published version of the manuscript.

**Funding:** The work has been partially supported by the Autonomous Province of Trento (ADP 2020).

**Acknowledgments:** We would like to thank the editor and the three anonymous reviewers for their suggestions and comments. Vittorio Capozzi acknowledges Francesco De Marzo of the Institute of Sciences of Food Production—CNR for the skilled technical support provided during the realization of this work.

**Conflicts of Interest:** The authors declare no conflict of interest.

## References

1. Carminati, D.; Perrone, A.; Neviani, E. Inhibition of *Clostridium sporogenes* growth in mascarpone cheese by co-inoculation with *Streptococcus thermophilus* under conditions of temperature abuse. *Food Microbiol.* **2001**, *18*, 571–579.

2. Franciosa, G.; Pourshaban, M.; Gianfranceschi, M.; Gattuso, A.; Fenicia, L.; Ferrini, A.M.; Mannoni, V.; De Luca, G.; Aureli, P. *Clostridium botulinum* spores and toxin in mascarpone cheese and other milk products. *J. Food Prot.* **1999**, *62*, 867–871. [PubMed]

3. Hinrichs, J. Mediterranean milk and milk products. *Eur. J. Nutr.* **2004**, *43*, i12–i17.

4. Del Prato, O.S. *Trattato di Tecnologia Casearia*; Edagricole: Bologna, Italy, 1998.

5. Donelly, C. *The Oxford Companion to Cheese*; Oxford University Press: New York, NY, USA, 2017; ISBN 978-0-19-933088-1.

6. Majchrzak, T.; Wojnowski, W.; Lubinska-Szczygeł, M.; Różańska, A.; Namieśnik, J.; Dymerski, T. PTR-MS and GC-MS as complementary techniques for analysis of volatiles: A tutorial review. *Anal. Chim. Acta* **2018**, *1035*, 1–13. [PubMed]

7. Dewulf, J.; Van Langenhove, H.; Wittmann, G. Analysis of volatile organic compounds using gas chromatography. *TrAC Trends Anal. Chem.* **2002**, *21*, 637–646.

8. Panseri, S.; Soncin, S.; Chiesa, L.M.; Biondi, P.A. A headspace solid-phase microextraction gas-chromatographic mass-spectrometric method (HS-SPME–GC/MS) to quantify hexanal in butter during storage as marker of lipid oxidation. *Food Chem.* **2011**, *127*, 886–889.

9. Biasioli, F.; Gasperi, F.; Yeretzian, C.; Märk, T.D. PTR-MS monitoring of VOCs and BVOCs in food science and technology. *TrAC Trends Anal. Chem.* **2011**, *30*, 968–977.

10. Jordan, A.; Haidacher, S.; Hanel, G.; Hartungen, E.; Märk, L.; Seehauser, H.; Schottkowsky, R.; Sulzer, P.; Märk, T.D. A high resolution and high sensitivity proton-transfer-reaction time-of-flight mass spectrometer (PTR-TOF-MS). *Int. J. Mass Spectrom.* **2009**, *286*, 122–128.

11. Capozzi, V.; Yener, S.; Khomenko, I.; Farneti, B.; Cappellin, L.; Gasperi, F.; Scampicchio, M.; Biasioli, F. PTR-ToF-MS Coupled with an Automated Sampling System and Tailored Data Analysis for Food Studies: Bioprocess Monitoring, Screening and Nose-space Analysis. *J. Vis. Exp.* **2017**, *123*, e54075.

12. Marsili, R. Flavors and off-flavors in dairy foods. In *Encyclopedia of Dairy Sciences*, 2nd ed.; Fuquay, J.W., Ed.; Academic Press: San Diego, CA, USA, 2011; pp. 533–551. ISBN 978-0-12-374407-4.

13. Mutarutwa, D.; Navarini, L.; Lonzarich, V.; Compagnone, D.; Pittia, P. GC-MS aroma characterization of vegetable matrices: Focus on 3-alkyl-2-methoxypyrazines. *J. Mass Spectrom.* **2018**, *53*, 871–881.

14. Delgado, F.J.; González-Crespo, J.; Cava, R.; Ramírez, R. Formation of the aroma of a raw goat milk cheese during maturation analysed by SPME–GC–MS. *Food Chem.* **2011**, *129*, 1156–1163. [PubMed]

15. Curioni, P.M.G.; Bosset, J.O. Key odorants in various cheese types as determined by gas chromatography-olfactometry. *Int. Dairy J.* **2002**, *12*, 959–984.

16. Jung, H.J.; Ganesan, P.; Lee, S.J.; Kwak, H.S. Comparative study of flavor in cholesterol-removed Gouda cheese and Gouda cheese during ripening. *J. Dairy Sci.* **2013**, *96*, 1972–1983. [PubMed]

17. Mallia, S.; Escher, F.; Rehberger, B.; Schlichtherle-Cerny, H. Aroma-active secondary oxidation products of butter. In Proceedings of the 3rd QLIF Congress: Improving Sustainability in Organic and Low Input Food Production Systems, Stuttgart, Germany, 20–23 March 2007.

18. Peterson, D.G.; Reineccius, G.A. Characterization of the volatile compounds that constitute fresh sweet cream butter aroma. *Flavour Fragr. J.* **2003**, *18*, 215–220.

19. Wang, W.; Zhang, L.; Li, Y. Production of Volatile Compounds in Reconstituted Milk Reduced-Fat Cheese and the Physicochemical Properties as Affected by Exopolysaccharide-Producing Strain. *Molecules* **2012**, *17*, 14393–14408. [PubMed]

20. Povolo, M.; Contarini, G. Comparison of solid-phase microextraction and purge-and-trap methods for the analysis of the volatile fraction of butter. *J. Chromatogr. A* **2003**, *985*, 117–125.

21. Qian, M.; Reineccius, G. Identification of aroma compounds in Parmigiano-Reggiano cheese by gas chromatography/olfactometry. *J. Dairy Sci.* **2002**, *85*, 1362–1369.

22. Gasperi, F.; Gallerani, G.; Boschetti, A.; Biasioli, F.; Monetti, A.; Boscaini, E.; Jordan, A.; Lindinger, W.; Iannotta, S. The mozzarella cheese flavour profile: A comparison between judge panel analysis and proton transfer reaction mass spectrometry. *J. Sci. Food Agric.* **2001**, *81*, 357–363.

23. Boscaini, E.; van Ruth, S.; Biasioli, F.; Gasperi, F.; Märk, T.D. Gas Chromatography–Olfactometry (GC-O) and Proton Transfer Reaction–Mass Spectrometry (PTR-MS) Analysis of the Flavor Profile of Grana Padano, Parmigiano Reggiano, and Grana Trentino Cheeses. *J. Agric. Food Chem.* **2003**, *51*, 1782–1790.

24. Gallardo-Escamilla, F.J.; Kelly, A.L.; Delahunty, C.M. Influence of Starter Culture on Flavor and Headspace Volatile Profiles of Fermented Whey and Whey Produced from Fermented Milk. *J. Dairy Sci.* **2005**, *88*, 3745–3753.

25. van Ruth, S.M.; Koot, A.; Akkermans, W.; Araghipour, N.; Rozijn, M.; Baltussen, M.; Wisthaler, A.; Märk, T.D.; Frankhuizen, R. Butter and butter oil classification by PTR-MS. *Eur. Food Res. Technol.* **2008**, *227*, 307–317.

26. Makhoul, S.; Yener, S.; Khomenko, I.; Capozzi, V.; Cappellin, L.; Aprea, E.; Scampicchio, M.; Gasperi, F.; Biasioli, F. Rapid non-invasive quality control of semi-finished products for the food industry by direct injection mass spectrometry headspace analysis: The case of milk powder, whey powder and anhydrous milk fat. *J. Mass Spectrom.* **2016**, *51*, 782–791. [PubMed]

27. Benozzi, E.; Romano, A.; Capozzi, V.; Makhoul, S.; Cappellin, L.; Khomenko, I.; Aprea, E.; Scampicchio, M.; Spano, G.; Märk, T.D.; et al. Monitoring of lactic fermentation driven by different starter cultures via direct injection mass spectrometric analysis of flavour-related volatile compounds. *Food Res. Int.* **2015**, *76*, 682–688. [PubMed]

28. Yépez, A.; Russo, P.; Spano, G.; Khomenko, I.; Biasioli, F.; Capozzi, V.; Aznar, R. In situ riboflavin fortification of different kefir-like cereal-based beverages using selected Andean LAB strains. *Food Microbiol.* **2019**, *77*, 61–68. [PubMed]

29. Romano, A.; Capozzi, V.; Spano, G.; Biasioli, F. Proton transfer reaction-mass spectrometry: Online and rapid determination of volatile organic compounds of microbial origin. *Appl. Microbiol. Biotechnol.* **2015**, *99*, 3787–3795.

30. Gallardo-Escamilla, F.J.; Kelly, A.L.; Delahunty, C.M. Sensory Characteristics and Related Volatile Flavor Compound Profiles of Different Types of Whey. *J. Dairy Sci.* **2005**, *88*, 2689–2699.

31. Burdock, G.A. *Fenaroli's Handbook of Flavor Ingredients*; CRC Press: Boca Raton, FL, USA, 2019; ISBN 978-1-00-069466-6.

32. Jia, W.; Wang, H.; Shi, L.; Zhang, F.; Fan, C.; Chen, X.; Chang, J.; Chu, X. High-throughput foodomics strategy for screening flavor components in dairy products using multiple mass spectrometry. *Food Chem.* **2019**, *279*, 1–11.

33. Puniya, A.K. *Fermented Milk and Dairy Products*; CRC Press: Boca Raton, FL, USA, 2015; ISBN 978-1-4665-7800-5.

34. Cattaneo, T.M.P.; Summa, C.; Bertolo, G.; Giangiacomo, R. Spreadability of Mascarpone cheese: Sensory and objective measurements. *Milchwissenschaft* **2005**, *60*, 399–402.

35. Su, W.-H.; He, H.-J.; Sun, D.-W. Non-Destructive and rapid evaluation of staple foods quality by using spectroscopic techniques: A review. *Crit. Rev. Food Sci. Nutr.* **2017**, *57*, 1039–1051.

36. van den Berg, F.; Lyndgaard, C.B.; Sørensen, K.M.; Engelsen, S.B. Process Analytical Technology in the food industry. *Trends Food Sci. Technol.* **2013**, *31*, 27–35.

37. Capozzi, V.; Makhoul, S.; Aprea, E.; Romano, A.; Cappellin, L.; Sanchez Jimena, A.; Spano, G.; Gasperi, F.; Scampicchio, M.; Biasioli, F. PTR-MS Characterization of VOCs Associated with Commercial Aromatic Bakery Yeasts of Wine and Beer Origin. *Molecules* **2016**, *21*, 483. [PubMed]

38. Campbell-Sills, H.; Capozzi, V.; Romano, A.; Cappellin, L.; Spano, G.; Breniaux, M.; Lucas, P.; Biasioli, F. Advances in wine analysis by PTR-ToF-MS: Optimization of the method and discrimination of wines from different geographical origins and fermented with different malolactic starters. *Int. J. Mass Spectrom.* **2016**, *397–398*, 42–51.

39. Bottiroli, R.; Pedrotti, M.; Aprea, E.; Biasioli, F.; Fogliano, V.; Gasperi, F. Application of PTR-TOF-MS for the quality assessment of lactose-free milk: Effect of storage time and employment of different lactase preparations. *J. Mass Spectrom.* **2020**. [CrossRef]

40. Cappellin, L.; Biasioli, F.; Fabris, A.; Schuhfried, E.; Soukoulis, C.; Märk, T.D.; Gasperi, F. Improved mass accuracy in PTR-TOF-MS: Another step towards better compound identification in PTR-MS. *Int. J. Mass Spectrom.* **2010**, *290*, 60–63.

41. Cappellin, L.; Biasioli, F.; Granitto, P.M.; Schuhfried, E.; Soukoulis, C.; Costa, F.; Märk, T.D.; Gasperi, F. On data analysis in PTR-TOF-MS: From raw spectra to data mining. *Sens. Actuators B Chem.* **2011**, *155*, 183–190.

42. Lindinger, W.; Hansel, A.; Jordan, A. On-line monitoring of volatile organic compounds at pptv levels by means of proton-transfer-reaction mass spectrometry (PTR-MS) medical applications, food control and environmental research. *Int. J. Mass Spectrom. Ion Process.* **1998**, *173*, 191–241.

**Sample Availability:** Samples of the compounds are not available from the authors.

*Review*

# Flavor and Texture Characteristics of 'Fuji' and Related Apple (*Malus domestica* L.) Cultivars, Focusing on the Rich Watercore

**Fukuyo Tanaka [1],*, Fumiyo Hayakawa [2] and Miho Tatsuki [3]**

[1]   Central Region Agricultural Research Center, National Agriculture and Food Research Organization (NARO), Tsukuba 305-8666, Japan

[2]   Food Research Institute, National Agriculture and Food Research Organization (NARO), Tsukuba 305-8642, Japan; fumiyoha@affrc.go.jp

[3]   Institute of Fruit Tree and Tea Science, National Agriculture and Food Research Organization (NARO), Tsukuba 305-8605, Japan; tatsuki@affrc.go.jp

*   Correspondence: fukuyot@affrc.go.jp; Tel.: +81-29-838-8814

Academic Editor: Eugenio Aprea
Received: 12 February 2020; Accepted: 29 February 2020; Published: 2 March 2020

**Abstract:** Watercore is a so-called physiological disorder of apple (*Malus domestica* L.) that commonly occurs in several well-known cultivars. It is associated with a rapid softening of the flesh that causes a marked changed in flavor and texture. In Asia, apples with watercore are preferred and considered a delicacy because of their enhanced sweet flavor. The 'Fuji' cultivar, the first cultivar with rich watercore that is free from texture deterioration, has played a key role in the development of the market for desirable watercored apples. This review aimed to summarize and highlight recent studies related to the physiology of watercore in apples with special focus on 'Fuji' and related cultivars.

**Keywords:** 'Fuji'; watercore; sweetness; flavor; texture; flesh browning disorder; apple

---

## 1. Introduction

The 'Fuji' cultivar has maintained a large share of the global apple production market over the last two decades [1]. Originally, 'Fuji' was selected from cross between 'Ralls Janet' and 'Delicious' in the Tohoku region of Japan and gained popularity because it is extremely juicy and crisp with a sweet flavor similar to that of 'Delicious' [2]. 'Fuji' is also susceptible to watercore development. Watercore is a phenomenon that presents as a translucent appearance at the core and/or flesh of the fruit, and it is caused by the intercellular spaces of the affected tissue being filled with fluid. It has been reported that watercore development is related to sugar metabolism during the maturation process and fruit mineral composition. Watercored apple is prone to several physiological disorders, such as browning and breakdown during storage [3–9]. Furthermore, strains of the 'Delicious' cultivar are also highly susceptible to watercore, which is typically accompanied by changes in texture traits, such as softening and mealiness. These undesirable characteristics that commonly occur in watercored 'Delicious' strains have caused watercoring in apples to be viewed negatively.

In spite of this, watercored 'Fuji' has gradually become desirable in Japan and other Asian countries, and the palatability of 'Fuji' and watercored apples has been identified in the last decade. Today, Japanese producers and consumers generally value watercored apple owing to its excellent fruit flavor, which occurs when it fully matures on the tree. In fact, watercored apples are often advertised using phrases such as aroma-rich and pineapple-like. Furthermore, the rich watercore trait has become a breeding target with the aim of increasing the sweet flavor in apple [10]. Aprea et al. [11] proposed that apple breeding programs must take into account factors such as volatile compounds, texture parameters, minor components, and information from sensory panels. However, perceived sweetness is difficult to be described because it is always perceived in combination with other sensory properties, which influence its evaluation. Sweetness perception is a complex and multisensory process, and only gustatory stimuli are insufficient to fully understand and predict it [12]. In this work, we focus on 'Fuji' and the related cultivars and review the mechanism of watercoring in apple palatability. We also assess various characters, such as flavor, texture, and genetic properties, employing integrated analysis of instrumental and sensory profiling.

## 2. Flavor Characteristics

### 2.1. Sensory Analysis

Although watercored apple has been extremely popular among Japanese consumers, there were little published data regarding the overall acceptance for watercored apple available. In order to characterize the unique flavor and overall acceptance, Tanaka et al. [13] conducted sensory analysis using watercored and nonwatercored 'Fuji' with 29 trained panelists. With respect to overall acceptance, watercored apples scored significantly higher than nonwatercored apples (Table 1). Overall intensities of aroma and taste and five sensory attributes were scored using a seven-point scale. Taste intensity was evaluated with nose clip. Overall aroma intensity and perception of sweet and fruity flavors were enhanced in watercored apple, whereas green and sour perception was enhanced in nonwatercored apple (Table 1). Overall taste intensity, in which the influence of aroma was eliminated by clips, was not significantly different, indicating that the contribution of aroma to the overall acceptance and characteristics of flavor was remarkably large in this case.

**Table 1.** Sensory evaluation for watercored and nonwatercored 'Fuji' apples.

| Sample Status | Overall Acceptability | Overall Intensity | | Sensory Attribute | | | | |
| --- | --- | --- | --- | --- | --- | --- | --- | --- |
| | | Aroma | Taste | Green | Fruity | Floral | Sweet | Sour |
| Nonwatercored | 3.0 | 4.0 | 4.1 | 4.3 | 4.0 | 3.1 | 3.9 | 4.0 |
| Watercored | 3.5 | 4.5 | 4.2 | 3.6 | 4.4 | 4.2 | 4.6 | 3.2 |
| Significance | ** | ** | ns | ** | * | *** | ** | *** |

Apple: Products of a commercial orchard, peeled, cored, and cut into bite-size pieces just before being served. Evaluation: a seven-point categorical scale (1–7); 29 panelists trained for quantitative destructive analysis, 10 females and 19 males. Taste was evaluated with nose clip to eliminate the influence of aroma. Significance: *, **, *** indicate significant differences at the level of $p < 0.05$, $p < 0.01$, or $p < 0.001$, respectively, using paired $t$-test; ns means not significant. Reproduced with permission from Tanaka et al. [13].

### 2.2. Analysis of Volatile and Water-Soluble Compounds

Volatile and water-soluble components were profiled for watercored and nonwatercored 'Fuji' and 'Koutoku', a progeny of 'Fuji' (Tables S1 and S2) [13]. In both cultivars, ethyl esters and methyl esters of fatty acids were detected in watercored fruit; their peak intensities were as large as several to several hundred times those of nonwatercored apple (Table S1). In addition, principal component analysis of the intensities of the 109 components suggested that the PC1 score was differentiated by cultivar, whereas the PC2 score was differentiated by the presence of watercore (Figure 1). The PC2 loadings suggested that ethyl esters, methyl esters, sorbitol, galactaric acid, erythronic acid, and dehydroascorbic acid were associated with watercore. Increase in sorbitol was consistent with previous reports [14–17]. This integrated profiling analysis suggested that an increase in methyl esters and ethyl esters is crucial to the attributes and desirability of watercored apple. Similar phenomena have been revealed by Dixon et al. [18] when, following a short-term exposure to hypoxic conditions, time courses of apple aroma components and odor units for 10 apple cultivars were analyzed, and their results indicated that odor unit values highly corresponded to ethyl ester levels.

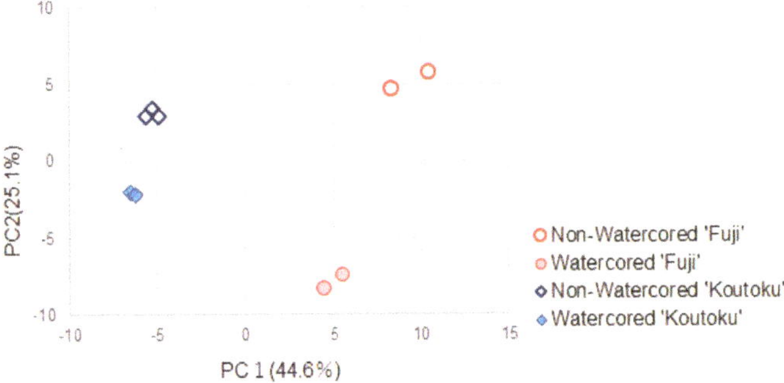


**Figure 1.** Principal component analysis score plots of the volatiles and solubles of fruit juice. PC1 and PC2 scores were discriminated by cultivars and watercore existence, respectively. Top 10 of PC2 loadings were (1) ethyl butanoate, (2) ethyl propanoate, (3) ethyl 2-methylbutanoate, (4) ethyl acetate, (5) ethyl hexanoate, (6) sorbitol, (7) galactaric acid, (8) methyl 2-methylbutanoate, (9) methyl acetate, and (10) ethyl tiglate. Reproduced with permission from Tanaka et al. [13].

Ethyl esters have been reported to have an apple-like, fruity, sweet aroma with an extremely low threshold value. For example, Komthong et al. [19] analyzed head-space volatiles of 'Fuji' using aroma extract dilution analysis and determined flavor dilution factor, which is the lowest dilution ratio of the volatile compounds. Then, methyl 2-methylbutanoate and ethyl 2-methylbutanoate were estimated and determined to be the most potent odorants in the volatiles based on their lowest threshold odor values. Moreover, we demonstrated that an increase in ethyl esters significantly enhanced the perception of apple-like sweetness by sensory evaluation using a series of 'Fuji' samples that had chemically modified aroma (Figure 2). Based on these data, ethyl esters appear to be potent, key flavor compounds in watercored apples.

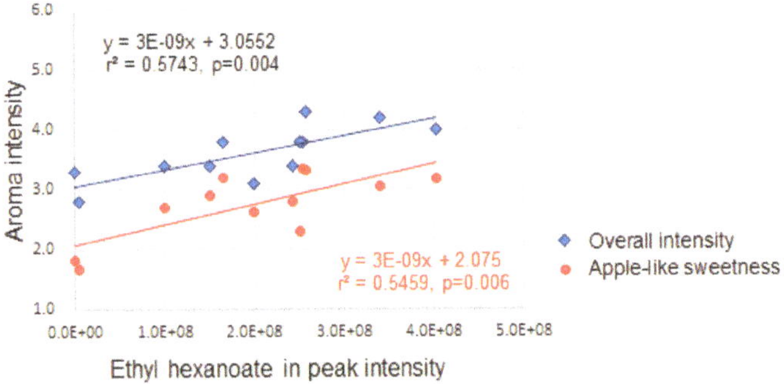


**Figure 2.** Ethyl hexanoate and aroma intensity in ester-enhanced 'Fuji' by incubation with ethanol mixture. Ethyl hexanoate is shown as a representative of ethyl esters because major ethyl esters of 'Fuji' correlate with one another in their peak intensities.

The difference among sugar and sorbitol contents, soluble solid contents (SSC), and gene expression related to sugars in watercored and nonwatercored tissues has been studied extensively [14,16,20,21]. Specifically, sorbitol accumulation in the watercored tissues was observed in several cases, whereas fructose, glucose, sucrose, total sugar contents, and SSC were only observed in a few cases [13,15,16,20]. Fructose, the sweetest sugar of the apple components [22,23], tended to be lower in watercored tissues. Sorbitol tends to be present at a low content and exhibits weaker perceptive sweetness compared with other sugars, even though it increases in content in watercored tissues. [23,24]. The comparison between watercored and nonwatercored tissues often found decreases in sweetness in the watercored tissues. According to Melad-Herreros [15] and Williams et al. [16], nonaffected tissues of watercored apples, often edible parts, scored higher for fructose and sucrose than the affected tissues (Table 2), indicating that watercored apples may in fact be sweeter. Harker et al. [25] stated that two apples needed to differ in °Brix (SSC) by more than 1 before evoking a change in response to a perceived sweet taste for the median panelist. A difference of 1 °Brix corresponds to 1% difference in sucrose. Table 2 presents the estimated sweetness using two equations that defined sucrose sweetness as 1 [26,27]. Our estimation (a) took sorbitol into account based on the estimation summarized by Kitahata and Machinami [22], whereas (b), known as the total sweetness index, did not [23]. Williams et al. [17] also found that the difference between nonwatercored and nonaffected tissues of watercored apples was nearly 1. In this case, there was a perceived sweetness difference between the edible part of a watercored and that of a nonwatercored apple at a near-threshold level. These findings are in agreement with the results of sensory evaluation (Table 1) of taste intensity, which found that while watercored apples were generally evaluated a little intense, they did not differ significantly from control samples. Given these findings, the significant difference in sweetness is likely affected by components other than sugars.

**Table 2.** Sugar profiles and estimated perceived sweetness of various apple cultivars.

| Cultivar | Watercore | Sugar Contents (g/100 g FW) | | | | Estimated Sweetness | | Ref. |
|---|---|---|---|---|---|---|---|---|
| | | Fructose | Glucose | Sucrose | Sorbitol | (a) | (b) | |
| 'Fuji' | absent | 5.7 | 3.1 | 1.5 | 0.4 | 12.0 | 12.3 | [14] [1] |
| | present (richest level) | 5.5 | 2.2 | 3.4 | 1.2 | 12.7 | 13.2 | |
| 'Fuji' | absent | 6.6 | 2.3 | 1.8 | 0.5 | 12.3 | 13.4 | |
| | present | 5.6 | 1.9 | 1.8 | 1.5 | 11.3 | 11.7 | |
| 'Gloster' | absent | 5.2 | 1.9 | 3.2 | 0.3 | 11.5 | 12.4 | [15] |
| | present | 3.8 | 1.7 | 2.6 | 0.9 | 9.2 | 9.6 | |
| 'Delicious' | absent | 6.4 | 1.7 | 2.2 | 0.2 | 11.9 | 13.1 | |
| | present | 5.1 | 1.5 | 2.0 | 1.0 | 10.2 | 10.7 | |
| 'Esperiega' | absent | 7.2 | 1.51 | 2.8 | 0.8 | 13.7 | 14.7 | [16] |
| | present (nonaffected site) | 6.9 | 2.0 | 3.2 | 1.5 | 14.4 | 15.0 | |
| | present (affected site) | 6.3 | 2.2 | 1.5 | 2.9 | 13.0 | 12.6 | |
| 'Winesap' | absent | 3.2 | 4.0 | 3.8 | 0.9 | 11.3 | 11.6 | |
| | present (nonaffected site) | 3.4 | 4.2 | 4.1 | 1.3 | 12.2 | 12.4 | [17] [1] |
| | present (affected site) | 3.0 | 3.7 | 3.8 | 1.8 | 11.4 | 11.1 | |

[1] Original sugar contents were converted to g/100 g FW. Estimated sweetness: (a) (1.0 [sucrose]) + (1.3 [fructose]) + (0.7 [glucose]) + (0.7 [sorbitol]) [22]; (b) = (1.0 [sucrose]) + (1.5 [fructose]) + (0.76 [glucose]) [23].

Recently, the importance of aroma components in the characteristics of flavor and preference in apple has been widely recognized. Aprea et al. [11] reported that sorbitol content correlated with perceived sweetness better than any other single sugar or total sugar content. Furthermore, their predictive model based on partial least squares regression included not only SSC but also volatile compounds and revealed that several volatiles are possibly contributing to flavor. Having a sweet taste is an important but difficult attribute to be predicted using objective measurements [25]. The contribution of sugars to the enhancement of perceived sweetness in watercored apple is likely limited, whereas the profile of aroma components varies widely and accounts for several of the unique flavor profiles. Aroma components between watercored and nonwatercored apples can be markedly different. For instance, our analysis revealed that the detected levels of most ethyl esters that created an aroma profile with characteristics similar to pineapple or ginjoshu (high-quality sake) were ten times their levels in nonwatercored apples (Table S1) [13,28–31]. Considering these profiles of flavor components and sensory attributes, the contribution of aroma components, such as ethyl esters, is crucial in producing the flavor characteristics in watercore-rich apples.

*2.3. Mechanism of Enhanced Ethyl Ester Synthesis in Watercored Apples*

Because ethyl esters are crucial in aroma and flavor profiles in apples, different analyses have already focused on the synthesis. Dixon and Hewett [18] reported that apple volatile compounds increased in ethanol and ethyl ester concentrations after exposure to hypoxic conditions. Specifically, the synthesis of ethyl esters was high in watercore-susceptible cultivars 'Red Delicious', synonymous with 'Delicious', and 'Fuji' and low in nonsusceptible cultivars 'Golden Delicious' and 'Cox's Orange Pippin'. It has also been reported that ethyl esters from apples subjected to controlled-atmosphere (CA) storage exhibited a temporary increase in ethanol and ethyl ester concentrations. Hypoxia likely activates anaerobic glycolysis and ethanol synthesis, causing an increase in ethyl ester production [32,33]. One study found that a decrease in respiration and an increase in ethanol and acetaldehyde concentrations in watercored tissues of 'Richard Delicious', a sport of 'Delicious', shared similarity with apples that were exposed to hypoxic conditions or CA-stored [3]. Furthermore, Tanaka et al. [13] analyzed oxygen distribution within a fruit and demonstrated low-oxygen status at the watercored position (Figure 3), whereas nonwatercored fruits were flat. These phenomena support the concept that ethyl ester synthesis is enhanced under hypoxic conditions within watercored tissues, resulting in distinctive, fermented flavor.

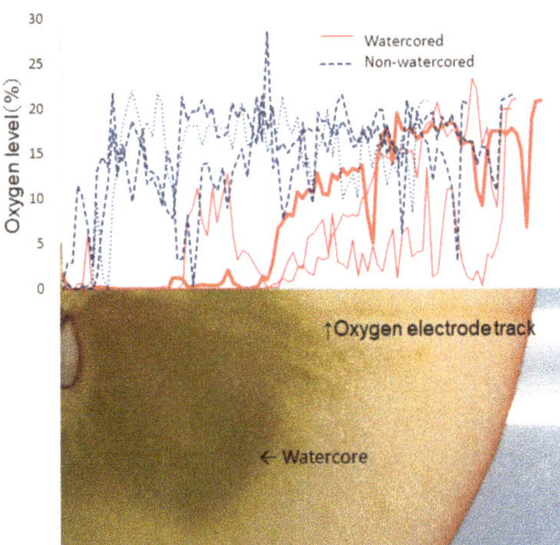


**Figure 3.** Oxygen distribution related to watercore of 'Fuji' apple. Three fruits were measured for each class. Thick red line corresponds to the photograph. Reproduced with permission from Tanaka et al. [13].

Questions also arise regarding discrimination of volatiles and related gene expression in a fruit associated with oxygen levels. This highly variable, cultivar-dependent response of apple cultivars to hypoxic conditions may also be associated with the physiological processes involved in the development of watercore, which has not been fully elucidated to date. To better understand the metabolism of the characteristic aroma profiles, a fusion analysis of molecular biology and metabolomics will be required.

## 3. Texture Characteristics

### 3.1. Apple Cultivars and Texture Measurement

Texture is a key factor that affects consumer preference of apple [34–36]. Texture comprises crispness, mealiness, juiciness, firmness, and other traits and has been reported to influence perceived sweetness [25,37]. Among the texture traits, crispness and juiciness are favorable for apple, whereas softness or mealiness are avoided [35]. Traditional watercore-susceptible cultivars were often accompanied by mealiness and rapid softening [5]. However, the occurrence of watercore and softening is under separate regulations, and cultivars have been developed with one but not the other [10,38]. Here, we review the studies on apple texture as it is related to watercore susceptibility.

Crispness is a sensory and integrated attribute defined as the amount and pitch of sound generated when the fruit is first bitten using the incisor [37,39,40], and it is often estimated from firmness because the two traits are highly and positively correlated [35]. Softening is usually caused by a reduction in firmness, which is typically measured using a penetrometer or sensory analysis. Cell shape, cell size, cell packing, and overall fruit anatomy as well as chemistry of the cell wall and membrane and the role of cell turgor affect firmness [41]. Among them, macromolecular network structures of cell walls, mainly comprising pectin, hemicellulose and cellulose, confer flesh cell rigidity, however, the structures are gradually lost by the cell-wall-modifying enzymes such as β-galactosidase, α-L-arabinofuranosidase, polygalacturonase, pectin methylesterase, and others. Ethylene reportedly stimulated these enzymes, subsequently causing flesh softening. Turgor reduction was also associated with firmness reduction [41,42]. Although cell membranes of apple are not typically associated with cell wall swelling and juiciness [41], so far as watercore is concerned, it may play roles in apple juiciness to some extent, as described below (Section 3.2).

Mealiness is defined as the amount of small, lumpy particles that become apparent during chewing in sensory analysis [37,39,43]. It is due to the loss of cell–cell adhesion or cell separation [37]. Iwanami et al. [38] investigated 23 cultivars and a breeding line under a time-course experiment to evaluate firmness and mealiness and divided them into four groups based on their results after 40 days of storage at 20 °C. The watercore-susceptible cultivars 'Starking Delicious' and 'Red Gold' were placed in the most rapid mealiness developing group, whereas 'Fuji' was the firmest and most nonmealy cultivar. This was consistent with other previous studies [44–46]. Iwanami et al. [47] also found that the softening performance of an apple cultivar during storage was highly dependent on the degree of mealiness and turgor reduction rate. The softening rates of all mealy cultivars were high; moreover, the softening rates of nonmealy cultivars were significantly correlated with the turgor reduction rates. In other words, nonmealy cultivars with slow turgor reduction can be expected to exhibit high storage performance. 'Fuji' had the lowest turgor reduction rate, which most likely contributed to its firmness and crispness. In addition, 'Starking Delicious', another sport of 'Delicious', surprisingly exhibited the slowest turgor reduction rate among the tested cultivars contrary to its trait of rapid softening. 'Fuji' seems to inherit the excellent trait of slow turgor reduction from the softening cultivar 'Delicious' and not from the slow softening 'Ralls Janet'. Differences in storage performance between 'Fuji' and the other 'Delicious' strains may mainly be due to differences in mealiness or nonmealiness.

A genetic contribution to watercore and mealiness in the 'Fuji'-related apples was demonstrated by Kunihisa et al. [10], who examined genomic dissection of 'Fuji' using 115 accessions of its descendants and parents. In that study, one quantitative trait loci (QTL) was detected for the following traits: degree of watercore and mealiness, acidity, and harvest day. The QTL for a high degree of watercore was detected in the middle of chromosome (chr)14, whereas the one for mealiness was detected at the middle of chr1. 'Fuji' has inherited haplotypes from both 'Delicious' and 'Ralls Janet'. The haplotype of 'Fuji' derived from 'Delicious' in the chr14 region dominantly causes watercore, whereas one in the chr1 region causes mealiness. For mealiness, another QTL associated with *MdPG1* was detected from different $F_1$ population [40]. So far, as 'Fuji' descendant, however, 90% of selected cultivars or superior breeding lines have inherited the haplotype of 'Fuji'-derived 'Ralls Janet' at the region of chr1 [10,48].

Sadamori, a leader of the 'Fuji' breeding team, recounted that most of the seedlings of 'Ralls Janet' and 'Delicious' generated sweet but mealy fruit in his memoir [49]. Among them (592 fruits), they found only two crisp and nonmealy fruits. One of them, which exhibited excellent flavor, was what eventually became 'Fuji' [49]. There was only a 0.3% frequency of nonmealy flesh from that cross; however, nonmealy phenotypes are more common in 'Fuji'-related accessions. Therefore, newly developed watercore-susceptible lines derived from 'Fuji' have an improved chance of possessing both excellent flavor and texture. Additional genetic information on the turgor reduction after harvest and its physiological understandings will help further improve and maintain the crispness of apples.

### 3.2. Watercore and Texture

Juiciness positively contributes to perceived freshness and is dependent on water content [50]. The water content of watercored apples is higher than that of nonwatercored apples, which is caused by the fluid within intercellular spaces or apoplast that causes watercore. Iwanami et al. [50] reported that both the water content of the whole fruit and apoplast tissues positively correlated with juiciness, affirming that watercored apples exhibit greater juiciness than nonwatercored apples. Although the report did not refer to watercore, the juiciest apple, 'Oyume', in their data is a cultivar that generally develops rich watercore. Maintaining the perception of freshness in apples, which are commonly stored for relatively long periods of time, is crucial for continued consumer appeal and requires appropriate storage conditions.

In order to establish a storage technique for high-quality watercored 'Fuji', Onodera et al. (2010) [51] investigated storability of apples that exhibited >30% watercoring. Time-course measurements of firmness and watercore degree were taken during 3 months of storage and 14 days of shelf time under regular atmosphere (RA). Both watercore degree and firmness decreased with time, and these exhibited significant positive correlation to each other (Figure 4). These results were in agreement with those of a previous report of Bowen and Watkins [14], which stated that flesh firmness at harvest initially tended to decrease with watercore scores and then significantly increased as watercore enhanced. These data suggest that highly watercored apples may maintain a firmer texture than lesser watercored apples for a few months after harvest. Further case examples are required.

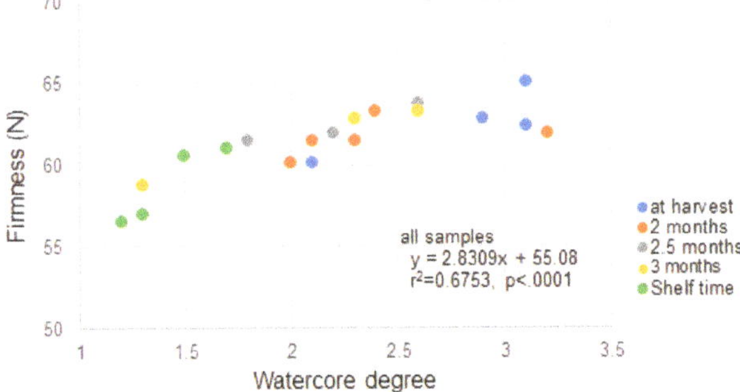


**Figure 4.** Watercore degree (0–4) and flesh firmness of 'Fuji' apple. Reproduced with permission from Onodera [51].

Being a major plant growth regulator and ripening hormone, ethylene is considered to be involved in the softening of apples [41,52]. Internal ethylene concentration (IEC) was measured in relation to watercore degree. Bowen and Watkins [13] reported that IEC increased with the watercore degree, whereas Argenta et al. [53] reported that IEC was higher in fruit with a low watercore score, and it decreased in fruits with a high watercore score compared with watercore-free. There have been few studies regarding the firmness of watercored apples under storage and its regulation, limiting what is currently known. As watercored apples gain popularity, further studies will likely greatly elucidate the relationship between firmness and watercoring.

### 4. Watercore During Storage

Watercored apples are likely to develop physiological disorders in the flesh, including watercore breakdown, internal browning, and various other disorders and, in some cases, worsen the degree of existing disorders [3–5,53–56]. These disorders often hinder the storability of apples and their use as a long-shelf life commodity. Watercore development is accompanied by photosynthetic carbohydrate accumulation in the fruit; consequently, as harvest is delayed, the degree of watercore increases [14,57,58]. Therefore, watercore-susceptible cultivars are often harvested long before maturity at the expense of sweet flavor.

Onodera et al. [51] investigated the storability of highly watercored 'Fuji' with or without 1-methylcyclopropene (1-MCP) treatment for 3 months. Watercore breakdown did not occur until 3 months after harvest, irrespective of 1-MCP treatment and temperature settings. Another experiment in Figure 5 presents a time course of watercore degree and incidence of watercore breakdown during shelf life. The storage conditions were set at −1 °C or 2 °C for an initial 2 months and at 5 °C for 9 days followed by 20 °C for 14 days. Watercore degrees gradually decreased in all treatments during the experiment, and the incidence of watercore breakdown was detected at 14 days after storage at 20 °C. These results indicate that watercored 'Fuji' can be stored for up to 3 months under RA with refrigeration.

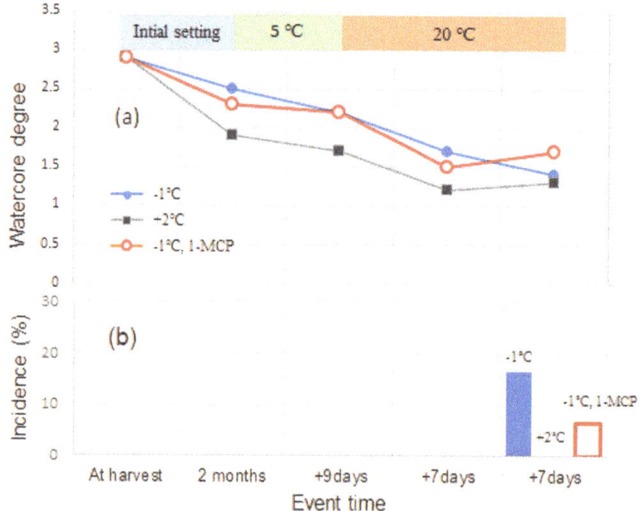


**Figure 5.** Time course of watercore degree (0–4) and watercore breakdown incidence (area %) in 'Fuji' apple. *n* = 30. (**a**) Watercore degree, (**b**) watercore breakdown incidence. Watercore breakdown was not observed in the apples that were initially stored at 2 °C. Reproduced with permission from Onodera [51].

Storage performance over a much longer duration than that reported by Onodera et al. [51] was reported by Kasai et al. [59] to determine which cultivars were resistant to physiological disorders and deterioration of flavor and texture. Apples of 30 cultivars were harvested at their commercial harvest time in the fall and stored under RA, CA, and 1-MCP treatment until mid-June of the next year at 0 °C followed by under RA at 20 °C for 5 days. The watercore degree and flesh browning disorder, which is regarded as a serious problem in watercored apples, were analyzed, and flesh browning disorder was found to occur in most cultivars irrespective of the presence or absence of watercore at harvest, except for 'Shuyo' and 'Ambitious'. Figure 6a presents the relationship between watercore scores at harvest and flesh browning disorder incidence after storage. Seven cultivars scored >2 in watercore, and most of them had high incidence of physiological disorders. However, the incidence in 'Fuji' was low for the score of watercore, which may have been due to the disappearance of watercore. The remaining watercore after storage and the incidence of flesh browning disorders are presented in Figure 6b. Flesh browning occurred irrespective of the remaining watercore score and storage condition; however, highly watercored remaining fruits exhibited severe flesh browning without exception. Watercore is not the only cause of the flesh browning disorder; however, prolonged, severe watercore greatly enhances the severe incidence in flesh.

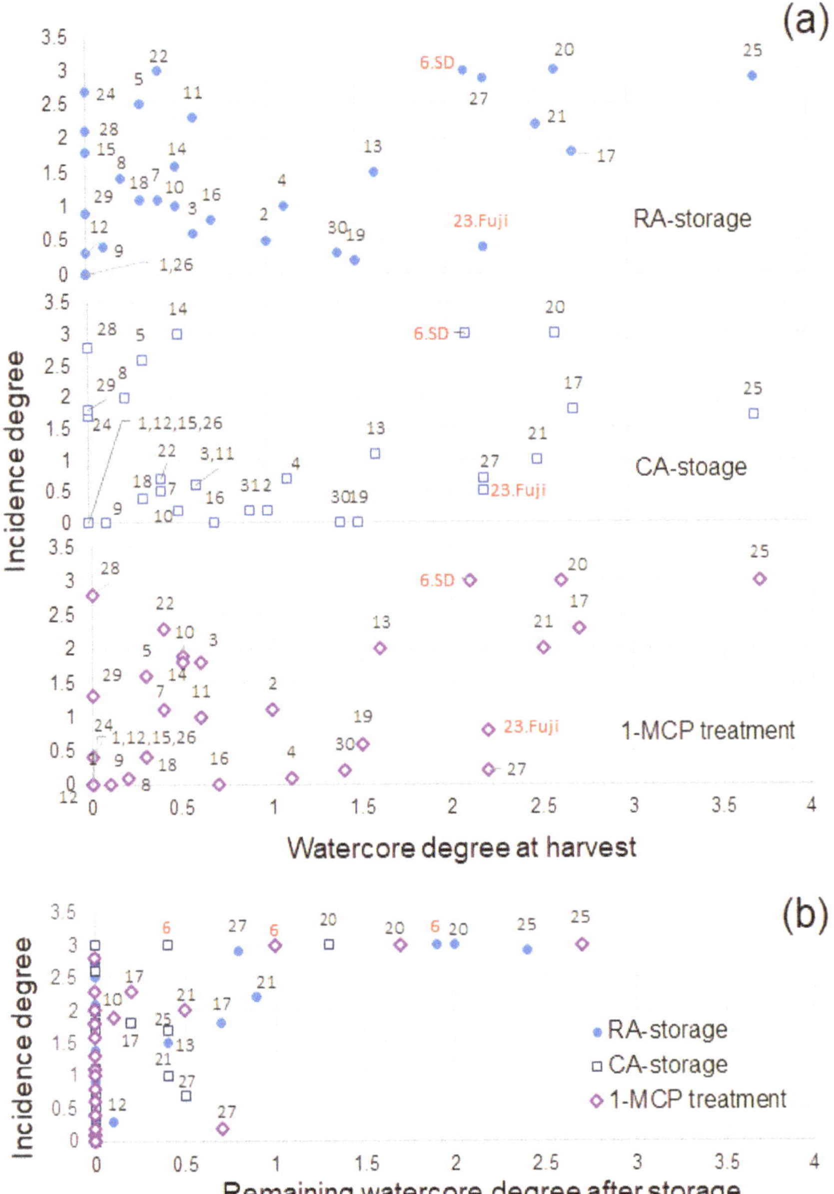



**Figure 6.** Watercore degree (0–4) and incidence of flesh browning disorders (0–3) in various apple cultivars. (**a**) Incidence degree to watercore degree at harvest; (**b**) Incidence degree to watercore degree after storage. $n = 10$. 1: 'Shuyo'; 2: 'Seimei'; 3: 'Shinano Sweet'; 4: 'Sekaiichi'; 5: 'Morinokagayaki'; 6: **'Starking Delicious' (SD)**; 7: 'Kitarou'; 8: 'Jona Gold'; 9: 'Koutaro'; 10: 'Yoko'; 11: 'Megumi'; 12: 'Aori 27'; 13: 'Aikanokaori'; 14: 'Mutsu'; 15: 'Shinano Gold'; 16: 'Mahe 7'; 17: 'Hokuto'; 18: 'Orin'; 19: 'Aori 15'; 20: 'Gunma Meigetsu'; 21: 'Koukou'; 22: 'Slim Red'; **23: 'Fuji'**; 24: 'Mellow'; 25: 'Koutoku'; 26: 'Ambitious'; 27: 'Romu 50'; 28: 'Granny Smith'; 29: 'Cripps Pink'; 30 'Aori 21'; 31: 'Fuji' (bagged). Reproduced with permission from Kasai et al. [59].

Storage longer than 4–5 months usually utilizes several treatments to suppress respiration and ethylene function, which results in an inhibition of aroma synthesis. This inhibits the generation of distinct, sweet aroma, which is the advantage of fresh watercore-rich apples and which cannot be produced after CA storage. In other words, watercored apples should be eaten within a few months of harvest or earlier, especially highly watercore-rich fruits.

Based on work from several previous studies, watercore development can be enhanced or inhibited using cultural techniques on watercore-susceptible cultivars. Watercore is promoted by low or high air temperatures during the preharvest period, large fruit, poor calcium concentration, high nitrogen and boron nutrition, a high leaf-to-fruit ratio, excessive fruit thinning, high or low light exposure, growth in volcanic ash soil, ethrel (ethephon) and gibberellin treatment, and girdling of the trunk and limbs [9]. Therefore, to develop rich watercore for a premium product, fruits are allowed to increase photosynthate accumulation by means of increasing the light received by the leaves and fruits and harvesting at full maturity. To maintain a long shelf life without the watercore physiological disorder, photosynthate accumulation in fruits is limited by earlier harvesting and fruit bagging. Apple producers choose one of these cultivation methods according to demand and their business policies.

## 5. Conclusions

Watercore in apple had been avoided for years due to the mealy texture and brown flesh incidence associated with it. Currently, however, watercore-rich apples are gaining popularity, mainly in Asian countries. 'Fuji', the first rich-watercored cultivar that is free from texture deterioration, greatly contributed to the paradigm shift. 'Fuji' resulted from a cross made in 1939, and though many decades have passed, the potential of 'Fuji' as a high-quality apple is still being shown by integration of diverse analytical methods, such as instrumental analysis and sensory, chemical, physiological, and genetic aspects. Still, there are many unresolved issues related to apple quality. Expanding the understanding of the nature and physiology of apple will continue to lead to improvements in apple quality by utilizing various concepts, approaches, and techniques.

**Supplementary Materials:** The following are available online. Table S1: Intensity of volatiles in watercored and nonwatercored 'Fuji' and 'Koutoku' apples, Table S2: Intensity of water-soluble compounds in watercored and nonwatercored 'Fuji' and 'Koutoku' apples.

**Funding:** This research was partly supported by grants from the Project of the Bio-oriented Technology Research Advancement Institution, NARO (the special scheme project on advanced research and development for next-generation technology).

**Acknowledgments:** We would like to thank Kasai, Statoshi of Aomori Prefectural Industrial Technology Research Center, Onodera, Reiko of Yamagata Agricultural Research Center, and Kunihisa, Miyuki of Institute of Fruit Tree and Tea Science, NARO for their contributions to preparation of the manuscript. We also thank Enago (www.enago.jp) for the English language review.

**Conflicts of Interest:** The authors declare no conflict of interest.

## References

1. O'Rourke, D. World production, trade, consumption and economic outlook for apples. In *Apples: Botany, Production, and Uses*; Ferree, D.C., Warrington, I.J., Eds.; CABI: Oxon, Wallingford, UK, 2003; pp. 15–29.
2. Sadamori, S.; Murakami, H.; Suzuki, H.; Isizuka, S. New breeding hybrids of apple. *Bull. Tohoku Natl. Agric. Exp. Stn.* **1955**, *4*, 125–128. (In Japanese)
3. Smagula, J.M.; Bramlage, W.; Southwick, R.; Marsh, H. Effects of watercore on respiration and mitochondrial activity in Richared Delicious apples. In *Proceedings of the American Society for Horticultural Science*; Amer Soc Horticultural Science: Geneva, NY, USA, 1968; pp. 753–761.
4. Faust, M.; Shear, C.; Williams, M.W. Disorders of carbohydrate metabolism of apples (watercore, internal breakdown, low temperature and carbon dioxide injuries). *Bot. Rev.* **1969**, *35*, 169–194. [CrossRef]
5. Sharples, R. A note on the occurrence of watercore breakdown in apples during 1966. *Plant Pathol.* **1967**, *16*, 119–120. [CrossRef]
6. Marlow, G.C.; Loescher, W.H. Watercore. *Hortic. Rev.* **1984**, *6*, 189–251.

7.  Kasai, S.; Arakawa, O. Antioxidant levels in watercore tissue in 'Fuji'apples during storage. *Postharvest Biol. Technol.* **2010**, *55*, 103–107. [CrossRef]

8.  Tanaka, F.; Tatsuki, M.; Matsubara, K.; Okazaki, K.; Yoshimura, M.; Kasai, S. Methyl ester generation associated with flesh browning in 'Fuji'apples after long storage under repressed ethylene function. *Postharvest Biol. Technol.* **2018**, *145*, 53–60. [CrossRef]

9.  Itai, A. Watercore in fruits. In *Abiotic Stress Biology in Horticultural Plants*; Springer: Tokyo, Japan, 2015; pp. 127–145.

10. Kunihisa, M.; Moriya, S.; Abe, K.; Okada, K.; Haji, T.; Hayashi, T.; Kawahara, Y.; Itoh, R.; Itoh, T.; Katayose, Y.; et al. Genomic dissection of a 'Fuji' apple cultivar: Re-sequencing, SNP marker development, definition of haplotypes, and QTL detection. *Breed. Sci.* **2016**, *66*, 499–515. [CrossRef]

11. Aprea, E.; Charles, M.; Endrizzi, I.; Laura Corollaro, M.; Betta, E.; Biasioli, F.; Gasperi, F. Sweet taste in apple: The role of sorbitol, individual sugars, organic acids and volatile compounds. *Sci. Rep.* **2017**, *7*, 44950. [CrossRef]

12. Charles, M.; Aprea, E.; Gasperi, F. Factors influencing sweet taste in apple. In *Bioactive Molecules in Food*; Mérillon, J.-M., Ramawat, K.G., Eds.; Springer International Publishing: Cham, Switzerland, 2018; pp. 1–22.

13. Tanaka, F.; Okazaki, K.; Kashimura, T.; Ohwaki, Y.; Tatsuki, M.; Sawada, A.; Ito, T.; Miyazawa, T. Profiles and physiological mechanisms of sensory attributes and favor components in watercored apple. *J. Jpn. Soc. Food Sci. Technol.* **2016**, *63*, 101–116. (In Japanese) [CrossRef]

14. Bowen, J.H.; Watkins, C.B. Fruit maturity, carbohydrate and mineral content relationships with watercore in 'Fuji'apples. *Postharvest Biol. Technol.* **1997**, *11*, 31–38. [CrossRef]

15. Zupan, A.; Mikulic-Petkovsek, M.; Stampar, F.; Veberic, R. Sugar and phenol content in apple with or without watercore. *J. Sci. Food Agric.* **2016**, *96*, 2845–2850. [CrossRef] [PubMed]

16. Melado-Herreros, A.; Munoz-García, M.-A.; Blanco, A.; Val, J.; Fernández-Valle, M.E.; Barreiro, P. Assessment of watercore development in apples with MRI: Effect of fruit location in the canopy. *Postharvest Biol. Technol.* **2013**, *86*, 125–133. [CrossRef]

17. Williams, M.W. Relationship of sugars and sorbitol to watercore in apples. *Proc. Am. Soc. Hortic. Sci.* **1966**, *88*, 67–75.

18. Dixon, J.; Hewett, E.W. Exposure to hypoxia conditions alters volatile concentrations of apple cultivars. *J. Sci. Food Agric.* **2000**, *81*, 22–29. [CrossRef]

19. Komthong, P.; Hayakawa, S.; Katoh, T.; Igura, N.; Shimoda, M. Determination of potent odorants in apple by headspace gas dilution analysis. *Lwt-Food Sci. Technol.* **2006**, *39*, 472–478. [CrossRef]

20. Gao, Z.; Jayanty, S.; Beaudry, R.; Loescher, W. Sorbitol transporter expression in apple sink tissues: Implications for fruit sugar accumulation and watercore development. *J. Am. Soc. Hortic. Sci.* **2005**, *130*, 261–268. [CrossRef]

21. Marlow, G.C.; Loescher, W.H. Sorbitol metabolism, the climacteric, and watercore in apples. *J. Am. Soc. Hortic. Sci.* **1985**, *110*, 676–680.

22. Kitahata, S.; Machinami, T. Sugar. In *Molecular Recognition of Taste and Aroma*; The Chemical Society of Japan, Ed.; Gakkai Shuppann Center: Tokyo, Japan, 1999; Volume 40, pp. 50–60. (In Japanese)

23. Magwaza, L.S.; Opara, U.L. Analytical methods for determination of sugars and sweetness of horticultural products—A review. *Sci. Hortic.* **2015**, *184*, 179–192. [CrossRef]

24. Ma, B.; Chen, J.; Zheng, H.; Fang, T.; Ogutu, C.; Li, S.; Han, Y.; Wu, B. Comparative assessment of sugar and malic acid composition in cultivated and wild apples. *Food Chem.* **2015**, *172*, 86–91. [CrossRef]

25. Harker, F.R.; Marsh, K.B.; Young, H.; Murray, S.H.; Gunson, F.A.; Walker, S.B. Sensory interpretation of instrumental measurements 2: Sweet and acid taste of apple fruit. *Postharvest Biol. Technol.* **2002**, *24*, 241–250. [CrossRef]

26. Baldwin, E.A.; Scott, J.W.; Einstein, M.A.; Malundo, T.M.M.; Carr, B.T.; Shewfelt, R.L.; Tandon, K.S. Relationship between sensory and Instrumental analysis for tomato flavor. *J. Am. Soc. Hortic. Sci.* **1998**, *123*, 906. [CrossRef]

27. Beckles, D.M. Factors affecting the postharvest soluble solids and sugar content of tomato (*Solanum lycopersicum* L.) fruit. *Postharvest Biol. Technol.* **2012**, *63*, 129–140. [CrossRef]

28. Tokitomo, Y.; Steinhaus, M.; Buttner, A.; Schieberle, P. Odor-active constituents in fresh pineapple (*Ananas comosus* [L.] Merr.) by quantitative and sensory evaluation. *Biosci. Biotechnol. Biochem.* **2005**, *69*, 1323–1330. [CrossRef] [PubMed]

29. Zheng, L.Y.; Sun, G.M.; Liu, Y.G.; Lv, L.L.; Yang, W.X.; Zhao, W.F.; Wei, C.B. Aroma volatile compounds from two fresh pineapple varieties in China. *Int. J. Mol. Sci.* **2012**, *13*, 7383–7392. [CrossRef] [PubMed]

30. Ichikawa, E.; Hosokawa, N.; Hata, Y.; Abe, Y.; Suginami, K.; Imayasu, S. Breeding of a sake yeast with improved ethyl caproate productivity. *Agric. Biol. Chem.* **1991**, *55*, 2153–2154.

31. Isogai, A.; Utsunomiya, H.; Kanda, R.; Iwata, H. Changes in the aroma compounds of sake during aging. *J. Agric. Food Chem.* **2005**, *53*, 4118–4123. [CrossRef] [PubMed]

32. Lee, J.; Mattheis, J.P.; Rudell, D.R. Antioxidant treatment alters metabolism associated with internal browning in 'Braeburn' apples during controlled atmosphere storage. *Postharvest Biol. Technol.* **2012**, *68*, 32–42. [CrossRef]

33. Lumpkin, C.; Fellman, J.K.; Rudell, D.R.; Mattheis, J.P. 'Fuji' apple (*Malus domestica* Borkh.) volatile production during high pCO$_2$ controlled atmosphere storage. *Postharvest Biol. Technol.* **2015**, *100*, 234–243. [CrossRef]

34. Hampson, C.; Quamme, H.; Hall, J.; MacDonald, R.; King, M.; Cliff, M. Sensory evaluation as a selection tool in apple breeding. *Euphytica* **2000**, *111*, 79–90. [CrossRef]

35. Iwanami, H. Breeding for fruit quality in apple. *Breed. Fruit Qual.* **2011**, 173–200.

36. Bowen, A.J.; Blake, A.; Tureček, J.; Amyotte, B. External preference mapping: A guide for a consumer-driven approach to apple breeding. *J. Sens. Stud.* **2018**, *34*, e12472. [CrossRef]

37. Harker, F.R.; Redgwell, R.J.; Hallett, I.C.; Murray, S.H.; Carter, G. Texture of fresh fruit. *Hortic. Rev.* **1997**, *20*, 121–224.

38. Iwanami, H.; Moriya, S.; Kotoda, N.; Takahashi, S.; Abe, K. Influence of mealiness on the firmness of apples after harvest. *HortScience* **2005**, *40*, 2091. [CrossRef]

39. Jowitt, R. The terminology of food texture. *J. Texture Stud.* **1974**, *5*, 351–358. [CrossRef]

40. Harker, F.R.; Johnston, J.W. Importance of texture in fruit and its interaction with flavour. In *Fruit and Vegetable Flavour*; Woodhead Publishing Limited: Cambridge, UK, 2008; pp. 132–149.

41. Johnston, J.W.; Hewett, E.W.; Hertog, M.L.A.T.M. Postharvest softening of apple (*Malus domestica*) fruit: A review. *New Zealand J. Crop Hortic. Sci.* **2002**, *30*, 145–160. [CrossRef]

42. Wei, J.; Ma, F.; Shi, S.; Qi, X.; Zhu, X.; Yuan, J. Changes and postharvest regulation of activity and gene expression of enzymes related to cell wall degradation in ripening apple fruit. *Postharvest Biol. Technol.* **2010**, *56*, 147–154. [CrossRef]

43. Meilgaard, M.; Civille, G.; Carr, B. *Sensory Evaluation Techniques*, 2nd ed.; CRC: Boca Raton, FL, USA, 1991.

44. Nara, K.; Kato, Y.; Motomura, Y. Involvement of terminal-arabinose and -galactose pectic compounds in mealiness of apple fruit during storage. *Postharvest Biol. Technol.* **2001**, *22*, 141–150. [CrossRef]

45. Iwanami, H.; Ishiguro, M.; Kotoda, N.; Takahashi, S.; Soejima, J. Optimal sampling strategies for evaluating fruit softening after harvest in apple breeding. *Euphytica* **2005**, *144*, 169–175. [CrossRef]

46. Wakasa, Y.; Kudo, H.; Ishikawa, R.; Akada, S.; Senda, M.; Niizeki, M.; Harada, T. Low expression of an endopolygalacturonase gene in apple fruit with long-term storage potential. *Postharvest Biol. Technol.* **2006**, *39*, 193–198. [CrossRef]

47. Iwanami, H.; Moriya, S.; Kotoda, N.; Abe, K. Turgor closely relates to postharvest fruit softening and can be a useful index to select a parent for producing cultivars with good storage potential in apple. *HortScience* **2008**, *43*, 1377. [CrossRef]

48. Moriya, S.; Kunihisa, M.; Okada, K.; Iwanami, H.; Iwata, H.; Minamikawa, M.; Katayose, Y.; Matsumoto, T.; Mori, S.; Sasaki, H.; et al. Identification of QTLs for flesh mealiness in apple (*Malus* × *domestica* Borkh.). *Hortic. J.* **2017**, *86*, 159–170. [CrossRef]

49. Sadamori, S. Memories on breeding of apple 'Fuji'. *J. Agric. Sci.* **1972**, *27*, 419–421. (In Japanese)

50. Iwanami, H.; Moriya, S.; Okada, K.; Abe, K.; Kawamorita, M.; Sasaki, M.; Moriya-Tanaka, Y.; Honda, C.; Hanada, T.; Wada, M. Instrumental measurements of juiciness and freshness to sell apples with a premium value. *Sci. Hortic.* **2017**, *214*, 66–75. [CrossRef]

51. Onodera, R.; Kudo, M.; Takahashi, K.; Nakamura, Y.; Hayama, H. Long term storage of water core apple 'Fuji'. *Tohoku Agric. Res.* **2010**, *63*, 111–112. (In Japanese)

52. Tatsuki, M. Ethylene biosynthesis and perception in fruit. *J. Jpn. Soc. Hortic. Sci.* **2010**, *79*, 315–326. [CrossRef]

53. Argenta, L.; Fan, X.; Mattheis, J. Impact of watercore on gas permeance and incidence of internal disorders in 'Fuji' apples. *Postharvest Biol. Technol.* **2002**, *24*, 113–122. [CrossRef]

54. O'Loughlin, J. Storage behaviour of 'Richard' and 'Starkrimson' delicious apples. *Sci. Hortic.* **1978**, *9*, 41–46.

55. Fukuda, H. Relationship of watercore and calcium to the incidence of internal storage disorders of 'Fuji' apple fruit. *J. Jpn. Soc. Hortic. Sci.* **1984**, *53*, 298–302. [CrossRef]

56. Argenta, L.C.; Fan, X.; Mattheis, J.P. Responses of 'Fuji' apples to short and long duration exposure to elevated $CO_2$ concentration. *Postharvest Biol. Technol.* **2002**, *24*, 13–24. [CrossRef]

57. Kato, K.; Sato, R. The ripening of apple fruits. II. Interrelations of respiration rate, $C_2H_4$ evolution rate and internal gas concentrations, and their relations to specific gravity or watercore during maturation and ripening. *J. Jpn. Soc. Hortic. Sci.* **1978**, *46*, 530–540. (In Japanese) [CrossRef]

58. Harker, F.R.; Watkins, C.B.; Brookfield, P.L.; Miller, M.J.; Reid, S.; Jackson, P.J.; Bieleski, R.L.; Bartley, T. Maturity and regional influences on watercore development and its postharvest disappearance in 'Fuji' apples. *J. Am. Soc. Hortic. Sci.* **1999**, *124*, 166. [CrossRef]

59. Kasai, S.; Kobayashi, T.; Kudo, T.; Goto, S. Appropriate combinations of apple cultivars and storage technologies for long-term storage. *Hortic. Res. (Jpn.)* **2019**, *18*, 173–184. (In Japanese) [CrossRef]

**Sample Availability:** Samples of the compounds are not available from the authors.

*Review*

# Volatile Flavor Compounds in Cheese as Affected by Ruminant Diet

Andrea Ianni [1], Francesca Bennato [1], Camillo Martino [2], Lisa Grotta [1] and Giuseppe Martino [1,*]

1   Faculty of Bioscience and Technology for Food, Agriculture and Environment, University of Teramo, Via R. Balzarini 1, 64100 Teramo, Italy; andreaianni@hotmail.it (A.I.); francescabennato82@gmail.com (F.B.); lgrotta@unite.it (L.G.)
2   Istituto Zooprofilattico Sperimentale dell'Abruzzo e del Molise "G. Caporale", Via Campo Boario, 64100 Teramo, Italy; c.martino@izs.it
*   Correspondence: gmartino@unite.it; Tel.: +39-0861-266950

Received: 11 December 2019; Accepted: 21 January 2020; Published: 22 January 2020

**Abstract:** Extensive research has been conducted concerning the determination and characterization of volatile compounds contributing to aroma and flavor in cheese. Considerable knowledge has been accumulated on the understanding of the mechanisms through which these compounds are formed during ripening, as well as on the optimization of the methodological approaches which lead to their detection. More recently, particular attention has been given to the aromatic properties of milk and cheeses obtained from lactating dairy ruminants fed experimental diets, characterized, for instance, by the addition of trace elements, natural supplements, or agricultural by-products rich in bioactive compounds. The purpose of this review is to summarize the major families of volatile compounds most commonly found in these types of dairy products at various ripening stages, describing in greater detail the role of animal diet in influencing the synthesis mechanisms most commonly responsible for cheese flavor determination. A large number of volatile compounds, including carboxylic acids, lactones, ketones, alcohols, and aldehydes, can be detected in cheese. The relative percentage of each compound depends on the biochemical processes that occur during ripening, and these are mainly mediated by endogenous enzymes and factors of bacterial origin whose function can be strongly influenced by the bioactive compounds taken by animals with the diet and released in milk through the mammary gland. Further evaluations on the interactions between volatile compounds and cheese matrix would be necessary in order to improve the knowledge on the synthesis mechanisms of such compounds; in addition to this, more should be done with respect to the determination of synergistic effects of flavor compounds, correlating such compounds to the aroma of dairy products.

**Keywords:** lactating ruminants; milk; cheese; volatile compound; lipolysis; proteolysis

---

## 1. Introduction

Chemical stability represents the fundamental characteristic of numerous processed foods. However, in the case of cheese, reference is made to a highly dynamic product from the biochemical point of view, especially in those cheeses subjected to ripening. During this period hundreds of volatile compounds (VOCs) can be produced, thus giving rise to flavors and odors that are characteristic of each cheese variety [1].

The main biochemical pathways that occur during the cheese ripening are represented by the metabolism of residual lactose, lactate, and citrate, lipolysis which is associated to the release of free fatty acids (FFAs), and proteolysis that is responsible for casein degradation to peptides with different molecular weights and free amino acids (FAAs). In addition, all the catabolic reactions against FFAs and peptides that give rise to a wide range of VOCs should be included [2].

In the last decades several studies have been conducted with the aim of investigating the specific mechanisms responsible for the production of sapid compounds in cheese during ripening. This approach was driven by the intention to obtain information on the flavor chemistry of many cheese varieties. An aspect to which less attention has been given regards the influence of the feeding strategies administered to ruminants on the volatile profile found in ripened dairy products. It is well known that by modifying animal diet, variations in the chemical-nutritional composition can be induced in milk. Consequently, the characteristics found in milk can be transferred in cheese, making available different substrates for the metabolic functions of starter or non-starter bacteria and for the activities of lipolytic or proteolytic enzymes of endogenous origin [3]. This means that volatile and sensory characteristics of ripened cheeses are largely defined by the technological approach and the initial chemical composition of the raw material [4]. The basic dietary factors that should be considered in ruminants for their effect on milk composition are represented by the fiber content, the ratio between forage and a concentrate (generally consisting of cereal and legume flours in addition to mineral and vitamin supplements), the carbohydrate composition of the concentrate, and the lipid amount, meal frequency, and intake [5]. Over time these aspects have been extensively characterized, especially with a view to obtaining a milk with a greater predisposition to be used for the production of manufactured products [6].

Numerous studies have focused their attention on the correlation between certain variations in the chemical composition of milk and the presence in the ruminant diet of specific classes of bioactive compounds, for instance polyphenols and terpenes, which can be mostly found in plants [7,8]. In this regard a mention should be made to the work of Walker et al. [9], who discussed the most relevant aspects able to induce effects on fatty acid composition of dairy cows' milk. High intake of starch is associated with increased de novo synthesis of fat in the mammary gland, with a consequent increase in the milk of saturated fatty acids (SFAs). In contrast, dietary intake of higher concentrations of polyunsaturated fatty acids (PUFAs) was demonstrated to be effective in inducing higher concentrations of unsaturated fatty acids (UFAs), including conjugated linoleic acid (CLA). An increased intake of starch-based concentrates is instead responsible for the reduction in milk fat concentration, a phenomenon which can be attributed to variations in the balance between lipogenic and glucogenic volatile FAs of ruminal origin. However, reduced fat levels in milk are presumably dependent also on the increased production in the rumen of long-chain FAs containing a *trans*-10 double bond, specifically C18:1 *trans*-10 and C18:2 *trans*-10 *cis*-12, in response to feeding strategies characterized by increased concentrations of PUFAs and/or starch.

In this context, we should include all studies in which ruminant diets have been integrated with agro-industrial by-products and the effects on the chemical-nutritional composition of milk and cheeses have been evaluated. For instance, the supplementation of dairy ewes' diet with an olive crude phenolic concentrate obtained from olive oil wastewater was demonstrated to be effective in inducing in milk an increase in concentration of polyunsaturated fatty acids [10]. A similar behavior was also observed by administering dairy cows with a diet enriched with dried grape pomace, the main by-product of the wine industry; in this study the authors also evidenced an improvement of the oxidative stability of ripened cheese [11]. In addition to this, grape pomace supplementation was also demonstrated to induce in cow's milk a significant increase in concentration of lactose and β-lactoglobulin, although no effects were found for α-lactalbumin, albumin, and caseins [12]. Although the consideration may be speculative, it is conceivable that such a finding may derive from the ability of bioactive compounds of dietary origin to influence bovine gene expression. Indeed, a recent study has shown that 75 days of dietary supplementation with dried grape pomace were effective in inducing variations in the whole-transcriptome of Friesian calves. In that case the authors specifically focused their attention on the pathway of cholesterol biosynthesis, and correlated the observed molecular variations with the reduction in both serum cholesterol and lipid oxidation in carcasses [13].

More recently, a fair number of papers have been published concerning the influence of the feeding strategy on the volatile profile of ripened dairy products obtained from lactating ruminants. The objective of this review is therefore to reorganize, as much as possible, the findings in this research area, in order to obtain a clearer view on the possible correspondences between the type of administered diet and variations in concentration of specific VOCs found in dairy products during ripening. The discussion will be performed on the individual classes of compounds (carboxylic acids, aldehydes, lactones, ketones, alcohols, esters, and phenolic compounds), also giving a nod to the relevance of specific VOCs in flavor perception and summarizing, if appropriate, the principal biochemical pathways by which flavor compounds are produced and that could be influenced by the presence of specific bioactive compounds of dietary origin.

## 2. Biochemical Mechanisms Responsible for the Production of Volatile Flavor Compounds in Dairy Products

The biochemical mechanisms that characterize cheese ripening can be grouped into primary and secondary events. Primary events are represented by the metabolism of residual lactose, lactate, and citrate, as well as lipolysis and proteolysis. These events are then followed by secondary biochemical mechanisms involved in the metabolism of fatty acids and amino acids, which directly contribute to the synthesis of many VOCs, credited as having a high capacity to influence the cheese flavor [1,2,14].

### 2.1. Metabolism of Residual Lactose, Lactate, and Citrate

Lactose is the most represented carbohydrate in milk and is converted to lactate during the cheesemaking by the lactic acid bacteria (LAB), inducing a decrease in pH. In turn, lactate can be further processed by LAB in order to release formate, acetaldehyde, ethanol, and acetate [1,2]. With regard to citrate, the residue remaining in the curd can be converted by citrate-positive LAB into acetate and lactate after cheesemaking. This event is also responsible for the production of additional volatile compounds such as acetoin, 2,3-butanediol, diacetyl, and 2-butanone [15].

### 2.2. Metabolism of Free Amino Acids (FAA)

The catabolism of FAAs represents the biochemical pathway mainly involved in the production of aldehydes, alcohols, carboxylic acids, amines, and sulfur compounds (Figure 1) [16,17].

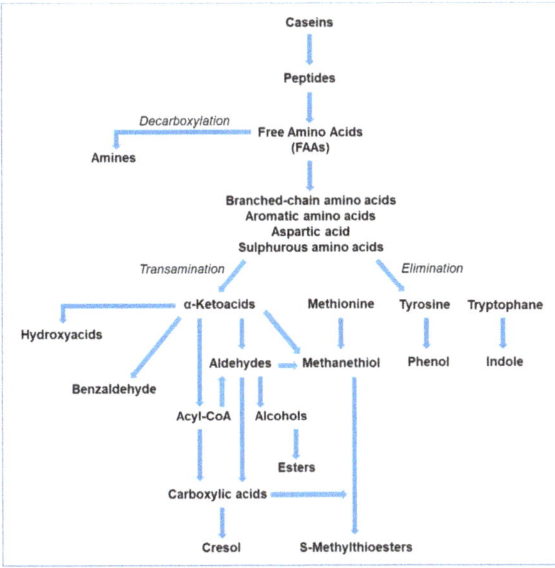


**Figure 1.** Schematic representation of the free amino acid (FAA) catabolism in cheese, modified from Bertuzzi et al. (2018).

The amino acid aminotransferase catalyzes a transamination reaction which leads to the conversion of aromatic amino acids, branched-chain amino acids, methionine, and aspartic acid into $\alpha$-ketoacids. These compounds are then further metabolized to branched-chain and aromatic aldehydes, acyl-CoA, hydroxy acids, and methanethiol [16,18]. The production of 2-methylpropanal, 2-methylbutanal, and 3-methylbutanal is respectively due to the transamination of valine, isoleucine, and leucine, while the transamination reaction in which the substrate is represented by aspartic acid, is responsible for the release of oxaloacetate, which is in turn further converted into acetoin, diacetyl, or 2,3-butanediol [15,19]. Recently, the pivotal role of the aspartic acid transamination was demonstrated in the production of diacetyl in *Lactobacillus paracasei* [20]. Previously, in *Lactococcus lactis* var. *maltigenes* the existence of specific enzymatic pathways responsible for the production of phenylacetaldehyde and methional was observed, as a result of phenylalanine and methionine reduction, respectively [21].

With regard to aromatic aldehydes, these compounds are mainly formed starting from $\alpha$-keto acids deriving from the benzaldehyde released by the spontaneous oxidation of tryptophan and phenylalanine. In this case it is therefore necessary to establish a condition causing a predisposition to a redox reaction, which is reported to be strongly influenced by the temperature, since an increase of this parameter involves catabolism acceleration [22]. Aldehydes represent the substrate of several dehydrogenases, which are able to convert such compounds to alcohols or to oxidize them into the corresponding carboxylic acids [16]. The metabolism of molds and yeast has been reported to be mainly involved in the biosynthesis of primary and aromatic alcohols, with the consequent release of corresponding carboxylic acids. In this regard, a study conducted by Yvon and Rijnen on *Geotrichum candidum* and yeasts isolated from Camembert allowed for the characterization of the mechanisms leading to the production of alcohols and carboxylic acids through FAA metabolism [23].

The excessive proteolysis in cheeses subjected to an uncontrolled ripening in terms of environmental conditions and duration leads to the formation of high concentrations of FAAs that can be decarboxylated, mainly by non-starter LAB, with the consequent release of biogenic amines, which are associated with poor flavor and potentially negative effects on consumer health. The most relevant biogenic amines are represented by histamine, tyramine, cadaverine, and putrescine, which are respectively synthesized starting from histidine, tyrosine, lysine, and ornithine [24].

In the context of FAA catabolism, a noteworthy aspect is also represented by the elimination reactions, which cleave the side chain of amino acids through a reaction catalyzed by a lyase. Over time substantial evidence has been collected about the fact that these reaction are associated to potential negative effects on flavor, as a consequence of the release of compounds such as p-cresol, phenethanol, and indole. This pathway also leads to the synthesis of methanethiol from methionine, which can be metabolized through a variety of different pathways. The major biosynthetic pathway in several strains is that of cystathionine, which involves the intervention of a cystathionine lyase. The further catabolism of methanethiol occurs through oxidative mechanisms performed by numerous LAB species, which are responsible for the production of dimethyldisulfide and dimethyltrisulfide. These compounds are reported to be characterized by low odor perception, thus markedly influencing the cheese flavor [25,26].

### 2.3. Metabolism of Free Fatty Acids (FFAs)

Lipolysis in dairy products is supported by the activity of lipases, microbial enzymes, enzymes of endogenous origin, and enzymes deriving from the added rennet pastes, which catalyze the triglyceride hydrolysis, with the consequent production of medium-chain (chain lengths up to 10 carbon atoms) and long-chain (chain lengths over than 10 carbon atoms) FFAs, di- and mono-glycerides, and glycerol [27].

The flavor properties of cheese are directly influenced by FFAs abundance and pH, and these parameters tend to influence each other. In presence of high pH values in cheese, the FFAs are reported to be less prone to the release of compounds capable of significantly influencing the flavor. Specifically, in this condition the FFAs are converted in non-volatile salts which induce the onset of unpleasant "soapy" aromatic notes. When pH is low, the FFAs are present in volatile form in the dairy matrix, and their excessive increase in concentration is generally effective in inducing a rancid taste [14].

As schematized in Figure 2, methyl ketones, secondary alcohols, straight-chain aldehydes, lactones, esters, and S-thioesters represent classes of VOCs partially deriving from the catabolism of FFAs, which therefore can contribute to the formation of cheese also indirectly as precursors of aromatic compounds [18,28]. FFAs can undergo oxidation, giving origin to β-ketoacids, which are converted to the corresponding methyl ketones through decarboxylation [17,27]. The biosynthetic pathway of methyl ketones is mainly associated to biochemical mechanisms performed by molds; however, hypotheses with regard to synthesis mechanisms induced by heating milk, or, alternatively, derived from a direct esterification of β-ketoacids [28] have been proposed. With regard to ketones, their possible overestimation in the volatile profile of dairy samples can occur as a consequence of the direct conversion of the β-ketoacids in the gas chromatograph inlet [29].

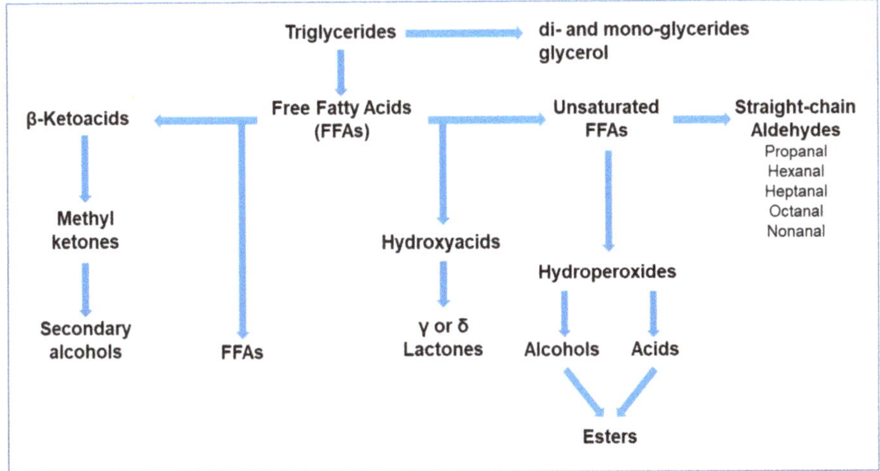


**Figure 2.** Schematic representation of free fatty acid (FFA) catabolism in cheese, modified from Bertuzzi et al. (2018).

The enzymatic reduction of ketones leads to the production of secondary alcohols, a mechanism mainly attributed to molds (such as *Penicillium* spp.) which are specifically responsible for the production of 2-pentanol, 2-heptanol, and 2-nonanol in blue veined cheeses [27]. Such compounds are reported to minimally contribute to the cheese flavor, although the 2-heptanol was identified as a strong odorant in Gorgonzola and Grana Padano cheese [26].

Unsaturated free fatty acids and esterified fatty acids can undergo an auto-oxidation process through non-enzymatic mechanisms, releasing straight-chain aldehydes, mainly propanal, hexanal, heptanal, octanal, and nonanal, that are responsible for the so defined "green grass-like" aroma [27].

The synthesis of esters can occur by esterification, mediated by esterases which use alcohols and carboxylic acids as substrates, or through alcoholysis, which involves the activity of acyltransferases and leads to the synthesis of esters from alcohols, acylglycerols, or acyl-CoA mainly derived from the metabolism of FAAs and FFAs. The transfer to alcohols of fatty acyl groups from acylglycerols or acyl-CoA derivatives represents the main biosynthetic mechanism adopted by LAB to obtain esters. These compounds are associated to pleasant fruity notes able to reduce the sharpness and bitterness that occur in dairy products in which an increase in concentration of FFAs and amines is observed [26]. During esterification or alcoholysis the production of S-methyl-thioesters may occur, a phenomenon mainly correlated to the presence of methanethiol, therefore strictly dependent on the metabolism of specific bacterial species such as *Micrococcaceae Brevibacterium linens*, and *Geotrichum candidum*.

S-methyl-thioesters can be alternatively released by the reaction between FFAs and methanethiol, and are commonly found on the surface of mold-ripened cheese and in blue veined cheeses, conferring strong odors with low threshold perception [30].

Lactones are produced by hydroxylated FFAs, which are integrated in milk triglycerides and released by reactions catalyzed by specific enzymes or induced by heating processes. In addition, hydroxylated FFAs can alternatively be produced by the catabolism of unsaturated fatty acids mediated by lipoxygenases and hydratases of microbial origin [28].

Unlike what has been reported for the other classes of compounds, phenols and terpenes can be identified in several varieties of dairy products, as a direct consequence of their presence in milk before the cheese-making. Phenolic compounds are mostly found in appreciable concentrations in goat and ewe milk, and while they are generally associated with pleasant aromatic notes, they tend to negatively affect the cheese flavor if present in excessive concentrations.

## 3. Major Volatile Flavor Compounds Found in Ripened Cheese and Influenced by Ruminant Diet

The lipolysis and the catabolism of fatty acids represent the most common biochemical mechanisms in cheese during ripening [27]. Therefore, the most represented family of VOCs in cheeses is usually that of carboxylic acids, generally composed of acids from C2 (acetic) to C10 or C12 (decanoic or dodecanoic) [31], followed by other classes of compounds such as aldehydes, lactones, ketones, alcohols, esters, and phenolic compounds. As summarized in Table 1, all these classes of compounds may undergo variations in quantity and composition, as a consequence of variations in the diet administered to ruminants.

**Table 1.** Summary of the most relevant variations found in different dairy products obtained from ruminants fed experimental diets.

| VOC Family | Dietary Supplement (Ruminant) | Type of Dairy Product | Effects | Ref. |
|---|---|---|---|---|
| Carboxylic acids | Dried grape pomace (Friesian cows) | Fresh and 28-day ripened Caciotta cheese | ↓ Acetic acid | [32] |
| | Nutrient-rich pasture (Simmental cows) | 12-month ripened Montasio cheese | ↓ Butanoic acid ↓ Hexanoic acid | [33] |
| | Dried grape pomace (Friesian cows) | 28-day ripened Caciotta cheese | | [34] |
| | Dried licorice root (Saanen goats) | Fresh and 30-day ripened Caciotta cheese | ↓ Hexanoic acid | [35] |
| | Organic zinc (Friesian cows) | 30-day ripened Caciotta cheese | ↑ Butanoic acid ↑ Hexanoic acid | [36] |
| | | 5-day stored Giuncata cheese | | [37] |
| | Extruded linseed (Saanen goats) | 60-day ripened Caciotta cheese | ↓ Dodecanoic acid | [38] |
| Aldehydes | Organic zinc (Friesian cows) | 120-day ripened Caciocavallo cheese | ↑ Nonanal | [39] |
| | Organic zinc (half-breed ewes) | 90-day ripened Pecorino cheese | ↑ Hexanal | [40] |
| | Organic selenium (Friesian cows) | 30-day ripened Caciotta cheese | ↓ Hexanal ↓ Heptanal | [41] |
| Lactones | Organic zinc (Friesian cows) | 120-day ripened Caciocavallo cheese | ↑ γ-nonalactone ↑ γ-dodecalactone ↑ δ-nonalactone ↑ δ-decalactone ↑ δ-dodecalactone ↑ δ-tetralactone | [39] |
| | | 30-day ripened Caciotta cheese | ↑ δ-octalactone ↑ δ-decalactone | [36] |

**Table 1.** *Cont.*

| VOC Family | Dietary Supplement (Ruminant) | Type of Dairy Product | Effects | Ref. |
|---|---|---|---|---|
| Lactones | Organic selenium (Friesian cows) | 30-day ripened Caciotta cheese |  | [41] |
|  | Dried olive pomace (Friesian cows) | 30-day ripened Caciotta cheese | ↑ γ-dodecalactone<br>↑ δ-octalactone | [42] |
| Ketones and Alcohols | Silages (Simmental cows) | 68-day, 200-day and 360-day ripened Montasio cheese | ↑ acetone ↑ 2-3-butanedione<br>↑ 2-butanone<br>↑ 2-hexanone<br>↑ 2-heptanone<br>↑ 2-methyl-1-butanol | [43] |
|  | Nutrient-rich vs nutrient-poor pasture (Simmental cows) | 60-day ripened Montasio cheese | ↑ 2-Propanone [1]<br>↑ 2-Hepta-none [1]<br>↑ 2-Undecanone [1]<br>↑ 3-Methyl-1-butanol [2] | [44] |
|  | Organic selenium (Friesian cows) | 120-day ripened Caciocavallo cheese | ↑ 2-pentanone<br>↑ 2-nonan-2-one<br>↓ Hexanol | [45] |
| Esters | TMR + native pasture (dairy cows) | Ragusano Cheese | Geranyl acetate [3]<br>[E]-Methyl-jasmonate [3] | [46] |
|  | Organic zinc (Friesian cows) | 120-day ripened Caciocavallo cheese | ↑ Ethyl butanoate<br>↑ Ethyl hexanoate<br>↑ Ethyl octanoate<br>↑ Ethyl nonanoate<br>↑ Ethyl decanoate<br>↑ Ethyl dodecanoate<br>↑ Ethyl tetradecanoate<br>↑ Ethyl hexadecanoate | [39] |
|  |  | 30-day ripened Caciotta cheese | ↑ Ethyl hexanoate<br>↑ Ethyl hexadecanoate | [36] |
| Phenolic compounds | Pasture (dairy cows) | Raw milk | ↑ Toluene | [47] |
|  | Crops (dairy cows) |  | ↑ Ptaquiloside<br>↑ Genistein<br>↑ Daidzein | [48] |

VOC = volatile compound; TMR = total mixed ration; ↑ = Increase in concentration; ↓ = Decrease in concentration. [1] Data referred to cheese obtained from cows fed the nutrient-poor pasture. [2] Data referred to cheese obtained from cows fed the nutrient-rich pasture. [3] Compounds only found in cheeses obtained from cows fed the experimental feeding strategy.

## 3.1. Acids

Acetic acid can be synthesized by the catabolism of lactose, citrate, and FAAs, can alternately be derived from propionic fermentation, and is associated to pungent, vinegary, and acidic notes [1,32]. Ianni et al. [34] showed a significant decrease in concentration of this compound in fresh and 28-day ripened cheeses obtained from lactating cows fed for 60 days with 5% dietary supplementation of grape pomace, the major by-product of the oenological industry. A plausible explanation for this finding probably lies in the fact that grape pomace induced in milk an increase in concentration of long-chain fatty acids, making less likely the release of short-chain free fatty acids. In this regard, the study of Harper et al. [49] is of note, in which milk fat was substituted with various vegetable lipids in Romano and Cheddar cheeses. During the ripening process of cheese slurries, low molecular weight free fatty acids were formed, although the vegetable fats did not contain these compounds. As also discussed by Urbach [29], this interesting behavior was not fully characterized by authors from a microbiological and biochemical point of view, and it is therefore possible that the behavior observed may in part have been determined by exogenous factors.

Butanoic and hexanoic acids are considered to be the primary cause of strong and, in some cases, unpleasant odors defined as cheesy, rancid, and sweaty, and their tendency to increase in concentration during ripening in hard cheeses has been widely observed and characterized [50,51]. In a recent study conducted by Aprea et al. [33] a significantly lower concentration of both compounds was observed in ripened Montasio cheese obtained from Italian Simmental cows grazing in a pasture characterized by a nutrient-rich vegetation type. A similar behavior was also observed in other studies in which the ruminant diet was supplemented with plant matrices, which is particularly interesting from the biological point of view, due to the high content of bioactive compounds. Specifically, the reduction of butanoic and hexanoic acids in ripened cheeses was obtained by enriching the diet of Saanen goats with 1% of dried licorice root for 60 days [35], and by administering dietary grape pomace supplementation in lactating Friesian cows [34]. In contrary to the above reported, an increase in butanoic and hexanoic acid was evidenced in dairy products obtained from Friesian cows given dietary zinc supplementation. The particularity in this case lies in the fact that this finding was observed both in a 30-day ripened Caciotta cheese and in a fresh Italian dairy product, the Giuncata cheese, which was analyzed after 5 days of storage at 4 °C [36,37]. The general increase in concentration of FFAs, such as butanoic and hexanoic acids, is commonly explained by the extent of starter cell autolysis during cheese ripening, with the consequent release of enzymes, especially lipases, that promote lipolysis by cleaving the ester linkage between the fatty acid and the glycerol of the triacylglycerol [27]. The breaking of the bacterial envelope and the release of enzymatic factors into the extracellular environment is mediated by peptidoglycan hydrolases, commonly named autolysins, which are characterized by an N-terminal domain, a central catalytic domain, and a C-terminal domain containing a binding motif for zinc, which therefore represent a valuable cofactor [52]. The increase of FFAs in the presence of zinc may therefore depend by the ability of the trace element to favor bacterial autolysis in cheese. In the case of licorice root and grape pomace the opposite phenomenon was instead observed, presumably due to the ability of bioactive compounds deriving from these matrices to slow down the lipolytic action. In this regard it could be taken into account that lipase activity in cheese is strongly influenced by the concentration and type of fatty acids present in the reaction environment [3]. Indeed, the dietary intake of both licorice and grape pomace induced in milk significant variations in the fatty acid profile, with a presumable effect especially on lipases of endogenous origin. This interpretation could also be applied to the just-mentioned studies based on the use of zinc with respect to the reduced production of short-chain FFAs in dairy products, and an increase in concentration of long-chain fatty acids in milk and specifically vaccenic (C18:1 *trans*-11), oleic (C18:1 *cis*-9), linoleic (C18:2 *cis*-9, *cis*-12), and rumenic (C18:2 *cis*-9, *trans*-11) acids.

With regard to the longer chain carboxylic acids, the picture seems to appear more complex. In a study conducted on ripened goat cheese, evidence was found of a tendency for octanoic, decanoic and dodecanoic acids to increase in concentration after short aging periods (about 12 weeks), reaching concentrations well above the threshold values of aroma perception [53]. Octanoic acid is considered to be the main "goaty" compound in dairy products, and is reported to exhibit a waxy aroma that strongly contributes to the flavor of hard goat cheeses. Also, decanoic and dodecanoic acids undoubtedly influence the overall flavor of hard cheeses, and their increase is generally associated to soapy flavor [54]. With regard to the effect of the ruminant diet on the concentration of these compounds in dairy products during ripening, the study conducted by Bennato et al. [38] is noteworthy, as a reduction of dodecanoic acid was observed in a 60-day ripened cheese obtained from goats given dietary supplementation with extruded linseed. This plant matrix did not induce changes in the chemical composition of milk; the only variation was represented by the increase in concentration of linolenic acid (C18:3 *cis*9, *cis*12, *cis*15), which is known to be particularly represented in linseed. As previously reported, an effect of different acidic compositions of milk in influencing the activity of endogenous lipases during cheese ripening [3] could be hypothesized.

### 3.2. Aldehydes

Aldehydes are strongly flavored compounds and are commonly associated in foods to aroma defects referred to as oxidative rancidity [1,27]. These compounds are mainly released by the catabolism of FAAs and, in turn, represent the substrate for specific dehydrogenases responsible for the production of alcohols and carboxylic acids [17]. Aldehydes can also derive from non-enzymatic auto-oxidation reactions which lead to the degradation of unsaturated fatty acids, both free and esterified. These reactions do not occur with high frequency, since the cheese is characterized by a reducing environment. However, this event is responsible for the release of straight-chain aldehydes, which are reported to be associated with pleasant flavor notes [28]. The dairy matrices particularly rich in polyunsaturated fatty acids are therefore more prone to encountering oxidative phenomena able to produce aldehydes.

Recently, the enrichment of the ruminant diet with trace elements, such as zinc and selenium, resulted effective in inducing in milk, and consequently in cheese, an increase in concentration of PUFAs [39–41]. With regard to zinc, the authors specifically observed an increase in desaturation of stearic acid, and this finding was at least in part attributed to the role of zinc as a cofactor for a protease involved in the expression of stearoyl coenzyme A desaturase (SCD) in the mammary gland. SCD is reported to be an endoplasmic reticulum-bound enzyme responsible for the $\Delta^9$-desaturation of saturated fatty acyl-CoAs. The gene expression of this enzyme is mediated by the sterol response element binding proteins (SREBPs) which are activated by a metalloprotease (Site-2 protease) that needs zinc to perform its catalytic function [55,56]. In these studies, the analysis of volatile profile in dairy products did not evidence significant variations in the amount of total aldehydes. Authors discussed this finding by advancing the hypothesis of a role of zinc and selenium in curbing the oxidative damage, a conclusion also supported by the reduction of lipid oxidation evaluated by measuring in cheese the thiobarbituric acid reactive substances (TBARS). In this regard, zinc has been reported to inhibit lipid peroxidation in biological systems by competing with prooxidant metals (i.e., Cu and Fe) for binding sites, thus decreasing their ability to transfer electrons in a particular environment [57]. In the case of selenium, its antioxidant property lies in the ability to act as a scavenger of reactive oxygen-based radicals, with a direct effect in opposing the lipid oxidation in biological systems [58]. Although no differences were found in the total aldehyde content, interesting differences were observed at the level of individual compounds. The dietary zinc supplementation induced a significant increase in concentration of nonanal and hexanal in 120-day ripened Caciocavallo cheese and in 90-day ripened Pecorino cheese, respectively [39,40], whereas the selenium supplementation administered to Friesian cows was effective in inducing a decrease in hexanal and heptanal in samples of 30-day ripened Caciotta cheese [41]. Therefore, in light of what has been reported, dietary zinc enrichment seems to induce a better effect on the aromatic properties of ripened dairy products, since nonanal and hexanal, unlike other aldehydes, are commonly associated with pleasant herbal and slightly fruity notes [28].

### 3.3. Lactones

The main precursors of lactones are represented by hydroxylated FFAs which are incorporated in milk fat triglycerides and released as a result of enzymatic lipolytic mechanisms or the heating process. Hydroxylated FFAs can also be produced by the activities of lipoxygenases or hydratases of microbial origin, within the catabolism of unsaturated fatty acids. A reaction of one-step transesterification is effective in synthesizing lactones from hydroxylated FFAs [28]. These mechanisms heavily and quite positively affect the cheese flavor, since lactones are associated with very pronounced fruity notes, although they have been found to also contribute in cheese to the buttery character [59]. The synthesis of lactones leads to the release of $\alpha$- and $\beta$-lactones that are reported to be highly reactive and unstable, while $\gamma$- and $\delta$-lactones are stable and have been identified in several dairy products. In Cheddar cheese, the concentration of lactones rapidly increased in the early stages of the ripening, reaching levels well above their thresholds of flavor perception. $\delta$-Octalactone was reported to be the most represented lactone in Parmigiano Reggiano cheese, while $\gamma$-decalactone, $\delta$-decalactone, $\gamma$-dodecalactone, and $\delta$-dodecalactone have been found in several French blue cheeses [60].

Also in this case, the studies previously cited have highlighted an active role of the ruminant diet in inducing a change in the relative concentration of this class of volatile compounds. In Caciocavallo cheese obtained from Friesian cows given zinc supplementation, there was evidence of different lactones: γ-nonalactone, γ-dodecalactone, δ-nonalactone, δ-decalactone, δ-dodecalactone, and δ-tetralactone. All these compounds went through a significant increase in concentration at the end of the 120 days of ripening [39]. No specific studies have been conducted on the effect of zinc in the biosynthetic pathway of lactones; however, as reported in the previous paragraphs, a role of the trace element in promoting the starter cells autolysis with consequent release of lipases in the dairy environment could be hypothesized [52]. This event has been reported to be responsible for the increase in concentration of FFAs, from which hydroxyacids, precursors of γ- and δ-lactones [28], are derived. The dietary zinc supplementation was also reported to induce an increase in concentration of lactones in samples of 30-day ripened Caciotta cheese, in which the compounds involved were however limited to δ-octalactone and δ-decalactone [36]. This phenomenon, involving δ-octalactone and δ-decalactone, has also been observed in samples of 30-day ripened Caciotta cheese obtained from lactating Friesian cows given dietary selenium supplementation [41]. This finding therefore suggests a common role of trace elements in favoring the biochemical mechanisms, especially of an enzymatic type, responsible for the synthesis of this class of VOCs.

Interestingly, lactones did not seem to undergo noteworthy variations in experimentations in which the diet of dairy goats and cows was supplemented with plant matrices such as linseed, licorice root, and grape by-products, rich in compounds credited of considerable interest from a biological point of view because of their well characterized anti-inflammatory and antioxidant properties [34,35,38]. An exception to this consideration is found in the study conducted by Castellani et al. [42], who administered to dairy cows a dietary supplementation of olive pomace, a by-product of the olive oil production, rich in fiber and unsaturated fatty acids. In samples of 30-day ripened Caciotta cheese, authors observed an increase in concentration of γ-dodecalactone and δ-octalactone, together with compounds belonging to other chemical classes such as 2-octenal and 1-hexanol.

### 3.4. Ketones and Alcohols

Ketones and alcohols mainly derive from biochemical mechanisms involving the lysis of triglycerides and the oxidation of saturated FFAs, with the consequent production of ketoacids that are decarboxylated to ketones which, in turn, can be reduced to obtain alcohols [61]. These compounds are mainly released by molds such as *Penicillium roqueforti* and *Penicillium camemberti*, which are responsible for typical odors that characterize the aroma of ripened blue veined cheeses [27]. In order to appreciate the potential contribution of these compounds to the cheese aroma, it could be useful to consider that in water, methyl ketones are reported to be characterized by perception thresholds that are quite low, ranging from 0.09 mg·100 g$^{-1}$ for 2-heptanone and 4.09 to 50.0 mg·100 g$^{-1}$ for 2-propanone [28].

In many studies on the volatile profile of cheeses, these classes of compounds are present in limited concentrations, precisely due to the fact that in the manufacturing of many dairy products molds are unwanted and strongly countered [29]. In addition to this, it should be mentioned that the concentration of ketones and alcohols does not seem to be particularly related to the degree of maturation of the cheeses, with a heterogeneous condition of complicated interpretation.

Stefanon and Procida [43] conducted a study aiming to evaluate the effects of including silage in dairy cow diet on the volatile profile of Montasio cheeses. During cheese ripening, significant variations were evidenced for ketones and mostly for the amount of total alcohols, with specific changes in concentration of ethanol, isobutanol, 1-penten-3-ol, and 2-methyl-1-butanol. Authors discussed these findings by assuming a direct effect of diet composition in affecting microbial and chemical fermentations in cheese during ripening rather than a transfer of selected compounds from milk. The Montasio cheese was also the subject of the research conducted by Bovolenta et al. [44], who performed evaluations on the volatile profile of cheese obtained by using raw milk coming from Italian Simmental cows grazing on two alpine pastures different for botanical composition. The "nutrient-poor pasture" resulted effective in inducing molding of the volatile profile of 60-day ripened cheese; in particular an overall increase in concentration of ketones, phenolic compounds, and terpenes was observed, with consequent slight effects noticed by panelists in the sensory analyses. With specific regard to terpenes, during cheese ripening differences were observed that the authors justified by taking into account the study of Belviso et al. [62] who demonstrated the ability of lactic acid bacteria isolated from cheese to influence the terpenoid biosynthesis.

An interesting behavior was recently observed in the volatile profile of Caciotta cheese obtained by enriching dairy cows diet with olive pomace. The dietary supplementation was effective in inducing a significant increase in FFAs, ester, and ketones in raw milk; however, following pasteurization and cheese-making, these differences disappeared both in the fresh and 30-day ripened dairy product [63]. Authors did not specifically investigate this phenomenon but hypothesized that the observed variations could at least in part derive from a change in the microbial population in pasteurized milk cheese, thus passing from a prevalence of lactic acid bacteria in raw milk to a greater concentration of the microbial genera used for pasteurized cheese manufacturing (*Lactococcus*, *Lactobacillus*, *Streptococcus*, and *Propionibacterium*).

Ianni et al. [45] compared the aromatic compounds of Caciocavallo cheeses obtained from Friesian cows fed a standard diet and a diet supplemented with selenium. Although the trace element did not induce differences in the chemical composition of milk and cheese, interesting variations were identified in the volatile profile of 120-day ripened cheeses, in which an increase in concentration of two methyl ketones (2-pentanone and 2-nonan-2-one) and a decrease of an alcohol (1-hexanol) were found. In this study changes in the family of ethyl esters were also highlighted, but no evaluations were executed on the hypothetical consumer acceptability of the experimental dairy product, since no sensory analyses were performed.

*3.5. Esters*

Esters represent a class of VOCs indirectly involved in the metabolism of FFAs [27]. Many of these compounds are reported to have low perception thresholds and are widely associated to a pleasant aroma characterized by sweet, fruity, and floral notes; furthermore, esters are appreciated for their role in stemming the bitterness and sharpness of cheeses, very often due to the high content of amines and FFAs [64].

Carpino et al. [46] analyzed the aroma-active compounds of Ragusano cheese obtained from dairy cows fed a total mixed ration (TMR) supplemented with native Sicilian pastures, in comparison with the same cheese obtained from cows fed only TMR. In samples of Ragusano cheese derived from native pasture 8 unique volatile flavor compounds were identified, among which 2 were esters, specifically geranyl acetate and [E]-methyl jasmonate. The latter compound represents a mediator of the physiological defense mechanisms adopted by plants subjected to stress induced by herbivorous insects. Specifically, when damage occurs plants produce VOCs reported to have detrimental effects on insect physiology [65]. The physical damage of the plant tissue entails the activation of the octadecanoid-lipoxygenase (LOX) pathway, responsible for the release of a wide range of lipid-derived VOCs. Therefore, it is conceivable that a small part of these compounds can be identified in raw milk and consequently in the cheese of ruminants fed with fresh pasture. The finding concerning the identification of unique odor-active esters was also found by analyzing the volatile profile of 60-day ripened goats' milk cheese obtained from animals fed a dietary supplementation of extruded linseed, a well characterized plant matrix rich in linolenic acid (C18:3 *cis*-9, *trans*-12, *trans*-15). These esters, only detected in the "experimental" ripened cheese, were specifically the butanoic acid pentyl ester, the butyric acid 2-ethylhexyl ester, and the isopentyl hexanoate [54].

The previously mentioned addition of zinc to the diet of lactating dairy cows resulted effective in inducing noteworthy variations in volatile esters in derived dairy products. In 120-day ripened Caciocavallo cheese a marked increase in concentration of all the detected ethyl esters was shown, specifically ethyl butanoate, ethyl hexanoate, ethyl octanoate, ethyl nonanoate, ethyl decanoate, ethyl dodecanoate, ethyl tetradecanoate, and ethyl hexadecanoate. Interestingly, the last two compounds resulted only present in ripened cheese samples obtained from zinc feeding [39]. As previously reported, these data allow the discussion of a role of zinc in inducing an increase in lipolytic activity on the triglycerides present in the dairy matrix [52], with a consequent increase in concentration in the reaction environment of FFAs, contributing to the determination of the volatile profile not only directly, but also giving rise to other families of compounds, including esters [28]. In a 30-day ripened Caciotta cheese, the dietary zinc supplementation induced an increase in concentration of only two compounds, ethyl hexanoate and ethyl hexadecanoate, whereas no variations in this class of compounds were evidenced in Giuncata cheese, a fresh dairy product, that was analyzed after 5 days of storage at 4 °C [36,37]. In light of what has been just reported, it seems plausible that the observed increase in concentration of volatile ethyl esters is related to the length of the maturing period, although this consideration should be properly verified. As a partial support to the discussion, in one study lactating ewes were administered a zinc-enriched diet. In addition, samples of Pecorino cheese matured for 90 days showed an increase in concentration of two ethyl esters, specifically ethyl butanoate and ethyl hexanoate. In this case a slight but still significant reduction in concentration of ethyl octanoate was also reported [40].

### 3.6. Phenolic Compounds

Phenolic compounds are secondary metabolites of plants to which interesting properties are attributed from the biological point of view [66]. For that reason, over time great interest has been given to the development of functional dairy products containing specific phenolic compounds, such as catechin, tannic acid, hesperetin, and flavones, or natural crude compounds, for instance grape extract, green tea extract, and dehydrated cranberry powder [67].

Their presence in animal products can also derive from the direct transfer of these compounds from green herbage, or the synthesis by rumen bacteria which are reported to be mainly responsible for the lignin breakdown into monomeric phenols, through a mechanism characterized by decomposition of benzyl ether bonds of lignin polymers under anaerobic conditions [68]. Previous studies focusing on the evaluation of meat quality evidenced the presence of specific phenolic compounds in ruminant fat as a consequence of the ingestion of higher percentages in green herbage than in grain-based diets; specifically, the identification of 4-methylphenol in ruminant fat was reported to be positively affected by grazing [69,70].

As reported by O'Connell and Fox [71], the majority of phenolic volatile compounds identified in milk and dairy products are strictly related to the diet administered to ruminants, although a proportion may represent the product of FFA catabolism, preferably exploiting tyrosine as a precursor. In another study, in which lactating Friesian cows received a diet enriched with olive pomace, in pasteurized milk cheeses an increase in phenolic compounds was observed, specifically phenylacetaldehyde and 2-phenylethyl alcohol, both derived from the catabolism of phenylalanine. Authors discussed the finding by assuming a non-enzymatic Strecker degradation of phenylalanine or by enzymatic transamination of phenylalanine as an imide that is subsequently degraded to give phenylacetaldehyde, that, in turn, undergoes reduction to produce 2-phenylethyl alcohol [72].

With specific regard to cow milk, a study conducted by Villeneuve et al. [47] showed higher concentrations of toluene in samples obtained from cows on pasture, in comparison with milk samples collected from animals fed hay and silage. Authors discussed this finding by advancing the hypothesis of a greater degradation of $\beta$-carotene in forages such as silage or hay subjected to wilting and sun curing following harvesting [73]. Therefore, authors concluded that cows on pasture presumably consumed more $\beta$-carotene, explaining the significant increase in milk of toluene concentration. To better understand this finding, the study conducted by Contarini et al. [74] should be taken into account, where the effect of different heat treatments on the volatile profile of milk by applying a dynamic headspace capillary gas chromatography coupled with multivariate statistical approach was studied. Also in this work, it was assumed that the identification of toluene in raw milk was the consequence of β-carotene degradation. Furthermore, it was evidenced that the identification of toluene in milk, together with 2-pentanone, 2-heptanone, pentanal, and 3-methylbutanal, was effective in discriminating in-bottle sterilized milk (in which these compounds are more greatly represented) from pasteurized samples.

Other studies confirmed that by feeding cattle with high levels of particular crops, other phenolic compounds may also be detected in ruminant milks, such as ptaquiloside, a norsesquiterpene from bracken (*Pteridium aquilinum*), or genistein and daidzein (derived from clover) [48].

## 4. Conclusions

In this review, the main biochemical mechanisms characterizing dairy products during ripening have been recalled, and the influence of different feeding strategies on the production and relative concentrations of various VOCs in fresh and ripened cheeses has been discussed.

Despite the large amount of research activity, to date the influence of certain dietary strategies on the quality of dairy products has not been well characterized, and there is a lack of findings useful to establish VOCs directly transferred from feeds to animal products that could be used for authenticity studies in order to discriminate milk samples or fresh and ripened dairy products. Furthermore, it should be also kept in mind that there is considerable variability induced by the cheese manufacturing process (heating, starter cultures type, ripening conditions), which could eliminate some of the VOCs present in milk. This remains an interesting challenge for researchers in the field of animal production.

**Author Contributions:** Conceptualization, A.I. and G.M.; investigation, A.I., F.B., C.M., and L.G.; resources, G.M.; writing—original draft preparation, A.I.; writing—review and editing, F.B. and C.M.; supervision, G.M. All authors have read and agreed to the published version of the manuscript.

**Funding:** This research received no external funding.



## References

1. McSweeney, P.L.; Sousa, M.J. Biochemical pathways for the production of flavour compounds in cheeses during ripening: A review. *Le Lait* **2000**, *80*, 293–324. [CrossRef]

2. McSweeney, P.L. Biochemistry of cheese ripening. *Int. J. Dairy Technol.* **2004**, *57*, 127–144. [CrossRef]

3. Fox, P.F.; Guinee, T.P.; Cogan, T.M.; McSweeney, P.L. Microbiology of cheese ripening. In *Fundamentals of Cheese Science*; Springer: Boston, MA, USA, 2017; pp. 333–390.

4. Buchin, S.; Delague, V.; Duboz, G.; Berdague, J.L.; Beuvier, E.; Pochet, S.; Grappin, R. Influence of pasteurization and fat composition of milk on the volatile compounds and flavor characteristics of a semi-hard cheese. *J. Dairy Sci.* **1998**, *81*, 3097–3108. [CrossRef]

5. Sutton, J.D. Altering milk composition by feeding. *J. Dairy Sci.* **1989**, *72*, 2801–2814. [CrossRef]

6. DePeters, E.J.; Cant, J.P. Nutritional factors influencing the nitrogen composition of bovine milk: A review. *J. Dairy Sci.* **1992**, *75*, 2043–2070. [CrossRef]

7. White, S.L.; Bertrand, J.A.; Wade, M.R.; Washburn, S.P.; Green, J.T., Jr.; Jenkins, T.C. Comparison of fatty acid content of milk from Jersey and Holstein cows consuming pasture or a total mixed ration. *J. Dairy Sci.* **2001**, *84*, 2295–2301. [CrossRef]

8. De Noni, I.; Battelli, G. Terpenes and fatty acid profiles of milk fat and "Bitto" cheese as affected by transhumance of cows on different mountain pastures. *Food Chem.* **2008**, *109*, 299–309. [CrossRef]

9. Walker, G.P.; Dunshea, F.R.; Doyle, P.T. Effects of nutrition and management on the production and composition of milk fat and protein: A review. *Aust. J. Agric. Res.* **2004**, *55*, 1009–1028. [CrossRef]

10. Cappucci, A.; Alves, S.P.; Bessa, R.J.; Buccioni, A.; Mannelli, F.; Pauselli, M.; Viti, C.; Pastorelli, R.; Roscini, V.; Serra, A.; et al. Effect of increasing amounts of olive crude phenolic concentrate in the diet of dairy ewes on rumen liquor and milk fatty acid composition. *J. Dairy Sci.* **2018**, *101*, 4992–5005. [CrossRef]

11. Ianni, A.; Di Maio, G.; Pittia, P.; Grotta, L.; Perpetuini, G.; Tofalo, R.; Cichelli, A.; Martino, G. Chemical–nutritional quality and oxidative stability of milk and dairy products obtained from Friesian cows fed with a dietary supplementation of dried grape pomace. *J. Sci. Food Agric.* **2019**, *99*, 3635–3643. [CrossRef]

12. Chedea, V.S.; Pelmus, R.S.; Lazar, C.; Pistol, G.C.; Calin, L.G.; Toma, S.M.; Dragomir, C.; Taranu, I. Effects of a diet containing dried grape pomace on blood metabolites and milk composition of dairy cows. *J. Sci. Food Agric.* **2017**, *97*, 2516–2523. [CrossRef] [PubMed]

13. Iannaccone, M.; Elgendy, R.; Giantin, M.; Martino, C.; Giansante, D.; Ianni, A.; Dacasto, M.; Martino, G. RNA sequencing-based whole-transcriptome analysis of friesian cattle fed with grape pomace-supplemented diet. *Animals* **2018**, *8*, 188. [CrossRef] [PubMed]

14. Ardö, Y.; McSweeney, P.L.; Magboul, A.A.; Upadhyay, V.K.; Fox, P.F. Biochemistry of cheese ripening: Proteolysis. In *Cheese*; Academic Press: Burlington, MA, USA, 2017; pp. 445–482.

15. Singh, T.K.; Drake, M.A.; Cadwallader, K.R. Flavor of Cheddar cheese: A chemical and sensory perspective. *Compr. Rev. Food Sci. Food Saf.* **2003**, *2*, 166–189. [CrossRef]

16. Ganesan, B.; Weimer, B.C. Amino acid catabolism and its relationship to cheese flavor outcomes. 2017. Biochemistry of cheese ripening: Proteolysis. In *Cheese: Chemistry, Physics, and Microbiology*, 4th ed.; McSweeney, P.L.H., Fox, P.F., Cotter, P.D., Everett, D.W., Eds.; Academic Press: San Diego, CA, USA, 2017; pp. 483–516.

17. Kilcawley, K.N. Cheese flavour. In *Fundamentals of Cheese Science*; Fox, P.F., Guinee, T.P., Cogan, T.M., McSweeney, P.L.H., Eds.; Springer: Boston, MA, USA, 2017; pp. 443–474.

18. Smit, G.; Smit, B.A.; Engels, W.J. Flavour formation by lactic acid bacteria and biochemical flavour profiling of cheese products. *FEMS Microbiol. Rev.* **2005**, *29*, 591–610. [CrossRef] [PubMed]

19. Ardö, Y. Flavour formation by amino acid catabolism. *Biotechnol. Adv.* **2006**, *24*, 238–242. [CrossRef] [PubMed]

20. Peralta, G.H.; Wolf, I.V.; Bergamini, C.V.; Perotti, M.C.; Hynes, E.R. Evaluation of volatile compounds produced by *Lactobacillus paracasei* I90 in a hard-cooked cheese model using solid-phase microextraction. *Dairy Sci. Technol.* **2014**, *94*, 73–81. [CrossRef]

21. Morgan, M. The chemistry of some microbially induced flavor defects in milk and dairy foods. *Biotechnol. Bioeng.* **1976**, *18*, 953–965. [CrossRef]

22. Klačanová, K.; Fodran, P.; Rosenberg, M. The possible production of natural flavours by amino acid degradation. *Chem. Mon.* **2010**, *141*, 823–828. [CrossRef]

23. Yvon, M.; Rijnen, L. Cheese flavour formation by amino acid catabolism. *Int. Dairy J.* **2001**, *11*, 185–201. [CrossRef]

24. Schirone, M.; Tofalo, R.; Perpetuini, G.; Manetta, A.; Di Gianvito, P.; Tittarelli, F.; Battistelli, N.; Corsetti, A.; Suzzi, G.; Martino, G. Influence of Iodine Feeding on Microbiological and Physico-Chemical Characteristics and Biogenic Amines Content in a Raw Ewes' Milk Cheese. *Foods* **2018**, *7*, 108. [CrossRef]

25. Forde, A.; Fitzgerald, G.F. Biotechnological approaches to the understanding and improvement of mature cheese flavour. *Curr. Opin. Biotechnol.* **2000**, *11*, 484–489. [CrossRef]

26. Curioni, P.M.G.; Bosset, J.O. Key odorants in various cheese types as determined by gas chromatography-olfactometry. *Int. Dairy J.* **2002**, *12*, 959–984. [CrossRef]

27. Collins, Y.F.; McSweeney, P.L.H.; Wilkinson, M.G. Lipolysis and free fatty acid catabolism in cheese: A review of current knowledge. *Int. Dairy J.* **2003**, *13*, 841–866. [CrossRef]

28. Bertuzzi, A.S.; McSweeney, P.L.; Rea, M.C.; Kilcawley, K.N. Detection of volatile compounds of cheese and their contribution to the flavor profile of surface-ripened cheese. *Compr. Rev. Food Sci. Food Saf.* **2018**, *17*, 371–390. [CrossRef]

29. Urbach, G. The flavour of milk and dairy products: II. Cheese: Contribution of volatile compounds. *Int. J. Dairy Technol.* **1997**, *50*, 79–89. [CrossRef]

30. Helinck, S.; Spinnler, H.E.; Parayre, S.; Dame-Cahagne, M.; Bonnarme, P. Enzymatic versus spontaneous S-methyl thioester synthesis in *Geotrichum candidum*. *FEMS Microbiol. Lett.* **2000**, *193*, 237–241. [CrossRef] [PubMed]

31. Faccia, M.; Trani, A.; Natrella, G.; Gambacorta, G. Chemical-sensory and volatile compound characterization of ricotta forte, a traditional fermented whey cheese. *J. Dairy Sci.* **2018**, *101*, 5751–5757. [CrossRef]

32. Beuvier, E.; Buchin, S. Raw milk cheese. In *Cheese: Chemistry, Physics and Microbiology*, 3rd ed.; Fox, P.F., McSweeney, P.L.H., Cogan, T.M., Guinee, T.P., Eds.; Elsevier Academic Press: London, UK, 2004; Volume 1, pp. 319–345.

33. Aprea, E.; Romanzin, A.; Corazzin, M.; Favotto, S.; Betta, E.; Gasperi, F.; Bovolenta, S. Effects of grazing cow diet on volatile compounds as well as physicochemical and sensory characteristics of 12-month-ripened Montasio cheese. *J. Dairy Sci.* **2016**, *99*, 6180–6190. [CrossRef]

34. Ianni, A.; Innosa, D.; Martino, C.; Bennato, F.; Martino, G. Compositional characteristics and aromatic profile of caciotta cheese obtained from Friesian cows fed with a dietary supplementation of dried grape pomace. *J. Dairy Sci.* **2019**, *102*, 1025–1032. [CrossRef]

35. Bennato, F.; Ianni, A.; Martino, C.; Di Luca, A.; Innosa, D.; Fusco, A.M.; Pomilio, F.; Martino, G. Dietary supplementation of Saanen goats with dried licorice root modifies chemical and textural properties of dairy products. *J. Dairy Sci.* **2020**, *103*, 52–62. [CrossRef]

36. Ianni, A.; Innosa, D.; Martino, C.; Grotta, L.; Bennato, F.; Martino, G. Zinc supplementation of Friesian cows: Effect on chemical-nutritional composition and aromatic profile of dairy products. *J. Dairy Sci.* **2019**, *102*, 2918–2927. [CrossRef] [PubMed]

37. Ianni, A.; Iannaccone, M.; Martino, C.; Innosa, D.; Grotta, L.; Bennato, F.; Martino, G. Zinc supplementation of dairy cows: Effects on chemical composition, nutritional quality and volatile profile of Giuncata cheese. *Int. Dairy J.* **2019**, *94*, 65–71. [CrossRef]

38. Bennato, F.; Ianni, A.; Innosa, D.; Grotta, L.; D'Onofrio, A.; Martino, G. Chemical-nutritional characteristics and aromatic profile of milk and related dairy products obtained from goats fed with extruded linseed. *Asian-Austral J. Anim. Sci.* **2019**, *33*, 148–156. [CrossRef] [PubMed]

39. Ianni, A.; Martino, C.; Innosa, D.; Bennato, F.; Grotta, L.; Martino, G. Zinc supplementation of lactating dairy cows: Effects on chemical-nutritional quality and volatile profile of Caciocavallo cheese. *Asian-Austral J. Anim. Sci.* **2019**, in press. [CrossRef] [PubMed]

40. Martino, C.; Ianni, A.; Grotta, L.; Pomilio, F.; Martino, G. Influence of zinc feeding on nutritional quality, oxidative stability and volatile profile of fresh and ripened ewes' milk cheese. *Foods* **2019**, *8*, 656. [CrossRef] [PubMed]

41. Ianni, A.; Martino, C.; Pomilio, F.; Di Luca, A.; Martino, G. Dietary selenium intake in lactating dairy cows modifies fatty acid composition and volatile profile of milk and 30-day-ripened caciotta cheese. *Eur. Food Res. Technol.* **2019**, *245*, 2113–2121. [CrossRef]

42. Castellani, F.; Vitali, A.; Bernardi, N.; Marone, E.; Grotta, L.; Martino, G. Lipolytic volatile compounds in dairy products derived from cows fed with dried olive pomace. *Eur. Food Res. Technol.* **2019**, *245*, 159–166. [CrossRef]

43. Stefanon, B.; Procida, G. Effects of including silage in the diet on volatile compound profiles in Montasio cheese and their modification during ripening. *J. Dairy Res.* **2004**, *71*, 58–65. [CrossRef]

44. Bovolenta, S.; Romanzin, A.; Corazzin, M.; Spanghero, M.; Aprea, E.; Gasperi, F.; Piasentier, E. Volatile compounds and sensory properties of Montasio cheese made from the milk of Simmental cows grazing on alpine pastures. *J. Dairy Sci.* **2014**, *97*, 7373–7385. [CrossRef]

45. Ianni, A.; Bennato, F.; Martino, C.; Innosa, D.; Grotta, L.; Martino, G. Effects of selenium supplementation on chemical composition and aromatic profiles of cow milk and its derived cheese. *J. Dairy Sci.* **2019**, *102*, 6853–6862. [CrossRef]

46. Carpino, S.; Mallia, S.; La Terra, S.; Melilli, C.; Licitra, G.; Acree, T.E.; Barbano, D.M.; Van Soest, P.J. Composition and aroma compounds of Ragusano cheese: Native pasture and total mixed rations. *J. Dairy Sci.* **2004**, *87*, 816–830. [CrossRef]

47. Villeneuve, M.P.; Lebeuf, Y.; Gervais, R.; Tremblay, G.F.; Vuillemard, J.C.; Fortin, J.; Chouinard, P.Y. Milk volatile organic compounds and fatty acid profile in cows fed timothy as hay, pasture, or silage. *J. Dairy Sci.* **2013**, *96*, 7181–7194. [CrossRef]

48. King, R.A.; Mano, M.M.; Head, R.J. Assessment of isoflavonoid concentrations in Australian bovine milk samples. *J. Dairy Res.* **1998**, *65*, 479–489. [CrossRef] [PubMed]

49. Harper, W.J.; Kristoffersen, T.; Wang, J.Y. Formation of free fatty acids during the ripening of fat modified cheese slurries. *Milchwissenschaft* **1978**, *33*, 604–608.

50. Zabaleta, L.; Gourrat, K.; Barron, L.J.R.; Albisu, M.; Guichard, E. Identification of odour-active compounds in ewes' raw milk commercial cheeses with sensory defects. *Int. Dairy J.* **2016**, *58*, 23–30. [CrossRef]

51. Fox, P.F.; McSweeney, P.L.; Cogan, T.M.; Guinee, T.P. *Cheese: Chemistry, Physics and Microbiology, General Aspects*; Elsevier: Oxford, UK, 2004; Volume 1.

52. Huard, C.; Miranda, G.; Wessner, F.; Bolotin, A.; Hansen, J.; Foster, S.J.; Chapot-Chartier, M.P. Characterization of AcmB, an N-acetylglucosaminidase autolysin from *Lactococcus lactis*. *Microbiology* **2003**, *149*, 695–705. [CrossRef] [PubMed]

53. Attaie, R.; Richter, R.L. Formation of volatile free fatty acids during ripening of Cheddar-like hard goat cheese. *J. Dairy Sci.* **1996**, *79*, 717–724. [CrossRef]

54. Woo, A.H.; Kollodge, S.; Lindsay, R.C. Quantification of major free fatty acids in several cheese varieties. *J. Dairy Sci.* **1984**, *67*, 874–878. [CrossRef]

55. Miyazaki, M.; Ntambi, J.M. Role of stearoyl-coenzyme A desaturase in lipid metabolism. *Prostaglandins Leukot. Essent. Fat. Acids* **2003**, *68*, 113–121. [CrossRef]

56. Smith, S.B.; Lunt, D.K.; Chung, K.Y.; Choi, C.B.; Tume, R.K.; Zembayashi, M. Adiposity, fatty acid composition, and delta-9 desaturase activity during growth in beef cattle. *Anim. Sci. J.* **2006**, *77*, 478–486. [CrossRef]

57. Bray, T.M.; Bettger, W.J. The physiological role of zinc as an antioxidant. *Free Radic. Biol. Med.* **1990**, *8*, 281–291. [CrossRef]

58. Zimmerman, M.T.; Bayse, C.A.; Ramoutar, R.R.; Brumaghim, J.L. Sulphur and selenium antioxidants: Challenging radical scavenging mechanisms and developing structure-activity relationships based on metal binding. *J. Inorg. Biochem.* **2015**, *145*, 30–40. [CrossRef] [PubMed]

59. Castada, H.Z.; Hanas, K.; Barringer, S.A. Swiss Cheese flavor variability based on correlations of volatile flavor compounds, descriptive sensory attributes, and consumer preference. *Foods* **2019**, *8*, 78. [CrossRef] [PubMed]

60. Thierry, A.; Collins, Y.F.; Mukdsi, M.A.; McSweeney, P.L.; Wilkinson, M.G.; Spinnler, H.E. Lipolysis and metabolism of fatty acids in cheese. In *Cheese*; Academic Press: Burlington, MA, USA, 2017; pp. 423–444.

61. Urbach, G. The chemical and biochemical basis of cheese and milk aroma. In *Microbiology and Biochemistry of Cheese and Fermented Milk*; Springer: Boston, MA, USA, 1997; pp. 253–298.

62. Belviso, S.; Giordano, M.; Dolci, P.; Zeppa, G. Degradation and biosynthesis of terpenoids by lactic acid bacteria isolated from cheese: First evidence. *Dairy Sci. Technol.* **2011**, *91*, 227. [CrossRef]

63. Castellani, F.; Vitali, A.; Bernardi, N.; Marone, E.; Palazzo, F.; Grotta, L.; Martino, G. Dietary supplementation with dried olive pomace in dairy cows modifies the composition of fatty acids and the aromatic profile in milk and related cheese. *J. Dairy Sci.* **2017**, *100*, 8658–8669. [CrossRef] [PubMed]

64. Liu, S.Q.; Holland, R.; Crow, V.L. Esters and their biosynthesis in fermented dairy products: A review. *Int. Dairy J.* **2004**, *14*, 923–945. [CrossRef]

65. Dicke, M.; Gols, R.; Ludeking, D.; Posthumus, M.A. Jasmonic acid and herbivory differentially induce carnivore-attracting plant volatiles in lima bean plants. *J. Chem. Ecol.* **1999**, *25*, 1907–1922. [CrossRef]

66. Pereira, D.; Valentão, P.; Pereira, J.; Andrade, P. Phenolics: From chemistry to biology. *Molecules* **2009**, *14*, 2202–2211. [CrossRef]

67. Han, J.; Britten, M.; St-Gelais, D.; Champagne, C.P.; Fustier, P.; Salmieri, S.; Lacroix, M. Polyphenolic compounds as functional ingredients in cheese. *Food Chem.* **2011**, *124*, 1589–1594. [CrossRef]

68. Kajikawa, H.; Kudo, H.; Kondo, T.; Jodai, K.; Honda, Y.; Kuwahara, M.; Watanabe, T. Degradation of benzyl ether bonds of lignin by ruminal microbes. *FEMS Microbiol. Lett.* **2000**, *187*, 15–20. [CrossRef]

69. Knudsen, K.E.B. Carbohydrate and lignin contents of plant materials used in animal feeding. *Anim. Feed Sci. Technol.* **1997**, *67*, 319–338. [CrossRef]

70. Raes, K.; Balcaen, A.; Dirinck, P.; De Winne, A.; Claeys, E.; Demeyer, D.; De Smet, S. Meat quality, fatty acid composition and flavour analysis in Belgian retail beef. *Meat Sci.* **2003**, *65*, 1237–1246. [CrossRef]

71. O'Connell, J.E.; Fox, P.F. Significance and applications of phenolic compounds in the production and quality of milk and dairy products: A review. *Int. Dairy J.* **2001**, *11*, 103–120. [CrossRef]

72. Castellani, F.; Bernardi, N.; Vitali, A.; Marone, E.; Grotta, L.; Martino, G. Proteolytic volatile compounds in milk and cheese of cows fed dried olive pomace supplementation. *J. Anim. Feed Sci.* **2018**, *27*, 361–365. [CrossRef]

73. Nozière, P.; Graulet, B.; Lucas, A.; Martin, B.; Grolier, P.; Doreau, M. Carotenoids for ruminants: From forages to dairy products. *Anim. Feed Sci. Technol.* **2006**, *131*, 418–450. [CrossRef]

74. Contarini, G.; Povolo, M.; Leardi, R.; Toppino, P.M. Influence of heat treatment on the volatile compounds of milk. *J. Agric. Food Chem.* **1997**, *45*, 3171–3177. [CrossRef]

MDPI

St. Alban-Anlage 66

4052 Basel

Switzerland

Tel. +41 61 683 77 34

Fax +41 61 302 89 18

www.mdpi.com

*Molecules* Editorial Office

E-mail: molecules@mdpi.com

www.mdpi.com/journal/molecules